Lecture Notes in Computer Science 9932

Commenced Publication in 1973
Founding and Former Series Editors:
Gerhard Goos, Juris Hartmanis, and Jan van Leeuwen

More information about this series at http://www.springer.com/series/7409

Feifei Li · Kyuseok Shim
Kai Zheng · Guanfeng Liu (Eds.)

Web Technologies and Applications

18th Asia-Pacific Web Conference, APWeb 2016
Suzhou, China, September 23–25, 2016
Proceedings, Part II

 Springer

Editors
Feifei Li
School of Computing
University of Utah
Salt Lake City, UT
USA

Kyuseok Shim
School of Electrical Engineering
Seoul National University
Seoul
Korea (Republic of)

Kai Zheng
Soochow University
Suzhou
China

Guanfeng Liu
Soochow University
Suzhou
China

ISSN 0302-9743   ISSN 1611-3349 (electronic)
Lecture Notes in Computer Science
ISBN 978-3-319-45816-8   ISBN 978-3-319-45817-5 (eBook)
DOI 10.1007/978-3-319-45817-5

Library of Congress Control Number: 2016949587

LNCS Sublibrary: SL3 – Information Systems and Applications, incl. Internet/Web, and HCI

Printed on acid-free paper

This Springer imprint is published by Springer Nature
The registered company is Springer International Publishing AG Switzerland

Message from the General Chairs and Program Committee Chairs

Welcome to APWeb 2016! This is the 18th Conference in the Asia Pacific Web Series. Since the first APWeb conference in 1998, APWeb has evolved over time to lead the frontier of data-driven information technology research. It has now firmly established itself as a leading Asia-Pacific focused international conference on research, development, and advanced applications on large-scale data management, Web and search technologies, and information processing. Previous APWeb conferences were held in Guangzhou (2015), Changsha (2014), Sydney (2013), Kunming (2012), Beijing (2011), Busan (2010), Suzhou (2009), Shenyang (2008), Huangshan (2007), Harbin (2006), Shanghai (2005), Hangzhou (2004), Xi'an (2003), Changsha (2001), Xi'an (2000), Hong Kong (1999), and Beijing (1998).

APWeb 2016 was held during September 23–25 in the beautiful and cultural city of Suzhou, China, a city proud of its history of 2500 years. The host organization of APWeb 2016 is Soochow University, one of the fastest developing universities in China.

As in previous years, the APWeb 2016 program featured the main conference with research papers, an industry track, tutorials, distinguished lectures, demos, and a panel. APWeb this year received 215 paper submissions to the main conference from North America, South America, Europe, Asia, and Africa. Each submitted paper underwent a rigorous review process by at least three independent referees from the Program Committee, with detailed review reports. Finally, 79 full papers, 24 short papers, and 17 demo papers were accepted and included in these proceedings. The conference this year had three satellite workshops:

- Second International Workshop on Web Data Mining and Applications (WDMA 2016)
- First International Workshop on Graph Analytics and Query Processing (GAP 2016)
- First International Workshop on Spatial-temporal Data Management and Analytics (SDMA 2016)

We were fortunate to have three world-leading scientists as our keynote speakers: Zhihua Zhou (Nanjing University, China), Cyrus Shahabi (University of Southern California, USA), and Yufei Tao (The University of Queensland, Australia). The Distinguished Lecture Series, co-chaired this year by Xiaokui Xiao (Nanyang Technology University, Singapore) and Xiaochun Yang (Northeastern University, China), invited active and high-impact researchers to discuss their work at APWeb. The two speakers this year were Chen Li (University of California, Irvine, USA) and Jianliang Xu (Hong Kong Baptist University, Hong Kong, China).

The success of APWeb 2016 would not have been possible without the hard work of a great team of people, including Workshop Co-chairs Rong Zhang (ECNU, China) and Wenjie Zhang (University of New South Wales, Australia); Tutorial Co-chairs, Wook-Shin Han (POSTECH, South Korea) and Weng-Chih Peng (National Jiao Tong University, Taiwan); Panel Chair, Ji-Rong Wen (Renmin University of China); Industrial Co-chairs, Luna Xin Dong (Google Research, USA) and Ying Yan (Microsoft Research, China); Demo Co-chairs, Zhifeng Bao (RMIT, Australia) and Xiangliang Zhang (KAUST, Saudi Arabia); Distinguished Lecture Series Co-chairs, Xiaokui Xiao (Nanyang Technological University, Singapore) and Xiaochun Yang (Northeastern University, China); Publication Chair, Guanfeng Liu (Soochow University, China); Social Media and Publicity Chair, Han Su (University of Southern California, USA); Sponsorship and Finance Chairs, An Liu and Lei Zhao (Soochow University, China); and Web Masters, Yan Zhao, and Yang Li (Soochow University, China).

We would also like to take this opportunity to extend our sincere gratitude to the Program Committee members and external reviewers. A special word of thanks goes to the Local Organization Chair, Zhixu Li (Soochow University, China) and his team of organizers and volunteers! Last but not least, we would like to thank all the sponsors, the APWeb Steering Committee led by Jeffrey Yu, and the host organization, Soochow University, for their support, help, and assistance in organizing this conference.

This year's APWeb conference was also the last APWeb conference in its current form. From next year, APWeb and WAIM will be officially combined into one conference, under the new name APWeb/WAIM Joint Conference on the Web and Big Data. These two conferences have many things in common, such as research topics, their regional focuses, and target audiences. In addition, they both share the same goal, to be a world-class research conference on World Wide Web, Internet, and Big Data research and applications with a clear focus on the Asia Pacific region. This year's APWeb was the end of a chapter that we are all proud of, and we are very excited about the new start!

We hope the participants enjoyed APWeb 2016 and Suzhou!

July 2016

General Chairs

Tamer Ozsu
Xiaofang Zhou

Program Committee Chairs

Feifei Li
Kyuseok Shim
Kai Zheng

Organization

General Co-chairs

Tamer Ozsu   University of Waterloo, Canada
Xiaofang Zhou   The University of Queensland, Australia and Soochow
                 University, China

Program Committee Co-chairs

Feifei Li   University of Utah, USA
Kyuseok Shim   Seoul National University, South Korea
Kai Zheng   Soochow University, China

Workshop Co-chairs

Atsuyuki Morishima   University of Tsukuba, Japan
Rong Zhang   ECNU, China
Wenjie Zhang   University of New South Wales, Australia

Tutorial Co-chairs

Wook-Shin Han   POSTECH, Korea, South Korea
Weng-Chih Peng   National Jiao Tong University, Taiwan
Yasushi Sakurai   Kumamoto University, Japan

Panel Chair

Ji-Rong Wen   Renmin University of China, China

Industrial Co-chairs

Luna Xin Dong   Google Research, USA
Ying Yan   Microsoft Research, China

Demo Co-chairs

Zhifeng Bao   RMIT, Australia
Xiangliang Zhang   KAUST, Saudi Arabia

Distinguished Lecture Series Co-chairs

Xiaokui Xiao   Nanyang Technological University, Singapore
Xiaochun Yang   Northeastern University, China

Research Students Symposium Co-chairs

Guoliang Li   Tsinghua University, China
Zi Huang   University of Queensland, Australia

Publication Chair

Guanfeng Liu   Soochow University, China

Social Media and Publicity Chair

Han Su   University of Southern California, USA

Sponsorship and Finance Chairs

An Liu   Soochow University, China
Lei Zhao   Soochow University, China

Local Organization Chairs

Zhixu Li   Soochow University, China
Yan Zhao   Soochow University, China

Web/Information Co-chair

Jun Jiang   Soochow University, China

Program Committee

Aimin Feng   Nanjing University of Aeronautics and Astronautics, China
Alex Thomo   University of Victoria, Canada
Alfredo Cuzzocrea   University of Calabria, Italy
Anirban Mondal   Xerox Research Centre, India
Aviv Segev   KAIST, Korea, South Korea
Baoning Niu   Taiyuan University of Technology, China
Bin Yang   Aalborg University, Denmark
Bingtian Dai   Singapore Management University, Singapore
Carson Kai-Sang Leung   University of Manitoba, Canada
Chaofeng Sha   Fudan University, China
Chen Lin   Xiamen University, China
Chengkai Li   University of Texas at Arlington, USA

Chih-Hua Tai	National Taipei University, Taiwan
Cuiping Li	Renmin University, China
Defu Lian	University of Electronic Science and Technology of China, China
Dejing Dou	University of Oregon, USA
Dhaval Patel	IIT-R, India
Dongxiang Zhang	National University of Singapore, Singapore
Feida Zhu	Singapore Management University, Singapore
Fuzhen Zhuang	ICT, Chinese Academy of Sciences, China
Fuzheng Zhang	Microsoft Research, Beijing, China
Gabriel Ghinita	University of Massachusetts at Boston, USA
Ganzhao Yuan	South China University of Technology, China
Guoliang Li	Tsinghua University, China
Haibo Hu	Hong Kong Baptist University, Hong Kong, China
Han Sun	University of Southern California, USA
Haruo Yokota	Tokyo Institute of Technology, Japan
Hiroaki Ohshima	Kyoto University, Japan
Hong Gao	Harbin Institute of Technology, China
Hong Chen	Renmin University, China
Hongyan Liu	Tsinghua University, China
Hongzhi Yin	University of Queensland, Australia
Hua Yuan	University of Electronic Science and Technology of China, China
Hua Wang	Victoria University, Australia
Hua Lu	Aalborg University, Denmark
James Cheng	The Chinese University of Hong Kong, Hong Kong, China
Jiajun Liu	Renmin University, China
Jialong Han	Nanyang Technological University, Singapore
Jian Li	Tsinghua University, China
Jianbin Qin	University of New South Wales, Australia
Jianbin Huang	Xidian University, China
Jiannan Wang	University of California, Berkerley, USA
Jianzhong Qi	University of Melbourne, Australia
Jimmy Huang	York University, Canada
Jinchuan Chen	Renmin University, China
Shi Gao	UCLA, USA
Jingfeng Guo	Yanshan University, China
Jizhou Luo	Harbin Institute of Technology, China
Ju Fan	National University of Singapore, Singapore
Jun Gao	Peking University, China
Junfeng Zhou	Yanshan University, China
Junjie Wu	Beihang University, China
Kai Zeng	UC Berkeley, USA
Ke Yi	Hong Kong University of Science and Technology, Hong Kong, China

Kun Ren	Yale University, USA
Leong Hou U	University of Macau, Macau, China
Liang Hong	Wuhan University, China
Lianghuai Yang	Zhejiang University of Technology, China
Lili Jiang	Max-Planck Institute, Germany
Ling Chen	University of Technology, Sydney, Australia
Man Lung Yiu	Hong Kong Polytechnical University, Hong Kong, China
Markus Endres	University of Augsburg, Germany
Meihui Zhang	Singapore University of Technology and Design, Singapore
Mihai Lupu	Vienna University of Technology, Austria
Mizuho Iwaihara	Waseda University, Japan
Mohammed Eunus Ali	Bangladesh University of Engineering and Technology, Bangladesh
Nicholas Jing Yuan	Microsoft Research, Beijing, China
Ning Yang	Sichuan University, China
Panos Kalnis	King Abdullah University of Science and Technology, Saudi Arabia
Peng Wang	Fudan University, China
Qi Liu	University of Science and Technology of China, China
Qingzhong Li	Shandong University, China
Quan Thanh Tho	Ho Chi Minh City University of Technology, Vietnam
Quan Zou	Xiamen University, China
Raymond Wong	Hong Kong University of Science and Technology, Hong Kong, China
Rui Zhang	University of Melbourne, Australia
Rui Chen	Hong Kong Baptist University, Hong Kong, China
Sanghyun Park	Yonsei University, South Korea
Sangkeun Lee	Oak Ridge National Laboratory, USA
Sang-Won Lee	Sungkyunkwan University, South Korea
Shengli Wu	Jiangsu University, China
Shinsuke Nakajima	Kyoto Sangyo University, Japan
Shuai Ma	Beihang University, China
Shuo Shang	China University of Petroleum-Beijing, China
Sourav Bhowmick	Nanyang Technological University, Singapore
Srikanta Bedathur	IBM Research India, India
Stavros Papadopoulos	Intel Labs, USA
Taketoshi Ushiama	Kyushu University, Japan
Tieyun Qian	Wuhan University, China
Ting Deng	Beihang University, China
Toshiyuki Amagasa	University of Tsukuba, Japan
Tru Hoang Cao	Ho Chi Minh City University of Technology, Vietnam
Vicent Zheng	Advanced Digital Sciences Center, Singapore
Wee Siong Ng	Institute for Infocomm Research, Singapore
Wei Wang	University of New South Wales, Australia
Wei Wang	Fudan University, China

Weining Qian	East China Normal University, China
Weiwei Sun	Fudan University, China
Weiwei Ni	Southeast University, China
Wen Zhang	Wuhan University, China
Wook-Shin Han	POSTECH, South Korea
Xiang Lian	University of Texas Rio Grande Valley, USA
Xiang Zhao	National University of Defence Technology, China
Xiangliang Zhang	King Abdullah University of Science and Technology, Saudi Arabia
Xiaochun Yang	Northeast University, China
Xiaofeng He	East China Normal University, China
Xiaowei Yang	South China University of Technology, China
Xike Xie	Aalborg University, Denmark
Xuanjing Huang	Fudan University, China
Xueqing Gong	East China Normal University, China
Yanghua Xiao	Fudan University, China
Yang-Sae Moon	Kangwon National University, South Korea
Yaokai Feng	Kyushu University, Japan
Yasuhiko Morimoto	Hiroshima University, Japan
Yi Zhuang	Zhejiang Gongshang University, China
Yijie Wang	National University of Defense Technology, China
Ying Zhao	Tsinghua University, China
Yinghui Wu	University of California, Santa Barbara, USA
Yong Zhang	Tsinghua University, China
Yoshiharu Ishikawa	Nagoya University, Japan
Yu Gu	Northeast University, China
Yuan Fang	Institute for Infocomm Research, Singapore
Yunjie Yao	East China Normal University, China
Yunjun Gao	Zhejiang University, China
Zakaria Maamar	Zayed University, United Arab Emirates
Zhaonian Zou	Harbin Institute of Technology, China
Zhengjia Fu	Advanced Digital Sciences Center, Singapore
Zhenying He	Fudan University, China
Zhiyong Peng	Wuhan University, China
Zhoujun Li	Beihang University, China
Zouhaier Brahmia	University of Sfax, Tunisia

Demo Track Program Committee

Chengbin Peng	KAUST, Saudi Arabia
Chuan Shi	Beijing University of Posts and Telecommunications, China
Feng Li	Microsoft Research, China
Huayu Wu	I2R, Singapore
Jianqiu Xu	NUAA, China
Jingbo Zhou	Baidu Research, USA

Ke Deng	RMIT University, Australia
Peng Zhang	University of Technology, Sydney, Australia
Pinghui Wang	Xian Jiaotong University, China
Qing Xie	Wuhan University of Technology, China
Rui Zhou	Victoria University, Australia
Sang-Wook Kim	Hanyang University, South Korea
Tao Jiang	Jiaxing University, China
Wei Lee Woon	Masdar Institute, United Arab Emirates
Xiangmin Zhou	RMIT University, Australia
Xiaoli Wang	Xiamen University, China
Xiaoying Wu	Wuhan University, China
Yong Zeng	Microsoft, USA
Yuchen Li	National University of Singapore, Singapore

Industry Track Program Committee

Liang Jeff Chen	Microsoft Research, China
Wee Hyong Tok	Microsoft Azure, China
Wenyuan Cai	Hypers, China
Sheng Huang	IBM China Research Lab, China
Chen Wang	RRX360, China

Contents – Part II

Research Full Paper: Streaming and Real-Time Data Analysis

Flexible and Adaptive Stream Join Algorithm. 3
  Junhua Fang, Xiaotong Wang, Rong Zhang, and Aoying Zhou

Finding Frequent Items in Time Decayed Data Streams. 17
  *Shanshan Wu, Huaizhong Lin, Leong Hou U, Yunjun Gao,
  and Dongming Lu*

A Target-Dependent Sentiment Analysis Method for Micro-blog Streams. . . . 30
  Yongheng Wang, Hui Gao, and Shaofeng Geng

Online Streaming Feature Selection Using Sampling Technique
and Correlations Between Features 43
  Hai-Tao Zheng and Haiyang Zhang

Real-Time Anomaly Detection over ECG Data Stream Based
on Component Spectrum 56
  Meng Wu, Zhen Qiu, Shenda Hong, and Hongyan Li

A Workload-Driven Vertical Partitioning Approach Based
on Streaming Framework. 68
  Hong Kang, Mengyu Guo, and Xiaojie Yuan

Research Full Paper: Recommendation System

Spica: A Path Bundling Model for Rational Route Recommendation 85
  Lei Lv, Yang Liu, and Xiaohui Yu

FHSM: Factored Hybrid Similarity Methods for Top-N
Recommender Systems 98
  Xin Xin, Dong Wang, Yue Ding, and Chen Lini

Personalized Resource Recommendation Based on Regular Tag
and User Operation 111
  Sisi Liu, Yongjian Liu, and Qing Xie

Academic Paper Recommendation Based on Community Detection
in Citation-Collaboration Networks 124
  Qisen Wang, Wenzhong Li, Xiao Zhang, and Sanglu Lu

Improving Recommendation Accuracy for Travelers by Exploiting
POI Correlations.   137
   Kai Zhang, Dapeng Zhao, and Xiaoling Wang

A Context-Aware Method for Top-k Recommendation in Smart TV   150
   Peng Liu, Jun Ma, Yongjin Wang, Lintao Ma, and Shanshan Huang

A Hybrid Method for POI Recommendation: Combining Check-In Count,
Geographical Information and Reviews   162
   *Xiefeng Xu, Pengpeng Zhao, Guanfeng Liu, Caidong Gu, Jiajie Xu,
   Jian Wu, and Zhiming Cui*

Improving Temporal Recommendation Accuracy and Diversity
via Long and Short-Term Preference Transfer and Fusion Models.   174
   Bei Zhang and Yong Feng

An Approach for Cross-Community Content Recommendation:
A Case Study on Docker   186
   Yang Yong, Li Ying, Tang Hongyan, Jia Tong, and Shao Wenlong

Research Full Paper: Data Quality and Privacy

Scalable Private Blocking Technique for Privacy-Preserving
Record Linkage   201
   Shumin Han, Derong Shen, Tiezheng Nie, Yue Kou, and Ge Yu

Repair Singleton IDs on the Fly   214
   Xingcan Cui, Xiaohui Yu, and De Guo

Modeling for Noisy Labels of Crowd Workers   227
   *Qian Yan, Hao Huang, Yunjun Gao, Chen Ying, Qingyang Hu,
   Tieyun Qian, and Qinming He*

Incomplete Data Classification Based on Multiple Views   239
   Ming Sun, Hongzhi Wang, Fanshan Meng, Jianzhong Li, and Hong Gao

Fuzzy Keywords Query   251
   *Shanshan Han, Hongzhi Wang, Hong Gao, Jianzhong Li,
   and Shenbin Huang*

EPLA: Efficient Personal Location Anonymity   263
   *Dapeng Zhao, Kai Zhang, Yuanyuan Jin, Xiaoling Wang,
   Patrick C.K. Hung, and Wendi Ji*

A Secure and Robust Covert Channel Based on Secret Sharing Scheme.   276
   Xiaorong Lu, Yang Wang, Liusheng Huang, Wei Yang, and Yao Shen

Discovering Approximate Functional Dependencies from Distributed
Big Data   289
    Weibang Li, Zhanhuai Li, Qun Chen, Tao Jiang, and Zhilei Yin

Research Full Paper: Query Optimization and Scalable Data Processing

Making Cold Data Identification Efficient in Non-volatile Memory Systems . . .   305
    Binbin Wang and Jiwu Shu

Preference Join on Heterogeneous Data   317
    Changping Wang, Chaokun Wang, Hao Wang, Jun Chen,
    and Xiaojun Ye

Accelerating Time Series Shapelets Discovery with Key Points.   330
    Zhenguo Zhang, Haiwei Zhang, Yanlong Wen, and Xiaojie Yuan

An Adaptive Partition-Based Caching Approach for Efficient Range
Queries on Key-Value Data   343
    Wei Ge, Min Chen, Chunfeng Yuan, and Yihua Huang

Handling Estimation Inaccuracy in Query Optimization   355
    Chiraz Moumen, Franck Morvan, and Abdelkader Hameurlain

Practical Study of Subclasses of Regular Expressions in DTD
and XML Schema.   368
    Yeting Li, Xiaolan Zhang, Feifei Peng, and Haiming Chen

Research Short Paper

Chinese Microblog Sentiment Analysis Based on Sentiment Features   385
    Weiwei Li, Yuqiang Li, and Yan Wang

FVBM: A Filter-Verification-Based Method for Finding Top-k Closeness
Centrality on Dynamic Social Networks.   389
    Yiyong Lin, Jinbo Zhang, Yuanxiang Ying, Shenda Hong,
    and Hongyan Li

Online Hot Topic Detection from Web News Based on Bursty
Term Identification   393
    Chao Wang, Xue Zhao, Ying Zhang, and Xiaojie Yuan

Grouped Team Formation in Social Networks.   398
    Ze Lv, Jianbin Huang, Yu Zhou, Heli Sun, and Xiaolin Jia

Profit Maximizing Route Recommendation for Vehicle Sharing Requests. . . .   402
    Zhiqiang Zhao, Jianbin Huang, Hua Gao, Heli Sun, and Xiaolin Jia

Ontology-Based Interactive Post-mining of Interesting Co-location Patterns. . . .   406
Xuguang Bao, Lizhen Wang, and Hongmei Chen

AALRSMF: An Adaptive Learning Rate Schedule for Matrix Factorization . . .   410
Feng Wei, Hao Guo, Shaoyin Cheng, and Fan Jiang

A Graph Clustering Algorithm for Citation Networks   414
Bo Zhang, Tiezheng Nie, Derong Shen, Yue Kou, Ge Yu, and Ziwei Zhou

A Distributed Frequent Itemsets Mining Algorithm Using Sparse Boolean
Matrix on Spark   419
Yonghong Luo, Zhifan Yang, Huike Shi, and Ying Zhang

A Simple Stochastic Gradient Variational Bayes for the Correlated
Topic Model.   424
Tomonari Masada and Atsuhiro Takasu

Reasoning with Large Scale OWL 2 EL Ontologies Based on MapReduce. . .   429
*Zhangquan Zhou, Guilin Qi, Chang Liu, Raghava Mutharaju,
and Pascal Hitzler*

Purchase and Redemption Prediction Based on Multi-task Gaussian Process
and Dimensionality Reduction   434
Chao Wang, Xiangrui Cai, Zhenguo Zhang, and Yanlong Wen

Similarity Recoverable, Format-Preserving String Encryption   439
Yijin Li and Wendy Hui Wang

RORS: Enhanced Rule-Based OWL Reasoning on Spark.   444
*Zhihui Liu, Zhiyong Feng, Xiaowang Zhang, Xin Wang,
and Guozheng Rao*

A Hadoop-Based Database Querying Approach for Non-expert Users   449
Yale Chai, Chao Wang, Yanlong Wen, and Xiaojie Yuan

A Collaborative Join Scheme on a MIC-Based Heterogeneous Platform.   454
Kailai Zhou, Hui Sun, Hong Chen, Tianzhen Wu, and Cuiping Li

Pairwise Expansion: A New Topdown Search for mCK Queries Problem
over Spatial Web   459
Yuan Qiu, Tadashi Ohmori, Takahiko Shintani, and Hideyuki Fujita

Mentioning the Optimal Users in the Appropriate Time on Twitter   464
*Zhaoyun Ding, Xueqing Zou, Yueyang Li, Su He, Jiajun Cheng,
Fengcai Qiao, and Hui Wang*

Historical Geo-Social Query Processing   469
Xiaoying Chen, Chong Zhang, Yanli Hu, Bin Ge, and Weidong Xiao

WS-Rank: Bringing Sentences into Graph for Keyword Extraction   474
Fan Yang, Yue-Sheng Zhu, and Yu-Jia Ma

Efficient Community Maintenance for Dynamic Social Networks   478
Hongchao Qin, Ye Yuan, Feida Zhu, and Guoren Wang

Open Sesame! Web Authentication Cracking via Mobile App Analysis   483
Hui Liu, Yuanyuan Zhang, Juanru Li, Hui Wang, and Dawu Gu

K-th Order Skyline Queries in Bicriteria Networks   488
Shunqing Jiang, Jiping Zheng, Jialiang Chen, and Wei Yu

A K-Motifs Discovery Approach for Large Time-Series Data Analysis   492
Yupeng Hu, Cun Ji, Ming Jing, and Xueqing Li

User Occupation Prediction on Microblogs.   497
Xia Lv, Peiquan Jin, and Lihua Yue

Industry Full Paper

Combo-Recommendation Based on Potential Relevance of Items   505
Yanhong Pan, Yanfei Zhang, and Rong Zhang

Demo Paper

OPGs-Rec: Organized-POI-Groups Based Recommendation   521
JiaPeng Li, Yanxia Xu, and Lei Zhao

A Demonstration of Encrypted Logistics Information System   525
Huakang Li, Xinwen Zhang, Yitao Yang, and Guozi Sun

PCMiner: An Extensible System for Analysing and Detecting Protein
Complexes   529
Danyang Xiao, Jia Zhu, Yong Tang, Lingxiao Chen, and Jingmin Wei

MASM: A Novel Movie Analysis System Based on Microblog   533
*Xingcheng Wu, Jia Zhu, Yong Tang, Rui Ding, Xueqin Lin,
and Chuanhua Xu*

A Text Retrieval System Based on Distributed Representations.   537
Zhe Zhao, Tao Liu, Jun Chen, Bofang Li, and Xiaoyong Du

A System for Searching Renting Houses Based on Relaxed
Query Answering   542
Jianfeng Du, Kunxun Qi, and Can Lin

TagTour: A Personalized Tourist Resource Recommendation System.   547
Tian Han, Yongjian Liu, and Qing Xie

A Chronic Disease Analysis System Based on Dirty Data Mining   552
    Ming Sun, Hongzhi Wang, Jianzhong Li, Hong Gao, and Shenbin Huang

ADDS: An Automated Disease Diagnosis-Aided System   556
    *Zhentuan Xu, Xiaoli Wang, Yating Chen, Yangbin Pan, Mengsang Wu,
    and Mengyuan Xiong*

Indoor Map Service System Based on Wechat Two-Dimensional Code   561
    Chen Guo and Qingwu Hu

Factorization Machine Based Business Credit Scoring by Leveraging
Internet Data   565
    Ge Zhu and Lin Li

OICRM: An Ontology-Based Interesting Co-location Rule Miner   570
    Xuguang Bao, Lizhen Wang, and Meijiao Wang

CB-CAS: A CAS-Based Cross-Browser SSO System   574
    Peng Gao, Yongjian Liu, and Qing Xie

A Demonstration of QA System Based on Knowledge Base   579
    Zhenjiang Dong, Hong Chen, Jingqiang Chen, Huakang Li, and Tao Li

An Alarming and Prediction System for Infections Disease Based
on Combined Models   583
    *Jiahong Li, Hongzhi Wang, Shengqiang Zhang, Xiangyu Gao, Ziqi Qu,
    and Shenbin Huang*

Co-location Detector: A System to Find Interesting Spatial
Co-locating Relationships   588
    Xuguang Bao, Lizhen Wang, and Qing Xiao

KEIPD: Knowledge Extraction and Inference System
for Personal Documents   592
    Zhaoyang Lv, Yuanyuan Liu, and Xiaohui Yu

Author Index   597

Contents – Part I

Research Full Paper: Spatio-temporal, Textual and Multimedia Data Management

Probabilistic Nearest Neighbor Query in Traffic-Aware Spatial Networks.... 3
  Shuo Shang, Zhewei Wei, Ji-Rong Wen, and Shunzhi Zhu

NERank: Bringing Order to Named Entities from Texts............... 15
  Chengyu Wang, Rong Zhang, Xiaofeng He, Guomin Zhou,
  and Aoying Zhou

FTS: A Practical Model for Feature-Based Trajectory Synthesis.......... 28
  Jiapeng Li, Wei Chen, An Liu, Zhixu Li, and Lei Zhao

Distributed Text Representation with Weighting Scheme Guidance
for Sentiment Analysis.................................... 41
  Zhe Zhao, Tao Liu, Xiaoyun Hou, Bofang Li, and Xiaoyong Du

A Real Time Wireless Interactive Multimedia System 53
  Hong Li, Wei Yang, Yang Xu, Jianxin Wang, and Liusheng Huang

Mining Co-locations from Continuously Distributed Uncertain Spatial Data ... 66
  Bozhong Liu, Ling Chen, Chunyang Liu, Chengqi Zhang,
  and Weidong Qiu

An Online Approach for Direction-Based Trajectory Compression
with Error Bound Guarantee................................ 79
  Bingqing Ke, Jie Shao, Yi Zhang, Dongxiang Zhang, and Yang Yang

A Data Grouping CNN Algorithm for Short-Term Traffic Flow Forecasting ... 92
  Donghai Yu, Yang Liu, and Xiaohui Yu

Efficient Evaluation of Shortest Travel-Time Path Queries in Road
Networks by Optimizing Waypoints in Route Requests
Through Spatial Mashups.................................. 104
  Detian Zhang, Chi-Yin Chow, Qing Li, and An Liu

Discovering Companion Vehicles from Live Streaming Traffic Data 116
  Chen Liu, Xiongbin Wang, Meiling Zhu, and Yanbo Han

Time-Constrained Sequenced Route Query in Indoor Spaces 129
  Wenyi Luo, Peiquan Jin, and Lihua Yue

Near-Duplicate Web Video Retrieval and Localization Using Improved
Edit Distance   141
   Hao Liu, Qingjie Zhao, Hao Wang, and Cong Zhang

Efficient Group Top-*k* Spatial Keyword Query Processing   153
   Kai Yao, Jianjun Li, Guohui Li, and Changyin Luo

Research Full Paper: Social Media Data Analysis

Learn to Recommend Local Event Using Heterogeneous Social Networks . . .   169
   *Shaoqing Wang, Zheng Wang, Cuiping Li, Kankan Zhao,
   and Hong Chen*

Top-*k* Temporal Keyword Query over Social Media Data   183
   Fan Xia, Chengcheng Yu, Weining Qian, and Aoying Zhou

When a Friend Online is More Than a Friend in Life: Intimate Relationship
Prediction in Microblogs   196
   Yunshi Lan, Mengqi Zhang, Feida Zhu, Jing Jiang, and Ee-Peng Lim

Community Inference with Bayesian Non-negative Matrix Factorization   208
   Xiaohua Shi and Hongtao Lu

Maximizing the Influence Ranking Under Limited Cost in Social Network. . .   220
   Xiaoguang Hong, Ziyan Liu, Zhaohui Peng, Zhiyong Chen, and Hui Li

Towards Efficient Influence Maximization for Evolving Social Networks. . . .   232
   Xiaodong Liu, Xiangke Liao, Shanshan Li, and Bin Lin

Detecting Community Pacemakers of Burst Topic in Twitter   245
   Guozhong Dong, Wu Yang, Feida Zhu, and Wei Wang

Dynamic User Attribute Discovery on Social Media   256
   Xiu Huang, Yang Yang, Yue Hu, Fumin Shen, and Jie Shao

The Competition of User Attentions Among Social Network Services:
A Social Evolutionary Game Approach   268
   Jingyuan Li, Yuanzhuo Wang, Yuan Lu, Xueqi Cheng, and Yan Ren

Mechanism Analysis of Competitive Information Synchronous
Dissemination in Social Networks   280
   Yuan Lu, Yuanzhuo Wang, Jianye Yu, Jingyuan Li, and Li Liu

A Topic-Specific Contextual Expert Finding Method in Social Network.   292
   Xiaoqin Xie, Yijia Li, Zhiqiang Zhang, Haiwei Pan, and Shuai Han

Budget Minimization with Time and Influence Constraints
in Social Network   304
   Peng Dou, Sizhen Du, and Guojie Song

B-mine: Frequent Pattern Mining and Its Application to Knowledge
Discovery from Social Networks.   316
   Fan Jiang, Carson K. Leung, and Hao Zhang

An Efficient Online Event Detection Method for Microblogs
via User Modeling   329
   Weijing Huang, Wei Chen, Lamei Zhang, and Tengjiao Wang

Man-O-Meter: Modeling and Assessing the Evolution of Language Usage
of Individuals on Microblogs   342
   Kuntal Dey, Saroj Kaushik, Hemank Lamba, and Seema Nagar

Research Full Paper: Modelling and Learning with Big Data

Forecasting Career Choice for College Students Based on Campus Big Data. . . .   359
   Min Nie, Lei Yang, Bin Ding, Hu Xia, Huachun Xu, and Defu Lian

Online Prediction for Forex with an Optimized Experts Selection Model   371
   *Jia Zhu, Jing Yang, Jing Xiao, Changqin Huang, Gansen Zhao,
   and Yong Tang*

Fast Rare Category Detection Using Nearest Centroid Neighborhood.   383
   *Song Wang, Hao Huang, Yunjun Gao, Tieyun Qian, Liang Hong,
   and Zhiyong Peng*

Latent Semantic Diagnosis in Traditional Chinese Medicine   395
   Wendi Ji, Ying Zhang, Xiaoling Wang, and Yiping Zhou

Correlation-Based Weighted K-Labelsets for Multi-label Classification.   408
   Jingyang Xu and Jun Ma

Classifying Relation via Bidirectional Recurrent Neural Network Based
on Local Information.   420
   Xiaoyun Hou, Zhe Zhao, Tao Liu, and Xiaoyong Du

Psychological Stress Detection from Online Shopping   431
   Liang Zhao, Hao Wang, Yuanyuan Xue, Qi Li, and Ling Feng

Confidence-Learning Based Collaborative Filtering with Heterogeneous
Implicit Feedbacks   444
   Jing Wang, Lanfen Lin, Heng Zhang, and Jiaqi Tu

Star-Scan: A Stable Clustering by Statistically Finding Centers and Noises . . .   456
   Nan Yang, Qing Liu, Yaping Li, Lin Xiao, and Xiaoqing Liu

Aggregating Crowd Wisdom with Instance Grouping Methods   468
  Li'ang Yin, Zhengbo Li, Jianhua Han, and Yong Yu

CoDS: Co-training with Domain Similarity for Cross-Domain Image
Sentiment Classification   480
  Linlin Zhang, Meng Chen, Xiaohui Yu, and Yang Liu

Feature Selection via Vectorizing Feature's Discriminative Information   493
  Jun Wang, Hengpeng Xu, and Jinmao Wei

A Label Inference Method Based on Maximal Entropy Random Walk
over Graphs   506
  Jing Pan, Yajun Yang, Qinghua Hu, and Hong Shi

An Adaptive kNN Using Listwise Approach for Implicit Feedback   519
  Bu-Xiao Wu, Jing Xiao, Jia Zhu, and Chen Ding

Quantifying the Effect of Sentiment on Topic Evolution
in Chinese Microblog   531
  Peng Fu, Zheng Lin, Hailun Lin, Fengcheng Yuan, Weiping Wang,
  and Dan Meng

Measuring Directional Semantic Similarity with Multi-features   543
  Bo Liu, Xuanhua Shi, and Hai Jin

Finding Latest Influential Research Papers Through Modeling Two
Views of Citation Links   555
  Lu Huang, Hongyan Liu, Jun He, and Xiaoyong Du

Multi-label Chinese Microblog Emotion Classification via Convolutional
Neural Network   567
  Yaqi Wang, Shi Feng, Daling Wang, Ge Yu, and Yifei Zhang

Mining Recent High Expected Weighted Itemsets
from Uncertain Databases   581
  Wensheng Gan, Jerry Chun-Wei Lin, Philippe Fournier-Viger,
  and Han-Chieh Chao

Context-Aware Chinese Microblog Sentiment Classification
with Bidirectional LSTM   594
  Yang Wang, Shi Feng, Daling Wang, Yifei Zhang, and Ge Yu

Author Index   607

Research Full Paper: Streaming and Real-Time Data Analysis

Flexible and Adaptive Stream Join Algorithm

Junhua Fang, Xiaotong Wang, Rong Zhang$^{(\boxtimes)}$, and Aoying Zhou

Institute for Data Science and Engineering, Software Engineering Institute,
East China Normal University, Shanghai, China
{jh.fang,xt.wang}@ecnu.cn, {rzhang,ayzhou}@sei.ecnu.edu.cn

Abstract. Flexibility and self-adaptivity are important to real-time join processing in a parallel shared-nothing environment. Join-Matrix is a high-performance model on distributed stream joins and supports arbitrary join predicates. It can handle data skew perfectly since it randomly routes tuples to cells with each steam corresponding to one side of the matrix. Designing of the partitioning scheme of the matrix is a determining factor to maximize system throughputs under the premise of economizing computing resources. In this paper, we propose a novel flexible and adaptive scheme partitioning algorithm for stream join operator, which ensures high throughput but with economical resource usages by allocating resources on demand. Specifically, a lightweight scheme generator, which requires the sample of each stream volume and processing resource quota of each physical machine, generates a join scheme; then a migration plan generator decides how to migrate data among machines under the consideration of minimizing migration cost while ensuring correctness. Extensive experiments are done on different kind of join workloads and show high competence comparing with baseline systems on benchmark.

1 Introduction

Explosively rapid growth in data exposes challenges to the applications on streaming processing, such as online financial analysis, traffic managements and environment monitoring systems. Data skew is a common phenomenon in those scenarios and leads to lingering processing in parallel shared-nothing environment. In this context, load balancing [9,11,15,17,18] is crucial for improving throughput by evanishing those lingering tasks which are overloaded. There have been work to solve the imbalance problem among tasks for different operations, such as summarization [4], aggregation [3,12] and join [8,16,18], which may group data by key for processing. One of the most challenging tasks above is to support theta joins [5,10,13] in a flexible and scalable manner under workload skewness.

For processing theta join on streams with skew data distribution, there are two kinds of popular processing models which are join-biclique [10] and join-matrix [5,13]. Lin et al. propose a join-biclique model [10] which organizes the clusters as a complete bipartite graph for joining big data streams. Join-matrix model supports distributed join processing with arbitrary join predicates

F. Li et al. (Eds.): APWeb 2016, Part II, LNCS 9932, pp. 3–16, 2016.
DOI: 10.1007/978-3-319-45817-5_1

perfectly. Recently, it has been studied on both MapReduce-like systems and distributed stream systems. Based on this model, Okcan [13] et al. introduce two partitioning schemes in a MapReduce job, namely 1-Bucket and M-Bucket. It performs well only when input or output dominates processing cost and it requires to get input statistics before optimization execution. DYNAMIC, a join operator designed in [5], adopts a grid-layout partitioning scheme on the matrix. Although it is resilient to data skew as taking a random distribution as routing policy for input tuples and it can perfectly handle any join predicates for it ensures each tuple of one stream to meet any tuple in the other stream. However, it suffers from inflexibility and huge amount of tuple duplication while scaling out or down. For example, *Dynamic* in [5] assumes the number of tasks in a matrix must be a power of two and scales out by splitting the states of every task to four tasks if a task stores a number of tuples exceeding specified memory capacity.

The matrix model incurs the following disadvantages: **(1)** The number of tasks is strictly decided by the number of cells in the matrix, which is calculated by multiplying the number of rows and columns of the matrix; **(2)** In the case of stream change, that is stream volume increasing or shrinking, adding or removing processing tasks must be consistent with matrix cells. Since the allocation of tasks is decided by the number of cells of the matrix, it greatly limits the flexibility of processing with the dynamics of stream, and may cause resource waste by generating more tasks than needed.

The technical contributions of this paper include: **(1)** We design a general strategy to generate the processing scheme for *theta* join operator under dynamic stream changes at runtime, which achieves scalability, effectiveness, and efficiency by a single shot. **(2)** We propose a light weight computation model to generate migration plan, which brings in minimal data transmission overhead. **(3)** We present a detailed theoretical analysis for the proposed model to prove the usability and correctness. **(4)** We implement our algorithms on Storm and give extensive experimental evaluation to our proposed techniques by comparing with existing work using abundant datasets.

The rest of this paper is organized as follows. Section 2 gives the overview and preliminaries. Section 3 presents our scheme generation algorithms to support load change. Section 4 proposes the migration plan generation algorithm. Section 5 presents empirical evaluations of our proposal. Section 6 reports on related work on stream join and load balance. Section 7 concludes the paper and addresses future research directions.

2   Preliminaries

2.1   Matrix Model

In order to make it easy for explanation, all notations used in the rest of this paper are summarized in Table 1. A partitioning scheme on the matrix model splits $R \bowtie S$ into a number of smaller parallel join processing units which are the cells in the matrix decided by rows and columns. Each cell holds partial subset

Table 1. Table of notations

Notations	Description	Notations	Description
$M/M_o/M_n$	Matrix/old matrix/new matrix	R, S	R, S stream
α, β	Number of rows and columns of M	i/j	i^{th} row, j^{th} column in old scheme
m_{ij}	Element of i-th row and j-th column	k/l	k^{th} row, l^{th} column in new shceme
r_i/s_j	Stream in i-th row or j-th column	mp	The migration plan
h_{ij}^R/h_{ij}^S	The sub-range of stream R/S that has been stored in m_{ij}	s_{kl}^R/s_{kl}^S	The sub-range of stream R/S that should be stored in m_{kl}
$V(V_h)$	Memory size of each task(half size $V_h = \frac{V}{2}$)	npi	The mapping of tasks between old and now scheme
$|o|$	The volume of set o	NP/MP	the set of npi/mp

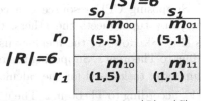

(a) Partition Scheme to Area: $R \bowtie S$ (b) Calculation Area of Partition:$|R| = |S| = 6$

Fig. 1. Example of partition scheme

of data from each stream, which is represented as a range $[b, e]$ to denote the begin b and end e points along the stream window. In Fig. 1(a), a join operation between two data streams R and S can be modeled as a matrix, each side of which corresponds to one stream. The calculation area can be represented by a rectangular with width $|R|$ and length $|S|$. A partitioning scheme splits the area into cells (r_i, s_j) $(0 \leq i \leq \alpha - 1, 0 \leq j \leq \beta - 1)$ of equal size representing stream volume as shown in Fig. 1(a).

2.2 Optimization Goal

Our optimization goal is to figure out the proper values for α and β to achieve the minimal resource usages. Supposing the maximum memory size for each task is V, we formulate our goal as an optimization problem defined as Eq. 1:

$$\min \alpha \cdot \beta, \quad \text{s.t.} \quad |R| \cdot \beta + |S| \cdot \alpha \leq \alpha \cdot \beta \cdot V; \quad \alpha \geq 1, \beta \geq 1. \tag{1}$$

3 Scheme Generation

3.1 Model Design

Since those subsets may be replicated along rows or columns, the values of α and β decide the memory consumption which is proportional to the sub-area's semi-perimeter valued as $|r_i| + |s_j|$ as in [5]. Given the area or the perimeter, we introduce the following two well known theories:

Theorem 1. *Given the area with a constant value, the square has the smallest perimeter among all the rectangles.*

Theorem 2. *Given the perimeter with a constant value, the square has the biggest area among all the rectangles.*

Based on these two theories, we have the following corollary on partitioning scheme:

Corollary 1. *If there exist α and β which can make $\frac{|R|}{\alpha} = \frac{|S|}{\beta} = V_h$, the consumption of processing resource for $R \bowtie S$ is minimal.*

Proof. Supposing CPU resource is a constant value in each task, in order to ensuring any tuple meets the others, the computation complexity for stream join is $|R| \cdot |S|$. However the memory usage will be minimized if $\frac{|R|}{\alpha} = \frac{|S|}{\beta} = V_h$ according to Theorem 1. Supposing the memory resource of each task is constant, the number of tasks used for the calculation (total area) is smallest when $\frac{|R|}{\alpha} = \frac{|S|}{\beta} = V_h$ according to Theorem 2. The network communication cost is relevant to memory for the volume of tuples stored equals to transmission volume. Based on the discussion above, we can draw a conclusion that Corollary 1 is established.

According to Corollary 1, if the volumes of two streams $|R|$ and $|S|$ can both be divisible by V_h, receiving tuples with quantity of V_h from both streams is a prefect solution to generate matrix scheme M with minimal resource usages. However, stream volumes may not be divisible by V_h. Given that the number of row (column) in matrix M must be an integer, we get the row number $\alpha = \lceil \frac{|R|}{V_h} \rceil$, and the column number $\beta = \lceil \frac{|S|}{V_h} \rceil$. Then the number of tasks N used in matrix M can be expressed as

$$N = \lceil \frac{|R|}{V_h} \rceil \cdot \lceil \frac{|S|}{V_h} \rceil \tag{2}$$

In those $\lceil \frac{|R|}{V_h} \rceil \cdot \lceil \frac{|S|}{V_h} \rceil$ cells, we primarily loading the front rows and columns, then, when the stream volume can not be evenly divided by V_h, it generates fragment data for the tasks (called fragment tasks) in the last row and the last column in matrix M. For example, given task memory $V = 10\,\text{GB}$, R stream volume $|R| = 6\,\text{GB}$ and S stream volume $|S| = 6\,\text{GB}$, its calculation area is shown in Fig. 1(b). Processing $R \bowtie S$ will take up 4 tasks for its matrix M with two rows and two columns. In M, $m_{00} = (5\,\text{GB}, 5\,\text{GB})$, $m_{01} = (5\,\text{GB}, 1\,\text{GB})$, $m_{10} = (1\,\text{GB}, 5\,\text{GB})$ and $m_{11} = (1\,\text{GB}, 1\,\text{GB})$. Then m_{01}, m_{10}, m_{11} are fragment tasks in that the sum of $|r_{ij}|$ and $|s_{ij}|$ in these tasks is smaller than memory V.

3.2   Generation Scheme

To find an optimal processing scheme, we differentiate the two streams as a primary stream volume P and a secondary stream volume D. Supposing we split P into P_γ subsets assigned to each task, we first ensure the memory usage for those subsets from P and keep the remaining memory for data split from D. Then for D, The split subset D_γ is calculated as $D_\gamma = \lceil \frac{D}{V - \frac{P}{P_\gamma}} \rceil$. We use N_c to represent the number of tasks and it can be calculated as Eq. 3:

$$N_c = P_\gamma \cdot D_\gamma = P_\gamma \cdot \lceil \frac{D}{V - \frac{P}{P_\gamma}} \rceil \tag{3}$$

As declared in Corollary 1, the number of tasks is minimized when $\frac{|R|}{\alpha} = \frac{|S|}{\beta} = V_h$. Then, to get the minimal number of tasks for processing, we make $P_\gamma \in \{\lceil \frac{|R|}{V_h} \rceil, \lfloor \frac{|R|}{V_h} \rfloor, \lceil \frac{|S|}{V_h} \rceil, \lfloor \frac{|S|}{V_h} \rfloor\}$ in Eq. 3 and take the minimal N_c for our scheme.

Theorem 3. *Given stream volumes $|R|$, $|S|$ and memory size V of task, using matrix model for $R \bowtie S$, the number of tasks generated by $P_\gamma * D_\gamma$ as Eq. 3 by selecting P_γ from $\{\lceil \frac{|R|}{V_h} \rceil, \lfloor \frac{|R|}{V_h} \rfloor, \lceil \frac{|S|}{V_h} \rceil, \lfloor \frac{|S|}{V_h} \rfloor\}$ is the minimal.*

Proof. We assume that there exists a matrix M' with row number α' and column number β' which can be used for $R \bowtie S$ and the number of tasks N' is smaller than N_c. In other words, there is a number P'_γ : $P'_\gamma \notin \{\lceil \frac{|R|}{V_h} \rceil, \lfloor \frac{|R|}{V_h} \rfloor, \lceil \frac{|S|}{V_h} \rceil, \lfloor \frac{|S|}{V_h} \rfloor\} \Rightarrow N' < N_c$. According to Corollary 1, $\frac{|R|}{\alpha'}$ is closer to V_h than $\frac{|R|}{\alpha}$, and $\frac{|S|}{\beta'}$ is also closer to V_h than $\frac{|S|}{\beta}$. However, it is impossible for $\frac{|R|}{\alpha'}$ and $\frac{|S|}{\beta'}$ to get closer to V_h simultaneously. Moreover, P_γ occupies all possible values that make $\frac{P}{P_\gamma}$ nearest to V_h. Hence, there is not any smaller N' existing.

4   Migration Plan Generation

4.1   Task-Load Mapping Generation

Supposing m_{ij} and m_{kl} corresponds to two area which are M_o in old schema and M_n in new schema respectively, and each area is one join processing task. We define a correlation coefficient λ_{kl}^{ij} for each pair of tasks corresponding to m_{ij} and m_{kl} respectively. λ_{kl}^{ij} is a measurement for the stream overlapping between m_{ij} and m_{kl} calculated as Eq. 4.

$$\lambda_{kl}^{ij} = (h_{ij}^R \cap s_{kl}^R) \cdot |R| + (h_{ij}^S \cap s_{kl}^S) \cdot |S| \tag{4}$$

A new indicant $npi =< m_{ij}, m_{kl}, \lambda_{kl}^{ij} >$ (task mapping item) is defined to represent the effort for migration $(|s_{kl}^R| + |s_{kl}^S| - \lambda_{kl}^{ij})$ when using the task in charge of m_{ij} for the data in m_{kl}. Our target is to minimize system migration cost during scheme change. The whole procedure of task pairing is described in Algorithm 1 and can be divided into two parts: (1) *part I* enumerates all the possible *npis* shown in line (2–5) in Algorithm 1; (2) *part II* generates task pairing relationship with the purpose of minimizing migration by selecting *npi* with the biggest λ_{kl}^{ij} into NP set. This NP set will generate the task-load mapping with the least migration according to Theorem 4.

Algorithm 1. Generation Task-Load Mapping

---

input: Old scheme M_o, New scheme M_n
output: Task mapping NP
1: Initialize $NP = Null$
2: **foreach** m_{ij} in Old scheme M_o **do**
3:     **foreach** m_{kl} in New scheme M_n **do**
4:         Calculate λ_{kl}^{ij} according to Eq. 4
5:         Add the item $< m_{ij}, m_{kl}, \lambda_{kl}^{ij} >$ to a temporary set NPI
6: **foreach** task mapping item with max λ_{kl}^{ij} in NPI **do**
7:     **if** m_{ij} or m_{kl} in npi not exist in NP **then**
8:         $< m_{ij}, m_{kl} > \rightarrow NP$
9:         Delete $< m_{ij}, m_{kl} >$ from NPI
10: **return** NP

Theorem 4. *Among task pairings between the old and new scheme, NP set produced by Algorithm 1 leads to the minimal migration cost.*

Proof. For *part I* in Algorithm 1, it generates *npis* with the size of $|M_o| \cdot |M_n|$. In other words, $\alpha_o \cdot \beta_o \cdot \alpha_n \cdot \beta_n$ items are generated where α_o and β_o are the number of row and column in old scheme M_o and α_n and β_n are the number of row and column in new scheme M_n. Obviously, $|NP|$ is the smaller one between $|M_o|$ and $|M_n|$ and each m_{ij} or m_{kl} appears in NP only once at most. Then we can conclude that the current maximal λ_{kl}^{ij} is independent of others. That is to say *partII* described as line (6–9) in Algorithm 1 produces the maximal cumulative sum of λ_{kl}^{ij}. It means there is the maximal volume of non-migrating data in NP, and also implies the task mapping NP leads to the minimal migration cost. Based on the discussion above, we can draw that Theorem 4 is established.

4.2   Migration Plan Generation

As described above, a migration plan defines how to migrate data among tasks when scheme changes. In order to make it easy for explanation, we describe data moving among tasks with Stream R, and it will be the same for Stream S. We use n_{kl}^{R} to denote the range of data in stream R that should be moved into area m_{kl}, calculated as Eq. 5:

$$n_{kl}^{R} = s_{kl}^{R} - (s_{kl}^{R} \cap h_{kl}^{R}) \tag{5}$$

Migration plan $mp = < m_{ij}, m_{kl}, N_{ij}^{R} >$ tells the data moving between two area m_{ij} in old scheme M_o and m_{kl} in new scheme M_n, with N_{ij}^{R} representing the data moving from area m_{ij} to area m_{kl} for R. We define two kinds of actions for moving: duplicating and migrating. Duplicating happens among tasks along the same row/column, otherwise, it is data migrating. On each m_{kl} in new scheme M_n, it deletes the migrated data from set $h_{kl}^R - s_{kl}^R$ for stream R, which is represented as $mp = < \odot, m_{kl}, h_{kl}^R - s_{kl}^R >$. All the calculations are the same for stream S.

Migration plan generation is described in Algorithm 2 and is divided into two steps as follows:

Step-1: Splitting stream data for matrix cells. According to matrix characteristics described in Sect. 2.1, it is easy for us to get the whole data set of stream R or S by combining the data from the first row or the first column in M_o. According to the new scheme M_n, we can divide the streams evenly to fill each cell as in line (1–8);

Step-2: Deleting migrated tuples. It deletes migrated data under the new scheme M_n in line (9–11).

A partitioning scheme changes from 2×1 to 2×2 as depicted in Fig. 2. In old scheme M_o, each area manages half volume of stream from R and total volume from S shown in Fig. 2(a): $h_{00}^R = [0, \frac{1}{2}]$, $h_{10}^R = [\frac{1}{2}, 1]$ and $h_{00}^S = [0, 1]$, $h_{10}^S = [0, 1]$. When the workload of streams increases, system may scale out by adding one more column with two tasks forming a 2×2 scheme as shown in Fig. 2(b). In this case, data partitions of R are unchanged where tasks in the first row still manage half volume ($s_{0j}^R = [0, \frac{1}{2}], j \in \{0, 1\}$) and tasks in the second row manage the other half ($s_{1j}^R = [\frac{1}{2}, 1], j \in \{0, 1\}$). Stream S should be split into two partitions for two columns, each of which manages $\frac{1}{2}$ range of data, that is $s_{i0}^S = [0, \frac{1}{2}]$, $s_{i1}^S = [\frac{1}{2}, 1]$, with $i \in \{0, 1\}$.

Algorithm 2. Migration Plan Generation

---

input: Old scheme M_o, New scheme M_n,
    Task mapping NP
output: Migration plan MP
1: **foreach** row i with column 0 in old scheme M_o **do**
2:     **foreach** m_{kl} in new scheme M_n **do**
3:         **if** $h_{i0}^R \cap n_{kl}^R \neq Null$ **then**
4:             $< m_{i0}, m_{kl}, h_{i0}^R \cap n_{kl}^R > \rightarrow$
    MP
5: **foreach** column j with row 0 in old scheme M_o **do**

6:     **foreach** m_{kl} in new scheme M_n **do**
7:         **if** $h_{0j}^S \cap n_{kl}^S \neq Null$ **then**
8:            $< m_{0j}, m_{kl}, h_{0j}^S \cap n_{kl}^S > \rightarrow$
    MP
9: **foreach** task m_{kl} in new scheme M_n **do**
10:     $< \odot, m_{kl}, h_{kl}^R - s_{kl}^R > \rightarrow MP$
11:     $< \odot, m_{kl}, h_{kl}^S - s_{kl}^S > \rightarrow MP$
12: **return** MP

According to the discussion in Sect. 4.1, NP is $\{< m_{00}^o, m_{00}^n >, < m_{10}^o, m_{10}^n >\}$ as shown in Fig. 2(b). In Fig. 2(b), we label the relevant task pairs between M_o and M_n by assigning tasks the same numbers. The tasks tagged with red *new* in m_{01} and m_{11} are new additive tasks. m_{01} needs data $n_{01}^R = [0, \frac{1}{2}]$

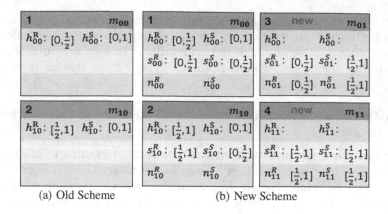

(a) Old Scheme   (b) New Scheme

Fig. 2. Example of scheme change

and $n_{01}^S = [\frac{1}{2}, 1]$; m_{11} needs data $n_{11}^R = [\frac{1}{2}, 1]$ and $n_{11}^S = [\frac{1}{2}, 1]$. According to Algorithm 2, s_{01}^R and s_{11}^R are generated by duplicating R data from m_{00} and m_{10}, respectively. m_{01} and m_{11} generate S by duplicating $[\frac{1}{2}, 1]$ from m_{00}. Since S has been reallocated according to discussion above, then the range $[\frac{1}{2}, 1]$ of S is deleting from m_{00} and m_{01}.

5 Evaluation

All of the approaches in our experiment are implemented and run on top of *Apache Storm* [1]. The overall workflow of the adjustment components for distributed stream join is as follows. At the end of each time interval (such as 5 s), the tasks report the information about current resource usage (such as memory load) to an *controller* module. Then the *controller* decides whether to change the processing scheme; If processing scheme needs change, *controller* first produces a new scheme (Sect. 3.2); Accordingly, it expects to explore the task-load mapping function for mapping tasks in an old scheme to ones in a new scheme (Sect. 4.1); Finally, it schedules the data migration among tasks (Sect. 4.2).

5.1 Experimental Setup

Environment: The *Storm* system (version 0.10.1) is deployed on a 21-instance HP blade cluster with CentOS 6.5 operating system. Each instance in the cluster is equipped with two Intel Xeon processors (E5335 at 2.00 GHz) with four cores.

Data Sets: We evaluate all the approaches using the existing benchmark TPC-H [2] and generate databases using the *dbgen* tool shipped with TPC-H benchmark. Before feeding to the stream system, we pre-generate and pre-process all the input data sets. Specifically, we adjust the degree of skew on the join attributes by defining skew parameter z for the Zipf function and we set $z = 1$ by default.

Queries: We conduct the experiments on three join queries, namely E_{Q_5}, B_{NCI} and B_{MR}, among which the first two are used in [5, 10]. E_{Q_5} is an equi-join which represents the most expensive operation in query Q_5 from TPC-H benchmark. B_{NCI} and B_{MR} are both band-joins, which are different in memory usage by different data selectivity on attribute *Quantity*.

E_{Q_5} : SELECT * FROM LINEITEM, REGION, NATION, SUPPLIER WHERE REGION.orderkey = LINEITEM.orderkey AND LINEITEM.suppkey = SUPPLIER.suppkey AND SUPPLIER.nationkey = NATION.nationkey

B_{NCI} : SELECT * FROM LINEITEM L1, LINEITEM L2   WHERE L1.orderkey - L2.orderkey$\leqslant$1 AND L1.shipmode = 'TRUK' AND L2.shipinstruck = 'NONE' AND L2.Quantity > 48

B_{MR} : SELECT *   FROM LINEITEM L1, LINEITEM L2   WHERE L1.orderkey - L2.orderkey $\leqslant$1 AND L1.shipmode = 'TRUK' AND L2.shipinstruck = 'NONE') AND L2.Quantity > 10

We implement both full-history joins and window-based joins, where full-history joins are used to verify system's scalability and window-based joins are used to validate algorithms' flexibility and self-adaptability.

Baseline Approaches: For the purpose of comparison, we implement four different distributed stream join algorithms: *MFM*, *Square*, *Dynamic* [5] and *Readj* [7]. *MFM* and *Square* are proposed in this paper. *MFM* denotes our flexible and adaptive algorithm that generates the scheme with less tasks according to Eq. 3. *Square* adopts a naive method to obtain the task number defined in Eq. 2. *Dynamic* [5] assumes the number of tasks in a matrix must be a power of two. If one stream doubles its volume, *Dynamic* adjusts matrix scheme by doubling the cells along the side corresponding to this stream. Meanwhile, it halves cells along the other side of the matrix. Besides, *Dynamic* scales out by splitting the states of every task to four tasks if a task stores a number of tuples exceeding specified memory capacity. *Readj* [7] is designed to minimize the load by redistributing tuples based on a hash function on keys.

Evaluation Metrics: We measure resource utilization and system performance through the following metrics: *Task number* is the total number of tasks used in system and each task is equipped with a constant quota of memory V; *Thoughtput* is the average number of tuples that processed by system per minute (time unit); *Migration volume* is the total amount of tuples migrated to other tasks during scheme changing; *Migration plan time* is the average time spent on generating a migration plan.

5.2 Scalability

To testify the scalability of our join algorithm, we set $V = 8 \cdot 10^5$ and continue load all $6 \cdot 10^6$ tuples into our system by executing B_{NCI}. Figure 3 shows increasing of task number and migration cost during loading data into the system. With increase of task number shown in Fig. 3(a), the memory utilization consumed by *Dynamic* also increases dramatically which is proportional to task number. The naive method *Square* consumes more memory compared to *MFM* since its task number increases a little bit more. Our algorithm MFM performs the best among those methods, which can scale out with minimal number of tasks and applies for resources on its real demand. Figure 3(b) illustrates the changes on migration cost with query B_{NCI} when loading the whole dataset into our system. Consistently, $Dynamic$ causes the highest migration cost than all other algorithms, because $Dynamic$ suffers from massive replications to maintain its matrix structure. Furthermore, $Square$ and MFM yield low migration volume in that they involve less tasks. From Fig. 3 we find that the migration volume increases along with data loading. This is because all matrix schemes progressively get larger.

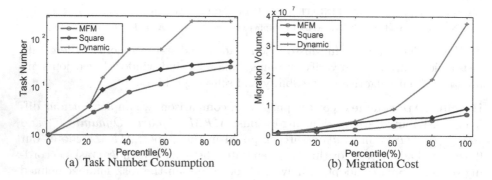

(a) Task Number Consumption   (b) Migration Cost

Fig. 3. Performance of full-history join with B_{NCI}

In addition, we examine the latency for generating migration plan and throughput for equi-join E_{Q_5} with different algorithms. For the purpose of load balance, we define the balance indicator θ_t for task instance d during time interval T_t as $\theta_t = |\frac{L_t(d) - \bar{L}_t}{L_t}|$, where $\bar{L}$ is the average load of all task instances. For this group of experiments, we set $\theta_t \leq 0.05$. Figure 4(a) provides the latency for generating migration plan. Obviously, the latency of $Readj$ is much larger than all other algorithms for $Readj$ is designed to minimize the load difference among tasks by redistributing data on keys with a hash function and it must recalculate the balance states for each scale-out processing. The other algorithms including $Dynamic$ use random distribution as routing policy, so they need not do calculation for balance scheduling. Figure 4(b) draws the throughput of each algorithm under different data skewness. Throughput of $Readj$ decreases with severer skewness because it spends more time for generating migration plan. Although tasks used by $Dynamic$ is much more than our methods, the throughput of ours is more than $Dynamic$ due to its massive migration cost.

5.3 Dynamic

This group of experiments shows the performance with window-based join, which bounds the memory consumption based on the window size. For this experiment,

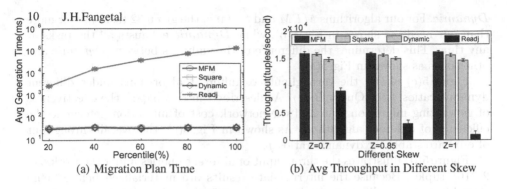

(a) Migration Plan Time

(b) Avg Throughput in Different Skew

Fig. 4. Performance of full-history with E_{Q_5}

we set window size as 5 min and the average input rate is about $1.8 \cdot 10^4$ tuples per second. We provide maximum 32 tasks for this group testing. The dynamics is simulated by altering the relative stream volume ratio $|R|/|S|$ between stream R and S [5] with the total volume $2 \cdot 10^7$ tuples, where the ratio fluctuates between f and $\frac{1}{f}$ with f defined as the fluctuation rate.

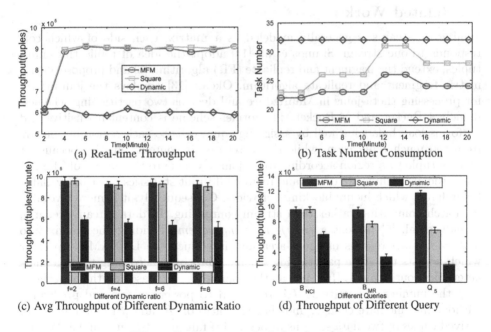

(a) Real-time Throughput

(b) Task Number Consumption

(c) Avg Throughput of Different Dynamic Ratio

(d) Throughput of Different Query

Fig. 5. Performance of window-based join

Figure 5(a) and (b) depict the throughput and number of tasks used for query B_{MR}. Figure 5(a) shows that our methods have better throughput compared to

Dynamic. For our algorithms MFM and *Square*, the given 32 tasks are far more than our needs as shown in Fig. 5(b), while *Dynamic* exhausts all the tasks at any time. This determines the difference of throughputs between *Dynamic* and other ones as shown in Fig. 5(a).

Figure 5(c) shows the throughputs of different algorithms under different dynamic ratios f for Query B_{NCI}. As described in this paper, the effectiveness of generating migration plan and the network cost of migration determine the efficiency of different algorithms. As shown in Fig. 5(c), the overall throughput of ours are stable for dynamic ratios f.

Figure 5(d) illustrates the throughput of different queries under the workload $2 \cdot 10^7$ tuples. Because the intermediate results are materialized before being stored in memory, different queries generate different volume of states which are to be stored in memory. Since E_{Q_5} is lack of filters on predicate, it should store all tuples within a window for join processing and then it requires more memory. In this way, throughput of *Dynamic* decreases dramatically due to its memory requirement for E_{Q_5}. For the two band-joins, B_{NCI} will have more throughput for its filters $Quantity > 48$ can filter out more tuples than $Quantity > 10$ in B_{MR}. This also indicates that B_{MR} requires more tasks than B_{NCI} and it has lower throughput when the total memory size is predefined.

6   Related Work

Joining on steams is generally modeled as a matrix, each side of which corresponds to one stream. Stamos et al. [14] adopt the idea of replicating input tuples, extend the fragment and replicate (FR) algorithm [6] and propose a symmetric fragment and replicate algorithm. Okcan [13] employs the join-matrix for processing theta-joins in MapReduce and designs two partitioning schemes, namely 1-Bucket and M-Bucket. The former scheme is content-insensitive and performs load balancing well by assigning equal cells to each region but suffers from too much replication, while the latter one is content-sensitive because it maps a tuple to a region according to its join key. Due to the nature of MapReduce, the algorithms are offline and require all input statistics must be available beforehand, which incurs blocking behaviors. Consequently, it is more favorable for batch computing rather than stream computing. In data stream scenario, Elseidy et al. [5] present a *(n,m)- mapping scheme* dividing the matrix into $J(J = n \times m)$ regions of equal area and introduce the DYNAMIC operator which adjusts the state partitioning scheme adaptively according to data characteristics continuously. However, all the approaches are based on the hypothesis that the number of partitions J is restricted to powers of two and predefined without intermediate change, and that the ratio of $|R|$ and $|S|$ (the number of arrived tuples of two data streams respectively) falls in between $\frac{1}{J}$ and J. What's more, the flexibility of the matrix structure is deteriorated when the matrix need to scale out (down).

7   Conclusion

In this paper, we focus on designing flexible and adaptive *theta* join algorithms to handle distributed join processing on stream system. Inspired by the matrix-based method which can ensure the correctness of join result and be immune to data skewness, we propose a new scheme change algorithm based on matrix model which inherits all advantages of traditional methods but improves them on scalability and effectiveness. We implement our design on Storm and compare it with the other state-of-art work to verify our idea. In future work, we will continue to design algorithms for *theta* join which may break through the matrix model aiming to make full usage to system resource.

Acknowledgments. This work is partially supported by National High Technology Research and Development Program of China (863 Project) No. 2015AA015307, National Science Foundation of China under grant (No. 61232002 and NO. 61332006), and National Science Foundation of Shanghai (No. 14ZR1412600). The corresponding author is Rong Zhang.

References

1. Apache Storm. http://storm.apache.org/
2. The TPC-H Benchmark. http://www.tpc.org/tpch
3. Nasir, M.A.U., De Francisci Morales, G., et al.: The power of both choices: practical load balancing for distributed stream processing engines. In: ICDE, pp. 137–148 (2015)
4. Cormode, G., Muthukrishnan, S.: An improved data stream summary: the count-min sketch and its applications. J. Algorithms **55**(1), 58–75 (2005)
5. Elseidy, M., Elguindy, A., Vitorovic, A., Koch, C.: Scalable and adaptive online joins. In: VLDB, pp. 441–452 (2014)
6. Epstein, R.S., Stonebraker, M., Wong, E.: Distributed query processing in a relational data base system. In: SIGMOD, pp. 169–180 (1978)
7. Gedik, B.: Partitioning functions for stateful data parallelism in stream processing. VLDB J. **23**(4), 517–539 (2014)
8. Huebsch, R., Garofalakis, M., Hellerstein, J., Stoica, I.: Advanced join strategies for large-scale distributed computation. In: VLDB, pp. 1484–1495 (2014)
9. Kwon, Y., Balazinska, M., et al.: Skewtune: mitigating skew in mapreduce applications. In: SIGMOD, pp. 25–36 (2012)
10. Lin, Q., Ooi, B.C., Wang, Z., Yu, C.: Scalable distributed stream join processing. In: SIGMOD, pp. 811–825 (2015)
11. Liu, B., Zhu, Y., Jbantova, M., et al.: A dynamically adaptive distributed system for processing complex continuous queries. In: VLDB, pp. 1338–1341 (2005)
12. Nasir, M.A.U., Serafini, M., et al.: When two choices are not enough: balancing at scale in distributed stream processing. In: ICDE (2016)
13. Okcan, A., Riedewald, M.: Processing theta-joins using mapreduce. In: SIGMOD, pp. 949–960 (2011)
14. Stamos, J.W., Young, H.C.: A symmetric and replicate algorithm for distributed joins. IEEE Trans. Parallel Distrib. Syst. **4**(12), 1345–1354 (1993)

15. Ufler, B., Augsten, N., Reiser, A., Kemper, A.: Load balancing in mapreduce based on scalable cardinality estimates. In: ICDE, pp. 522–533 (2012)
16. Vitorovic, A., ElSeidy, M., Koch, C.: Load balancing and skew resilience for parallel joins. In: ICDE (2016)
17. Xing, Y., Hwang, J., Cetintemel, U., Zdonik, S.: Providing resiliency to load variations in distributed stream processing. In: VLDB, pp. 775–786 (2006)
18. Xu, Y., Kostamaa, P., Zhou, X., Chen, L.: Handling data skew in parallel joins in shared-nothing systems. In: SIGMOD, pp. 1043–1052 (2008)

Finding Frequent Items in Time Decayed Data Streams

Shanshan Wu[1], Huaizhong Lin[1(✉)], Leong Hou U[2],
Yunjun Gao[1], and Dongming Lu[1]

[1] College of Computer Science and Technology,
Zhejiang University, Hangzhou, China
{wuss,linhz,gaoyj,ldm}@zju.edu.cn
[2] Department of Computer and Information Science,
University of Macau, Macau, China
ryanlhu@umac.mo

Abstract. Identifying frequently occurring items is a basic building block in many data stream applications. A great deal of work for efficiently identifying frequent items has been studied on the landmark and sliding window models. In this work, we revisit this problem on a new streaming model based on time decay, where the importance of every arrival item is decreased over the time. To address the importance changes over the time, we propose a new heap structure, named Quasi-heap, which maintains the item order using a lazy update mechanism. Two approximation algorithms, Space Saving with Quasi-heap (SSQ) and Filtered Space Saving with Quasi-heap (FSSQ), are proposed to find the frequently occurring items based on the Quasi-heap structure. Extensive experiments demonstrate the superiority of proposed algorithms in terms of both efficiency (i.e., response time) and effectiveness (i.e., accuracy).

1 Introduction

A data stream is a massive unbounded sequence of items continuously received at a rapid rate and it appears in a variety of applications, such as network monitoring, financial monitoring, web logging, etc. Substantial analytical studies have been devoted to the data streams, such as clustering [6], classification [17], and mining frequent patterns [3]. Finding frequent items [8,11,14,16,18] has received considerable attentions in the data stream analytical tasks. This problem has been served as an important building block for different data stream mining problems, such as mining frequent itemsets [3] and computing the entropy of a data stream [2].

In typical data stream scenarios, the item arrival rate is very high so that not every received item can be kept in the main memory. Thereby, the solutions normally scan every arrival item once (i.e., sequential access) and drop unpromising items (e.g., less frequently occurring items) when the main memory becomes full. Complying with these constraints, the prior studies mostly focus on how to answer the data stream problem approximately with an error bound.

© Springer International Publishing Switzerland 2016
F. Li et al. (Eds.): APWeb 2016, Part II, LNCS 9932, pp. 17–29, 2016.
DOI: 10.1007/978-3-319-45817-5_2

Early solutions [8,10,11,14,16] of this problem are developed based on two traditional streaming models, the landmark (i.e., the frequent items can be any item in the entire stream) and the sliding window (i.e., the frequent items can only be the items of the current window). While the landmark model preserves better data completeness, it ignores the importance of newly arrival items. On the other hand, the sliding window model partially addresses the item freshness but the items not occurring in the current window are completely ignored.

To address these, finding the frequent items in *time decayed* data streams has received substantial attention over the past decade [4,5,9,12,15,19]. Under the *time decay* model, the weight of a received item is decreased over the time and the frequent items are then computed based on the time decayed weights. This model preserves better completeness (i.e., every item is considered) and item freshness (i.e., the recent items are more important) than the prior streaming models.

In the landmark and the sliding window models, the count of an item always increases by one when the item comes from the stream. Hence, given an ordered list of the frequent items (maintained by a linked list or a heap structure), we can easily maintain the order consistency by swapping the affected item with its neighbor items. Under the time decay model, the weight of every item is updated over the time subject to the decay function. The order maintenance may become costly since the updated item may swap with multiple items in the ordered list.

To address this challenge, we propose a new heap structure, named Quasi-heap, which maintains the order of the items in a heap by a lazy manner. Based on the Quasi-heap, two approximation algorithms are studied to solve the frequent item problem on time decayed data streams. We briefly list our main contributions as follows.

- We propose a new heap structure, named Quasi-heap, to maintain the frequent items based on their time decayed counts.
- We invent two approximation algorithms, Space Saving with Quasi-heap (based on Space Saving [16]) and Filtered Space Saving with Quasi-heap (based on Filtered Space Saving [10]). Our improved algorithms answer the frequent item problem with reasonable memory overhead and a guaranteed error bound.
- Extensive experiments are conducted to demonstrate the superiority of our algorithms in terms of the running time and the estimation accuracy.

The remainder of this paper is organized as follows. A survey of related work is presented in Sect. 2. Section 3 formulates the frequent item problem in time decayed streams. Section 4 discusses and analyzes the proposed Quasi-heap. Sections 5 and 6 depict two improved algorithms SSQ and FSSQ, respectively. Section 7 evaluates the proposed algorithms, and we conclude this paper in Sect. 8.

2 Related Work

2.1 Landmark Model and Sliding Window Model

In the landmark and the sliding window models, there are a great deal of work proposed to find the frequent items from a data stream. These work can be classified into two main categories [7], counter-based and sketch-based. The counter-based algorithms are deterministic, which only monitor a subset of the items from a data stream. These algorithms maintain a set of counters to track the frequent items over the subset. Space Saving (SS) [16], Lossy Counting [14], and Frequent [11] are the representative ones in this category. The sketch-based algorithms which use a set of array counters to estimate the frequencies of the items. Different from the counter-based algorithms, each item is projected into a set of corresponding sketches by some hash functions. The frequency of an item is estimated from the counters of its corresponding sketches. To minimize the collision probability of the hash functions, we can increase the granularity of the sketch (i.e., more counters are used). However, this will lead to huge memory consumption. CountSketch [2], Count-Min Sketch [8], and Filtered Space Saving (FSS) [10] are the representative algorithms in this category.

However, these work either treat the stale and the fresh data the same (i.e., the landmark model) or remove the stale item by a subjective window length (i.e., the sliding window model). In real world applications, it is more desirable if the frequent item problem not only considers every arrival item but also treats the fresh items more important than the stale items.

2.2 Time Decay Model

Finding frequent items in a time decayed data stream has received substantial attention from the community recently [4,5,9,12,15,19]. Zhang et al. [19] proposed two ϵ-approximation algorithms called Frequent-Estimating (FE) and FE with Heap (FEH). FE updates the frequent item result for an item arrival in $O(\epsilon^{-1})$ time by a linked list and FEH updates the result in $O(\log \epsilon^{-1})$ by a heap structure. Chen et al. [5] proposed another ϵ-approximation algorithm, called Frequent-item Counting (FC), which takes $O(1)$ time to maintain the answer for each arrival item by a hash function. Mei and Chen [15] proposed to estimate the frequencies of items by multiple hash functions. However, their work did not give the analysis of the memory consumption and the estimation accuracy.

Recent developments attempted to improve the estimation accuracy by either exploiting the decay function or employing a new data structure. Lim et al. [12] proposed a new ϵ-approximation algorithm, TwMinSwap, which takes $O(\epsilon^{-1})$ time to process each arrival item. The basic idea is to drop the minimum item (with the smallest counter) when the memory becomes full, where the counters are updated over the time by multiplying the decay rate. λ-HCount algorithm [4] employs a double linked list to record the frequent items and improves the frequency estimation accuracy by multiple hash functions. The items monitored in the double linked list are arranged in the descending order of their recently updated time. Since the items are organized in a double linked queue structure, the algorithm can reallocate an item entry to the end of the list in $O(1)$ time.

All the above algorithms are based on a backward decay function where the item importance is decreased over the time. The main challenge under the backward decay function is that the weights of the existing items are constantly changed. To address this, Cormode et al. [9] studied an alternative decay function that is a monotone increasing function to the *age* of an item (i.e., the subtraction of the arrival time and the origin time). In this model, the item weight is fixed when the *age* of an item is decided. In other words, the weights of the existing items become stable and the problem of finding the frequent item becomes easier. However, the *forward* weight of an item will become very large (due to the *age*) if the system has been running for a long time. One possible solution is to reset the origin time periodically but it needs extra effort to recompute the frequent items. The effectiveness of this model on the frequent item problem is unknown.

For clarity, in this work we focus on finding the frequent items based on the backward decay function as it is widely adopted in the prior studies.

3 Definitions and Preliminaries

Table 1 summarizes the notation to be used in this paper. We use a standard stream model with discrete timestamps and only one item arrives at every timestamp. The current data stream $D_n = <I_1, I_2, ..., I_n>$, where $I_t \in \{a_1, ..., a_D\}$.

Table 1. Summary of notation

Notation	Description	Notation	Description
D_n	The data stream up to time n	I_t	The received item at time t
D	The number of distinct items	$\{a_1, ..., a_D\}$	The set of distinct items
ϕ	Frequency threshold	ϵ	Error tolerance parameter
m	The length of the monitored list	τ	Time decay rate
$C_t(a_i)$	The decayed count of a_i at time t	$c_t(a_i)$	The estimate value of $C_t(a_i)$

While processing a long stream, it is reasonable to treat a recent item more important than an old item. In this work we adopt a backward time decay model that is used to gradually decrease the effect of the obsolete items.

Definition 1 (Decayed count of an item, $C_n(a_i)$ [3]). *Given a time decay rate τ ($0 < \tau \le 1$), $C_n(a_i)$ is the decayed count of an item a_i at time n, i.e.,*

$$C_n(a_i) = C_{n-1}(a_i) \times \tau + W_n(a_i) \tag{1}$$

where $C_1(a_i) = W_1(a_i)$ and $W_k(a_i)$ is a function that indicates the arrival of an item a_i at the timestamp k. Specifically, if $I_k = a_i$, $W_k(a_i) = 1$; otherwise, $W_k(a_i) = 0$.

Definition 2 defines the frequent item in this work subject to a frequency threshold ϕ. Specifically, an item a_i is in the result set if and only if its *normalized decayed count* is higher than ϕ and the normalization factor is the sum of all items $|D_n|$ $(= \frac{1-\tau^n}{1-\tau}$, which approaches to $\frac{1}{1-\tau}$ when $n \to \infty)$.

Definition 2 (ϕ-frequent item). *Given a stream D_n, and a frequency threshold ϕ, $0 < \phi \le 1$. If the decayed count $C_n(a_i)$ is higher than $\phi \times |D_n|$ $(|D_n|$ is the sum of decayed counts of all items in D_n), then a_i is a ϕ-frequent item.*

We need huge memory to maintain the ϕ-frequent items exactly in a long running data stream, where the space complexity is $\Omega(D)$ when the number of distinct items in the stream is D [7]. Thereby, the typical solutions focus on answering this problem approximately subject to an error bound ϵ. Formally, the approximation version of the problem is defined as follows.

Problem 1. (ϵ-Approximate Decayed Frequent items). Given a data stream D_n, the ϵ-approximate frequent item set contains all items where their decayed counts are higher than $(\phi - \epsilon) \times |D_n|$.

4 Quasi-heap

To find the frequent items in a data stream, a group of counters is used to record the candidate items. We assume that there are m counters available (subject to the memory budget) and these counters are organized into a linked list or a heap structure. To address the memory budget, the prior studies [16] replace the smallest item by the newly arrival item when the memory becomes full.

According to Definition 1, the decayed count of an item a_i is increased when a_i is received from the stream. Thereby, the position of a_i in the counters should be changed in order to keep the order correct. The complexity of each update takes $O(m)$ (for the linked list) and $O(\log m)$ (for the heap structure), which is definitely too time consuming in data stream environments. Hence, we propose a new data structure called Quasi-heap which aims at postponing their sorting operations when the decayed count of an existing item is increased. In other words, we allow certain order inconsistency in the Quasi-heap structure. An example of the Quasi-heap is given in the following Example 1.

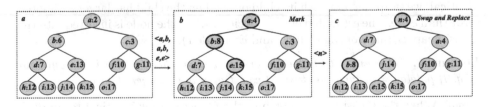

Fig. 1. A running example of Quasi-heap

Example 1. (An example of Quasi-heap). In Fig. 1, each node consists of the item name and its decayed count. We adopt $\tau = 1$ for ease of presentation. Figure 1*a* shows a Quasi-heap that contains 12 items where the order is identical to that of the ordinary heap. Upon receiving the next sequence $< a, b, a, b, e, e >$, the corresponding counts of a, b, e are increased. Instead of running heapify to maintain the heap structure, we only mark these items as *delayed* (e.g., these items marked by thick lines in Fig. 1*b*) since they are the old items in the Quasi-heap. While receiving a new item n, we start to run the heapify partially to those *delayed* nodes starting from the root. After the heapify process, c becomes the root node as it is the smallest item in the Quasi-heap. And then, c is replaced by n (cf. Fig. 1*c*) where the *estimated count* of n is set to 4 (i.e., the count of c + 1) as followed the suggestion of other counter-based algorithms [16].

According to the discussion in Example 1, the main operation, *delayedSorting*, is to execute heapify partially on the Quasi-heap so that the minimum node can be properly identified and removed.

Algorithm 1. *delayedSorting*

1: $c.error \leftarrow c.error \times \tau^{(t-c.ut)}$; $c.cnt \leftarrow c.cnt \times \tau^{(t-c.ut)}$; $c.ut \leftarrow t$;
2: **if** $c.delay = 0$ or c is a leaf node **then** return;
3: *delayedSorting* (each child counter of c);
4: **if** $c.cnt > sml.cnt$ **then**   ▷ sml is the smaller of the two child counters
5:      swap c and sml and $sml.delay \leftarrow 1$;
6: **if** $c.cnt = sml.cnt$ **then**
7:        **if** $sml.error > c.error$ **then**
8:            swap c and sml and $sml.delay \leftarrow 1$;
9: $c.delay = 0$;

Algorithm 1 describes the *delayedSorting* operation in detail. The information of an item is updated in line 1. If the delayed flag of an item is not marked, in line 2, then its count must be the minimum count in its subtree according to Lemma 1 (being discussed shortly). If the delayed flag of an item is marked, the order of this item may be inaccurate so that we need to execute the *delayedSorting* operation on its each child (line 3). After the recursive calls, if the count of the root is larger than that of its children, we swap the root with its child in lines 4–5. If the counts are identical, we swap the root with its child only when the child has larger estimated error than c (i.e., the estimated error is decided when the node is inserted into the Quasi-heap, cf. line 7 of Algorithm 2 and Example 1).

Lemma 1. *Let p and q be two counters in a Quasi-heap and p is an ancestor of q. If $p.delay = 0$, then the decayed count of p is no larger than q.*

Proof. When the Quasi-heap is not full, the Quasi-heap is identical to an ordinary heap so that $p.cnt \leq q.cnt$ due to the heapify.

When the Quasi-heap becomes full, there are two cases. If p is not received from the data stream again, then $p.cnt \leq q.cnt$ is still held no matter whether q

has been updated or not due to the monotonicity of the decayed count function (cf. Eq. 1). If p has been received from the data stream again, the *delay* flag of p must be set to 1. The *delay* flag is reset to 0 only when the subtree of p is refined by the heapify (cf. Algorithm 1) so that $p.cnt \leq q.cnt$ is still held.   □

Time Complexity Analysis: When the newly arrival item is in the Quasi-heap, we only update the decayed count of this item and mark the delay flag. Hence, processing an existing item takes $O(1)$ time. When the newly arrival item is not in the Quasi-heap, we replace the minimum item of the Quasi-heap by this new item. To find the minimum item in the Quasi-heap, we execute Algorithm 1 to ensure the correctness of the order. The cost of Algorithm 1 is $O(m)$ as it may traverse the entire tree in the worst case. However, this case is very rare to happen in real datasets. In addition, a frequent item is likely kept in the Quasi-heap (as their counts are high) and it is more frequently received form the data stream than other items. The response time of processing the existing items is dramatically reduced from $O(\log m)$ to $O(1)$. In our experiments, the Quasi-heap can reduce response time up to 80 % as compared with the ordinary heap.

5 Space Saving Algorithm with Quasi-heap

We study a counter-based algorithm, SSQ (that is based on the SS algorithm [16]), to find the ϵ-approximate decayed frequent items. Algorithm 2 depicts the SSQ in detail. If the new arrival item c is already in the Quasi-heap (lines 2–3), we update its statistics and mark the *delay* flag as 1. Otherwise, we first check whether the Quasi-heap is full or not. If the Quasi-heap is not full (lines 9–11), we simply execute heapify to maintain the consistence of the Quasi-heap. Otherwise, we run the *delayedSorting* from the root of the Quasi-heap and replace the *refined* root by the new item c. Similar to the SS algorithm, the estimated count of a new item c is derived from the count of the removal item r.

Algorithm 2. *SSQ* Algorithm
___
1: **for** each coming item c at timestamp t **do**
2:     **if** c is tracked in Quasi-heap **then**
3:         $c.error \leftarrow c.error \times \tau^{t-c.ut}$; $c.cnt \leftarrow c.cnt \times \tau^{t-c.ut}+1$; $c.ut \leftarrow t$; $c.dealy \leftarrow 1$
4:     **if** c is not tracked in Quasi-heap **then**
5:         **if** Quasi-heap is full **then**
6:             $delayedSorting(r)$ ▷ r is the root of Quasi-heap
7:             $c.error \leftarrow r.cnt \times \tau^{t-r.ut}$; $c.cnt \leftarrow r.cnt \times \tau^{t-r.ut}+1$; $c.ut \leftarrow t$;
8:             $c.dealy \leftarrow 1$; and replace r by c
9:         **if** Quasi-heap is not full **then**
10:            create a new counter c; $c.error \leftarrow 0$; $c.cnt \leftarrow 1$; $c.ut \leftarrow t$; $c.dealy \leftarrow 0$
11:            insert and maintain c in the Quasi-heap
___

We present two properties of SSQ algorithm, which are based on the properties proposed in SS [16] and FE [19] with some minor modifications.

Lemma 2. *Among all m counters, the minimum counter value $\mu \leq \frac{1-\tau^n}{m(1-\tau)}$.*

Proof. The sum of estimated counts of all m items in the monitored list is equal to the sum of decayed counts of all n items in the data stream, i.e., $\Sigma_i c_n(a_i) = \frac{1-\tau^n}{1-\tau}$. It is following that $m\mu \leq \Sigma_i c_n(a_i)$, so $\mu \leq \frac{1-\tau^n}{m(1-\tau)}$. $\qquad\square$

Based on the Lemma 2, SSQ can use confined space (i.e., $m = \lceil 1/\epsilon \rceil$) to find ϵ-approximate frequent items by securing the error ratio at most $\epsilon \times |D_n|$ [16].

Theorem 1 (No False Negative). *For any item a_i with the decayed count $C_n(a_i)$ greater than μ is present in the Quasi-heap.*

Proof. We prove the theorem by contradiction. Assume at the current time t, an item a_i with decayed count $C_t(a_i) > \mu$ is not in the Quasi-heap. Then, a_i must be evicted sometime in the past. Suppose a_i was last evicted at time unit t', then its decayed count $C_{t'}(a_i) = C_t(a_i)/\tau^{t-t'}$, which is larger than $\mu/\tau^{t-t'}$ (based on the condition $C_t(a_i) > \mu$). Let $\mu_{t'}$ be the minimal count at t', then $\mu_{t'} \leq \mu/\tau^{t-t'}$. We can get $C_{t'}(a_i) = C_t(a_i)/\tau^{t-t'} > \mu/\tau^{t-t'} \geq \mu_{t'}$. So, clearly, $C_{t'}(a_i) \geq \mu_{t'}$. This fact means that the estimated count of the item a_i was greater than the minimum counter value when it was evicted at time unit t'. This contradicts the fact that the SSQ algorithm evicts the item with the minimum counter value. $\qquad\square$

6   Filtered Space Saving Algorithm with Quasi-Heap

In this section, we propose a sketch-based algorithm, FSS with Quasi-Heap (FSSQ) (that is based on the FSS algorithm [10]), to find the frequent items. The FSSQ algorithm employs two data structures, (1) the Quasi-heap and (2) a sketch (i.e., a two-dimensional array with width w and depth d). The sketch used in this work is similar to that of the Count-Min algorithm [7]. Each entry of the sketch is composed of an estimated count and the time of last update, denoted as (cnt, ut). To update the value of the sketch entries, we need d pairwise-independent hash functions: $h_1, ..., h_d : \{1, 2, ..., D\} \rightarrow \{1, 2, ..., w\}$. FSSQ improves the estimation accuracy since the count of a new item is estimated by d sketch entries instead of the minimum item in the Quasi-heap (cf. SSQ).

Algorithm 3 depicts the FSSQ, whose idea is similar to that of Algorithm 2 except the situation that a new item is not tracked in Quasi-heap and Quasi-heap becomes full, hence we omit the similar parts. We first run *delayedSorting* from the root (line 4) in order to find the minimum item. Next we update the corresponding sketch entries by c (lines 6–7), and estimate its minimum value among d corresponding sketch entries (line 8). If the estimated minimum count is larger than the root of Quasi-heap, then we replace the root by the new item c (lines 12–13) and update the corresponding sketch entries by the evicted item r (lines 10–11).

The properties of the FSSQ are given in Lemmas 3–5 and Theorem 2.

Algorithm 3. *FSSQ* Algorithm

1: **for** each coming item c at timestamp t **do**
2:     **if** c is not tracked in the Quasi-heap **then**
3:         **if** Quasi-heap is full **then**
4:             $delayedSorting(r)$ ▷ r is the root of Quasi-heap
5:             Let $s[x, y]$ be the counter in the sketch entry (x, y);
6:             **for** $j = 1, \cdots, d$ **do**   ▷ d is the depth of the sketch
7:                 $s[j, h_j(c)].cnt \leftarrow s[j, h_j(c)].cnt \times \tau^{t-s[j,h_j(c)].ut}+1$; $s[j, h_j(c)].ut \leftarrow t$;
8:             $est \leftarrow min_{1 \leq j \leq d} s[j, h_j(c)].cnt\}$;
9:             **if** $est > r.cnt$ **then**
10:                 **for** $j = 1, \cdots, d$ **do**
11:                     $s[j, h_j(r)].cnt \leftarrow r.cnt \times \tau^{t-r.ut}$; $s[j, h_j(r)].ut \leftarrow t$;
12:                 $c.error \leftarrow r.cnt \times \tau^{t-r.ut}$; $c.cnt \leftarrow r.cnt \times \tau^{t-r.ut} + 1$; $c.ut \leftarrow t$;
13:                 $c.dealy \leftarrow 1$; and replace r by c

---

Lemma 3. *At any moment, for each item a_i, the minimum count μ in the Quasi-heap is no less than the minimum entry that the hash function values of a_i associates, i.e., $\mu \geq min_{1 \leq j \leq d} s[j, h_j(a_i)]$.*

Proof. In the initialization phase, all the hits are reflected in Quasi-heap, and all the entries in sketch have value of 0. Hence, the conclusion is trivially true.

When the Quasi-heap has been full. For a new coming item, if it is being monitored in the Quasi-heap, its counter in the Quasi-heap increases and all the entries in the sketch remain unchanged. If a new coming item is not being monitored in the Quasi-heap, the increment of the entry in the sketch may lead to the situation that the minimum entries become larger than μ. However, at this case, a replacement is taken place. Hence, the conclusion is still true.   □

Lemma 4. *For any item in the Quasi-heap, its overestimated error is no greater than μ.*

Proof. For any item a_i in the Quasi-heap, its maximum overestimated error is always assigned the minimum decayed count in the entries that a_i associates, when a replacement takes place. This value is no greater than μ according to Lemma 3, so the maximum overestimated error is no greater than μ.   □

Lemma 5. *Assume $w = \lceil e/\epsilon \rceil$ and $d = \lceil ln(1/\delta) \rceil$, in which e is the Euler's constant, i.e., the base of natural logarithms. For any item in the Quasi-heap, its count error is no greater than $\epsilon/(1 - \tau)$ with probability at least $1 - \delta$.*

Proof. We omit the proof due to the space limit.

Theorem 2. *For any item with $C_n(a_i) > \mu$ is present in the Quasi-heap.*

Proof. We first prove that a_i is present in the Quasi-heap. If the last arrival time of a_i is t_1 time unit ago, then the estimated count of a_i in the sketch t_1 time units ago is no less than $C_n(a_i)/\tau^{t_1}$. Thereby, $min_{1 \leq j \leq d} s[j, h_j(a_i)]/\tau^{t_1} \geq C_n(a_i)\tau^{t_1} > \mu/\tau^{t_1}$. This means that a_i was inserted in the monitored counters t_1 time units ago. We also need to show there is no false negative result. The proof is identical to that of Theorem 1 if a_i is present in the Quasi-heap.   □

7   Experimental Study

In this section, we empirically evaluated the efficiency and effectiveness of SSQ and FSSQ using both real and synthetic datasets. We compared proposed solutions with the state-of-the-art solutions, TwMinSwap [12] (counter-based) and λ-HCount [4] (sketch-based). All methods were implemented in C++ and compiled using the Microsoft Visual Studio 2012 compiler. All experiments were conducted on a 3.20 GHz Pentium PC machine with 8 GB main memory running Windows 7 Professional Edition.

7.1   Experimental Settings

The synthetic datasets are generated based on Zipfian distributions. The Zipfian parameter value is from 0.8 to 2 (the default value is 1.0). We also used two real datasets widely evaluated in the data stream research [13]. The Kosarak is an anonymized click-stream on a Hungarian online news portal[1]. The Retail contains retail market basket data from an anonymous Belgian store [1]. In our experiments, we consider every single item in sequential order. Kosarak (Retail) is of all the items 8019015 (908576) and distinct items 41270 (16470).

We verify the efficiency with respect to **Time** (Each algorithm is run for 20 times and the average response time is reported), **Precision** (The fraction of the items identified by the algorithm that are actually frequent), and **Recall** (All the frequent items are detected due to the overly estimated count (cf. Theorem 1)).

7.2   Experimental Results

To verify the scalability, we varied one parameter in each set of experiments while setting other parameters to their default values. ϕ is from 0.0001 to 0.01 (the default value is 0.001), m is from 1000 to 4000 (the default value is 1000) and τ is from 0.97 to 1.00 (the default value is 0.995).

Performance Overview: In terms of the response time, SSQ yields better performance than state-of-art solutions, due to the Quasi-heap data structure proposed which can save the response time by up to 80 %. In terms of the precision, FSSQ performs the best among all methods due to the sketch structure which can get a better bound on the estimation count error.

Figure 2 shows the response time and the precision by varying the cardinality of the items. Notably, the response time of all methods increases as n becomes larger. We find that the counter-based algorithms are faster than the sketch-based algorithms; however, the counter-based algorithms is less accurate than the sketch-based algorithms as discussed in Sect. 6.

It is obvious that the response time decreases when the data becomes more skewed (cf. Fig. 3(a)). The cost of processing an existing item in the Quasi-heap is less than that of processing a new item in the Quasi-heap. The probability of receiving an existing item becomes higher when the data is more skewed.

---

[1] Frequent Itemset Mining Dataset Repository http://fimi.cs.helsinki.fi/data/.

Figures 4, 5 and 6 show the experiments conducted ϕ, τ, and memory, respectively. Due to the space limit, each figure only reports one set of experiments as all methods preform similarly on three datasets. For instance, SSQ is 35% − 40% faster than TwMinSwap on average for ϕ and τ on all three datsets. Figure 6 shows that the precision increases when we have more space.

7.3   Effectiveness of Quasi-heap

We also performed an experiment to investigate the advantage of the Quasi-heap (SSQ) as compared to the ordinary heap (SS). We reported the number of comparisons performed in the heap and the response time. Figure 7(a) shows that the performance between the Quasi-heap and the ordinary heap. For example, when the stream size is 10^7 and the data skew is 3.0, the number of comparisons in the Quasi-heap is $2k$ which is 118 times smaller than $236k$ in the ordinary heap. From Fig. 7(b), we can find that the Quasi-heap reduces the response time by up to 80% since huge amount of unpromising comparisons are delayed by the *delaySorting* method.

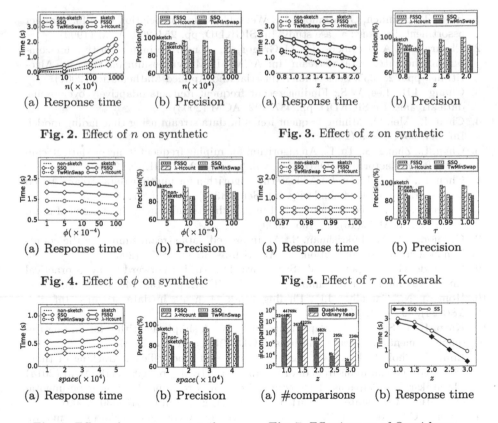

(a) Response time   (b) Precision   (a) Response time   (b) Precision

Fig. 2. Effect of n on synthetic   **Fig. 3.** Effect of z on synthetic

(a) Response time   (b) Precision   (a) Response time   (b) Precision

Fig. 4. Effect of ϕ on synthetic   **Fig. 5.** Effect of τ on Kosarak

(a) Response time   (b) Precision   (a) #comparisons   (b) Response time

Fig. 6. Effect of memory on retail   **Fig. 7.** Effectiveness of Quasi-heap

8 Conclusion and Future Work

In this paper, we focused on the problem of finding frequent items in data streams with a time decay model. In order to reduce the maintenance cost of the ordinary heap, we proposed a Quasi-heap data structure with a delayed sorting operation and invented two algorithms based on it. We extensively evaluated our methods on three datasets. In the future, we plan to extend the proposed methods in a distributed environment especially for data whose volume does not fit into memory of a stand-alone machine.

Acknowledgement. This work was supported by the public key plan of Zhejiang Province (2014C23005), National Science and Technology Supporting plan (2013BAH62F02 and 2013BAH27F01), China mobile research fund of education ministry (mcm20130671), the cultural relic protection science and technology project of Zhejiang Province, University of Macau RC (MYRG2014-00106-FST), and NSFC of China (61502548).

References

1. Brijs, T., Swinnen, G., Vanhoof, K., Wets, G.: Using association rules for product assortment decisions: a case study. In: SIGKDD, pp. 254–260. ACM (1999)
2. Chakrabarti, A., Cormode, G., McGregor, A.: A near-optimal algorithm for computing the entropy of a stream. In: ACM-SIAM Symposium on Discrete Algorithms, pp. 328–335. Society for Industrial and Applied Mathematics (2007)
3. Chang, J.H., Lee, W.S.: Finding recent frequent itemsets adaptively over online data streams. In: SIGKDD, pp. 487–492. ACM (2003)
4. Chen, L., Mei, Q.: Mining frequent items in data stream using time fading model. Inf. Sci. **257**, 54–69 (2014)
5. Chen, L., Zhang, S., Tu, L.: An algorithm for mining frequent items on data stream using fading factor. In: COMPSAC, vol. 2, pp. 172–177. IEEE (2009)
6. Chen, L., Zou, L.J., Tu, L.: A clustering algorithm for multiple data streams based on spectral component similarity. Inf. Sci. **183**(1), 35–47 (2012)
7. Cormode, G., Hadjieleftheriou, M.: Finding the frequent items in streams of data. Commun. ACM **52**(10), 97–105 (2009)
8. Cormode, G., Muthukrishnan, S.: An improved data stream summary: the count-min sketch and its applications. J. Algorithms **55**(1), 58–75 (2005)
9. Cormode, G., Shkapenyuk, V., Srivastava, D., Xu, B.: Forward decay: a practical time decay model for streaming systems. In: ICDE, pp. 138–149. IEEE (2009)
10. Homem, N., Carvalho, J.P.: Finding top-k elements in data streams. Inf. Sci. **180**(24), 4958–4974 (2010)
11. Karp, R.M., Shenker, S., Papadimitriou, C.H.: A simple algorithm for finding frequent elements in streams and bags. TODS **28**(1), 51–55 (2003)
12. Lim, Y., Choi, J., Kang, U.: Fast, accurate, and space-efficient tracking of time-weighted frequent items from data streams. In: CIKM, pp. 1109–1118. ACM (2014)
13. Manerikar, N., Palpanas, T.: Frequent items in streaming data: an experimental evaluation of the state-of-the-art. Data Knowl. Eng. **68**(4), 415–430 (2009)
14. Manku, G.S., Motwani, R.: Approximate frequency counts over data streams. In: PVLDB, pp. 346–357. VLDB Endowment (2002)

15. Mei, Q.L., Chen, L.: An algorithm for mining frequent stream data items using hash function and fading factor. In: Applied Mechanics and Materials, vol. 130, pp. 2661–2665. Trans Tech Publications (2012)
16. Metwally, A., Agrawal, D., Abbadi, A.E.: An integrated efficient solution for computing frequent and top-k elements in data streams. TODS **31**(3), 1095–1133 (2006)
17. Shaker, A., Senge, R., Hüllermeier, E.: Evolving fuzzy pattern trees for binary classification on data streams. Inf. Sci. **220**, 34–45 (2013)
18. Tong, Y., Zhang, X., Chen, L.: Tracking frequent items over distributed probabilistic data. World Wide Web **19**(4), 1–26 (2015)
19. Zhang, S., Chen, L., Tu, L.: Frequent items mining on data stream based on time fading factor. In: AICI, vol. 4, pp. 336–340. IEEE (2009)

A Target-Dependent Sentiment Analysis Method for Micro-blog Streams

Yongheng Wang[✉], Hui Gao, and Shaofeng Geng

College of Information Science and Engineering, Hunan University,
Changsha 410082, China
wyh@hnu.edu.cn, 1030278455@qq.com, sfgeng@163.com

Abstract. Sentiment analysis technique is useful for companies to analyze customer's opinion about products and to find potential customers. Most of the target-dependent sentiment analysis methods can not get acceptable accuracy. Recently some new sentiment analysis methods using Recursive Neural Networks (RNN) are promising but they are not target-dependent. In this paper we propose a target-dependent sentiment analysis method for micro-blog streams based on RNN. We use cluster-based data partitioning to get higher accuracy with limited labeled samples. A tree pruning method is proposed to remove irrelevant parts from the syntax tree. The original recursive neural network model is extended to support target-dependent sentiment analysis better. Experimental results on two corpuses with different targets show that the performance of our method is better than previous methods.

1 Introduction

Micro-blog, such as Twitter and Sina micro-blog, is one of the most popular applications in the age of Web 2.0 which allow users to publish short messages to share their opinions with others. Since its emergence, micro-blog has attracted a lot of people to record their lives, discuss hot topics, express and share their opinions. It has become an important resource of mining people's opinions and sentiments [1]. Micro-blog provides effective data to support customer satisfaction survey, public sentiment detection, sociological research, etc. Sentiment analysis technique can be used to classify the texts automatically according to the sentiments expressed by the authors. Recently one of the useful applications is to manage potential customers through sentiment analysis of micro-blog streams. By monitoring the online micro-blog streams and analyzing the customers' opinions about products, a company can find the defect of their products and find potential customers. Comparing to the traditional textual data, micro-blog posts have some new characters such as limited size, often containing spelling and grammatical mistakes, using variety of emoticons and colloquial expressions [2], etc. These characters make it more difficult to determine the sentiment polarity of micro-blog.

This work is supported by the National Natural Science Foundation of China under grant 61371116.

F. Li et al. (Eds.): APWeb 2016, Part II, LNCS 9932, pp. 30–42, 2016.
DOI: 10.1007/978-3-319-45817-5_3

Micro-blog posts often contain diverse evaluation objects, some of which are irrelevant to the given targets. However, the state-of-the-art sentiment analysis methods basically work in a target-independent way that all the sentiment expressions appearing in the posts will be regarded as being opinionated about the target, no matter it is right or not, which will lead to some mistakes [3]. Recently new sentiment analysis methods using Recursive Neural Networks (RNN) [4, 5] which are proposed by Stanford university show good performance. Deep learning with RNN is proved to be a promising direction for sentiment analysis. However, these methods have some limitations. A lot of labeled samples are needed to train the model. All nodes in the input trees should be labeled manually with one of the five sentiment levels, which is very expensive. Another limitation is that the RNN model is target-independent which cannot be used directly for target-dependent sentiment analysis. In this paper, we propose a target-dependent sentiment analysis method based on RNN (TSA-RNN). We use a cluster based data partitioning method to get higher accuracy with limited labeled samples. A tree pruning method is proposed to remove irrelevant parts from the syntax tree before constructing the RNN. The RNN model is also extended to support target-dependent sentiment analysis better.

2   Related Works

The current sentiment analysis methods mainly include lexicon based and machine learning based methods. Base on some sentiment dictionaries, the lexicon based methods use the dominant polarity of the opinion words to determine the polarity of text. Fiaidhi et al. used three scoring scheme to score the tweets, including subtracting positives from negatives, TF-IDF weighted scheme and Latent Dirichlet allocation (LDA) [6]. Yang et al. utilized HowNet and current opinion lexicon to construct a more comprehensive lexicon for internet textual data [7]. However, sentiment analysis for micro-blog usually gets poor performance since they contain many kinds of special emotional words and expressions that do not exist in the general opinion lexicons.

The machine learning based methods treat sentiment classification simply as a special case of text classification. These methods need large sets of labeled data to train sentiment classifiers. Pang et al. selected unigram, bigram, part-of-speech (POS) tag, and the positions of opinion words as features [8]. Naïve Bayesian (NB), Maximum Entropy (ME), and Support Vector Machine (SVM) are used to build the classifiers. Davidov et al. utilized Twitter tags and smileys as sentiment labels which avoided the need of manual annotation [9]. Their method allows identification and classification of diverse sentiment types of tweets. Recently, Socher et al. proposed a RNN-based sentiment analysis method [4, 5]. Their RNN model is trained using a sentiment treebank which includes fine grained sentiment labels for large amount of phrases in the parse trees. Their model outperforms all previous methods on several metrics.

Target-dependent methods classify the posts according to the sentiments expressed towards the given targets. Nasukawa and Yi applied the semantic analysis with syntactic parser and sentiment lexicon to identify the semantic relationships between sentiment expressions and subjects [10]. The polarity of sentiment expressions are determined according to the rules made manually. This kind of methods cannot analyze complicate sentences and have low recall because it is hard to create a comprehensive rule set. Jiang et al. proposed a target-dependent twitter sentiment classification method by using both target-dependent and context-aware approaches [11]. The target-dependent approach refers to incorporating syntactic features to decide whether or not the sentiment is about the given target. The context-aware approach considers related posts. Dong et al. proposed an adaptive recursive neural network model (AdaRNN) for target-dependent twitter sentiment classification [12]. They used dependency tree to construct the RNN and extended the RNN model by using different composition matrices depending on the linguistic tags and the combined vectors. Compared to their work, we extended the RNN model to be target-dependent while the composition matrices of AdaRNN are not target-dependent. We also used cluster based data partition to get high accuracy with limited labeled samples.

3   Sentiment Analysis with Recursive Neural Networks

In the RNN model [4, 5, 13], phrases and words are represented as D-dimensional vectors. It performs vector compositions based on binary trees, and compute parent vectors in a bottom up fashion using different types of compositionality functions recursively.

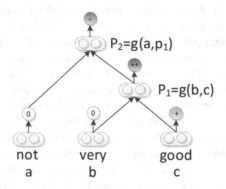

Fig. 1. System architecture

As illustrated in Fig. 1, when a phrase "not very good" is parsed into the binary tree, each word in leaf node is represented as a vector. The representation of "very good" is obtained by the composition of "very" and "good", and the representation of trigram "not very good" is recursively obtained by the composition of "not" and "very good". The model uses the following equation to compute the parent vectors.

$$v_P = f(g(v_l, v_r)) = f(W \begin{bmatrix} v_l \\ v_r \end{bmatrix} + b) \tag{1}$$

where v_P is the parent vector, v_l and v_r are the vectors of left and right child, g is the composition function, f = tanh is a standard element-wise nonlinearity, $W \in \Re^{D \times 2D}$ is the composition matrix, and b is the bias vector. Each parent vector $v_{p,i}$ is given to the same softmax classifier of Eq. (2) to compute its label probabilities.

$$s^{v_{p,i}} = soft\max(W_s v_{p,i}) \qquad (2)$$

where $Ws \in \Re^{5 \times D}$ is the sentiment classification matrix (sentiment is classified into 5 levels: very negative, negative, neutral, positive and very positive).

This basic model has been extended to Matrix-Vector RNN (MV-RNN) [13] which represents every word and longer phrase in a parse tree as both a vector and a matrix, and Recursive Neural Tensor Network (RNTN) [5] which uses the same, tensor-based composition function for all nodes.

4   TSA-RNN Method

4.1   TSA-RNN Overview

The overall architecture of TSA-RNN is shown in Fig. 2. The historical Micro-blogs are clustered offline according to the syntax tree structure similarity. Samples of a certain number are selected randomly from each cluster and labeled manually. The labeled samples are then used to train the target-dependent RNN model. The Micro-blogs in online streams are parsed into syntax tree. Irrelevant parts in the tree are analyzed and pruned with the help of domain ontology and syntactic paths library. The final binary tree is inputted into the target-dependent RNN to predict semantic label. The online Micro-blog streams are also clustered using evolving clustering algorithm. If a new cluster is found and the size of the cluster exceeds a threshold value, we can consider label the samples of this cluster and retrain the model.

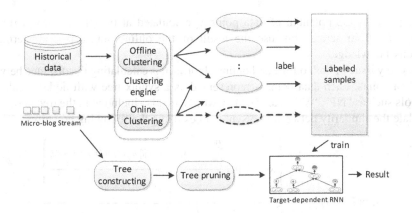

Fig. 2. The overall architecture of TSA-RNN

4.2   Cluster-Based Data Partitioning

Labeling the samples is quite an expensive task since all nodes in the tree must be labeled manually. In order to get better accuracy using the RNN model with limited labeled samples, the training samples should distribute evenly on the data space. Since the TSA-RNN model is constructed based on the syntax tree and it is target-dependent, we cluster the Micro-blogs using syntax tree structure and some key words as features. The key words are set manually which include the words that can affect the sentiment analysis for various targets, e.g. "than", "however", "but", etc.

We use subtractive clustering [14] as offline clustering algorithm to cluster historical data. Data points themselves are considered as candidate focal points (cluster centers). For every data point a potential value (density) is calculated using the following equation

$$pi(zi) = \sum_{j=1}^{N} e^{-\alpha \|z_i - z_j\|^2}, \ \alpha = 4/r_a^2 \tag{3}$$

where $p_i(z_i)$ is the potential value for data point z_i, N is the number of data points, and r_a is a positive constant. Thus, the measure of potential for a data point is a function of its distances to all other data points. The point with highest potential value is selected as the first cluster center. Then the potential of all data points are reduced by an amount that is dependent on their distance to the cluster center. The next cluster center is the point with the remaining maximum potential, and so on.

To cluster online Micro-blog streams, we use the evolving clustering (eClustering) algorithm [15, 16] which is built upon subtractive clustering. We defined the potential in eClustering as:

$$p_k(z_k) = \frac{1}{1 + (\sum_{i=1}^{k-1} \|z_i - z_k\|^2)/(k-1)} \tag{4}$$

where $p_k(z_k)$ is the potential of data point z_k calculated at time k. We don't run the eClustering from scratch but use the result of the offline subtractive clustering as previous knowledge.

The key issue of both offline and online algorithm is calculating the distance between two data points. Each data point is represented as a syntax tree with nodes labeled by symbols such as "NP", "VP", etc. The key words are also included in the tree nodes. We calculate the similarity between two syntax trees using the following equation:

$$sim(T_1, T_2) = \frac{\sum_{i=1}^{|T|} \left\{ size(T[i]) \frac{\sum_{j=1}^{|T[i]|} [\alpha + (1-\alpha) \cdot s_k(T[i].node[j])] \cdot ht(T[i].node[j])}{\sum_{j=1}^{|T[i]|} ht(T[i].node[j])} \right\}}{\sum_{i=1}^{|T|} size(T[i])} \tag{5}$$

where T[] is the set of common subtrees between two syntax trees T_1 and T_2, size(T[i]) is the nodes number of subtree T[i], function ht(x) gets the height of node x in the original syntax tree, $\alpha \in (0,1)$ is a weight factor, function $s_k(x)$ calculates the similarity between the key words from T_1 and T_2 on common node x. Equation (5) makes the similarity value to be proportional to the number of common subtrees, the size of common subtrees, the height of the nodes of common subtrees, and the key words similarity of common subtrees. We use the method proposed by Lin et al. [17] to get all common subtrees. The subtrees with small size (smaller than a threshold value) are ignored. The distance between two syntax trees T_1 and T_2 is defined as $1/sim(T_1,T_2)$.

4.3 Tree Pruning

Syntax tree usually include noise which makes the cost of classification to be high. Focusing on target-dependent sentiment analysis, we propose a pruning algorithm to prune the target-irrelevant appraisal expressions from syntax trees.

Some targets are not included directly in the text. For example, when we analysis smart phone "Samsung", "Android" and "screen" might be hidden targets which are related to the main target. In order to handle this issue, we create domain ontologies for the product opinion analysis. An example is shown in Fig. 3. We used a semi-automated method based on Formal Concept Analysis (FCA) to create the domain ontology. The formal concepts and relations are extracted based on FCA. We then select and adjust them by hand to create the domain ontology.

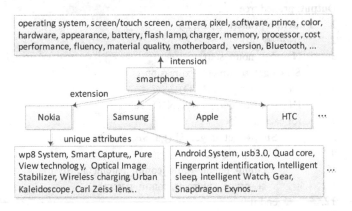

Fig. 3. An example domain ontology for smartphone

In order to prune useless part of syntax tree, we need to detect effective appraisal expressions which consist of the sentiment expressions and their targets [18]. The syntactic path between the sentiment expression and its target refers to the syntactic structure connects these two nodes in the syntax tree. In Fig. 4, the syntactic path between sentiment expression "negative comment" and its target "Jingdong" is

NN→NP→IP→VP→VP→NP→NN. By counting the syntactic paths on large scale corpus, we can find that the effective paths occur much more frequently than the invalid ones. Generally, the targets are nouns which we can detect through POS tagging, and the sentiment expressions can be found by opinion lexicons. After extracting the syntactic paths from syntax tree, we need to carry out the generalization that makes paths with small difference aggregate to one path. This step aggregates the same adjacent elements in a path, e.g., the path shown above can be generalized as NN→NP→IP→VP→NP→NN. We select the most frequently-appearing paths to build the syntactic path library.

The tree pruning algorithm is shown in Algorithm 1. In lines 3–4 noun and sentiment expression sets are obtained by POS tagging and sentiment lexicon. In lines 5–6, objects in set N but not in the ontology are put into set IN which indicates objects to be pruned. Objects (together with their attributes) in the ontology but have competitive relation with the target are also put into set IN. In lines 7–13, we get all sentiment expression – target pairs and match them with the syntactic path library. If the path is not match or the frequency of matched path is smaller than a threshold value, the path is regarded as an invalid path. We find the common parent node comm_parent of the sentiment expression SE_p and its target t, and then prune the subtrees of comm_parent that contains SE_p or t. After that, if there is not any subtree rooted at comm_parent, prune comm_parent from the syntactic tree as well.

Algorithm 1. Prune irrelevant parts from syntax tree

Input: t: syntax tree, O: domain ontology, Ta: target, L: path library
Output: pruned tree
1.   **Function** prune(t, O, Ta, L)
2.   IN ← ϕ
3.   N ← get_noun_set(t)
4.   SE ← get_sentiment_expression(t)
5.   IN +← filter_noun(N, O)
6.   IN +← filter_compititive(N, O, Ta)
7.   for each n in IN
8.   SE_p←get_target_SE(t,n, SE, Ta, L)
9.   if SE_p ≠ ϕ then
10.   comm_parent←find_comm_parent(t,n, SE_p)
11.   t←prune_subtree(t, comm_parent)
12.   end if
13.   end for
14.   return t

A tree pruning example is shown in Fig. 4 where we assume "Huawei" is the target. We can find that the sentiment expression set is {not bad, negative comment}, and the target set is {Huawei, JingDong}. With the help of domain ontology, we can find that "JingDong" is an object irrelevant to the target and should be pruned. By matching the syntactic path library, we discover that the path between "Jingdong" and "negative

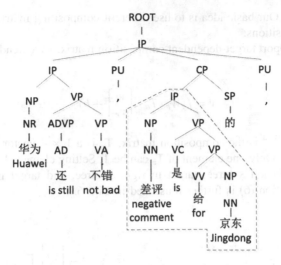

Fig. 4. An example of syntax tree and syntactic path

comment" is effective but the path between "Jingdong" and "not bad" is not. We prune the corresponding structure from the syntax tree following Algorithm 1 and mark it by dashed line in Fig. 4.

4.4 Target-Dependent Recursive Neural Etwork

The original RNN model uses one global matrix W to combine the vectors. But in target-dependent sentiment analysis, the composition matrix might be related to the targets. For the example shown in Fig. 5, we get different sentiment label when the

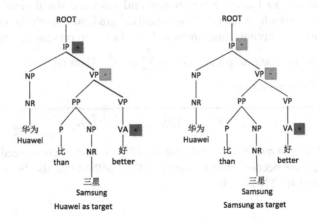

Fig. 5. Sentiment tags for different targets

target is different. Our basic idea is to use different compositing matrices according to different target positions.

In order to support target-dependent composition matrix, we extend Eq. (1) into the following:

$$v = f(W_t \cdot T_a \cdot \begin{bmatrix} v_l \\ v_r \end{bmatrix} + b) \tag{6}$$

where $W_t \in \Re^{D \times 2D \times 4}$ is the composition matrix, T_a is a 1×4 vector which indicates the target position. Only one element of T_a can be 1. Setting element 1–4 to 1 indicates no target, target in left subtree, target in right subtree, and target in both subtrees respectively. Equation (6) is further illustrated in Fig. 6.

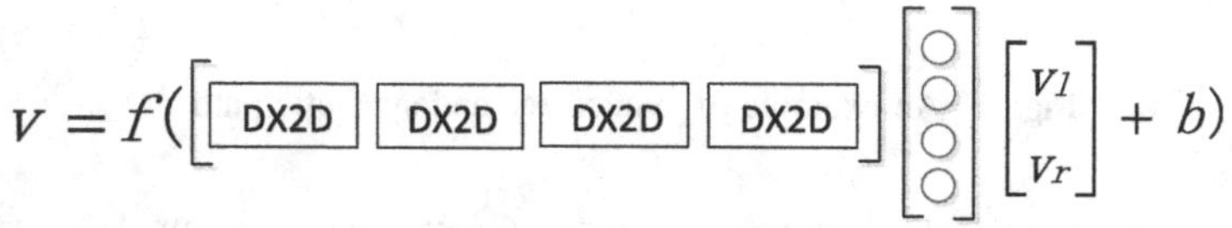

$$V = f\left(\begin{bmatrix} \boxed{DX2D} & \boxed{DX2D} & \boxed{DX2D} & \boxed{DX2D} \end{bmatrix} \begin{bmatrix} \bigcirc \\ \bigcirc \\ \bigcirc \\ \bigcirc \end{bmatrix} \begin{bmatrix} V_l \\ V_r \end{bmatrix} + b\right)$$

Fig. 6. Target-dependent composition of vectors

In order to train the model, let $\theta = \{W_t, b\}$ be the set of model parameters (the softmax parameter W_s can be learned separately). The cost function is defined as:

$$J(\theta) = \frac{1}{N} \sum_{i=1}^{N} \left[\frac{1}{2} \|f(\theta, x_i, y_i) - y_i\|^2 \right] + \frac{\lambda}{2} \|\theta\|^2 \tag{7}$$

where N is the number of samples and the second term is the regularization penalty. We use gradient descent to learn the parameters and calculate the derivative by using backpropagation through structure. The adaptive gradient algorithm (AdaGrad) is employed to solve the optimization problem [19]. Let $g_{t,i}$ represent the subgradient of the i'th parameter at time stamp t and $G_{t,i} = \sum_{s=1}^{t} g_{s,i} g_{s,i}^T$. Then the parameter update for the i'th parameter at time t is:

$$\theta_{t,i} = \theta_{t-1,i} - \alpha [diag(G_{t,i})]^{-1/2} g_{t,i} \tag{8}$$

where α is the learning rate. Since the diagonal of $G_{t,i}$ is used, we only need to store few values and the update becomes fast to compute. Our method can also be used to extend the MV-RNN and RNTN models.

5 Experimental Evaluations

We first built the syntactic path library. We utilized NLPIR Chinese segmentation system[1] for word segmentation and POS tagging, and Stanford Parser[2] for syntactic tree building. The opinion lexicon we used is the NTUSD lexicon from Taiwan University, which contains 2810 positive opinion words and 8276 negative ones. Besides, we put 50 frequently-used positive internet words and 53 negative ones into the lexicon. The corpus we used to build the syntactic path library contains 36,042 micro-blog posts grabbed from Tencent[3] Micro-blog, which covers many fields like cars, smartphone, shopping websites, and so on. The five most frequently-appearing paths and their occurring-frequency are shown in Table 1.

Table 1. Five most frequently-appearing paths in the syntactic path library

Serial number	Syntactic path	Frequency
1	NN→NP→IP→VP→VV	17,267
2	NN→NP→NN	14,087
3	NN→NP→VP→VV	14,068
4	NN→NP→ADJP→JJ	11,056
5	NN→NP→IP→VP→VA	9,311

We selected two keywords as the targets, i.e. "Huawei" and "WeiPinHui" (a Chinese online shopping website)[4], and grabbed posts containing the targets from Tencent Micro-blog. The number of posts for each target is 30,000. We selected 3,500 posts randomly for each target and manually classify each post as positive or negative towards the target. The corpus with target "Huawei" contains 1,643 positive posts and 1,379 negative posts, and corpus with target "WeiPinHui" contains 1,412 positive posts and 1583 negative posts. We select 2000 samples as training set, 300 samples as development set and the rest as test set. Seven methods are evaluated as following:

(1) SVM-convolution: SVM classifier with convolution kernel and syntactic tree features.
(2) SVM-composite: SVM classifier using unigram and syntactic tree as features, and composite kernel as the following:

$$K = (1 - \alpha) \cdot Tree\_\mathrm{ker}\,nel + \alpha \cdot Vector\_\mathrm{ker}\,nel \tag{9}$$

where $\alpha \in (0, 1)$ is a weight factor. We only present the best result for α.

---

[1] http://ictclas.nlpir.org.

[2] http://nlp.stanford.edu/software/lex-parser.shtml.

[3] http://t.qq.com.

[4] http://www.vip.com.

(3) RNN: the original RNN method.
(4) AdaRNN: the adaptive recursive neural network classifier proposed by Dong et al. [12]. We implemented the "AdaRNN-comb" method in their paper. Since some detailed information about the implementation of AdaRNN is lacked, we implemented the method based on our understanding.
(5) TSA-RNN-basic: TSA-RNN without tree pruning and cluster-based sample selecting.
(6) TSA-RNN-prune: TSA-RNN with tree pruning but without cluster-based sample selecting.
(7) TSA-RNN: TSA-RNN with both tree pruning and cluster-based sample selecting. Using this method, we clustered the total samples and got 5 and 8 clusters for target "Huawei" and "WeiPinHui" respectively. Then 2000 samples are selected from the clusters evenly for each target.

Table 2. The evaluation results for 6 methods. A: Accuracy, P: Precision, R: Recall, F: F-score

methods	Huawei							WeiPinHui						
	positive			negative			A	positive			negative			A
	P	R	F	P	R	F		P	R	F	P	R	F	
SVM-convolution	65.8	55.3	60.1	52.7	74.2	61.6	60.9	63.5	62.7	63.1	63	65.3	64.1	63.6
SVM-composite	67.2	59.3	63.0	62.9	69.7	66.1	64.6	65.7	64.8	65.2	65.7	68.1	66.9	66.1
RNN	71.8	69.6	70.7	70.2	70.5	70.3	70.5	70.5	71.9	71.2	69.4	71.9	70.6	70.9
AdaRNN	79.5	78.1	78.8	78.7	79.2	78.9	78.9	78.8	76.7	77.7	77.4	76.9	77.1	77.4
TSA-RNN-basic	80.2	82.8	81.5	80.4	83.3	81.8	81.7	82.8	80.5	81.6	80	81.1	80.5	81.1
TSA-RNN-prune	82.2	83.3	82.7	82.9	83.7	83.3	83.0	83.1	81.8	82.4	82.4	81.2	81.8	82.1
TSA-RNN	84.4	84.8	84.6	84.2	85.9	85.0	84.8	85.2	83.8	84.5	83.6	83.5	83.5	84.0

The evaluation result is shown in Table 2. From the table we can draw the following conclusions: (1) Our method achieves best performance both in positive and negative Micro-blog sets. The accuracy, precision, recall, and F-score of our method are all the highest among all methods. This result shows that the TSA-RNN model, tree pruning method and cluster-based sample selection method work well for target-dependent sentiment analysis of Micro-blogs. (2) The traditional SVM-based classifiers get low perofrmance when used for target-dependent sentiment analysis. Composite kernel-based method achieved better performance since it can capture both structured and flat features. (3) Although the original RNN classifier can get high performance for target-independent sentiment analysis, it cannot get acceptable performance for target-dependent sentiment analysis. The reason is that it uses a global composition matrix without considering targets. The TSA-RNN model uses target-related composition martix which improves the performance. (4) AdaRNN is a target-dependent sentiment classifier based on RNN. Its accuracy is better than SVM and original RNN since it adaptively propagates the sentiments of words to target depending on the context and syntactic relationships between them. (5) The basic method of TSA-RNN outperforms AdaRNN since we extended the RNN model to be target-dependent while the composition matrices of AdaRNN are not target-dependent.

(6) The tree pruning method improves the performance of TSA-RNN since it can remove some irrelevant parts which makes the following work easier. (7) The cluster-based sample selecting method can improve the performance of TSA-RNN since it makes the training samples be distributed evenly on the data space.

6 Conclusion and Future Work

Micro-blogs are important resource for sentiment analysis but it is quite difficult to determine target-dependent polarity. RNN is a promising model for sentiment analysis and it gets high accuracy for target-independent sentiment analysis. In this paper we proposed an extension to the RNN method which includes TSA-RNN model, cluster-based training sample selection and tree pruning. The experimental results shown that this method works well for target-dependent sentiment analysis of Micro-blog stream.

Our current extension to the RNN model is still simple. We believe the deep relationship between targets and sentiment expressions are still not modeled well. In the future we will develop new method to further improve the TSA-RNN model.

References

1. Pak, A., Paroubek, P.: Twitter as a corpus for sentiment analysis and opinion mining. In: Proceedings of the 7th International Conference on Language Resources and Evaluation (LREC 2010), Valletta, Malta, 17-23 May 2010
2. Aisopos, F., Papadakis, G., Tserpes, K., et al.: Content vs. context for sentiment analysis: a comparative analysis over microblogs. In: Proceedings of the 23rd ACM conference on Hypertext and social media (HT 2012), Milwaukee, USA, pp. 187–196, July 2012
3. Lixing, X., Ming, Z., Maosong, S.: Hierarchical structure based hybrid approach to sentiment analysis of Chinese micro blog and its feature extraction. J. Chin. Inf. Process. 26(1), 73–83 (2012). (in Chinese)
4. Socher, R., Pennington, J., Huang, E.H., et al.: Semi-supervised recursive autoencoders for predicting sentiment distributions. In: Proceedings of the Conference on Empirical Methods in Natural Language Processing (EMNLP 2011), Edinburgh, UK, 27-29 July 2011
5. Socher, R., Perelygin, A., Wu, J., et al.: Recursive deep models for semantic compositionality over a sentiment treebank. In: Proceedings of the Conference on Empirical Methods in Natural Language Processing (EMNLP 2013), Seattle, USA, 18-21 October 2013
6. Fiaidhi, J., Mohammed, O., Mohammed, S., et al.: Opinion mining over twitterspace: Classifying tweets programmatically using the R approach. In: Proceedings of the 7th International Conference on Digital Information Management, Macau, China, August 2012
7. Yang, C., Feng, S., Wang, D., et al.: Analysis on web public opinion orientation based on extending sentiment lexicon. J. Chin. Comput. Syst. 31(4), 691–695 (2010)
8. Pang, B., Lee, L., Vaithyanathan, S.: Thumbs up? Sentiment classification using machine learning techniques. In: Proceedings of Conference on Empirical Methods in Natural Language Processing, Philadelphia, pp. 79–86 (2002)

9. Davidov, D., Tsur, O., Rappoport, A.: Enhanced sentiment learning using twitter hashtags and smileys. In: Proceedings of the 23rd International Conference on Computational Linguistics, Beijing, pp. 241–249, August 2010

10. Nasukawa, T., Yi, J.: Sentiment analysis: capturing favorability using natural language processing. In: Proceedings of the 2nd International Conference on Knowledge Capture, Sanibel Island, pp. 70–77, October 2003

11. Jiang, L., Yu, M., Zhou, M.: Target-dependent twitter sentiment classification. In: Proceedings of the 49th Annual Meeting of the Association for Computational Linguistics: Human Language Technologies (ACL-HLT 2011), Portland, 19-24 June 2011, pp. 151–160 (2011)

12. Dong, L., Wei, F., Tan, C., et al.: Adaptive recursive neural network for target-dependent twitter sentiment classification. In: Proceedings of the 52nd Annual Meeting of the Association for Computational Linguistics (ACL 2014), Baltimore, 22-27 June 2014, pp. 49–54 (2014)

13. Socher, R., Huval, B., Manning, C.D., et al.: Semantic compositionality through recursive matrix-vector spaces. In: Proceedings of the Conference on Empirical Methods in Natural Language Processing and Computational Natural Language Learning (EMNLP-CoNLL 2012), Jeju, 12-14 July 2012

14. Chiu, S.: Fuzzy model identification based on cluster estimation. J. Intell. Fuzzy Syst. 2(3), 267–278 (1994)

15. Angelov, P., Zhou, X.: Evolving fuzzy system from data stream in real time. In: Proceedings of the 2006 International Symposium on Evolving Fuzzy Systems (EFS 2006), Lake District 7-9 September 2006, pp. 29–35 (2006)

16. Angelov, P.: Evolving Takagi-Sugeno fuzzy systems from data streams (eTS+). In: Angelov, P., Filev, D., Kasabov, N. (eds.) Evolving intelligent systems: methodology and applications, pp. 21–50. Wiley and IEEE Press, New York (2010)

17. Lin, Z., Wang, H., Mcclean, S., et al.: All common embedded subtrees for measuring tree similarity. In: Proceedings of the 2008 International Symposium on Computational Intelligence and Design (ISCID 2008), Wuhan, 17 October 2008, pp. 29–32 (2008)

18. Zhao, Y., Qin, B., Che, W., et al.: Appraisal expressions recognition based on syntactic path. J. Softw. 22(5), 887–898 (2011)

19. Duchi, J., Hazan, E., Singer, Y.: Adaptive subgradient methods for online learning and stochastic optimization. J. Mach. Learn. Res. 12, 2121–2159 (2011)

Online Streaming Feature Selection
Using Sampling Technique and Correlations
Between Features

Hai-Tao Zheng$^{(\boxtimes)}$ and Haiyang Zhang

Graduate School at Shenzhen, Tsinghua University, Tsinghua Campus,
The University Town, Shenzhen 518055, People's Republic of China
zheng.haitao@sz.tsinghua.edu.cn

Abstract. Feature selection is an important topic in data mining and machine learning, and has been extensively studied in many literature. In real-world applications, the dimensionality is extremely high, in millions, and keeps growing. Unlike traditional batch learning methods, online learning is more efficient for real-world applications. We define streaming features as features that flow in one by one over time whereas the number of training examples remains fixed. This is in contrast with the other kind of online learning methods that only deal with sequentially added instances. The key challenge for current online streaming feature selection is the large feature space, possibly of unknown or infinite size. To select a small number of features in an online manner more effectively, we propose a novel algorithm using sampling techniques and correlations between features. We evaluate the performance of the proposed algorithms for online streaming feature selection on several public datasets, and demonstrate their applications to real-world problems as image classification in computer vision. From Experiments, we can see that our algorithm consistently surpassed the baseline algorithms for all the situations.

Keywords: Online streaming feature selection · Sampling techniques · Correlations · Binary classification

1 Introduction

Feature selection is an important topic in data mining and machine learning. It has been extensively studied in many literatures [1–3]. In classification problems, it is the process of selecting a subset of the terms occurring in the training set and using only this subset as features. By removing irrelevant and redundant features, feature selection can alleviate the effect of the curse of dimensionality, improve the generalization performance, accelerate the learning process, and enhance the interpretive performance of the model, finally improve the performance of prediction models [4].

Most existing studies of feature selection are restricted to batch learning, which assumes the feature selection task is conducted in an off-line learning fashion and all the features of training instances are given a priori. But in real-world applications, not all features can be present in advance. For example, many image processing processes involve a search of potential features for machine learning algorithms to fulfill the

© Springer International Publishing Switzerland 2016
F. Li et al. (Eds.): APWeb 2016, Part II, LNCS 9932, pp. 43–55, 2016.
DOI: 10.1007/978-3-319-45817-5_4

pattern recognition goal, but image features are often expensive to generate and store and therefore may exist in a streaming format [5]. Another example is feature selection in bioinformatics, where acquiring the full set of features for every instance is expensive because of the high cost in conducting wet lab experiments [7].

Unlike the existing feature selection studies, we study the problem of Online Feature Selection (OFS). There are two research directions for online feature selection: One assumes that the number of features on training data is fixed while the number of instances increases over time, such as the OFS algorithm [7] and the RFOFS algorithm [4], which performs feature selection upon each training instance's arrival. The other online feature selection assumes that the number of training instances is fixed while the number of features increases over time, such as the Fast-OSFS [8] and alpha-investing algorithms [9]. We call such a method online streaming feature selection (OSFS).

Although there have been a variety of OSFS algorithms, they still meet difficulty in computational cost when the dimensionality is extremely high or the data is highly class-imbalanced. In this paper, we focus on the research of how to improve efficiency of OSFS with extremely high-dimensional or highly class-imbalance data.

In this paper, to tackle the challenges of efficiency, our contributions are as follows:

1. We deeply study the Fast-OFS algorithm and its three key steps. We optimize these steps by the theoretical analysis of the comparisons of features.
2. We develop Online Streaming Feature Selection using Sampling Techniques and Correlations between Features (STCF). The STCF algorithm improves efficiency of OSFS with extremely high-dimensional data by comparison of features and sampling techniques.
3. We validate the effectiveness and efficiency of STCF by conducting an extensive set of experiments. Our method has superior performance over the state-of-the-art online streaming feature selection methods on data sets of extremely high dimensionality.

The rest of the paper is organized as follows. Section 2 reviews related work. Section 3 presents the problem of current algorithms and our STCF algorithm. Section 4 discusses our empirical studies. At last, Sect. 5 concludes this work and our future work.

2 Related Work

Feature selection (FS) has been the focus of interest for quite some time and has been studied extensively in the literatures of data mining and machine learning [1, 10]. The existing FS algorithms generally can be grouped into three categories: super-vised, unsupervised, and semi-supervised FS. Supervised FS [11, 15] selects features according to labeled training data. When there is no label information available, unsupervised feature selection [12] attempts to select the important features which preserve the original data similarity or manifold structures. Semi-supervised feature selection methods [13], as its name says, exploit both labeled and unlabeled data information. The existing super-vised FS methods can be further divided into three groups, depending on how they combine the feature selection search with the construction of the classification model: Filter methods, wrapper methods, and embedded methods approaches. Filter methods

[11, 14] choose important features by measuring the correlation between individual features and output class labels, without involving any learning algorithm; wrapper methods [15] rely on a predetermined learning algorithm to decide a subset of important features. Embedded methods [16] aim to integrate the feature selection process into the model training process. Feature selection has found many applications [10], including bioinformatics, text analysis and image annotation. Our OFS technique generally belongs to supervised FS.

Contrast to the batch methods, online methods has become a hot research topic due to the efficiency of the algorithms and effectiveness in real-world applications. There are two research directions for online feature selection. One assumes that the number of features on training data is fixed while the number of instances increases over time [17]. A classical online learning method is the well-known Perceptron algorithm [18]. Recently, a lot of new online learning algorithms have been presented, for example, the Passive-Aggressive algorithm [19] and the OFS algorithm [7]. In 2015, we presented the RFOFS algorithm [4], which performs feature selection more accurate and more efficient upon each training instance's arrival.

Different from OFS, online streaming feature selection assumes that the number of training instances is fixed while the number of features increases over time. Perkins et al. firstly proposed the Grafting algorithm [18], a method based on gradient descent approach. Zhou et al. presented Alpha-investing [9] which sequentially considers new features as additions to a predictive model. However, Alpha-investing requires the prior information of the original feature set and never evaluates the redundancy among the selected features as time goes [8]. To tackle the drawbacks, Wu et al. presented the OSFS algorithm and its faster version, the Fast-OSFS algorithm [8]. To handle online feature selection with grouped features, Wang et al. proposed the OGFS (Online Group Feature Selection) algorithm [8]. However, the computational cost inherent in those algorithms may still be very expensive when the dimensionality is extremely high. Wu et al. proposed SAOFS (scalable and accurate online feature selection) and Group-SAOFS [6] to improve efficiency of OSFS by considering pairwise correlation bound between features. However, it cannot effectively deal with highly class-imbalanced data.

3 Online Streaming Feature Selection Using Sampling Techniques and Correlations Between Features

3.1 Problem Setting

In this section, we introduce the online feature selection with streaming features, named online streaming feature selection (OSFS). We formally define types of features.

In OSFS, feature dimensions may grow over time and may even extend to an infinite size and each feature is required to be processed online upon its arrival.

In general, a training data set is defined as $\{(x_i, y_i), 1 \leq i \leq N\}$, where N is the number of training instances, $x_i \in \mathbb{R}^d$ is a vector of d dimension, contains d features, Y is the class attribute that has two distinct class labels, $y_i \in \{-1, 1\}$, and a feature set F on dataset is defined by $F = \{F_1, F_2, \ldots, F_d\}$.

To characterize feature relevance, an input feature can be categorized into three disjoint groups, namely, strongly relevant, weakly relevant or irrelevant, among the groups, weakly relevant features can be further divided into redundant features and non-redundant features [15, 19].

Definition 3.1 Irrelevance: A feature F_i is irrelevant to Y iff it is neither strongly nor weakly relevant, and

$$\forall S \subseteq F - \{F_i\} \text{ s.t. } P(Y|F_i, S) = P(Y|S)$$

Weakly relevant features can be further divided into redundant features and non-redundant features based on a Markov blanket criterion [20].

Definition 3.2 Markov blanket: Denoting $M_i \subset F$ a subset of features, if for the given M_i the following property

$$\forall T \in F - M_i \text{ s.t. } P(Y|M_i, T) = P(Y|M_i)$$

holds, then M_i is a Markov blanket for C (MB(C) for short).

Definition 3.3 Redundant features: A feature F_i is redundant to the class attribute C, if and only if it is weakly relevant to Y and has a Markov blanket $MB(F_i)$, that is a subset of the Markov blanket MB(C).

The Markov blanket of the class attribute Y is the optimal feature subset which contains all the weakly relevant but non-redundant features and strongly relevant features. Thus our task is to determine whether a feature is in the Markov blanket.

3.2   Baseline Methods

There is a framework for feature selection with streaming features that contains two major steps: (1) online relevance analysis that discards irrelevant features and retains relevant ones; and (2) online redundancy analysis which eliminates redundant features from the features selected so far.

Fast-OSFS divides the online redundancy analysis (step 2 in framework) into two parts: The first part is a redundancy analysis which aims to remove the new relevant but redundant feature; the second part is triggered by whether the new feature is redundant, if the new feature is not redundant, validates each originally existing feature in OF, and checks whether any of these features has become redundant after the inclusion of new feature.

Fast-OSFS uses the G^2 test to denote the conditional independence or dependence, and then identify irrelevant and redundant features. Assuming p is the p-value returned by the G^2 test and α is a given significance level, Fast-OSFS judge independence by comparing p and α. There are three key steps for OSFS:

- Online Relevance Analysis: Determine the relevance of F_i to Y: if equation

$$P(Y|F_i) = P(Y) \tag{1}$$

holds, F_i is irrelevant to Y, thus F_i should be discarded;
- Online Redundancy Analysis 1: Determine whether F_i and Y are conditionally independent given S_{i-1}: if equation

$$P(Y|S_{i-1}, F_i) = P(Y|S_{i-1}) \tag{2}$$

holds, F_i has a Markov blanket in S_{i-1}, thus F_i should be discarded, S_{i-1} is the feature subset selected at time t_{i-1}.
- Online Redundancy Analysis 2: After F_i is added to OF, discard the redundant features in OF: we can get OF from

$$S_i = \mathrm{argmax}_{\zeta \subseteq \{S_{i-1}, F_i\}} P(Y|\zeta) \tag{3}$$

To judge whether a feature T in $\{S_{i-1}, F_i\}$ is redundant by using a Markov blanket, it is necessary to check all the subset of the $\{S_{i-1}, F_i\}$ (the total number of subsets is $2^{|\{S_{i-1}, F_i\}|}$). It is computationally prohibitive. Therefore, we need to simplify the computation process, find a suitable solution for Eqs. (1) to (3).

3.3   The STCF Algorithm: Using Correlations Between Features

In this section, we propose solutions for Eqs. (1) to (3) using: (1) correlations between features and (2) correlations between feature and class attribute. We use $0 \le CR(X, Y) \le 1$ to express the correlations between vector X and vector Y, it can be computed by mutual information, chi-square test or fisher's z-test and so on, if $CR(X, Y) = 0$, X is irrelevant to Y.

Solution to Eq. (1). Assming S_{i-1} is the selected feature subset at time t_{i-1}, and at time t_i, a new feature F_i arrives, if

$$CR(F_i, Y) > \varepsilon(0 \le \varepsilon < 1) \tag{4}$$

holds, F_i is said to be a relevant feature to Y; otherwise, F_i is discarded as an irrelevant feature and will never be considered again.

Solution to Eq. (2). After Eq. (1), if F_i is relevant, at time t_i, how can we determine whether F_i is redundant given S_{i-1}? If F_i is redundant given S_{i-1}, that is $CR(F_i, Y|S_{i-1}) = 0$. It means there is one feature or combined features which make F_i redundant. In our work, we focus on the situation of a single feature make F_i redundant. With this hypothesis, we solve Eq. (2) with the following lemma.

Lemma: Assming S_{i-1} is the selected feature subset at time t_{i-1}, and at time t_i, a new feature F_i arrives, if $\exists T \in S_{i-1}$, such that $CR(F_i, Y|T) = 0$, then for $CR(F_i, Y)$, $CR(F_i, T)$ and $CR(T, Y)$, $CR(F_i, Y)$ is the minimum value.

Proof: We may use mutual information to compute the CR value:

$$I(X, Y) = H(X) - H(X|Y)$$

The entropy of feature Y is defined as

$$H(X) = -\sum_{x_i \in X} P(x_i) \log_2 P(x_i)$$

And the entropy of Y after observing values of another feature Z is defined as

$$H(X|Y) = -\sum_{y_i \in Y} P(y_i) \sum_{x_i \in X} P(x_i|y_i) \log_2 P(x_i|y_i)$$

From above equations, we get the following,

$$I(F_i, Y) + I(F_i, T|Y) = I(F_i, T) + I(F_i, Y|T)$$

Then if $I(F_i, Y|T) = 0$ holds, we get the following,

$$I(F_i, Y) + I(F_i, T|Y) = I(F_i, T) \tag{5}$$

Thus, we found

$$CR(F_i, Y) \leq CR(F_i, T) \tag{6}$$

Then we prove $I(F_i, Y) \leq I(T, Y)$:
From computation of I(X, Y), we get the following,

$$I(F_i, T) + I(T, Y|F_i) = I(T, Y) + I(F_i, T|Y)$$

Then with Eq. (5), we get the following,

$$I(F_i, Y) + I(T, Y|F_i) = I(T, Y)$$

Thus, we found

$$CR(F_i, Y) \leq CR(T, Y) \tag{7}$$

By Eqs. (6) and (7), the Lemma is proven.

We deal with Eq. (2) as follows: when a new feature F_i at time t_i, if $\exists T \in S_{i-1}$, for $CR(F_i, Y), CR(F_i, T)$ and $CR(T, Y), CR(F_i, Y)$ is the minimum value, then F_i is discarded; otherwise, F_i is added to S_{i-1}.

Solution to Eq. (3)

Once F_i is added to S_{i-1} at time t_i, we will check which features within S_{i-1} can be removed due to the new inclusion of F_i. If $\exists T \in S_{i-1}$ such that $CR(T, Y|F_i) = 0$, then feature T can be removed from S_{i-1}.

Similar to Eqs. (6) and (7), if $CR(T, Y|F_i) = 0$, we can get,

$$CR(T, Y) \leq CR(F_i, Y) \tag{8}$$

$$CR(T, Y) \leq CR(T, F_i) \tag{9}$$

Accordingly, the solution to Eq. (3) is as follows: with the feature F_i added to feature subset S_{i-1} at time t_i and $S_i = S_{i-1} + F_i$, if $\exists T \in S_i - F_i$ for $CR(F_i, Y), CR(F_i, T)$ and $CR(T, Y), CR(T, Y)$ is the minimum value, then T can be removed from S_i.

3.4   The STCF Algorithm: Using Sampling Technique

Using the analysis of Sect. 3.3, we can greatly improve the efficiency of OSFS algorithm. To more effectively deal with highly class-imbalanced data, we need to determine: (1) whether the data is class-imbalanced data? (2) how to make data class-balanced? (3) how to get the best result?

For problem 1, we should distinguish relative imbalance and absolute imbalance. If dataset is relative imbalance, mutual information has been more considered with fewer data. Thus we only consider the absolute imbalance dataset, that is, absolute number of rare category is too small. For problem 2, we can use multiple sampling technique to building a subset which is class-balanced. After sampling and feature selection, we can get the best result by voting or ranking.

Using the analysis of Sect. 3.3 and above discussion, we propose the Online Streaming Feature Selection using Sampling Techniques and Correlations between Features (STCF algorithm) in detail, as shown in Algorithms 1 and 2. Algorithm 1 shows the sampling process, and Algorithm 2 shows the feature selection process. The feature selection process of STCF algorithm is implemented as follows: at time t_i, as a new feature F_i arrives, if $CR(F_i, Y) \leq \varepsilon$ holds at Step 5, then F_i is discarded as an irrelevant feature and STCF waits for a next coming feature; if not, at Step 11, STCF evaluates whether F_i should be kept given the current feature set S_{i-1}. If $\exists T \in S_{i-1}$ such that Eqs. (6) and (7) holds, we discard F_i and never consider it again. Once F_i is added to S_{i-1} at time t_i, S_{i-1} will be checked whether some features within S_{i-1} can be removed due to the new inclusion of F_i. At Step 16, if $\exists T \in S_{i-1}$ such that Eqs. (8) and (9) holds, T is removed.

ALGORITHM 1: The STCF Algorithm (sampling and voting process)

---

1. Input:
   Data: dataset, n*d matrix;
   Y: class attribute.
2. if Data is **class-imbalanced data, sampling to k class-balanced subset.**
3. **repeat**
4.   /* generate d random values from 1 to d */
5.   Id_list = randperm(d);
6.   for t=1:length(Id_list)
7.   id=Id_list(t);
8.   input feature Data(:, id);
9.   excute the online streaming feature selection process;
10.   end for
11. until k subset are all handled
12. voting

ALGORITHM 2: The STCF Algorithm (feature selection process)

1: Input:
   F_i: predictive features;
   Y: the class attribute;
   ε: a relevance threshold (0~1);
   S_{i-1}: the selected feature set at time t_{i-1};
   Output: S_i : the selected feature set at time t_i;
2: repeat
3:   get a new feature F_i at time t_i;
4:   /*Solve Eq.(1)*/
5:   if $CR(F_i, Y) \leq \varepsilon$ then
6:     Discard F_i;
7:     Go to Step 21;
8:   end if
9:   for each feature T ∈ S_{i-1} do
10:     /*Solve Eq.(2)*/
11:     if $CR(F_i, Y) \leq CR(F_i, T)$ && $CR(F_i, Y) \leq CR(T, Y)$ then
12:       Discard F_i;
13:       Go to Step 21;
14:     end if
15:     /*Solve Eq.(3)*/
16:     if $CR(T, Y) \leq CR(F_i, Y)$ && $CR(T, Y) \leq CR(T, F_i)$ then
17:       $S_{i-1} = S_{i-1} - T$;
18:     end if
19:   end for
20:   $S_i = S_{i-1} + F_i$;
21: until no features are available
22: Output S_i ;

---

4 Experiments

4.1 Datasets

We test the proposed algorithms on a number of publicly available benchmarking datasets. All of the datasets can be downloaded either from LIBSVM website[1] or UCI machine learning repository[2]. Besides these data sets, we also adopt two high-dimensional datasets and an image classification dataset: (i) webspam; (ii) kdd2010; (iii) CIFAR-10 image dataset[3]. We set the count of training samples to 10000 and the count of testing samples to 20000 if total samples' count > 30000.

Table 1 shows the statistics of the datasets used in our following experiments.

Table 1. List of datasets in our experiments

Dataset	Dimensions	Training samples	Testing samples
madelon	500	2000	600
epsilon	2000	10000	20000
gisette	5000	6000	1000
leukemia	7129	38	34
real-sim	20958	10000	20000
rcv1	47236	10000	20000
webspam	16,609,143	10000	20000
kdd2010	29,890,095	10000	20000
CIFAR-10	3072	2000	1992

4.2 Experimental Setup and Baseline Algorithms

We compare our proposed algorithm against the following three baselines:

1. Alpha-investing Algorithm; [9]
2. Fast-OSFS Algorithm; [8]
3. A scalable and accurate online feature selection for big data (SAOFS); [6]

To make a fair comparison, all algorithms adopt the same experimental settings. All the experiments were run over 10 times, each time with a random permutation of features in a dataset. Our Algorithm is computed by Fisher's z-test or G^2 test, ε in STCF and SAOFS is set to 0; The significance level is set to 0.01. Our experiment was run in matlab R2014b.

---

[1] http://www.csie.ntu.edu.tw/~cjlin/libsvmtools/.

[2] http://www.ics.uci.edu/~mlearn/MLRepository.html.

[3] http://www.cs.toronto.edu/kriz/cifar.html.

4.3   Experiment I: Prediction Accuracy, Running Time, the Number of Selected Features

Table 2 are the prediction accuracies of STCF against SAOFS, Fast-OSFS and Alpha-investing using the SVM classifier. From the table, we can see our algorithm's accuracy mostly surpass the other three algorithms. This is because we concerned the class-imbalanced data and features' sequential order.

Table 2. Prediction accuracy (SVM)

Dataset	STCF	SAOFS	Fast-OSFS	Alpha-investing
madelon	**0.653**	0.530	0.523	0.502
epsilon	0.840	**0.853**	0.792	0.623
gisette	0.874	0.873	**0.883**	0.750
leukemia	**0.941**	0.853	0.650	0.529
real-sim	**0.708**	0.631	0.540	0.560
rcv1	**0.842**	0.825	0.746	0.732
webspam	**0.913**	0.885	–	–
kdd2010	**0.864**	0.802	–	–

Table 3 is the running time of the feature selection process of STCF against SAOFS, Fast-OSFS and Alpha-investing. From the table, we can see our algorithm and SAOFS both greatly improve the running efficiency of online streaming feature selection when the dimensionality is extremely high. Even in the last two high dimension dataset, Fast-OFS and Alpha-investing run more than 3 days.

Table 3. Time cost (seconds)

Dataset	STCF	SAOFS	Fast-OSFS	Alpha-investing
madelon	**0.1**	0.2	0.3	0.3
epsilon	**19**	22	640	671
gisette	**3**	8	854	936
leukemia	**0.7**	2	1	0.8
real-sim	75	**61**	3805	4023
rcv1	**83**	94	>1 day	>1 day
webspam	**2905**	3058	–	–
kdd2010	2437	**2149**	–	–

Table 4 is number of selected features of STCF against SAOFS, Fast-OSFS and Alpha-investing. From the table, we can see features' count of our algorithm is generally similar to Fast-OSFS and less than SAOFS, that is because SAOFS algorithm's redundant judgments is sometimes worse than the other algorithms.

Table 4. Number of selected features

Dataset	STCF	SAOFS	Fast-OSFS	Alpha-investing
madelon	**3**	**3**	**3**	4
epsilon	**8**	16	12	9
gisette	4	22	6	**2**
leukemia	5	17	5	**2**
real-sim	49	171	53	**48**
rcv1	**55**	76	65	486
webspam	**164**	180	–	–
kdd2010	63	**51**	–	–

Through above three table, we can conclude the following:

- STCF vs Fast-OSFS and Alpha-investing: we observe that STCF is very competitive with these two algorithms. We spend shorter time to select feature subset and have a higher prediction accuracy rate. Fast-OSFS and Alpha-investing is very expensive in computation and even prohibitive on some data sets, while STCF can get result very fast. And for the number of selected features, STCF is close to Fast-OSFS.
- STCF vs SAOFS: SAOFS is focus on shorten the running time of algorithms, so its time cost is lower than Fast-OSFS and Alpha-investing. However, our method has the fastest speed in feature selection, even faster than SAOFS. And our method has higher accuracy rate and less features in selected feature subset. Thus, STCF surpass SAOFS algorithm in every aspects.

4.4   Experiment II: Applications to Image Classification

The experimental results are shown in Table 5. From table we can see our STCF algorithm performs significantly better than the other three approaches on CIFAR-10, which demonstrates the effectiveness of our algorithm.

Table 5. Image

Measurement	STCF	SAOFS	Fast-OSFS	Alpha-investing
Accuracy	0.78	0.75	0.68	0.65
Running time (seconds)	2	6	153	226
number of selected features	28	26	38	59

5   Conclusion and Future Work

In this paper, we utilized Correlations between Features to tackle online streaming feature selection with extremely high dimensionality. We utilized sampling techniques to handle class-imbalanced data and improve accuracy. Then we evaluate the performance of the proposed algorithms STCF for online streaming feature selection on several public datasets, two high-dimensional datasets and an image classification dataset. From the experiment results, we can see our algorithm is significantly more efficient and effective than the other algorithms.

There is one feature or combined features which make feature redundant. However, in our work, we only consider the situation of a single feature. In the future, we will find efficient and effective methods to consider positive feature interactions between features, or extend our framework to solve other problems, such as online multi-class classification.

Acknowledgement. This research is supported by National Natural Science Foundation of China (Grant No. 61375054 and 61402045), Natural Science Foundation of Guangdong Province (Grant No. 2014A030313745), Tsinghua University Initiative Scientific Research Program (Grant No. 20131089256), and Cross fund of Graduate School at Shenzhen, Tsinghua University (Grant No. JC20140001).

References

1. Dash, M., Liu, H.: Feature selection for classification. Intell. Data Anal. **1**(1–4), 131–156 (1997)
2. Saeys, Y., Inza, I., Larrañaga, P.: A review of feature selection techniques in bioinformatics. Bioinformatics **23**(19), 2507–2517 (2007)
3. Yu, L., Liu, H.: Feature selection for high-dimensional data: a fast correlation-based filter solution. In: ICML, pp. 856–863 (2003)
4. Zheng, H.-T., Zhang, H.: Online feature selection based on passive-aggressive algorithm with retaining features. In: Cheng, R., Cui, B., Zhang, Z., Cai, R., Xu, J. (eds.) APWeb 2015. LNCS, vol. 9313, pp. 707–719. Springer, Heidelberg (2015). doi:10.1007/978-3-319-25255-1_58
5. Glocer, K., Eads, D., Theiler, J.: Online feature selection for pixel classification. In: ICML, pp. 249–256 (2005)
6. Yu, K., Wu, X., Ding, W., Pei, J.: Towards scalable and accurate online feature selection for big data. In: IEEE ICDM 2014, pp. 660–669. IEEE (2014)
7. Wang, J., Zhao, P., Hoi, S.C.H., Jin, R.: Online feature selection and its applications. In: TKDE (2013), pp. 1–14 (2013)
8. Wu, X., Yu, K., Wang, H., Ding, W.: Online streaming feature selection. In: ICML, pp. 1159–1166 (2010)
9. Zhou, J., Foster, D.P., Stine, R., Ungar, L.H.: Streaming feature selection using Alpha-investing. KDD 2005, pp. 384–393 (2005)
10. Guyon, I., Elisseeff, A.: An introduction to variable and feature selection. J. Mach. Learn. Res. **3**, 1157–1182 (2003)
11. Dash, M., Gopalkrishnan, V.: Distance based feature selection for clustering microarray data. In: Haritsa, J.R., Kotagiri, R., Pudi, V. (eds.) DASFAA 2008. LNCS, vol. 4947, pp. 512–519. Springer, Heidelberg (2008)
12. Zhao, Z., Liu, H.: Spectral feature selection for supervised and unsupervised learning. In: ICML, pp. 1151–1157 (2007)

13. Ren, J., Qiu, Z., Fan, W., Cheng, H., Yu, P.S.: Forward semi-supervised feature selection. In: Washio, T., Suzuki, E., Ting, K.M., Inokuchi, A. (eds.) PAKDD 2008. LNCS (LNAI), vol. 5012, pp. 970–976. Springer, Heidelberg (2008)

14. Yu, L., Liu, H.: Feature selection for high-dimensional data: a fast correlation-based filter solution. In: ICML, pp. 856–863 (2003)

15. Kohavi, R., John, G.H.: Wrappers for feature subset selection. Artif. Intell. **97**(1–2), 273–324 (1997)

16. Xu, Z., Jin, R., Ye, J., Lyu, M.R., King, I.: Non-monotonic feature selection. In: ICML, p. 144 (2009)

17. Hoi, S.C.H., Wang, J., Zhao, P., Jin, R.: Online feature selection for mining big data. In: Proceedings of the 1st International Workshop on Big Data, Streams and Heterogeneous Source Mining: Algorithms, Systems, Programming Models and Applications, pp. 93–100. ACM (2012)

18. Perkins, S., Theiler, J.: Online feature selection using grafting. In: ICML 2003, pp. 592–599 (2003)

19. Koller, D., Sahami, M.: Toward optimal feature selection. In: ICML 1995, pp. 284–292 (1995)

20. Yu, K., Wu, X., Ding, W., Pei, J.: Scalable and accurate online feature selection for big data. arXiv:1511.09263v1 [cs.LG], (2015)

21. Zhou, J., Foster, D.P., Stine, R.A., Ungar, L.H.: Streamwise feature selection. J. Mach. Learn. Res. **7**, 1861–1885 (2006)

22. Lei, Yu., Liu, H.: Efficient feature selection via analysis of relevance and redundancy. J. Mach. Learn. Res. **5**, 1205–1224 (2004)

Real-Time Anomaly Detection over ECG Data Stream Based on Component Spectrum

Meng Wu[1,2], Zhen Qiu[1,2], Shenda Hong[1,2], and Hongyan Li[1,2(✉)]

[1] Key Laboratory of Machine Perception, Peking University,
Ministry of Education, Beijing 100871, China
[2] School of Electronics Engineering and Computer Science,
Peking University, Beijing 100871, China
lihy@cis.pku.edu.cn

Abstract. Anomaly detection is a popular research in the age of Big Data. As a typical application scenario, anomaly detection over ECG data stream is confronted with particular difficulties including high real-time requirement and poor data quality. In this article, a novel method based on component spectrum is presented to provide a practicable solution for the problem. Experiments on real data show that the proposed method achieves high sensitivity, high specificity and low false alarm rate.

Keywords: Anomaly detection · Component spectrum · ECG data stream

1 Introduction

In the age of Big Data, development and wide application of data acquisition systems make it possible to produce data continuously under many circumstances. In clinical field, medical monitors of Electrocardiograph (ECG) are commonly used in hospitals. Meanwhile, plenty of portable instruments can produce ECG data in the form of data streams with patients in any place. ECG data streams reflect health state of patients in real-time. Therefore, detecting anomaly automatically can be of great practical value.

Recently, researches on detecting anomalies and outliers have drawn increasing attention [1]. As regard of ECG data streams, however, detecting anomalies is confronting with four major challenges.

Poor Data Quality. ECG data is often noisy because of ambient influence and sensor error [2]. Take MIT-BIH Arrhythmia data [3] for example, serious noises appear every 6.08 min per channel on average. Noise can cause false alarms in monitoring. A statistic of monitoring on patients after cardiac surgery shows that only 10 % of 1307 alarms are proper [4]. Meanwhile, noises are unpredictable and overlap with ECG data in frequency domain. Thus, it is hard to remove them thoroughly without signal attenuation while processing in real time.

Pseudo-periodicity. ECG data are pseudo-periodic [5] since they actually vary along with the periodic cardiac electrophysiological activities, where the data

F. Li et al. (Eds.): APWeb 2016, Part II, LNCS 9932, pp. 56–67, 2016.
DOI: 10.1007/978-3-319-45817-5_5

present a repetitive pattern within a certain time interval. However, the length and amplitude of different cycles can be slightly various and also unpredictable. In this article, we use the concept of component spectrum to depict the relatively fixed typical components in various cycles, and view the data as the combination of the typical components. The method focuses on the relatively fixed typical components in pseudo-periodic data and separates noise at the same time.

High Data Volume. While processing data streams, methods should be scalable [6]. In application scenarios like observation of chronic disease patients or intensive care, the monitoring process can last several days or even months. A single low frequency ECG monitor of 125 Hz generates over 10 million data points per day. Therefore, the total data would be too voluminous to be stored completely, which means we can not afford to scan the data more than once.

High Real-Time Requirement. In clinical field, many acute diseases reflected by ECG data need to be treated immediately. For example, with regard to cardiac arrest, Hugh E. Stephenson analyzed 1200 cases in 1953 and came to the conclusion that the optimal opportunity to rescue cardiac arrest patients is within 4 min after the attack [7]. Thus, it is of significance that the anomaly detection task over ECG data streams be carried out in real time.

In this article, we propose a supervised method for online anomaly detection based on component spectrum, which consists of three main modules: a component spectrum representation module (CSRM), a latent class clustering module (LCCM) and a classification-based anomaly detection module (CADM).

Firstly, CSRM mines the typical components in ECG data streams and obtains component spectrum on this basis. As mentioned above, noises are unpredictable and appear randomly in ECG data streams. Oppositely, there are basic waves appear pseudo-periodically which are relatively invariant in morphology in spite of different physiological statuses. Thus, this module tends to acquire a set of relatively invariant components since noises are irregular.

Secondly, LCCM discovers the latent classes of data based on their different component spectrums. As known in clinical field, different physiological statuses correspond to different basic waves or different ways of combination. We introduce the concept of latent class to denote physiological status. And in this context, different latent classes correspond to different component spectrums.

Lastly, CADM is trained to decide whether a kind of latent class should be regarded as anomalous.

For example, in Fig. 1, the ECG data before the red vertical dotted line shows normal sinus rhythm, while the data after shows the sign of ventricular flutter, a serious kind of arrhythmia. Our method can discover latent classes, and determine whether or not trigger the alarm for anomalies in real time.

The rest of this article is organized as follows: In Sect. 2, we discuss the related work. Next, we describe the problem and the overall framework, followed by introductions to every module, experimental evaluation and conclusion.

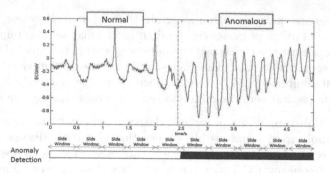

Fig. 1. An example of ECG data stream. The anomaly detection algorithm outputs whether the data of the current sliding window is anomalous. In this figure, the anomalous part is painted blue. (Color figure online)

2   Related Work

In this section, we will discuss methods of anomaly detection over data stream in two categories: supervised and unsupervised.

Supervised Anomaly Detection over Data Stream. Most researches falling in this category basically follows the three procedures of feature extraction, feature selection and classifier construction [8,9]. Feature extraction of pseudo-period data stream is mainly based on segmentation which requires preprocessing. Moreover, techniques like piecewise linear approximation [10] are widely used in the procedure of segmentation. A framework for detecting anomalies in uncertain pseudo-periodic time series is presented in [5], where the influence of noise is taken into account. However, the preprocessing of noise reduction and segmentation can be time consuming and error causing. Thus, these methods are not designed to be performed in real time.

In the mean time, online processing of anomaly detection also draws arisen attention recently. Some methods concerning on anomaly of ECG data are operated by recognizing the morphology of normal sinus beat [11], where expert acknowledge is needed to determine the appropriate range for all the features. This kind of methods are vulnerable to occasionally occurred noises. As introduced in [12,13], other methods are typically operated by finding a normal region containing some non-anomalous training samples while points outside the region are determined as anomalies. These methods are also not robust enough. J. Wang et al. proposed a method combining Symbolic Aggregate approXimation (SAX) and topic model [14], but the step of SAX imports error and the method may lose useful sequential information. Inspired by the idea of representation [15], we propose component spectrum representation to depict the appearance of typical components separately without losing important information.

Unsupervised Anomaly Detection over Data Stream. The unsupervised methods are mainly based on clustering, among which k-means is the most commonly used [16]. Conventional methods detecting outliers based on Euclidean distance [17], which are unsuitable since the waves would change slightly in pseudo-periodic data stream. In this case, dynamic time warping technique is used to match the waves with different length by stretching the data along the time shaft [18], while correlation distance matches waves with different amplitude. However, outliers detected by these method are more likely to be noises than anomalies in ECG data. The distance of an anomaly to a normal ECG cycle may still be shorter than that of two normal ECG cycles in different form in spite of the stretch techniques.

3   Framework of Proposed Method

In this section, we first provide a description of the problem and our goal, and then present the framework of the proposed method.

In this article, ECG data stream refers to the ECG data sampled by medical monitors in fixed frequency which came in form of data streams. They are manifested as potentially infinite pseudo-periodic time series data with irregularly occurred additive noise.

Given a ECG data stream, the aim is to decide whether the current data is anomalous. As a supervised method, the anomalies to find here are data that can characterize some kind of diseases.

As mentioned above, our method consists of a component spectrum representation module (CSRM), a latent class clustering module (LCCM) and a classification-based anomaly detection module (CADM). The parameters of the three modules are trained beforehand using the historical data of the same kind, so the method can be processed online.

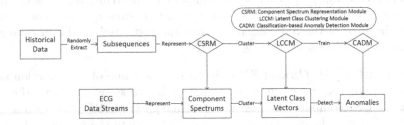

Fig. 2. The framework of the proposed method.

Figure 2 provides an overview of the proposed method. The upper half of the figure shows the three procedures to train the model. First, we randomly extract a set of fixed length subsequences from historical data. Then, the subsequences are used to train the CSRM which produces component spectrum. The trained

CSRM contains a set of components which can be used to represent any subsequence as its component spectrum. Next, we cluster the component spectrums and get parameters of different latent classes in LCCM. Last, the latent classes vectors are used to train whether the subsequences are anomalies in CADM. Meanwhile, the lower half of the figure shows the correlative online processing procedures of component spectrum representation, class vector obtaining and anomaly detection by classification.

4   Model Training and Online Processing

In this section, we introduce the proposed method in detail and explain the structure of each module of the framework.

4.1   Component Spectrum Representation

The goal of CSRM is to represent the subsequences as the linear combination of a small amount of typical components. And the training task here is to acquire the component set.

Definition 1 (Component Spectrum). For any pseudo-periodic data sequence ss with length l, if a set of components $CS = \{cs_i\}_{i=1}^{m}$ each with length l meet the condition that $ss = sp * CS + \varepsilon$ where ε is a small value, then we call CS the component set and sp the component spectrum of ss.

Formally, given n subsequences $SS = \{ss_i\}_{i=1}^{n}$ each with length l, assume the number of component set CS is m, then we can formulate the loss function as

$$L(CS, SP) = \sum_{i=1}^{n} ||ss_i - CS * sp_i||_2^2, s.t. \forall i, ||sp_i||_0 < k, \tag{1}$$

where $SP = \{sp_i\}_{i=1}^{n}$ is the set of component spectrum, and k is the maximum number of components used to represent a subsequence. The formula depicts the error of the representation while the restricted condition constrains the number of nonzero values in $sp_i, i \in [1, n]$.

Aiming to acquire CS and SP which make the value of loss function small enough, we can use the iterative algorithm with two steps: (1) given the component set CS, compute the best component spectrum SP which satisfy the sparsity; (2) given the component spectrum SP, update the component set CS to decrease the loss. Notice that the form of the loss function is similar to the loss function in dictionary learning algorithms. Thus, we apply a related technique in [19] and refine the loss function as

$$L(CS, SP) = ||SS - CS * SP||_F^2, s.t. SP \odot M = 0, \tag{2}$$

where $\odot$ denotes entry-wise (Schur) multiplication of two equally-sized matrices; and M is a mask matrix which satisfies M(i, j) = 1 if SP(i, j) = 0, and

$M(i, j) = 0$ otherwise. The restricted condition of this loss function forces all the zero entries in A to remain intact [19]. Readers may refer to the reference [19] for detailed computational process of the algorithm.

After the acquisition of the component set CS, we normalize the set. The normalized set NCS satisfies the condition of $||abs(cs_i)||_1 = 1, i \in [1, n]$, where $abs(cs_i)$ denotes the matrix where the value of any element is the absolute value of cs_i.

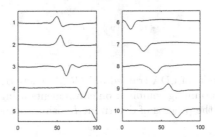

Fig. 3. Examples of the learned component set.

Due to the pseudo-periodicity of ECG data streams, we can randomly select plenty of subsequences from some standard ECG databases. Examples of the learned component set can be found in Fig. 3. An interesting phenomenon is that the acquired typical components look like some typical waves of ECG data. The left five components labeled as 1 to 5 look approximately like the QRS waves in ECG, only in different coordinate positions, while the right five components labeled as 6 to 10 look like the T waves in different coordinate positions. Notice that the coefficient of linear combination can be negative, so some of the waves are vertical inverted. This suggests the component set learned in this module accords with expert knowledge in clinical field.

4.2   Class Clustering

The task of LCCM is clustering the component spectrums to associate them with different phenotypes, and obtaining the latent class vector as the quantized association strength.

In the procedure of component spectrum representation, components of noise and other waves are separated. However, noises can not be wiped off directly by their morphology now that they varies in morphology. In order to reduce the influence of occasional noises, more than one consecutive subsequences analyzed in CSRM are considered here in a sliding window.

It is known that different phenotypes of ECG data root in different statuses of cardiac electrophysiological activities. If we divide the statuses into several types, then each type is associated with a particular combination mode of a series of typical components. Hence, we use the idea of topic model for reference, and view the generation process of ECG data as follows: firstly, one or more type of cardiac electrophysiological activities statuses (each associated with a latent class of data) are chosen in a certain chance; secondly, the associated components are chosen in a certain frequency. Latent class vector lc of a sliding window denotes probability of each phenotype's appearance in the sliding window.

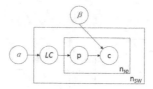

Fig. 4. The graphical model of LDA trained in LCCM, where α, β is are parameters of LDA. The process of determining combination components using β, current phenotype p repeats n_{sp} times and the process of determine lc using α repeats n_{SW} times to get SW.

According to CSRM, we have m normalized typical components in total. In the training step, we randomly select a set of sliding windows SW each with length L from some standard ECG databases. We train a latent Dirichlet allocation (LDA) model [20] to simulate the process, and optimize the aforementioned idea by replacing the showing up frequency of components with the sum of absolute value in component spectrums within a sliding window. The graphical model of LDA described above is shown in Fig. 4. Moreover, the strategy of downsampling is applied to improve performance since the normal samples are much more than the anomalous ones.

4.3   Classifier Training

The CADM is designed to solve the problem of anomaly detection, and can be seen as a binary classification problem based on latent class vectors. Since our method is independent of the classification methods, we simply apply a Gaussion-kernel support vector machine (SVM) here due to its high applicability and set the parameter to be selected automatically using a heuristic procedure.

To obtain the classification label, we preprocess the frequently-used annotations of ECG data streams and classify them to a normal or anomalous label. Table 1 shows the classification.

4.4   Online Processing

The main steps of online processing can be seen in Fig. 2, which are partition by a sliding window protocol, component spectrum representation, latent class vector obtaining and supervised anomaly detection. Algorithm 1 summarizes the procedure in form of pseudo-code.

Table 1. Classification of phenotypes in ECG.

Normal	Anomalous
Normal beat	(Aberrated) atrial premature beat
Paced beat	Premature ventricular contraction
Signal quality change	Fusion of ventricular and normal beat
Isolated QRS-like artifact	Nodal (junctional) premature beat
Comment annotation	Non-conducted P-wave (blocked APB)
Measurement annotation	Premature or ectopic supraventricular beat
P-wave peak	Ventricular or nodal (junctional) escape beat
T-wave peak	Fusion of paced and normal beat
Rythm change	R-on-T premature ventricular contraction
U-wave peak	Systole/Diastole
	Left or right bundle branch block
	Non-conducted pacer spike
	Ventricular flutter/fibrillation
	Atrial or supraventricular espace beat

Algorithm 1. Real-time Anomaly Detection over ECG Data Stream

Input: the ECG data stream S to be analyzed, the length of components l, the length of sliding window L, the component set CS, the latent class parameter α, β, the classification parameter $Model$

Output: The classification result A of current sliding window

1: **for** each incoming subsequence SW of S with length L **do**
2:     $SS \leftarrow$ reshape SW as a l by L/l matrix;
3:     Use CS to get the component spectrum SP;
4:     $sumsp \leftarrow 0$;
5:     **for all** $sp_i \in SP$ **do**
6:         $sumsp \leftarrow sumsp + abs(sp_i)$;
7:     **end for**
8:     Use α, β to get the latent class vector lc of $sumsp$;
9:     Use $Model$ to predict A;
10: **end for**

---

5   Experimental Evaluation

In this section, we demonstrate the effectiveness of our method over real data. The performance is evaluated in different aspects.

5.1   Experimental Setup

The experiments were conducted on an Intel Core i3-2120 3.30 GHz with 8 GB of memory, running Window 7. The data we used were from the MIT-BIH

Arrhythmia Database [3], a real database obtained by the Beth Israel Hospital Arrhythmia Laboratory. The data was sampled with the frequency of 360 Hz, along with the manual marked annotations corresponding to the annotation types in Table 1. The database contains 23 records which are chosen randomly from both patients and inpatients, and 25 records which are selected intentionally to include a variety of rare but clinically important phenomena.

5.2 Accuracy of Component Spectrum Representation

For the beginning of the experiments, we want to ensure that the component spectrum representation does not import an unbearable amount of error. As the fact that our method intends to be used in practical and trained beforehand, the training and online processing should be conducted based on different source data. Thus, we train the component set on the 25 records selected intentionally and observe the effectiveness using the 23 records randomly chosen. The subsequence length $l = 100$ and the sparsity parameter $k = l/8$ are set here.

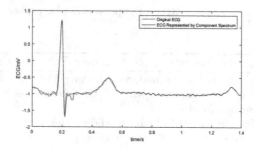

Fig. 5. An example of component spectrum representation.

The metric we use here is the root of mean square error (RMSE). Notice that the RMSE of the training set varies when parameters such as length, number of iterations are set as different value. In the experiment, RMSE of the training set is 3.829 % while RMSE of the test set is 4.587 %, which shows a strong generalization ability. An example of component spectrum representation can be find in Fig. 5. We can see the error is small enough not to affect the detection result.

5.3 Methods Comparison

We compare our method with a classical unsupervised method in [16] denoted as $k-means$ and the supervised method in [14] denoted as $SAX + LDA$ for short. Since a normal heart beat is usually a little less than a second, we set the sliding window length $L = 400$ through the experiments, which is about 1.11 times the sampling frequency. As noted in [14], the optimal performance is achieved when parameter $a = 3\text{--}4$ and $w = 6\text{--}8$. In the experiment, a is set to be 4 as it performed better than $a = 3$, while w is set to be 5 and 10 as it should divide L with no remainder. In our method, the number of latent classes is set to 25 due to the number of phenotypes. And we tried different value of subsequence length l. Note that the goal of component spectrum obtaining is separating typical components in the subsequences, so l is unnecessary and time-consuming to be set too large. As for parameter k in $k - means$, it is also set to be 25. Each experiment was repeated 5 times and the average value of the result is given in Fig. 6.

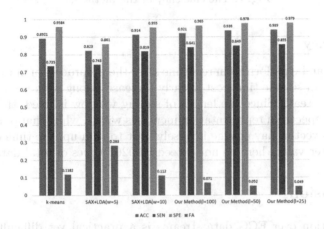

Fig. 6. The performance comparison of different methods.

We use matrices of ACC, SEN, SPE, and FA to validate the classification performance.

(1) $Accuracy\ (ACC) : (TP + TN)/(TP + TN + FP + FN)$;
(2) $Sensitivity\ (SEN) : TP/(TP + FN)$;
(3) $Specificity\ (SPE) : TN/(TN + FP)$;
(4) $FalseAlarmRate\ (FA) : FP/(TP + FP)$;

Where TP denotes true positive, TN denotes true negative; FP denotes false positive and FN denotes false negative.

As shown in Fig. 6, our method achieves higher accuracy, higher sensitivity, higher specificity and lower false alarm rate. To some degree, our method can be seen as $SAX + LDA$ where component spectrum representation is used instead of SAX. As a result of the experiment, our method outperforms $SAX + LDA$, so that component spectrum representation performs better here. A possible explanation would be that the subsequence length in our method is longer and thus more sequential information is included.

Fig. 7. The efficiency of the method.

5.4 Efficiency

Our method can be performed in real time after the preprocess of model training. In Fig. 7, we can see the time cost linearly increases along with the total length of ECG data stream. Since the length of sliding window is fixed, the time cost of component spectrum representation increases with l, while that of acquisition of latent class vector may change inversely. But to sum up, the time costs of all three parameter values here do not exceed 0.00267 times of total data length.

6 Conclusion

Outlier detection over ECG data streams is a practical yet difficult task. We analyze the challenges of the task and introduce an effective supervised method which can operate in real time and deal with the challenges well. We propose a novel concept of component spectrum of time series data. The decomposition procedure makes it possible to separate the typical wave components and noise components. The proposed method is operated over real data and it has showed effectiveness and low false alarm rate without the help of expert knowledge.

For future work, a possible direction would be to find clustering models according with the semantic of the data stream better. Also, the method can be applied to scenarios like other kinds of physiological data streams.

Acknowledgments. This work was supported by Natural Science Foundation of China (No. 61170003).

References

1. Cao, L., Yang, D., Wang, Q., Yu, Y., Wang, J., Rundensteiner, E.A.: Scalable distance-based outlier detection over high-volume data streams. In: International Conference on Data Engineering, pp. 76–87. IEEE (2014)
2. Jiang, X., Xie, C.: Home health telemonitoring system and data analyzing of physical parameters. In: Computer Engineering and Applications, pp. 213–215 (2011). (in Chinese)

3. Goldberger, A.L., Amaral, L.A.N., Glass, L., Hausdorff, J.M., Ivanov, P.C., Mark, R.G., et al.: Physiobank physiotoolkit. physionet: components of a new research resource for complex physiologic signals. Circulation 215–220 (2000)
4. Tsien, C.L.: Reducing false alarms in the intensive care unit: a systematic comparison of four algorithms. In: Proceedings: A Conference of the American Medical Informatics Association. AMIA Annual Fall Symposium, vol. 4, p. 894 (1997)
5. Ma, J., Sun, L., Wang, H., Zhang, Y., Aickelin, U.: Supervised anomaly detection in uncertain pseudoperiodic data streams. ACM Trans. Internet Technol. 16(1), 1–20 (2016)
6. Liu, S., Qu, Q., Chen, L., Ni, L.M.: SMC: a practical schema for privacy-preserved data sharing over distributed data streams. IEEE Trans. Big Data 1(2), 68–81 (2015)
7. Stephenson Jr., H.E., Reid, L.C., Hinton, J.W.: Some common denominators in 1200 cases of cardiac arrest. Ann. Surg. 137(5), 731–744 (1953)
8. Wang, J.S., Chiang, W.C., Hsu, Y.L., Yang, Y.T.C.: ECG arrhythmia classification using a probabilistic neural network with a feature reduction method. Neurocomputing 116, 38–45 (2013)
9. Doğan, B., Korürek, M.: A new ECG beat clustering method based on kernelized fuzzy c-means and hybrid ant colony optimization for continuous domains. Appl. Soft Comput. 12(11), 3442–3451 (2012)
10. Qi, J., Zhang, R., Ramamohanarao, K., Wang, H., Wen, Z., Wu, D.: Indexable online time series segmentation with error bound guarantee. World Wide Web-internet Web Inf. Syst., 1–43 (2013)
11. Ngo, D.H., Veeravalli, B.: Design of a real-time morphology-based anomaly detection method from ECG streams. In: International Conference on Bioinformatics and Biomedicine, pp. 829–836. IEEE (2015)
12. Aggarwal, C.C.: Outlier Analysis. Springer Publishing Company, New York (2015)
13. Chandola, V., Banerjee, A., Kumar, V.: Anomaly detection: a survey. ACM Comput. Surv. 41(3), 75–79 (2009)
14. Wang, J., Sun, X., She, M.F.H., Kouzani, A., Nahavandi, S.: Unsupervised mining of long time series based on latent topic model. Neurocomputing 103, 93–103 (2013)
15. Qu, Q., Qiu, J., Sun, C., Wang, Y.: Graph-based knowledge representation model and pattern retrieval. In: International Conference on Fuzzy Systems and Knowledge Discovery, vol. 5, pp. 541–545 (2008)
16. Arthur, D., Vassilvitskii, S.: k-means++: the advantages of careful seeding. In: ACM-SIAM Symposium on Discrete Algorithms, vol. 11, pp. 1027–1035 (2007)
17. Breunig, M.M., Kriegel, H.P., Ng, R.T., Sander, J.: LOF: identifying density-based local outliers. In: Proceedings of the 2000 ACM SIGMOD International Conference on Management of Data, pp. 93–104 (2000)
18. Shabib, A., Narang, A., Niddodi, C.P., Das, M.: Parallelization of searching and mining time series data using dynamic time warping. In: International Conference on Advances in Computing, Communications and Informatics. IEEE (2015)
19. Smith, L.N., Elad, M.: Improving dictionary learning: multiple dictionary updates and coefficient reuse. IEEE Sig. Process. Lett. 20(1), 79–82 (2013)
20. Blei, D.M., Ng, A.Y., Jordan, M.I.: Latent dirichlet allocation. J. Mach. Learn. Res. 3, 993–1022 (2003)

A Workload-Driven Vertical Partitioning Approach Based on Streaming Framework

Hong Kang, Mengyu Guo, and Xiaojie Yuan[✉]

College of Computer and Control Engineering, Nankai University, Tianjin, China
yuanxiaojie@dbis.nankai.edu.cn

Abstract. In cloud computing environment, data information is growing exponentially. Which raises new challenges in efficient distributed
data storage and management for large scale OLTP and OLAP applications. Horizontal and vertical database partitioning can improve the
performance and manageability for shared-nothing systems which are
popular in nowadays. However, the existing partitioning techniques can't
deal with dynamic information efficiently and can't get the real-time partitioning strategies. In this paper, we present *WSPA*: a workload-driven
stream vertical partitioning approach based on streaming framework.
We construct an affinity matrix to get the mapping information from a
workload and cluster attributes according to the attribute affinity, then
obtain the optimal partitioning scheme by a cost model. The experimental results show that *WSPA* has good partitioning quality and lower time
complexity than existing vertical partitioning method. It is an efficient
partitioning method for processing the dynamic and large scale queries.

Keywords: Database partitioning · Streaming framework · Workload-
driven

1 Introduction

Horizontal and vertical partitioning are important means to improve the performance and manageability, for both single system servers (e.g. [1]) and shared-
nothing systems (e.g. [2,3]). Horizontal partitioning allows data to be partitioned
into disjoint sets of rows that are physically stored on different machine nodes.
It is possible to reduce the query time where each node can scan its partitions in parallel [4,5]. On the other hand, vertical partitioning allows a table to
be partitioned into disjoint sets of columns. Partitioning technologies have been
researched heavily in the past [4,6–8]. Nowadays, researchers proposed workload-
based horizontal partitioning [9] and vertical partitioning [10] with the objective
of finding the configuration producing the minimum workload cost.

1.1 Problems with Partitioning in Cloud Computing Environment

In cloud computing environment, data information is growing exponentially and
changing too fast [12–14]. The dynamic big data should be processed in the

© Springer International Publishing Switzerland 2016
F. Li et al. (Eds.): APWeb 2016, Part II, LNCS 9932, pp. 68–81, 2016.
DOI: 10.1007/978-3-319-45817-5_6

fastest and most efficient way so as to obtain useful knowledge from it. Unfortunately, The existing partitioning approaches [11, 15–17] are not the real time partitioning methods. Assume the workload changes over time, e.g. dynamic changes in the workload due to a lot of new database applications, or a sharp increasing number of queries. In these situations, the existing partitioning strategies should be revisited to improve query efficiency. Further, the query contents (attributes or tuples) from the requesters are also not be predicted. When the current partitioning approaches attempt to build data models based on the unpredicted query contents and to find the optimal partitioning strategy, the cost will be very expensive. This is especially problematic if the database system has to handle bursts and peaks. For instance consider (i) in the field of financial, banking system grasping the real-time data from the various subsystems, monitoring the global state and providing decision support timely; (ii) real timely analyzing the state information of users in the social networking web sites, where schemas are often characterized by many n-to-n relationships; (iii) an shared-nothing OLAP system having to cope with new query patterns. In these types of applications, it is not acceptable for users to wait for the partitioning tool to analyze the workload in a much amount of time. If the system stalls due to a peak workload, the application provider may can't react quickly when problems appear or lose precious commercial opportunities.

1.2   Research Goals and Contributions

Our goal is to research a real-time partitioning approach which is able to cope with the dynamic, large scale and unpredicted analytical or transactional workloads in the cloud computing environment. The DBAs can reorganize the database and migrate the data according to the partitioning scheme. Moreover, the partitioning system should have the good scalability and high availability when handle the dynamic and heavy workloads. In this paper, we present *WSPA* (Workload-driven Stream Vertical Partitioning Approach), which processes the dynamic and large-scale queries by integrating with a streaming framework. The contributions of this paper are:

- We express the existing partitioning strategies and describe the limitations and weaknesses of them (Sect. 2).
- We introduce the preliminaries of our algorithm, put forward the conception of optimal affinity sequence and candidate split vector (Sect. 3).
- We present a real time vertical partitioning algorithm *WSPA*. Then we put forward an improved algorithm *WSPA-P* which takes advantage of the parallel computing mechanism of streaming framework (Sect. 4).
- We show an extensive evaluation of our algorithm over TPCC, TPCE, TPCH benchmarks and a real database CCEMIS. We compare the partition quality, partition time and iteration times with an existing partitioning approach O^2P, then analyze the experimental results (Sect. 5).

2   Related Work

David J. DeWitt and others put forward the thought of data partitioning [5], and proved that using data partition can significantly improve the retrieval performance of the database. Vertical partitioning started with early approaches of heuristic based partitioning [18] of data files. Database researchers pointed out that the number of ways to partition vertically, for x attributes, is given by bell number $B(x)$. To find the optimal solution is to enumerate all bell numbers. The complexity of this approach is $O(x^x)$, making it infeasible for large databases. To address this, the state-of-the-art work in vertical partitioning [19] develops the notion of attributes affinity, quantifying attribute co-occurrence in a given set of transactions. A follow up work [17] presents graphical algorithms to improve the complexity of their approach. Next, researchers proposed transaction based vertical partitioning [20], arguing that since transactions have more semantic meaning than attributes, it makes more sense to partition attributes according to a set of transactions. A recent work proposed an online vertical partitioning algorithm O^2P [10] which can take partitioning decisions automatically. This approach first monitors the queries using a slide time window, then executes the partitioning algorithm over a fixed time interval. The core idea of O^2P is computing the affinity between every pair of attributes and clustering them [17], then using greedy strategy to calculate the cost of every possible split line to get the best partitioning scheme. The split line between every pair of attributes will be calculated in one iteration. However, this approach copes with the static queries in an online way. It can't deal with stream queries and get the real time partitioning scheme on the arrival of each query. Moreover, this approach can't have good scalability when handle the dynamic and heavy workloads.

3   Preliminaries

3.1   Definition

In this section, we describe the definitions used in *WSPA*, then express the idea of our partitioning approach.

Definition 1. A *workload* W_i is a stream of queries $\{q_0, q_1, q_2, .., q_{i-1}, q_i\}$, seen till time i, where $0 < ... < i - 1 < i$. Further, in order to obtain the partitioning information from the workload, we need to model the attributes which are accessed by a query.

Definition 2. An *attributes list* $L(q_i)$ is the list of attributes which are accessed by q_i, where $q_i \in W_i$.

In order to map the attributes relationships among these queries, we also introduce attribute affinity as a measure of pair-wise attribute similarity. We compute affinities between every pair of attributes, then cluster them such that high affinity pairs are as close in neighborhood as possible. To compute affinity between different attributes, we also need to know their access patterns.

A visit function $Visit(q,a)$ denotes whether or not query q references attribute a. $Visit(q,a) = 1$ if q references a and 0 otherwise.

Definition 3. The *attributes affinity* $A(a_i,a_j)$ of two attributes a_i and a_j is defined as follow:

$$A(a_i, a_j) = \sum_q Visit(q, a_i) \bullet Visit(q, a_j) \qquad (1)$$

$A(a_i,a_j)$ is calculated as the sum of queries that access attributes a_i and a_j simultaneously, it represents the similarity of pair-wise attributes. Given attributes a_i, a_j and a_k, If $A(a_i,a_j) > A(a_i,a_k)$, then the relevance of a_i and a_j is stronger than that of a_i and a_k. Since the attributes affinity represents the relevance between every pair of attributes, we dynamically tuning the attribute sequence, and cluster the most relevant attributes. The new order of the attributes is an optimal attributes affinity sequence.

Definition 4. An *optimal attributes affinity sequence OpS* is a maximum associated sequence of the attributes which are accessed by q_i, where $q_i \in W_i$.

Every time a new query arrived, the attribute affinity between every pair of attributes is calculated, then the new OpS is dynamically created. In order to get the optimal split vector, we analyze the possible spit lines to partition the OpS. So that, we create the candidate split vector, which captures the logical candidate partitioning schemes over a given OpS, *i.e.*

Definition 5. A *candidate split vector CSV* is a row vector of possible split lines to partition the OpS, where a split line s_j denotes whether or not there is a split line between a_j and a_{j+1}. If $s_j = 1$ then there is a split line between a_j and a_{j+1} and 0 otherwise. For example, if $CSV = [0,0,0,1,0,0,1,0,0]$ and $OpS = \{a_0, a_1, a_2, a_3, a_4, a_5, a_6, a_7, a_8\}$, there are two split lines at position 3 and 6 of OpS, i.e. $\{a_0, a_1, a_2, a_3|a_4, a_5, a_6|a_7, a_8\}$.

The **Workload-driven Stream Vertical Partitioning Problem** will be expressed as follows: Given a query q_i, workload W_i and optimal attributes affinity sequence OpS_i at time i, find the optimal split vector s^{min} that minimize the execution cost, *i.e.*

$$s^{min} = \arg \min_{s \in CSV_i} (Cost(q_i, W_i, OpS_i)) \qquad (2)$$

The complexity of the above problem depends on the number of the candidate split lines included in CSV. Once the s^{min} is created, the partitioning scheme is updated immediately and outputted.

4 Partitioning Algorithm

In this section we describe the work flow of our approach $WSPA$. Moreover, we come up with a method $WSPA$-P which realizes parallel computing by integrating with the streaming framework.

4.1 Workload Mapping

In order to get the mapping information from a workload W_i, we define an affinity matrix M. The rows and columns of the matrix represent the attributes accessed by W_i, each element in the matrix $M[i,j] = A(a_i, a_j)$. When a query q_i arrived, the algorithm dynamically update the affinity matrix. If another new query q_{i+1} accesses to attributes already exist in M, the attribute affinities among these attributes are updated according to the query q_{i+1}. If q_{i+1} accesses to the new attributes which do not exist in M, the new attributes will be added to the matrix as an isolate node on the right.

Example 1. We list 2 queries that access to table $Order\_Line$ of TPCC. When q_1 arrived, the matrix is expressed as Matrix 1. When q_2 arrived, there is a new attribute $delivery\_d$ and an attribute $number$ which is already in Matrix 1. Then we add the new attribute $delivery\_d$ on the right of attributes, update the attribute affinity according to the accessed attributes $number$ and $delivery\_d$, the results are displayed in the gray cells of Matrix 2. Figure 1 shows the overall process of workload mapping.

4.2 Get Optimal Attributes Affinity Sequence

Given an affinity matrix, the quality of affinity clustering is defined as Aff, $i.e.$

$$Aff = \sum_{i-1}^{x}\sum_{j-1}^{x}A(a_i, a_j) \times [A(a_i, a_{j-1}) + A(a_i, a_{j+1})] \; i, j > 0 \qquad (3)$$

To get OpS, we respectively put the second attribute on the left and right of the first accessed attribute. The problem to obtain OpS can be transformed into a solution of the maximum value of Aff. When Aff has the maximum value, the corresponding positions are the optimal positions of the attributes.

Example 2. Consider Matrix 2 in Example 1. To get OpS, we put $deliver\_d$ on the right and left of $number$. Since $Aff\_right = 8 < Aff\_left = 12$, such that $OpS = \{delivery\_d, number, quantity\}$.

q_1: Select number, quantity from Order_Line	q_2: Select delivery_d, number from Order_Line

Matrix 1

	number	quantity
number	1	1
quantity	1	1

Matrix 2

	number	quantity	delivery_d
number	2	1	1
quantity	1	1	0
delivery_d	1	0	1

Fig. 1. Workload mapping

4.3   Get Candidate Split Vector

Suppose $K(q_i)$ is a collection of the attributes in $L(q_i)$ out of order, OpS_i is the optimal affinity sequence after q_i arrived. To get partitioning unit of OpS_i, we compare $K(q_i)$ with OpS_i, if there is a subset of OpS_i equals to $K(q_i)$, then $K(q_i)$ is a partitioning unit, and the split lines on both sides of $K(q_i)$ will be added to CSV_i.

Example 3. There are 5 queries arrived one by one. The list of attributes accessed by each query is list as follows: $L(q_1) = \{a_1, a_3, a_5\}$, $L(q_2) = \{a_1, a_5, a_6, a_4\}$, $L(q_3) = \{a_7\}$, $L(q_4) = \{a_1, a_3, a_5, a_4\}$, $L(q_5) = \{a_2, a_3, a_4, a_8\}$. When q_2 comes, we first update the affinity matrix shown in Fig. 2(a) according to the method proposed in 4.2. The result is shown as Fig. 2(b). So that we get the $OpS_2 = \{a_6, a_4, a_5, a_1, a_3\}$. Now we have two partitioning units of attributes $K(q_1) = \{a_1, a_3, a_5\}$ and $K(q_2) = \{a_1, a_5, a_6, a_4\}$. We put split lines on both sides of the units. So that CSV_2 is $[0\ 1\ 0\ 1]$.

	a_1	a_3	a_5
a_1	1	1	1
a_3	1	1	1
a_5	1	1	1

(a)

	a_6	a_4	a_5	a_1	a_3
a_6	1	1	1	1	0
a_4	1	1	1	1	0
a_5	1	1	2	2	1
a_1	1	1	2	2	1
a_3	0	0	1	1	1

(b)

Fig. 2. Get OpS_2

The overall process is shown in Fig. 3:

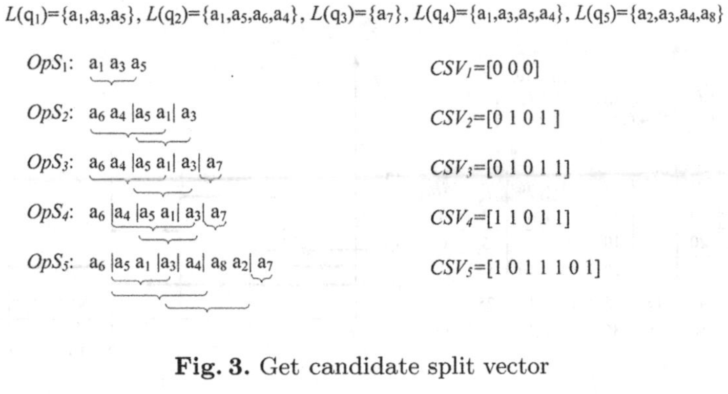

$L(q_1)=\{a_1,a_3,a_5\}$, $L(q_2)=\{a_1,a_5,a_6,a_4\}$, $L(q_3)=\{a_7\}$, $L(q_4)=\{a_1,a_3,a_5,a_4\}$, $L(q_5)=\{a_2,a_3,a_4,a_8\}$

OpS_1: $a_1\ a_3\ a_5$ $CSV_1=[0\ 0\ 0]$

OpS_2: $a_6\ a_4\ |a_5\ a_1|\ a_3$ $CSV_2=[0\ 1\ 0\ 1\ ]$

OpS_3: $a_6\ a_4\ |a_5\ a_1|\ a_3|\ a_7$ $CSV_3=[0\ 1\ 0\ 1\ 1]$

OpS_4: $a_6\ |a_4\ |a_5\ a_1|\ a_3|\ a_7$ $CSV_4=[1\ 1\ 0\ 1\ 1]$

OpS_5: $a_6\ |a_5\ a_1\ |a_3|\ a_4|\ a_8\ a_2|\ a_7$ $CSV_5=[1\ 0\ 1\ 1\ 1\ 0\ 1]$

Fig. 3. Get candidate split vector

4.4   Vertical Partitioning Using Cost Model

For each candidate split line in CSV, we will choose the best split line by computing the cost using a cost model. The split line with the minimum cost is the best split line. $L(q_i)$ is a list of attributes which q_i accessed. $C(q_i)$ is the access number of q_i. A possible partition line splits the OpS into two sets L and U-L, where U is the intersected attributes list, and U is also a subsequence of OpS, i.e.

$$U = \bigcup_{L(q_i) \cap L(q_j) \neq \Phi} L(q_j), \quad where \quad q_i, q_j \in W_i \quad and \quad 0 < j \leq i \tag{4}$$

Given $q_i, q_m, q_n \in W_i$, the cost model is defined as follows:

$$Cost(q_i, W_i, OpS_i) = | \sum_{L(q_m) \subset L} C(q_m) \times |L| - \sum_{L(q_n) \subset U-L} C(q_n) \times |U - L| | \tag{5}$$

Example 4. Figure 4 shows the attributes accessed by query q_i $(0 < i < 9)$ from t_0 to t_4, (a) is the query access situation, (b) shows the relationships among the accessed attributes at time t_3. Suppose that $CSV = [0, 1, 1, 0, 0, 0, 0, 0, 0]$, the possible split positions are 1 and 2. Since Cost2 < Cost1, partition at 2 is better than partition at 1, the best partition line is: $\{5, 1, 7, |, 2, 8, 3, 9, 10, 4, 6\}$.

4.5   Get Partitioning Scheme

Considering a query q_i arrived, then a new optimal affinity sequence OpS_i is created. If there is no intersection between $L(q_i)$ and the previous attributes list $L(q_j)$ $(0 < j < i)$ which exist in OpS_{i-1}, then q_i creates a new partition. So that a split line is added to partitioning scheme S. On the contrary, if there are intersections between $L(q_i)$ and the previous attributes lists $L(q_j)$, we need to use cost model for each split line in CSV_i to calculate the partitioning cost, then obtain the best split line and add it to S. Observe that the attributes lists haven't been broken in the current OpS will have the same candidate split lines

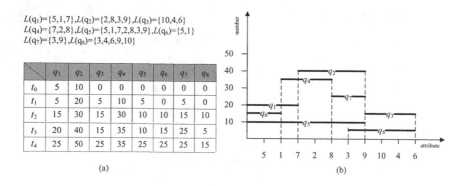

Fig. 4. Vertical partitioning using cost model

in the previous *OpS*. So we consider these candidate split lines for partitioning. Further, if the attributes lists are broken in the current *OpS*, the relevant split lines need to be removed from the current *CSV*. Algorithm1 shows the dynamic programming based enumeration in *WSPA*. First, *WSPA* gets the attributes list $L(q_i)$, updates the matrix and obtains the optical attributes affinity sequence *OpS* (Lines 1–3). If $L(q_i)$ creates a new partition then add split line to S and *WSPA* returns (Lines 4–8). Otherwise, it creates the candidate split vector *CSV* (Lines 10–18), chooses one split line having the lowest cost and add it to S (Lines 19–22). At last, return S (Line 23). The detail pseudo-code is shown in Algorithm 1.

Input: query q_i, CSV_{i-1}
Output: Partitioning Scheme S
1 $Update(Matrix)$;
2 $OpS_i = maxAff(right, left, prev)$;
3 $L(q_i) = GetAttributesList(q_i)$;
4 **if** $L(q_i) \bigcap OpS_i = \Phi$ **then**
5     $SinglePartList = getSplitLine(L(q_i))$;
6     $S = S + SinglePartList$;
7     $return\ S$;
8 **end**
9 **else**
10   **for** $i = 0$ *to* $size(CSV_{i-1}))$ **do**
11   **if** $CSV_{i-1}.get(i) \bigcap (OpS_i) = \Phi$ **then**
12          $splitline = getSplitLine(CSV_{i-1}, i)$;
13          $CSV_i = remove(CSV_{i-1}, splitline)$;
14   **end**
15   **end**
16    $NewSplitline = getNewSplitLine(L(q_i))$;
17    $CSV_i = add(CSV_i, Newsplitline)$;
18 **end**
19 **for** $i = 0$ *to* $size(CSV_i)$ **do**
20    $s^{min} = argmin_{s \in CSV_i}(Cost(q_i, W_i, OpS_i))$;
21 **end**
22 $S = S + s^{min}$;
23 $return\ S$;

Algorithm 1. Partitioning Algorithm

Theorem 1. *WSPA produces the correct greedy result.*

Proof. The best split line given in $(i+1)^{th}$ query of Algorithm 1 can be written as:

$$s_{i+1}^{min} = \underset{s \in CSV_{i+1}}{\operatorname{argmin}} (Cost(q_{i+1}, W_{i+1}, OpS_{i+1}))$$

$$= \underset{s \in L(q_j), \forall L(q_j) \subset OpS_{i+1}}{\operatorname{argmin}} (Cost(q_{i+1}, W_{i+1}, OpS_{i+1}))$$

$$= \underset{\forall L(q_j) \subset OpS_{i+1}}{\operatorname{argmin}} (\underset{s \in L(q_j)}{\operatorname{argmin}}(Cost(q_{i+1}, W_{i+1}, OpS_{i+1})))$$

$$= \underset{\forall L(q_j) \subset OpS_{i+1}}{\operatorname{argmin}} s_{i+1}^{min, L(q_j)}$$

That means to find the best split line we just need to compare the best split lines in each of the partitions which created by q_j, where $q_j \in W_{i+1}$. Hence, we can retain the best split lines from a partition for future comparison.

Theorem 2. *The runtime complexity of WSPA is O(n).*

Algorithm 1 shows that the complexity of *WSPA* depends on the length of *CSV*. Suppose the length of *CSV* is O(n), the complexity of the first part (Line 1–3) is O(n), the complexity of the second part (Line 4–8) is O(n), the complexity of the third part (Line 10–15) is O(n), the complexity of the forth part (line 19–21) is O(n). Hence, the complexity of *WSPA* is O(n).

4.6   WSPA-P

WSPA-P uses parallel computing to reduce the execution time. When calculating the affinity of matrix M, we calculate every row of the matrix in different computing units of streaming framework simultaneously, then add all intermediate results together to get the final result of M.

4.7   Algorithm Characteristics

In this section, we illustrate the salient features of our method:

(1) **Optimal Affinity Sequence Updated Dynamically.** The *OpS* can be updated dynamically according to the affinity information carried in a new query. Further, the *OpS* is incrementally updated. Usually, we only need to update the part of previous *OpS* which affected by the intersected attributes lists of q_i.
(2) **Unpredicted Workload Adaption.** Different from the traditional partitioning technology, the algorithm produces the correct greedy result based on the *OpS*. Which means we don't need to know the detail of the workload before partitioning. Therefore our approach can deal with the unpredicted workload.
(3) **Real-time Processing.** Through the integration with streaming framework, *WSPA* is able to process the workload in real time, the partitioning scheme is obtained immediately. Further, we use the parallel computing mechanism of streaming framework to obtain the *OpS*. This approach can improve the efficiency of *WSPA*.

Table 1. Algorithm comparison between $WSPA$ and O^2P

	WSPA	O^2P
Best time complexity	O(n)	O(n)
Worst time complexity	O(n)	$O(n^2)$
Real-time processing	Yes	No
Horizontal scaling	Yes	No
Workload	Dynamic/Static	Static
Table structure	Unpredicted	Known

(4) **High-throughput Adaption and Horizontal Scaling.** $WSPA$ can adapt to the high throughput by integrating streaming framework with Flume and Kafka. Further, based on the streaming framework, we can realize horizontal scaling by adding physical nodes easily. We compare the following properties: best time complexity, worst time complexity, real-time processing, horizontal scaling, workload type, table structure of $WSPA$ with O^2P. The result is list in Table 1.

5   Experiment

In this section, we firstly compare partition performance and partitioning time with O^2P, then evaluate the performance of our algorithm.

5.1   Data Sets

In this experiment, we use different types of data bases as the background of the experiment. The databases are created from TPCC, TPCE, TPCH and CCEMIS individually. TPCC, TPCE and TPCH are the popular OLAP and OLTP benchmarks. CCEMIS is a real database of Nankai University. Based on these databases, we made the SQL queries as the data sets to carry out the experiments. The average length of a query is 1.6 KB. A workload which size is 10 MB includes 6,000 queries. We change the size of workload (10M to 2G), the proportion of OLTP (0 to 1), the query generation rate(0.5 G/s to 5 G/s) to create the different data sets.

5.2   Comparison with O^2P

5.2.1   Iteration Times

The data sets used are TPCC, TPCE, TPCH and CCEMIS with a workload size of 100M, query number of 60,000, OLTP proportion of 0.5. Figure 5(a) shows the iteration times of $WSPA$, $WSPA$-P and O^2P.

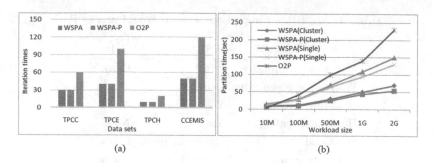

Fig. 5. Iteration times and partitioning time comparison with O^2P

The result shows that the iteration times of O^2P are more than *WSPA* and *WSPA-P*. Because O^2P keeps a slide time window and analyzes queries over a fixed time interval. The workload is divided into several pieces, and the queries included in each piece are still static. However, *WSPA* and *WSPA-P* get real-time result after each of the queries arrived, and the possible split lines exist in *CSV*. The best and worst complexity of *WSPA*, *WSPA-P* is $O(n)$.

5.2.2 Partitioning Time

The data sets we use are TPCC, TPCE, TPCH and CCEMIS with OLTP proportion of 0.5. The size of workload is increasing from 10 MB (6,000 queries) to 2 GB (1,200,000 queries). We compare the partitioning time of O^2P, *WSPA*(Cluster), *WSPA*(Single), *WSPA-P*(Cluster) and *WSPA-P*(Single). The results are shown in Fig. 5(b). The analysis results show that when the data scale is bigger, the advantages of *WSPA* and *WSPA-P* gradually enhance. Compared with O^2P, when the workload size is increasing continuously, *WSPA* and *WSPA-P* can assign computing tasks to different nodes through dynamical horizontal scaling. However, O^2P can't extend well, it can only run in a single node.

5.2.3 Evaluating Partition Quality

In order to assess the quality of the partitioning produced by our algorithm, we experimented on a variety of workloads created from TPCC, TPCE, TPCH and CCEMIS. The workload size is 100 MB, and the OLTP proportion ranges from 0 to 1. We compare the quality of the partitioning in terms of the number of distributed transactions. The results are summarized in Fig. 6.

We can see that (1) When handling workload with different OLTP/OLAP proportions, *WSPA* and *WSPA-P* have almost the same partitioning quality as O^2P. (2) On a more serious note, *WSPA* and *WSPA-P* can cope with the unpredicted workload and obtain the partitioning scheme. O^2P, however, has to know the table structure before partitioning. Hence, *WSPA* and *WSPA-P* have better applicability.

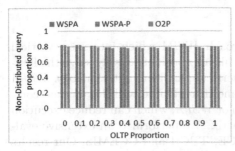 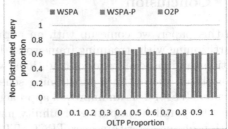

Fig. 6. Partition quality of WSPA on TPCE and TPCH

5.3   Horizontal Scaling Evaluation

We evaluate the horizontal scaling ability of *WSPA* by increasing workload size. As shown in Fig. 5(b), when the scale of data is small, the advantages of Storm clusters is not obvious due to the system initialization time, intermediately generation and transmission time. When the workload becomes larger than 20M and the above, compared to the single machine, the advantages of cluster is gradually reflected. Moreover, when we add new machines to the storm cluster, the partitioning time reduced markedly. As shown in Fig. 7(a), when the node number is added to 7, the partition time is reduced by nearly half.

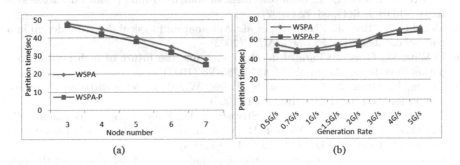

(a)   (b)

Fig. 7. Horizontal scaling and high-throughput adaption evaluation of WSPA

5.4   Evaluating High-Throughput Adaption Ability

In order to evaluate the ability of the high-throughput adaption, we use the dynamic query generator to create the streaming query. The generation rate is from 0.5 GB/s (3 million queries/s) to 5 GB/s (30 million queries/s). The result is shown in Fig. 7(b). By analyzing the result, we find that the generation rate increased by 1000 %. However, at the same time, partitioning time only increased by average 30 % and fluctuates in a small range. That means *WSPA* has the good ability to adapt to the high-throughput.

6   Conclusion

In this paper, we come up with an approach *WSPA* which integrates vertical partitioning with streaming framework. This approach effectively solve the limitations and disadvantages of traditional vertical partitioning technology. *WSPA* can obtain the optimal partitioning scheme in real time with lower time complexity. *WSPA-P* is an improved approach of *WSPA*, which can reduce execution time by parallel computing of affinity matrix. We performed an extensive evaluation of our algorithm over TPCC, TPCE, TPCH and CCEMIS. Our results show that *WSPA* and *WSPA-P* are faster than earlier approaches and still produce good quality partitioning results. Additionally, our results show that over dynamic and heavy workloads, *WSPA* and *WSPA-P* have the good scalability and high availability.

Acknowledgments. This work is partially supported by Tianjin Application Foundation and Advanced Technology Research Project No. 14JCYBJC15500 and Special Fund for the Doctoral Program of Higher Education No. 20130031120029.

References

1. DeWitt, D.J., Ghandeharizadeh, S., Schneider, D., Bricker, A., Hsiao, H.-I., Rasmussen, R.: The Gamma Database Machine Project
2. Dewitt, D.J., Metha, M.: Data placement in shared-nothing parallel database systems. VLDB J. **6**(1), 53–72 (1997)
3. Megiddo, N., Rao, J., Zhang, C., et al.: Automating physical database design in a parallel database, pp. 558–569. ACM (2002)
4. DeWitt, D., Gray, J.: Parallel database systems: the future of high performance database systems. Comm. ACM **35**(6), 85–98 (1992)
5. Chang, F., Dean, J., Ghemawat, S., Hsieh, W.: Bigtable: a distributed storage system for structured data. In: OSDI (2006)
6. Cooper, B.F., Ramakrishnan, R., Srivastava, U., Silberstein, A., Bohannon, P., Jacobsen, H.-A., Puz, N., Weaver, D., Yerneni, R.: PNUTS: Yahoo!s hosted data serving platform. PVLDB **1**(2), 1277–1288 (2008)
7. Curino, C., Jones, E., Zhang, Y., Wu, E., Madden, S.: Relationalcloud: the case for a database service. New England Database Summit (2010)
8. Ghandeharizadeh, S., DeWitt, D.J.: Hybrid-range partitioning strategy: a new declustering strategy for multiprocessor databases machines. In: VLDB (1990)
9. Curino, C., Jones, E., Zhang, Y., et al.: Schism: a workload-drive approach to database replication and partitioning. VLDB **3**(1/2), 48–57 (2010)
10. Jindal, A., Dittrich, J.: Relax and let the database do the partitioning online. In: Castellanos, M., Dayal, U., Lehner, W. (eds.) Enabling Real-Time Business Intelligence. LNBIP, vol. 126, pp. 65–80. Springer, Heidelberg (2012)
11. Grund, M., Krger, J., Plattner, H., Zeier, A., CudreMauroux, P., Madden, S.: HYRISEA—main memory hybrid storage engine. Bull. Tech. Committee Data Eng. **4**(1), 105–116 (2010)
12. Lynch, C.: Big data: how do your data grow. Nature **455**(7209), 28–29 (2008)
13. Li, G.J., Cheng, X.Q.: Research status and scientific thinking of big data. Bull. Chin. Acad. Sci. **27**(6), 647–657 (2012)

14. Wang, Y.Z., Jin, X.L., Cheng, X.Q.: Network big data: present and future. Chin. J. Comput. **36**(6), 1125–1138 (2013). (in Chinese with English abstract)
15. Agrawal, S., Chu, E., Narasayya, V.: Automatic physical design tuning: workload as a sequence. In: SIGMOD (2006)
16. Navathe, S., et al.: Vertical partitioning algorithms for database design. ACM TODS **9**(4), 680–710 (1984)
17. Navathe, S., Ra, M.: Vertical partitioning for database design: a graphical algorithm. In: SIGMOD (1989)
18. Pujol, J.M., Siganos, G., Erramilli, V., Rodriguez, P.: Scaling online social networks without pains. In: NetDB (2009)
19. Zilio, D.C.: Physical database design decision algorithms and concurrent reorganization for parallel database systems. Ph.D. thesis (1998)
20. Chu, W.W., Ieong, I.T.: A transaction-based approach to vertical partitioning for relational database systems. IEEE TSE **19**(8), 804–812 (1993)

Research Full Paper:
Recommendation System

Spica: A Path Bundling Model for Rational Route Recommendation

Lei Lv[1], Yang Liu[1], and Xiaohui Yu[1,2(✉)]

[1] School of Computer Science and Technology, Shandong University,
Jinan 250101, China
llsdu13@gmail.com, {yliu,xyu}@sdu.edu.cn
[2] School of Information Technology, York University,
Toronto, ON M3J 1P3, Canada

Abstract. This paper presents Spica, a path bundling model for rational route recommendation leveraging the intelligence and experience of the past driving records. In this model, the traffic surveillance system is employed to probe the traffic rhythm of a city and vehicle traveling record's intelligence is used to choose driving directions in the real world. We propose Joint Technique (JT) to build Time-Dependent Joint (TDJ) graph and model the dynamic traffic pattern so as to provide the rational fastest route to a given destination at a given starting time. Then we estimate the travel time in different time slots and based on TDJ graph, we propose Time-Dependent Heuristic Algorithm (TDHA) to compute the rational recommended routes. We build our model based on a real-world trajectory data set generated by totally 44,593,706 passage records in a period of a week, and evaluate the performance of our model by conducting extensive experiments. The recommended routes are effective and JT gives evidence of its rationality over previous ways.

Keywords: Joint technique · Driving directions · Rational route recommendation · Trajectory

1 Introduction

In the last decades, urban areas have witnessed an explosive growth in the number of vehicles, which makes the urban traffic congestion a pandemic illness and demands for more resources in metropolis and small city. In this case, finding highly efficient driving directions has become a daily activity and a fast driving route saves not only the time of a driver but also energy consumption (as most gas is wasted during the congestion waiting time).

Route planning is a part of people's daily life. Basically, driver traverses a route depends on total of three aspects: (1) the physical feature of a route, which includes distance, capacity (lanes) and the number of traffic lights as well as direction turns; (2) the time-dependent traffic flow on the route; and (3) a driver's driving behavior. Also driver's driving behaviors are varied according to their own progressing driving skills and experiences. For example, when driving

© Springer International Publishing Switzerland 2016
F. Li et al. (Eds.): APWeb 2016, Part II, LNCS 9932, pp. 85–97, 2016.
DOI: 10.1007/978-3-319-45817-5_7

on an unfamiliar road, a driver has to pay more attention to the lane signs, the guideposts and the other vehicles, hence drives relatively slowly. Therefore, a highly efficient driving direction model should take into consideration these three aspects: routes, traffic, and drivers.

By using taxi-GPS traces, recent literatures have made a considerable effort to find fastest driving routes. Though taxi traces can help better understand the urban mobility and taxi drivers can usually find out the fastest path to send passengers to a destination based on their knowledge, it becomes clear that there are three obviously problems existed in these taxi-trace methods. Firstly, taxi movements can not reflect the entire traffic condition. The occurrence of taxies on roads may be skewed from that of the entire set of vehicles. For example, observing more taxis on a road segment does not mean more occurrences of other vehicles. Secondly, taxi drivers have not always a pure destination, which makes their movements meaningless. Such as hanging out in the road to find a passenger and even greedy drivers will overcharge passengers by deliberately taking unnecessary detours. Thirdly, the most important one is, these methods give a road segment as a single-valued travel time function which results in wrong travel time links (later explicitly discussed). Generally, big cities have a huge number of vehicles traversing in urban areas. To understand the urban mobility sufficiently, we can make full use of analysis of not only taxi movements but also the other types of vehicles through the traffic surveillance system. By analyzing the global vehicle traces, the above problems can be avoided.

In this paper, we propose Spica, a path bundling model for rational route recommendation. The time-stamped vehicle trajectories are recorded by surveillance cameras to probe the traffic rhythm of a city in the physical world. The main idea of Spica is that it uses a new technique named *Joint Technique* to link the adjacent road segments about travel time and recommends the fastest driving routes by learning from all vehicles' past traveling records. Here are the main steps. First, we take this new technique JT to build TDJ graph and model the dynamic traffic pattern. Second, we estimate the travel time in different time slots among TDJ graph. Finally, based on TDJ graph, we use TDHA to compute the rational recommended routes.

Our main contributions of this paper can be summarized as follows.

- To the best of our knowledge, JT method gives a new angle of route planning and is important for time-dependent route rationalization. The previously time-dependent algorithm is imprudent in *route linking problem* (see Sect. 4.1).
- For the first time, we tackle the problem of route planning on traffic surveillance data set. Unlike taxi GPS traces, traffic surveillance data set can provide more enriched traffic information for us to analyze the route planning problem especially around the crossroad.
- We propose Spica to do route planning, this new model is not complex but effective.
- Extensive experiments are performed on a real data set (a traffic surveillance data set) and the results show the effectiveness of Spica and JT demonstrates evidence of its rationality over previous ways.

The rest of this paper is organized as follows. Section 2 reviews related work. Section 3 provides a brief description of preliminary preparation. The method of building TDJ graph is presented in Sect. 4. Section 5 describes TDHA and in Sect. 6, the performance evaluation is discussed. Finally, the discussion of the conclusions and the future vision are presented in Sect. 7.

2 Related Work

2.1 Shortest and Fastest Path

In [10,12,15], there are several well-known algorithms have been developed to find the shortest paths on a graph. the Dijkstra algorithm and the A* algorithm are both commonly used for efficiently searching for shortest paths. They presented landmark-based approaches to approximate the shortest path distances with lower computation costs.

In stead of finding the shortest paths on a graph, searching for the fastest routes in road networks is more useful for driving. The problem of finding the fastest paths on road networks has been investigated in [6–8,11]. In road networks, each road segment has its own historical speed patterns. In [7], Kanoulas et al. proposed a novel approach for finding the fastest path on urban area with these speed patterns. In addition, in [6], Gonzalez et al. introduced the problem of planning fastest path on urban area considering the type of roads and traffic patterns. However, in road networks, traffic status changes every time. Thus, the problem of finding the fastest paths on dynamic road networks is further studied in [8,11]. These studies developed efficiently models to monitor the fastest paths before arriving at the destination

The time-dependent fastest path problem is first studied in [1]. Dreyfus [4] proposed a straightforward generalization of Dijkstra algorithm but he did not notice that it does not work for a non-FIFO network [9]. Under the FIFO assumption, paper [2] provides a generalization of Dijkstra algorithm that can solve the problem with the same time complexity as the static fastest path problem. Demiryurek et al. presented a good case study comparing existing approaches for the TDFP problem on real-world networks [3]. Furthermore, in [17], Yuan et al. introduced a novel system named T-Drive to find the fastest paths based on the taxi drivers' intelligence. In this work, they focus on obtaining routes with lower travel costs.

2.2 Taxi-GPS Based Studies

Taxi drivers are usually experienced in finding the fastest (quickest) route to a destination based on their knowledge. In these years, lots of literatures [5,13,14, 16–19] came out to do research based on taxi-GPS traces. Paper [16,17] make use of taxi drivers' experiences to guide the fast driving routes while [18,19], and Veloso [13,14] analyze taxi-GPS traces to explore the relationships between pick-up and drop-off locations and visualize the spatiotemporal variation of taxi services. Ge et al. developed a taxi driving fraud detection system to investigate taxi driving fraud (eg. detours) in [5].

3 Preliminary Preparation

In this section, we define a few terms that are required for the subsequent discussion.

Definition 1. Road segment. *A road segment r is a one-way or bidirectional edge that has two terminal points ($r.s, r.e$, refer to the surveillance camera locations in this paper). If a road segment is one-way, r can only be traversed from $r.s$ to $r.e$; otherwise, people can travel from both terminal points, i.e., $r.s \rightarrow r.e$, or $r.e \rightarrow r.s$.*

Definition 2. Road network. *A road network G_r is a directed graph, $G_r = (V_r, E_r)$, where V_r is a set of nodes representing the terminal points of road segments, and E_r is a set of edges denoting road segments. The time needed for traversing an edge is dynamic during the time of day.*

Definition 3. Route. *A route R is a set of connected road segments, i.e., $R : r_1 \rightarrow r_2 \rightarrow ... \rightarrow r_n$, where $r_i.e = r_{i+1}.s$, $(1 \leq i < n)$. The start point and the end point of a route can be represented as $R.s = r_1.s$ and $R.e = r_n.e$.*

Definition 4. Vehicle trajectory. *A vehicle trajectory T_r is a sequence of camera locations pertaining to one trip. Each trajectory point $p \in P_r$ consists of a time stamp p.t (in minute). i.e., $T_r : p_1 \rightarrow p_2 \rightarrow ... \rightarrow p_n$, where $0 < p_{i+1}.t - p_i.t < \Delta T$, $(1 \leq i < n)$. P_r is a set of camera locations and ΔT defines the maximum travel time of a road segment.*

Definition 5. Joint edge. *A joint edge e_j has two travel time $e_j.t_1$ and $e_j.t_2$ that traversed two adjacent road segments ($e_j.r_1 : p_a \rightarrow p_b, e_j.r_2 : p_b \rightarrow p_c$) continuously and we estimate the travel time in Sect. 4.3. Each joint edge results from a series of real vehicle traces.*

Definition 6. Joint graph. *A joint graph $G_j = (V_j, E_j)$ is a directed graph that consists of a set of road segments as vertex set V_j appeared in vehicle trajectory data set and a set of joint edges as edge set E_j.*

4 Time-Dependent Joint Graph

This section first describes the motivation and construction of the time-dependent joint graph, then details the travel time estimation of joint edges.

4.1 Motivation of Joint Graph

Modern city especially the big city has lots of multi-lane roads such as six-lane roads and eight-lane roads, with quick lane, with slow lane, with through lane, and with left-turn lane and right-turn lane, etc. Different lanes often have different driving speed traversed by a vehicle and different driving speed means different travel time. However, previous time-dependent model have all treated

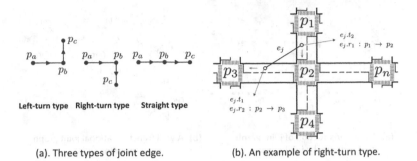

(a). Three types of joint edge.   (b). An example of right-turn type.

Fig. 1. Description of joint edge.

travel time on each road segment as conditional independence (such as [16,17]) which brings up some impossible driving phenomenon. For example, if a car travels much faster in right-turn lane than in through lane, which probably means a traffic congestion occurred in through lane. In this case, the vehicle near to the crossroad stays in solid line of through lane can not change its lane into right-turn lane for an express speed, but rather goes straight on congested condition and in this paper, we call this problem *route linking problem*.

Therefore, in order to ensure the driving rationality, for fear that a wrong travel time plan would appear in a recommended route, we put forward a new conception *joint edge* to link the travel time of those adjacent road segments rather than give each edge a single-valued function based on time of day independently. Figure 1 shows the *joint edge* in detail and the evaluation of joint edge will be specifically discussed in Sect. 6.4. We can see in Fig. 1(a) that joint edge has three types in total. Figure 1(b) shows an example of right-turn type of joint edge.

4.2   Building the TDJ

In practice, the traffic surveillance system can not cover any corner of the urban area but the main crossroad and some road segments. That is, we cannot recommend routes in a considerable detailed way based on vehicle trajectories.

In our method, we first extract vehicle trajectories representing individual trips from the traffic surveillance raw data set. Then we calculate travel time from preprocessed vehicle trajectories and divide them into joint edges. Finally, we use these road segments and obtained joint edges to build TDJ.

We observe from the vehicle trajectories that different weekdays (e.g., Tuesday and Friday) almost share similar traffic patterns while the weekdays and weekends have different patterns. Therefore, we build two different joint graphs for weekdays and weekends, respectively. Figure 2 illustrates a 24-joint-edge graph's transformation and r refers to road segment (see Definition 1).

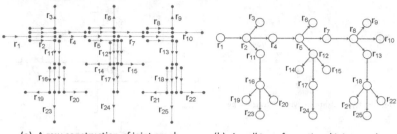

(a). A raw construction of joint graph that is drawn by joint edges.

(b). A well transformational joint graph corresponding to (a).

Fig. 2. An example of building TDJ.

4.3 Travel Time Estimation

In this part, we estimate the travel time distribution for each kind of joint edge in every time interval for all day.

Since the road network is dynamic (see Definition 2), we can use neither the same nor a predefined time partition method for all the joint edges. Moreover, our work is not a simple travel time estimation about each single road segment but the estimation for the joint edge (see Definition 5) and the reason has been showed in Sect. 4.1.

For the simplicity, we choose Δt minutes as the minimum time interval unit which is studied in evaluation part and use the fastest vehicle trajectory to estimate travel time of joint edges where $p_a.t$ is in the internal $[t, t + \Delta t)$ (see Eq. 1). Because of the diversity of drivers' personal behavior, skill and preferences, we define a weighting parameter δ as the experienced parameter to evaluate a driver's behavior in the following section.

$$e_j.t_1 = min\{p_b^1.t - p_a^1.t, p_b^2.t - p_a^2.t, \ \dots \ , p_b^n.t - p_a^n.t\};$$
$$e_j.t_2 = min\{p_c^1.t - p_b^1.t, p_c^2.t - p_b^2.t, \ \dots \ , p_c^n.t - p_b^n.t\}; \tag{1}$$
$$p_a^i.t \in [t, t + \Delta t), 0 < i \le n.$$

n is the total number of trajectories generated by traversing camera locations $\{p_a, p_b, p_c\}$ that $p_a.t$ is in the internal $[t, t + \Delta t)$.

4.4 Differentiate Drivers' Experiences

We use the parameter δ to describe a typical driver's driving proficiency who traverses a joint edge's locations $\{p_a, p_b, p_c\}$ continuously and use $\bar{\delta}$ to define a driver's experience traversing a certain trajectory $\{p_1, p_2, ..., p_n\}$. For example, $\bar{\delta} = 0.8$ means that the driver can outperform 80 % the fastest traversing a corresponding trajectory in terms of travel time under the same external condition. τ is the order of a time interval unit in a day.

$$\delta = \frac{1}{2} \cdot \left(\frac{e_j.t_1}{p_b.t - p_a.t} + \frac{e_j.t_2}{p_c.t - p_b.t} \right);$$

$$p_a.t \in [t, t + \Delta t), 0 < i \le n.$$

$$(2)$$

$$\bar{\delta}^\tau = \frac{1}{N} \sum_{i=1}^{N} \delta^{\tau_i} \quad \left(0 \le \tau < \left\lceil \frac{1440}{\Delta t} \right\rceil \right)$$

$$(3)$$

5   Time-Dependent Heuristic Algorithm

Unlike the static graph, the travel time of a road segment in TDJ graph changes all the day. In this case, we can not simply use those static shortest path algorithms such as dijkstra, bellman-ford and floyd algorithm but a heuristic search algorithm. This section introduces TDHA, which is just like a A* algorithm and includes two stages: heuristic evaluation function calculation and time-dependent breadth-first search algorithm (TBFS).

5.1   Heuristic Evaluation Function

To prepare for the TBFS, in this stage we calculate a heuristic evaluation function in order to reduce the searching costs. Firstly, we define a heuristic evaluation function $dPre_{p_i,p_j}^\tau$ as the time-distance of each pair of departure and destination (p_i, p_j) in τ^{th} internal (see Sect. 4.4) and initialize this value to 0. Then, scan each trajectory to update $dPre_{p_i,p_j}^\tau$ value in each τth time interval to the minimum value except zero. Algorithm 1 describes this initialized procedure in detail.

Algorithm 1. Heuristic Evaluation Function

1: Firstly, initialize each $dPre_{p_i,p_j}^\tau$ to zero, $0 < i < j \le n$;
2: **for** each vehicle trajectory $T_r : p_1 \to ... \to p_i \to ... \to p_j \to ... \to p_n$ **do**
3:     **for** $i = 1, ..., n - 1$ **do**
4:         **for** $j = i + 1, ..., n$ **do**
5:             $\tau = \left\lfloor \frac{1440}{p_i.t} \right\rfloor$;
6:             **if** $dPre_{p_i,p_j}^\tau > 0$ **then**
7:                 $dPre_{p_i,p_j}^\tau = min\{dPre_{p_i,p_j}^\tau, p_j.t - p_i.t\}$;
8:             **else**
9:                 $dPre_{p_i,p_j}^\tau = p_j.t - p_i.t$;
10:         **end if**
11:         **end for**
12:     **end for**
13: **end for**

5.2   Time-Dependent Breadth-First Search Algorithm

In the stage, for a given routing query (a departure p^{start} and a destination p^{end}) at specific time t_s, we propose TBFS algorithm (see Algorithm 3) which is a variant breadth-first search algorithm to find out the best route planning from p^{start} to p^{end}. Its initialization part is shown in Algorithm 2.

Algorithm 2. TBFS Initialization

1: TBFS $(G_j = (P_r, V_j, E_j), p^{start})$;
2: // program initialization;
3: **for** each $p \in P_r$ except p^{start} **do**
4:      $distance[p] = infinite$;
5: **end for**
6: i = 0, k = 0;
7: $prefix[0] = 0, parent[0] = p^{start}$;
8: **for** each joint edge $e \in E_j$ **do**
9:      **if** $p^{start} = p_a^e$ **then**
10:        $distance[p_b^e] = e.t_1$;
11:        $ENQUEUE(Q_r, e.r_1(p_a^e, p_b^e))$;
12:        $k = k + 1$;
13:        $prefix[k] = i, parent[k] = p_b^e$;
14:     **end if**
15: **end for**

G_j is a joint graph described in Definition 6, P_r is defined in Definition 4, each vertex in V_j described in Definition 1 and each joint edge in E_j refers to a road segment (p_a, p_b, p_c) which is described in Definition 5. There are three arrays $distance$, $prefix$ and $parent$. Array $distance$ records the value of single-source time-dependent shortest path from p^{start}, array $prefix$ keeps the index of prior node during the TBFS procedure and array $parent$ owns the number of the current node's father node. Algorithm 2 is prepared for the initialization of these three arrays. Then we execute TBFS procedure by Algorithm 3 and use $dPre$ to optimize its time complexity.

5.3   Route Generation

After TBFS procedure, we get results of three arrays $distance$, $prefix$ and $parent$ and a backtracking method is used to generate the final route. Firstly, we find out the p^{end}'s index that appeared for the first time in array $parent$ backward and push it into a stack. Then we backtrack its parent node according to $prefix$ until the current index becomes zero, and push their parent node into stack one by one. Finally, we can get the recommended route by popping the stack from the departure p^{start} and the destination p^{end}.

Algorithm 3. TBFS Algorithm

1: TBFS $(G_j = (P_r, V_j, E_j), p^{start})$;
2: // calculate the time-dependent shortest path;
3: **while** $not(empty(Q_r))$ **do**
4:   $i = i + 1$;
5:   $r = DEQUEUE(Q_r)$;
6:   **for** each joint edge $e(p_a, p_b, p_c)$ **do**
7:     **if** $\left\lfloor \frac{1440}{p_a^r.t} \right\rfloor = \left\lfloor \frac{1440}{p_a^e.t} \right\rfloor$ **and** (p_a^r, p_b^r) equals to (p_a^e, p_b^e) **then**
8:       $\tau_1 = \left\lfloor \frac{1440}{p_b^e.t} \right\rfloor$, $\tau_2 = \left\lfloor \frac{1440}{p_c^e.t} \right\rfloor$;
9:       **if** $dPre_{p_b^e, p^{end}}^{\tau_1} < dPre_{p_c^e, p^{end}}^{\tau_2}$ **then**
10:          continue;
11:       **end if**
12:       **if** $distance[p_c^e] > distance[p_b^e] + e.t_2$ **then**
13:          $distance[p_c^e] = distance[p_b^e] + e.t_2$;
14:          $ENQUEUE(Q_r, e.r_2(p_a^e, p_b^e))$;
15:          $k = k + 1$;
16:          $prefix[k] = i, parent[k] = p_c^e$;
17:       **end if**
18:     **end if**
19:   **end for**
20: **end while**

---

6   Performance Evaluation

The experiments are conducted on Dell OPTIPLEX 990, a PC with a 3.1 GHz Intel i5-2400 processor and 4 GB RAM. Extensive experiments have been made to evaluate the performance of the proposed Spica by using a real data set generated by the traffic surveillance system. In this section, we will first explain the data set and experimental settings, followed by the evaluation metrics to measure the performance and then, show the experimental results including JT method's effectiveness and model's superiority.

6.1   Study Area

Our data set is a traffic surveillance dataset of a major city in China and consists of real vehicle passage records from March 9, 2015 to March 15, 2015 that collected from the traffic surveillance system of a major metropolitan area. The data set contains 44,593,706 passage records and involves totally 900 camera locations on almost all cross roads and we divide it into two parts, resource set and prediction set. Moreover, the resource set covers the data from March 9 to March 12 on weekdays and March 14 on weekends. The prediction sets are traffic records on March 13 and March 15, respectively.

6.2   Data Pre-processing

We pre-process the data set to form the trajectories, decomposing them into joint edges in each time internal Δt on resource set (see Sect. 6.4). After data pre-processing and travel time estimation (see Sect. 4.3), we get a total of 4,415,329 trajectories and are ready to make the rational route recommendation.

6.3   Goodness-of-Fit Statistics

we present a **Match Rate** (M_R) to prove the reasonable accuracy of JT method and two evaluation statistics to evaluate the effectiveness of the routes recommended by different methods (as A) and the real route (as B), α **Rate** (R_α) and β **Rate** (R_β).

(1) M_R is a way to measure the traveling time matching accuracy in *route linking problem* and is calculated as

$$M_R = \frac{Number(real\ trajectories)}{Number(trajectories\ of travel\ time\ possible\ links)} \tag{4}$$

(2) R_α represents how many routes recommended by method A are faster than that of method B and is calculated as

$$R_\alpha = \frac{Number(A's\ travel\ time < B's\ travel\ time)}{Number(trajectories\ of\ prediction\ set)} \tag{5}$$

(3) R_β reflects to what extent the routes suggested by A are faster than the B's and is calculated as

$$R_\beta = \frac{B's\ travel\ time - A's\ travel\ time}{Number(B's\ travel\ time)} \tag{6}$$

6.4   Evaluation of Spica

Different value of time internal Δt means different number of joint edges and different time costs. Figure 3 shows the total number of joint edges and time costs that change along with different value of time internal Δt. According to Fig. 3(a) and (b), we choose $\Delta t = 5$ to make rational route recommendation.

We make predictions for all trajectories of prediction set and put T-Drive as the comparison method [17].

Firstly, based on Sect. 4.1's theory, the travel time is probably different on adjacent road segments, we vary the number (1 to 5) of travel time clusters [16, 17] on adjacent road segments and the length of trajectory when the number of cluster is 2 to assess M_R value. We have studied from previous literatures that they consider a driver travels in the same travel time level always corresponding to the same travel time cluster. Figure 4(a) and (b) show these two situations, respectively. We can see in the picture that whether more clusters and longer trajectories would both lead to less matching rates by using T-Drive way. On the contrary, JT can definitely keep accuracy of link problem as well.

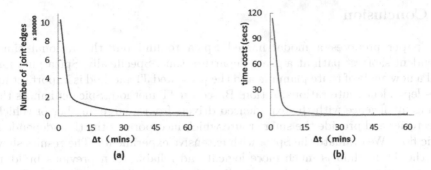

Fig. 3. The total number of joint edges and time costs change along with different value of time internal Δt

Secondly, we evaluate Spica by R_α and R_β and choose the real route as baseline. For the differentiate drivers' experience, we calculate each $\bar{\delta}$ of trajectories and times the value of recommended travel time to obtain the final travel time result. Because of the similarity of experimental results, we only give out the result of weekdays for short. Figure 4(c) shows the R_α in different star time interval τ of a day while Fig. 4(d) shows the proportion of routes under different R_β and both compared with T-Drive. These two figures show that though T-Drive could get the better β rate, Spica seems not bad and has the almost equal α rate. These experimental results prove the efficiency of Spica and the routes recommended by Spica are worked.

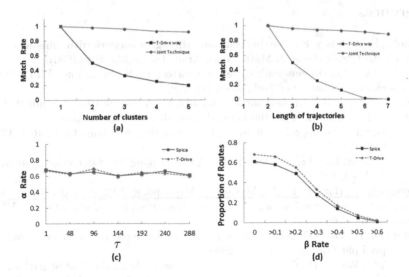

Fig. 4. The influence of *route linking problem* by metric M_R and comparison of overall performance of Spica measured by R_α and R_β.

7 Conclusion

This paper proposes a model named Spica to find out the rational time-dependent shortest path at a given departure time. Specifically, Spica puts forward a new angle of route planning and the presented JT method is important for time-dependent route rationalization. Based on JT method, Spica enhances the driving intelligence with the experienced drivers from a large number of vehicle trajectories and provide the smart route which incorporates the time-dependent traffic flow. We evaluate the Spica with extensive experiments. The results show that the JT method is much more logical and reliable than previous building ways and our model is effective in finding the practically fastest routes. Based on the TDHA algorithm, our model is not a time-consuming but a feasible job.

For the further work, the idea of *joint edge* can be extended and attached more importance. It is possible that we can use more kind of mobile data which is more fine-grained to describe trajectory of a moving object on the road. There are more efficient routing model can be built in the future. These problems deserve further study.

Acknowledgment. This work was supported in part by the National Basic Research 973 Program of China under Grant No. 2015CB352502, the National Natural Science Foundation of China under Grant Nos. 61272092 and 61572289, the Natural Science Foundation of Shandong Province of China under Grant Nos. ZR2012FZ004 and ZR2015FM002, the Science and Technology Development Program of Shandong Province of China under Grant No. 2014GGE27178, and the NSERC Discovery Grants.

References

1. Cooke, K.L., Halsey, E.: The shortest route through a network with time-dependent internodal transit times. J. Math. Anal. Appl. **14**(3), 493–498 (1966)
2. Dean, B.C.: Continuous-time dynamic shortest path algorithms. Ph.D. thesis, Massachusetts Institute of Technology (1999)
3. Demiryurek, U., Banaei-Kashani, F., Shahabi, C.: A case for time-dependent shortest path computation in spatial networks, pp. 474–477. ACM (2010)
4. Dreyfus, S.E.: An appraisal of some shortest-path algorithms. Oper. Res. **17**(3), 395–412 (1969)
5. Ge, Y., Xiong, H., Liu, C., Zhou, Z.-H.: A taxi driving fraud detection system, pp. 181–190. IEEE (2011)
6. Gonzalez, H., Han, J., Li, X., Myslinska, M., Sondag, J.P.: Adaptive fastest path computation on a road network: a traffic mining approach, pp. 794–805. VLDB Endowment (2007)
7. Kanoulas, E., Yang, D., Xia, T., Zhang, D.: Finding fastest paths on a road network with speed patterns, p. 10. IEEE (2006)
8. Lee, C.-C., Wu, Y.-H., Chen, A.L.P.: Continuous evaluation of fastest path queries on road networks. In: Papadias, D., Zhang, D., Kollios, G. (eds.) SSTD 2007. LNCS, vol. 4605, pp. 20–37. Springer, Heidelberg (2007)
9. Orda, A., Rom, R.: Shortest-path and minimum-delay algorithms in networks with time-dependent edge-length. J. ACM (JACM) **37**(3), 607–625 (1990)

10. Potamias, M., Bonchi, F., Castillo, C., Gionis, A.: Fast shortest path distance estimation in large networks, pp. 867–876. ACM (2009)
11. Tian, Y., Lee, K.C., Lee, W.-C.: Monitoring minimum cost paths on road networks, pp. 217–226. ACM (2009)
12. Tretyakov, K., Armas-Cervantes, A., García-Bañuelos, L., Vilo, J., Dumas, M.: Fast fully dynamic landmark-based estimation of shortest path distances in very large graphs, pp. 1785–1794. ACM (2011)
13. Veloso, M., Phithakkitnukoon, S., Bento, C.: Urban mobility study using taxi traces, pp. 23–30. ACM (2011)
14. Veloso, M., Phithakkitnukoon, S., Bento, C., Fonseca, N., Olivier, P.: Exploratory study of urban flow using taxi traces (2011)
15. Wu, L., Xiao, X., Deng, D., Cong, G., Zhu, A.D., Zhou, S.: Shortest path and distance queries on road networks: an experimental evaluation. Proc. VLDB Endowment 5(5), 406–417 (2012)
16. Yuan, J., Zheng, Y., Xie, X., Sun, G.: Driving with knowledge from the physical world, pp. 316–324. ACM (2011)
17. Jing Yuan, Y., Zheng, X.X., Sun, G.: T-drive: enhancing driving directions with taxi drivers' intelligence. IEEE Trans. Knowl. Data Eng. 25(1), 220–232 (2013)
18. Yuan, J., Zheng, Y., Zhang, L., Xie, X. and Sun, G.: Where to find my next passenger, pp. 109–118. ACM (2011)
19. Zheng, X., Liang, X., Ke, X.: Where to wait for a taxi? pp. 149–156. ACM (2012)

FHSM: Factored Hybrid Similarity Methods for Top-N Recommender Systems

Xin Xin[✉], Dong Wang, Yue Ding, and Chen Lini

School of Software, Shanghai Jiao Tong University,
800 Dongchuan Road, Shanghai, China
xin.xsjtu@yahoo.com, {wangdong,dingyue}@sjtu.edu.cn,
chenln.sjtu@yahoo.com

Abstract. Collaborative filtering (CF)-based methods in recommender systems believe that the user's preference of an item is the aggregation of the similar items or users. However, conventional item-based or user-based CF methods only consider either the item similarity or the user similarity. In this paper, we present hybrid-based methods for generating top-N recommendations in which both the item-item and user-user similarities are captured by the dot product of two low dimensional latent factor matrices. These matrices are learned using a stochastic gradient descent (SGD) algorithm to minimize two different loss functions, one is the squared error loss function and the other is the logistic loss function. A comprehensive set of experiments on multiple datasets is conducted to evaluate the performance of the proposed methods. The experimental results demonstrate the factored hybrid similarity methods (FHSM) achieve a superior recommendation quality in comparison with state-of-the-art methods.

Keywords: Recommender systems · Collaborative filtering · Low-rank · Matrix factorization

1 Introduction

Recommender systems [7] (RS) have been extremely common in variety of applications. Top-N recommendation as the majority form of RS aims to generate a ranked list of items to help users identify the interesting things [20]. Over the past decades, plenty of work has been done to develop new approaches to top-N recommender systems. These approaches can be broadly classified into three categories: CF-based methods, content-methods and hybrid methods [7]. CF-based methods have been a hot research point because of the high scalability and accuracy. In particular, CF-based methods can be classified into neighborhood-based methods and model-based methods [10]. Neighborhood-based methods compute the similarities between the users/items using the co-rating information and the unknown ratings are estimated as the aggregation of the most similar users/items. Typically, cosine similarity, correlation coefficient and other predefined similarity functions are adopted in these methods. Model-based algorithms

© Springer International Publishing Switzerland 2016
F. Li et al. (Eds.): APWeb 2016, Part II, LNCS 9932, pp. 98–110, 2016.
DOI: 10.1007/978-3-319-45817-5_8

use the collection of feedback information to learn a model, which is then used to perform recommendations. For example, matrix factorization [9] (MF) methods present users and items in a common latent space, then the user-item similarities can be captured in the latent space and utilized to generate recommendations.

Recently, a novel data-driven item-based CF method called SLIM [1] has been proposed. Unlike the conventional item-based CF method, SLIM derives the item similarity directly from data rather than any predefined similarity functions. A sparse aggregation coefficients matrix is learned and utilized to generate the unknown ratings. SLIM has been shown to achieve better performance than the conventional CF-based methods. However, SLIM only considers the co-rated items by one user and information of other users is ignored. In fact, there may be sets of users with similar preferences, leading to the assumption that incorporating information of other users can potentially generate better results.

In this paper, we propose FHSM, which incorporates both the information of items co-rated by one user and users who rated the same items. Precisely, FHSM captures both the item-item and user-user similarities through the dot product of two pairs of low-rank matrices. The experimental results on multiple datasets show that FHSM performs better than other state-of-the-art algorithms. Moreover, the result confirms the assumption that involving data-driven user similarities into recommendation can improve the performance of RS.

The key contributions of this paper are the followings:

1. We involve both item-item and user-user similarities in a unified framework and get better recommendation.
2. We present the item-item and user-user similarities in a latent space and estimate the proposed model using the SGD algorithm.
3. We estimate the proposed model using both squared error and logistic error in multiple datasets.

The rest of the paper is organized as follows. Section 2 introduces the notations used in this paper. Section 3 presents the related work. Section 4 describes the detail of the proposed methods. In Sect. 5, we illustrate the experimental results. Finally, we conclude the work in Sect. 6.

2 Notations

In this paper, all vectors are represented by bold lower case letters (e.g., $\mathbf{p}$,$\mathbf{q}$). All matrices are in bold upper case (e.g., $\mathbf{S}$,$\mathbf{T}$). The ith row of a matrix $\mathbf{W}$ is represented by $\mathbf{w}_i$ and $\mathbf{w}_j^T$ denotes the jth column. The estimated values is denoted by having a hat over it (e.g., $\hat{r}$). Gothic letters $\mathcal{C}$ and $\mathcal{D}$ are used to denote the user set and item set whose respective cardinalities are n and m ($|\mathcal{C}|=n$ and $|\mathcal{D}|=m$). Matrix $\mathbf{R}$ which is a binary matrix is utilized to represent the user-item transaction matrix. u and i are used to denote individual users and items. r_{ui} is used to denote the specific entry (u,i) in $\mathbf{R}$ which represents the implicit feedback information for user u on item i. $r_{ui} = 1$ means user u has interaction with item i and $r_{ui}=0$ means that u doesn't provide the feedback information of i. $\mathcal{L}(\cdot)$ denotes the loss function and regularization terms are represented by $Reg(\cdot)$.

3   Review of Relevant Research

The proposed methods in this paper are mainly motivated by combination of the basic belief of neighborhood-based CF methods and the model-based latent factor.

As one of the most successful methods, neighborhood-based CF methods first identify a set of similar items or users, then recommendations are generated based on those similar entities. It can be further classified into item-based methods and user-based methods. The basic belief of neighborhood-based methods is user u's preference of item i is the aggregation of similar items/users. A lot of predefined similarity measures [11] (e.g., cos similarity, Pearson correlation) are used to calculate the similarity among users/items. Neighborhood-based CF methods generate recommendation quickly but it suffers from low accuracy since there is essentially no knowledge learned from user-item interaction.

The other class of CF methods is model-based. Many model-based methods are based on MF. The key idea is to factorize the user-item matrix to represent the users' preferences and items' characteristics in a common latent space. The result is estimated by the dot product of the latent factors. Cremonesi et al. proposed PureSVD [15] in which users' and items' features are represented by the singular vectors of user-item matrix. Koren proposed the well-known SVD++[10] which models users by both explicit and implicit feedback. PMF [8] aims to generate recommendation using the maximum a posteriori (MAP) method.

Recently, another model-based method, SLIM [1] is proposed. It adopts the belief of neighbourhood-based method. However, SLIM doesn't involve the predefined similarity measures, it utilizes a kind of data-driven similarity. Precisely, SLIM approximates the matrix $\mathbf{R}$ as follows:

$$\widehat{\mathbf{R}} = \mathbf{RS} \tag{1}$$

$\mathbf{S}$ is a $m \times m$ sparse matrix of aggregation coefficients which can be considered as item-item similarity. SLIM directly learns the similarity matrix from data by solving a ℓ_1-norm and ℓ_2-norm regularized optimization problem. The objective function is as follows:

$$\min_{S} \frac{1}{2}\|\mathbf{R} - \mathbf{RS}\|_F^2 + \frac{\beta}{2}\|\mathbf{S}\|_F^2 + \lambda\|\mathbf{S}\|_1 \tag{2}$$
$$\text{subject to } \mathbf{S} \geq 0, diag(\mathbf{S}) = 0$$

where $\|\mathbf{S}\|_F$ is the Frobenius norm of $\mathbf{S}$ and $\|\mathbf{S}\|_1$ is the ℓ_1-norm which induces the sparsity. The constrain $\text{diag}(\mathbf{S}) = 0$ aims to ensure that r_{ui} will not be used to predict $\hat{r}_{ui}$. SLIM uses the non-negative constrain to ensure the positive aggregation of items but further research [2] claims that this constrain can be removed. SLIM adopts the aggregation belief of neighborhood-based CF and utilizes a learning process to get knowledge. Plenty of research has been done based on SLIM. SSLIM [3] integrates the side information. LorSLIM [5] involves the nuclear-norm to induce the low-rank property of **S**. HOSLIM [4] uses the potential higher-order information to generate better recommendation.

Motivated by the latent factor used in matrix factorization, Kabbur et al. proposed FISM [2] in which the aggregation matrix **S** is factored into two low-rank matrix **P** and **Q**. As a result, the item similarity can be captured in a latent space. The prediction rule of r_{ui} is as follows:

$$\hat{r}_{ui} = b_u + b_i + (n_u^+ - 1)^{-\alpha} \sum_{j \in \mathcal{R}_u^+ \setminus \{i\}} \mathbf{p}_j \mathbf{q}_i^T \qquad (3)$$

where b_u and b_i are user and item biases. The dot product of **P** and **Q** can be considered as the item similarity. FISM learns the parameters by solving the ℓ_2-norm optimization problem. Besides, Kerypis et al. also proposed UFSM [6] to solve the cold-start problems in recommendation. The basic idea is also using a learning process to get the aggregation of similar items.

However, all of these data-driven methods only take item similarity into account. Other users' information is ignored. Although in traditional neighborhood-based CF, item-based methods outperform the user-based ones, in the case of data-driven similarity, it can't be asserted that user similarity is useless. On the contrary, incorporating information of user similarity potentially generates better results.

This paper attempts to integrate both the item and user similarity in a unified model. The similarities are captured in the latent space using the SGD algorithm. The problem is solved by minimising the structural risk. Recommendations are generated by selecting the top-N items in the ranked list.

4   The Approach

Motivated by the basic belief that the rating score is the aggregation of similar items and similar users. We formulate the estimated value of r_{ui} as follows:

$$\hat{r}_{ui} = \sum_{j \in \mathcal{R}_u^+ \setminus \{i\}} \text{sim}(i, j) + \sum_{j \in \mathcal{R}_i^+ \setminus \{u\}} \text{sim}(u, j) \qquad (4)$$

where $\mathcal{R}_u^+$ denotes the item set rated by user u and $\mathcal{R}_i^+$ represents the users who have rated item i. Like the diagonal constrain in SLIM, we exclude user u and item i to ensure r_{ui} is not used to predict $\hat{r}_{ui}$. To capture these similarities in latent space, two pairs of low-rank matrices are involved. The similarities are represented in the dot product of the low-rank matrices. So, the prediction rule becomes:

$$\hat{r}_{ui} = \sum_{j \in \mathcal{R}_u^+ \setminus \{i\}} \mathbf{p}_j \mathbf{q}_i^T + \sum_{j \in \mathcal{R}_i^+ \setminus \{u\}} \mathbf{s}_j \mathbf{t}_u^T \qquad (5)$$

where $\mathcal{R}_u^+$ denotes the items viewed by user u and $\mathcal{R}_i^+$ means users who have interacted with item i. To control the number of neighborhoods, two parameters

α_1 and α_2 are introduced. These parameters can be considered as degree of agreement between similar entities. Besides, we can figure out that user bias won't affect the rank of items for a user but item bias will. The reason is that the user bias remains the same for a specific user. As a result, the proposed model only involves the item bias. So the final prediction rule of r_{ui} is:

$$\hat{r}_{ui} = b_i + (n_u^+ - 1)^{-\alpha_1} \sum_{j \in \mathcal{R}_u^+ \backslash \{i\}} \mathbf{p}_j \mathbf{q}_i^T + (n_i^+ - 1)^{-\alpha_2} \sum_{j \in \mathcal{R}_i^+ \backslash \{u\}} \mathbf{s}_j \mathbf{t}_u^T \qquad (6)$$

Recently, LorSLIM [5] has been proposed in which the low-rank property is introduced. LorSLIM uses the nuclear-norm to ensure the low-rank structure of the coefficient matrix. The motivation is that only a few latent variables account for the similarity. The low-rank structure also needs to be introduced in FHSM. However, we don't need to involve the nuclear-norm because the coefficient matrix has been factored into two low-rank matrices and the low-rank structure is guaranteed spontaneously.

To learn these parameters in the prediction rule, we minimize the structural risk using SGD algorithm [16]. In the next two parts, we introduce two different types of FHSM models in which two different loss functions are utilized.

4.1   FHSMrmse

The first loss function aims to minimize the squared loss error [16]. Precisely, the loss function is as follows:

$$\mathcal{L}(\cdot) = \sum_{u \in \mathcal{C}} \sum_{i \in \mathcal{D}} (r_{ui} - \hat{r}_{ui})^2 \qquad (7)$$

Note that unlike other rating prediction methods (e.g. SVD++, PMF) which only calculate the loss in the known ratings in **R**. FHSM takes all the entries into account. The reason is that FHSM aims to generate the rank of items but to perform the rating prediction. The drawback is taking all the entries will increase the computation complexity. However, Kabbur [2] has claimed that this complexity can be remitted by sampling from zero entries. The objective function of FHSMrmse is shown as follows:

$$\min_{\mathbf{P},\mathbf{Q},\mathbf{S},\mathbf{T}} \frac{1}{2} \sum_{u \in \mathcal{C}} \sum_{i \in \mathcal{D}} (r_{ui} - \hat{r}_{ui})^2 + \frac{\beta_1}{2} (\|\mathbf{P}\|_F^2 + \|\mathbf{Q}\|_F^2) + \frac{\beta_2}{2} (\|\mathbf{S}\|_F^2 + \|\mathbf{T}\|_F^2) + \frac{\lambda}{2} \|\mathbf{b_i}\|_2^2$$

$$(8)$$

where λ, β_1 and β_2 are the regularization weights for item bias vector, item latent factor matrices and user latent factor matrices. The optimization is performed using SGD algorithms. The detail is shown in Algorithm 1. The sample process is involved to reduce the computation complexity. The process first counts the number of non-zero entries (nnz) and then $\rho \cdot nnz$ zero entries are sampled randomly.

Algorithm 1. FHSMrmse learning process using SGD

1: $\eta \leftarrow$ learning rate
2: $\beta_1, \beta_2, \lambda \leftarrow$ regularization weights
3: $\rho \leftarrow$ sample factor
4: $iter \leftarrow 0$
5: Init $\mathbf{P}, \mathbf{Q}, \mathbf{S}, \mathbf{T}$
6: **while** $iter < maxIter$ or error on validation set decreases **do**
7: $\quad \mathcal{R}' \leftarrow \mathbf{R} \cup SampleZeroEntries(\mathbf{R}, \rho)$
8: $\quad$ **for all** $r_{ui} \in \mathcal{R}'$ **do**
9: $\quad\quad \mathbf{x}_1 \leftarrow (n_u^+ - 1)^{-\alpha_1} \sum_{j \in \mathcal{R}_u^+ \setminus \{i\}} \mathbf{P}_j$
10: $\quad\quad \mathbf{x}_2 \leftarrow (n_i^+ - 1)^{-\alpha_2} \sum_{j \in \mathcal{R}_i^+ \setminus \{u\}} \mathbf{s}_j$
11: $\quad\quad \hat{r}_{ui} \leftarrow b_i + \mathbf{q}_i^T \mathbf{x}_1 + \mathbf{t}_u^T \mathbf{x}_2$
12: $\quad\quad e_{ui} \leftarrow r_{ui} - \hat{r}_{ui}$
13: $\quad\quad b_i \leftarrow b_i + \eta(e_{ui} - \lambda b_i)$
14: $\quad\quad \mathbf{q}_i \leftarrow \mathbf{q}_i + \eta(e_{ui}\mathbf{x}_1 - \beta_1 \mathbf{q}_i)$
15: $\quad\quad \mathbf{t}_u \leftarrow \mathbf{t}_u + \eta(e_{ui}\mathbf{x}_2 - \beta_2 \mathbf{t}_u)$
16: $\quad\quad$ **for all** $j \in \mathcal{R}_u^+ \setminus \{i\}$ **do**
17: $\quad\quad\quad \mathbf{p}_j \leftarrow \mathbf{p}_j + \eta(e_{ui}(n_u^+ - 1)^{-\alpha_1} \mathbf{q}_i - \beta_1 \mathbf{p}_j)$
18: $\quad\quad$ **end for**;
19: $\quad\quad$ **for all** $j \in \mathcal{R}_i^+ \setminus \{u\}$ **do**
20: $\quad\quad\quad \mathbf{s}_j \leftarrow \mathbf{s}_j + \eta(e_{ui}(n_i^+ - 1)^{-\alpha_2} \mathbf{t}_u - \beta_2 \mathbf{s}_j)$
21: $\quad\quad$ **end for**;
22: $\quad$ **end for**;
23: $\quad iter \leftarrow iter + 1$
24: **end while**
25: **return** $\mathbf{P}, \mathbf{Q}, \mathbf{S}, \mathbf{T}, \mathbf{b_i}$

4.2 FHSMlogi

The second loss function comes from logistic regression [16]. When tackling with binary data, logistic regression is a consciously choice. Another choice is a ranking loss error based on Bayesian Personalized Ranking (BPR)[14] which is motivated by the fact that top-N recommendation aims to perform ranking other than rating prediction. But experiments in FISM show that this ranking loss function doesn't achieve a better performance. As a result, we don't involve this ranking loss error here. To tackle with binary feedback, we involve the logistic loss error which is shown as follows:

$$\mathcal{L}(\cdot) = -\sum_{u \in \mathcal{C}} \sum_{i \in \mathcal{D}} (r_{ui} \log(\frac{1}{1 + e^{-\hat{r}_{ui}}}) + (1 - r_{ui}) \log(1 - \frac{1}{1 + e^{-\hat{r}_{ui}}})) \qquad (9)$$

Note that like FHSMrmse, the loss error is also computed on all the entries. To reduce the computation complexity, the sample process is involved, too. The objection function of FHSMlogi is:

$$\min_{\mathbf{P,Q,S,T}} -\sum_{u\in\mathcal{C}}\sum_{i\in\mathcal{D}}(r_{ui}\log(\frac{1}{1+e^{-\hat{r}_{ui}}})+(1-r_{ui})\log(1-\frac{1}{1+e^{-\hat{r}_{ui}}}))$$
$$+\frac{\beta_1}{2}(\|\mathbf{P}\|_F^2+\|\mathbf{Q}\|_F^2)+\frac{\beta_2}{2}(\|\mathbf{S}\|_F^2+\|\mathbf{T}\|_F^2)+\frac{\lambda}{2}\|\mathbf{b_i}\|_2^2 \tag{10}$$

where the terms mean the same with FHSMrmse. The optimization is performed using SGD algorithm. Algorithm 2 shows the detail procedure.

Algorithm 2. FHSMlogi learning process using SGD

1: $\eta \leftarrow$ learning rate
2: $\beta_1, \beta_2, \lambda \leftarrow$ regularization weights
3: $\rho \leftarrow$ sample factor
4: $iter \leftarrow 0$
5: Init $\mathbf{P,Q,S,T}$
6: **while** $iter < maxIter$ or error on validation set decreases **do**
7:      $\mathcal{R}' \leftarrow \mathbf{R}\cup SampleZeroEntries(\mathbf{R},\rho)$
8:      **for all** $r_{ui} \in \mathcal{R}'$ **do**
9:          $\mathbf{x}_1 \leftarrow (n_u^+-1)^{-\alpha_1}\sum_{j\in\mathcal{R}_u^+\setminus\{i\}}\mathbf{P}_j$
10:         $\mathbf{x}_2 \leftarrow (n_i^+-1)^{-\alpha_2}\sum_{j\in\mathcal{R}_i^+\setminus\{u\}}\mathbf{s}_j$
11:         $\hat{r}_{ui} \leftarrow b_i + \mathbf{q}_i^T\mathbf{x}_1 + \mathbf{t}_u^T\mathbf{x}_2$
12:         $\varepsilon_{ui} \leftarrow r_{ui} - \dfrac{e^{\hat{r}_{ui}}}{e^{\hat{r}_{ui}}+1}$
13:         $b_i \leftarrow b_i + \eta(\varepsilon_{ui}-\lambda b_i)$
14:         $\mathbf{q}_i \leftarrow \mathbf{q}_i + \eta(\varepsilon_{ui}\mathbf{x}_1 - \beta_1\mathbf{q}_i)$
15:         $\mathbf{t}_u \leftarrow \mathbf{t}_u + \eta(\varepsilon_{ui}\mathbf{x}_2 - \beta_2\mathbf{t}_u)$
16:         **for all** $j \in \mathcal{R}_u^+\setminus\{i\}$ **do**
17:             $\mathbf{P}_j \leftarrow \mathbf{P}_j + \eta(\varepsilon_{ui}(n_u^+-1)^{-\alpha_1}\mathbf{q}_i - \beta_1\mathbf{P}_j)$
18:         **end for;**
19:         **for all** $j \in \mathcal{R}_i^+\setminus\{u\}$ **do**
20:             $\mathbf{s}_j \leftarrow \mathbf{s}_j + \eta(\varepsilon_{ui}(n_i^+-1)^{-\alpha_2}\mathbf{t}_u - \beta_2\mathbf{s}_j)$
21:         **end for;**
22:     **end for;**
23:     $iter \leftarrow iter + 1$
24: **end while**
25: **return P,Q,S,T,b$_i$**

---

5   Experimental Evaluation

5.1   Data Sets

The experiments are carried out on four real datasets to evaluate the proposed methods. All the ratings in datasets are converted to binary indications to represent the implicit feedback. The four datasets can be classified into two categories according to the size.

Table 1. Datasets

Dataset	#users	#items	#trns	rsize	csize	Density
FilmTrust	1508	2071	35497	23.54	17.14	1.14 %
ML100K	943	1682	100000	106.04	59.45	6.30 %
Yahoo	7558	3951	282075	37.32	71.39	0.94 %
Netflix	6079	5641	429339	70.63	76.11	1.25 %

The "#users", "#items" and "#trns" columns are the number of users, number of items and number of transactions, respectively, in each dataset. The "rsize" and "csize" columns are the average number of transactions for each user and for each item (i.e., row and column density of the user-item matrix), respectively, in each dataset. The "density" column is the density of each daset (i.e., density = #trns/(#users×#items)).

The first category contains FilmTrust and ML100K which are in the smaller size. Particularly, FilmTrust contains the rating information of 1508 users and 2071 movies. The total number of ratings is 35497. It also contains the trust information between users but we don't utilize the information. ML100K is obtained from the MovieLens research project which contains 943 users and 1682 movies.

The other category contains two lager datasets: Netflix and Yahoo. The Netflix dataset which contains the feedback information of 6079 users and 5641 movies is a subset extracted from the Netflix Prize dataset. Finally, the Yahoo dataset is obtained from Yahoo Music.

The detail characteristics of the four datasets are presented in Table 1.

5.2   Evaluation Methodology

We applied the 5-fold cross validation to evaluate the performance of the proposed methods. In each run, each dataset is split into a training set and a testing set by randomly selecting 80 % of the non-zero entries of each user as training set and the rest 20 % as testing set. The performance is evaluated by comparing the recommendation list of each user and the items of that user in testing set.

We adopt two well-applied measures as the evaluation metrics. The first one is Recall at N (Rec@N). Given the ranked recommendation list of size N for a user, Rec@N measures how many of the items liked by the user appeared in the list. Rec@N is computed for each user and the average is reported. The second one is Precision at N (Prec@N) which measures how many items are correct in the recommendation list. The average policy remains the same with Rec@N. The detail of these two metrics is shown as follows:

$$\text{Rec@N} = \frac{|\text{Recommended list} \cap \text{Testing set}|}{|\text{Testing set}|} \tag{11}$$

$$\text{Prec@N} = \frac{|\text{Recommended list} \cap \text{Testing set}|}{N} \tag{12}$$

5.3   Compared Methods

The performance of FHSM is compared with 5 other recommendation algorithms, including two neighborhood-based CF methods ItemKNN(cos) and ItemKNN(jac)[11], in which cosine similarity and Jaccard similarity are utilized, three model-based ranking methods BPRMF [14], SLIM [1] and FISM [2]. About FISM, we adopt FISMrmse other than FISMauc because experiments has shown that FISMrmse has a better performance [2]. These methods constitute the current state-of-the-art for top-N recommendation task. Therefore, they form a good set of methods to evaluate the proposed model.

5.4   Experiment Results

Table 2 shows the performance of different top-N recommendation algorithms. These results shows that FHSM has better performance than the compared methods. It conforms to the assumption that integrating user similarity can improve the recommendation quality. When it comes to loss functions, we can see that FHSMlogi has a slightly better performance than FHSMrmse. The reason is that logistics loss is more suitable to tackle with binary feedbacks. Besides, we can see that the recall in FilmTrust is much higher than other datasets. The reason is that the rsize of FilmTrust is smaller. It is obvious that all of these methods achieve a higher precision in FilmTrust. The reason is that FilmTrust also contains the user's social information. Although the social information is not involved by these methods, the users and movies in FilmTrust may have some internal relationships. Among the compared methods, we can see that FISM is substantially better than others because it combines the data-driven similarity and the low-rank latent spaces. It is obvious that all of these methods outperform ItemKNN because the similarity measures in ItemKNN is predefined and don't involve too much knowledge. The other reason is that all the entries in these datasets are converted into binary entries and the similarity measures used in ItemKNN are not so efficient in that case.

Figure 1 presents the performance of the methods (in terms of recall) for different top-N values (i.e., 5, 10, 15 and 20) for all four datasets. The results show that FHSM achieves a better performance than the compared methods in all top-N values, except $N = 15$ for FilmTrust. Actually, the advantage is not so obvious when the top-N value is large. However, the items which are in the tail of the recommendation list is not so important. As a result, the performance when the top-N value is small is more important and FHSM performs better in that case.

Figure 2 shows the impact (in terms of Rec@10) of α and latent factors in ML100K. The result of FHSMrmse is reported. FHSMlogi is almost in the same case. It is obvious that FHSM achieves the best performance when α_1 and α_2 are both around 0.6. When α_1 and α_2 increase to 1, the performance decreases dramatically. In fact, when the value of α increases, the agreement of similar entities becomes more and more strict. However, when α is too large, the agreement is

Table 2. Comparison of different Top-N recommendation algorithms

Dataset	FilmTrust				Prec@10	Rec@10	ML100K				Prec@10	Rec@10
	Params						Params					
ItemKNN(cos)	50	-	-	-	0.3397	0.5912	100	-	-	-	0.2676	0.1698
ItemKNN(jac)	50	-	-	-	0.3413	0.5952	100	-	-	-	0.2660	0.1701
BPRMF	0.1	300	-	-	0.3480	0.6341	0.01	200	-	-	0.3599	0.2398
SLIM	0.5	0.1	-	-	0.3599	0.6555	0.2	0.2	-	-	0.3469	0.2236
FISM	1e-4	96	1e-4	0.01	0.3542	0.6367	1e-3	300	2e-5	0.01	0.3715	0.2475
FHSMrmse	100	100	1e-4	1e-4	<u>0.3701</u>	<u>0.6618</u>	100	25	2e-5	1e-4	<u>0.3841</u>	<u>0.2560</u>
FHSMlogi	100	100	1e-4	1e-4	<u>0.3704</u>	<u>0.6639</u>	100	25	2e-5	1e-4	<u>0.3849</u>	<u>0.2566</u>
Dataset	Yahoo				Prec@10	Rec@10	Netflix				Prec@10	Rec@10
	Params						Params					
ItemKNN(cos)	100	-	-	-	0.0682	0.0891	100	-	-	-	0.1158	0.1160
ItemKNN(jac)	100	-	-	-	0.0678	0.0886	100	-	-	-	0.1162	0.1167
BPRMF	0.01	600	-	-	0.0946	0.1273	0.01	600	-	-	0.1325	0.1330
SLIM	1.2	0.1	-	-	0.1004	0.1328	0.8	0.1	-	-	0.1438	0.1450
FISM	1e-3	400	2e-5	0.01	0.1030	0.1368	1e-4	600	2e-5	0.01	0.1504	0.1507
FHSMrmse	150	150	2e-5	2e-5	<u>0.1162</u>	<u>0.1489</u>	400	400	2e-5	2e-5	<u>0.1598</u>	<u>0.1601</u>
FHSMlogi	150	150	2e-5	2e-5	<u>0.1170</u>	<u>0.1493</u>	400	400	2e-5	2e-5	<u>0.1603</u>	<u>0.1606</u>

The parameters represented for each method are as follows, respectively: ItemKNN(cos) and ItemKNN(jac): the number of neighbors; BPRMF: the regularization weight λ and the dimension of the latent space; SLIM: the regularization weights of ℓ_1-norm and ℓ_2-norm; FISM: the learning rate η, the dimension of the latent space, the regularization weights of item matrices and item bias; FHSM: the dimension of the item latent space (k_1), the dimension of the user latent space (k_2), the regularization weights of item matrices (β_1) and user matrices (β_2). Columns corresponding to Prec@10 and Rec@10 present the precision and recall when the size of recommendation list is 10, respectively.

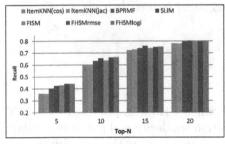

(a) FilmTrust

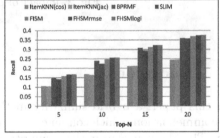

(b) ML100K

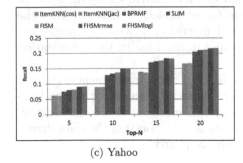

(c) Yahoo

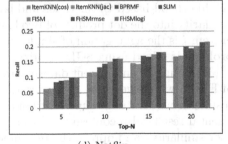

(d) Netflix

Fig. 1. Recommendation for different Top-N values

too strict to perform better recommendation. The results in Fig. 2 also demonstrate that FHSM achieves the best performance when k_1 is around 100 and k_2 is around 20. With the increase of latent factors, the recall firstly increases and then decreases slowly. The reason is more factors can capture the similarity in a higher dimension but too much factors break the low-rank property of similarity matrix which results in the decrease of recall. It is obvious that in ML100K, the proper factor of user similarity (k_2) is smaller than the item one (k_1). The reason is that in ML100K the number of users is smaller than the number of items, so we can not assert that the user similarity is determined by fewer factors than the item similarity because it depends on the datasets.

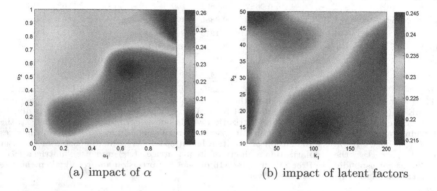

(a) impact of α (b) impact of latent factors

Fig. 2. impact of α and latent factors

6   Conclusions

In this paper, a hybrid similarity method called FHSM is proposed for top-N recommendation, which combines the item-item similarity and user-user similarity. The idea comes from the assumption that incorporating the data-driven user similarity can improve the recommendation quality. Motivated by the ideas of neighborhood-based CF methods and latent factor models, FHSM factorizes the similarity into the dot product of two low-rank matrices and a user u's preference of item i is the aggregation of similar users and items. Finally, the optimization problem is solved using SGD algorithm.

Experiments on different datasets are conducted to compare the performance of FHSM with other state-of-the-art top-N recommendation algorithms and the results demonstrate the superiority of the proposed methods. Besides, the experimental results also confirm the assumption that incorporating the data-driven user similarity can improve the recommendation quality.

References

1. Ning, X., Karypis, G.: Slim: sparse linear methods for top-n recommender systems. In: 2011 IEEE 11th International Conference on Data Mining (ICDM). IEEE (2011)
2. Kabbur, S., Ning, X., Karypis, G.: FISM: factored item similarity models for top-n recommender systems. In: Proceedings of the 19th ACM SIGKDD International Conference on Knowledge Discovery and Data Mining. ACM (2013)
3. Ning, X., Karypis, G.: Sparse linear methods with side information for top-n recommendations. In: Proceedings of the Sixth ACM Conference on Recommender Systems. ACM (2012)
4. Christakopoulou, E., Karypis, G.: HOSLIM: higher-order sparse linear method for top-n recommender systems. In: Tseng, V.S., Ho, T.B., Zhou, Z.-H., Chen, A.L.P., Kao, H.-Y. (eds.) PAKDD 2014. LNCS, vol. 8444, pp. 38–49. Springer International Publishing, Heidelberg (2014)
5. Cheng, Y., Li'ang, Y., Yong, Y.: LorSLIM: low rank sparse linear methods for top-n recommendations. In: 2014 IEEE International Conference on Data Mining (ICDM). IEEE (2014)
6. Elbadrawy, A., Karypis, G.: User-specific feature-based similarity models for top-n recommendation of new items. ACM Trans. Intell. Syst. Technol. (TIST) **6**(3), 33 (2015)
7. Adomavicius, G., Tuzhilin, A.: Toward the next generation of recommender systems: a survey of the state-of-the-art and possible extensions. IEEE Trans. Knowl. Data Eng. **17**(6), 734–749 (2005)
8. Salakhutdinov, R., Andriy M.: Probabilistic matrix factorization. NIPS (2011)
9. Koren, Y., Bell, R., Volinsky, C.: Matrix factorization techniques for recommender systems. Computer **8**, 30–37 (2009)
10. Koren, Y.: Factorization meets the neighborhood: a multifaceted collaborative filtering model. In: Proceedings of the 14th ACM SIGKDD International Conference on Knowledge Discovery and Data Mining. ACM (2008)
11. Sarwar, B., et al.: Item-based collaborative filtering recommendation algorithms. In: Proceedings of the 10th International Conference on World Wide Web, pp. 285–295. ACM (2011)
12. Delgado, J., Ishii, N.: Memory-based weighted-majority prediction for recommender systems In: Research and Development in Information Retrieval (1999)
13. Ning, X., Karypis, G.: Recent advances in recommender systems and future directions. In: Kryszkiewicz, M., Bandyopadhyay, S., Rybinski, H., Pal, S.K. (eds.) PReMI 2015. LNCS, vol. 9124, pp. 3–9. Springer International Publishing, Heidelberg (2015)
14. Rendle, S., et al.: BPR: Bayesian personalized ranking from implicit feedback. In: Proceedings of the Twenty-Fifth Conference on Uncertainty in Artificial Intelligence. AUAI Press (2009)
15. Cremonesi, P., Koren, Y., Turrin, R.: Performance of recommender algorithms on top-n recommendation tasks. In: Proceedings of the Fourth ACM Conference on Recommender Systems. ACM (2010)
16. Boyd, S., Lieven, V.: Convex Optimization. Cambridge University Press, Cambridge (2004)
17. Paterek, A.: Improving regularized singular value decomposition for collaborative filtering. In: Proceedings of KDD Cup and Workshop, vol. 2007 (2007)
18. Hu, Y., Yehuda, K., Chris V.: Collaborative filtering for implicit feedback datasets. In: 2008 Eighth IEEE International Conference on Data Mining, ICDM 2008. IEEE (2008)

19. Kabbur, S., Karypis, G.: NLMF: Nonlinear matrix factorization methods for top-n recommender systems. In: 2014 IEEE International Conference on Data Mining Workshop (ICDMW). IEEE (2014)
20. Ricci, F., et al.: Recommender Systems Handbook. Springer, Heidelberg (2011)

Personalized Resource Recommendation Based on Regular Tag and User Operation

Sisi Liu, Yongjian Liu, and Qing Xie$^{(\boxtimes)}$

School of Computer Science and Technology,
Wuhan University of Technology, Wuhan 430070, China
{liuss,felixxq}@whut.edu.cn, liuyj626@163.com

Abstract. In conventional tag-based recommendation system, the sparsity and impurity of social tag data significantly increase the complexity of data processing and affect the accuracy of recommendation. To address these problems, we consider from the perspective of resource provider and propose a resource recommendation framework based on regular tags and user operation feedbacks. Based on these concepts, we design the user feature representation integrating the information of regular tags, user operations and time factor, so as to precisely discover the user preference on different tags. The personalized recommendation algorithm is designed based on collaborative filtering mechanism by analyzing the general preference modeling of different users. We conduct the experimental evaluation on a real recommendation system with extensive user and tag data. Compared with traditional user-based collaborative filtering and the social-tag-based collaborative filtering, our approach can effectively alleviate the sparsity problem of tag data and user rating data, and our proposed user feature is more accurate to improve the performance of the recommendation system.

Keywords: Regular tag · User operation · User preference model · Collaborative filtering · Recommendation system

1 Introduction

In the information era, the recommendation system [14] has been designed to recommend relevant items to different users, by analyzing the characteristics of users and items. It has been applied in many successful system instances, e.g., Amazon product recommendation system [9], Netflix [1] and MovieLens [5].

With the development of Web 2.0, users are no longer satisfied with the information retrieved by keyword-based approaches, but tend to require the personalized information services according to their own preferences. Therefore, the tagging technique is designed, e.g., *Folksonomy* system [11], to enable users freely creating and using tags to describe the resources on the Web, and also sharing the tags with other users. In this way, the correlation between resources and the interaction between users can be effectively enhanced, since the tags

© Springer International Publishing Switzerland 2016
F. Li et al. (Eds.): APWeb 2016, Part II, LNCS 9932, pp. 111–123, 2016.
DOI: 10.1007/978-3-319-45817-5_9

describing the information of resources and users are more flexible and accurate to indicate the resource characteristics and the user interest preferences.

As the development of tagging technique, the concept of personalized information recommendation based on tags is proposed, e.g., tag-based recommendation system. Mishne first designed a simple automatic tag assignment system [12], which compared and clustered the user's blog information to generate a list of tags, and then filtered and sorted these tags into a result set to recommend to the user. Firan et al. [7] suggested that social tags can not only represent the resources on the Web, but also indicate the preferences of users. Since then, more and more researchers applied the tagging technique in the field of personalized recommendation, and achieved effective performance. For example, Chen et al. [3] mapped the tags to the users and contents for filtering, and proposed a hybrid recommendation approach considering the correlation of tags and the social network; Ma et al. [10] proposed a collaborative recommendation system based on social network and tags, and trust network was introduced to improve the recommendation confidence. However, the tag-based recommendation system has to face the problem of data sparsity, which significantly affects the efficiency and effectiveness of recommendation. Motivated by this situation, we propose an innovative tag-based resource recommendation solution in this work.

In tag-based recommendation system, the data sparsity results from the potentially unlimited set of *social tags* by free user tagging. This situation also produces noisy tags and leads to the problems of tag redundancy and semantic fuzziness, which will increase the burden of recommendation process and reduce the accuracy. Therefore, we consider from the perspective of resource providers in practical applications, and design the recommendation algorithm based on *regular tags* and user behaviors, so as to effectively reduce the tag noise and data sparsity. The regular tags are those created and maintained by the official resource providers, which are more accurate and strict to describe the intrinsic attributes of resources. The user behaviors are those operations when users browse the resources, which can be recognized as the feedbacks to resource providers and reflect the user preferences.

Generally, most recommendation systems are based on the tagging records of the user group and the item group to make the recommendation using collaborative filtering [9]. However, these systems are short in scalability due to the sparse tag matrix, and in the process of recommendation, they ignore the personalized characteristics of individual users. In this work, we propose to build the "implicit" user-item scoring model, by integrating the regular tags and the user operations. We synthesize the resource tag characteristics, the user operations and the time factor to represent the user feature, and the user operations are utilized to weight the user-tag matrix, so as to embed the user preference. Based on the user representation and regular tags, we analyze the user preference on the existing resources and potential new resources. In the process of recommendation, the preference scores on new resource items for target user and his similar users are calculated to form a ranking list, and collaborative filtering is employed to propose the final recommendation.

We summarize the contributions of this work as follows:

- We design the recommendation system from the perspective of resource provider, and propose regular tags to generate the standard tag system;
- We propose a new model to represent the user feature, which integrates the tag characteristics, the user operation and the time factor.
- The proposed approach is evaluated by practical system with extensive real dataset.

The following part of this paper is organized as below: In Sect. 2, we introduce the related works about tag-based recommendation approaches. Section 3 interprets our personalized recommendation strategy based on regular tags in details. The experimental study is provided in Sect. 4, and we conclude our work in Sect. 5.

2   Related Works

Generally, the tag-based recommendation systems suggest the resources to users by analyzing the tags and the rating scores assigned to the resources. A large number of approaches to improve the recommendation systems focus on the problem of data sparsity, including the social tag data and potential user rating data on the resources. Pan et al. [13] proposed to expand tag neighbors by calculating tag similarity and investigate the spectral clustering algorithm to filter out noisy and redundant tags, in this way to improve the recommendation accuracy. Yuan et al. [15] proposed a collaborative filtering recommendation algorithm based on a temporal interest evolution model and social tag prediction. The optimized tags are used to model the relationship between users, tags and resources, and the recommendation is made by community discovery and maximum tag voting. Durao et al. [6] proposed to extend the basic similarity calculation with external factors such as tag popularity, tag representativeness and the affinity between user and tag, so as to study and evaluate the recommendation system.

In addition, to alleviate the sparsity problem of user rating data, many algorithms propose to conduct the user characteristics from some indirect information derived from the user rating scores. For example, the number of user browsing and searching for resources is used as the user rating [8]. Cheng et al. [4] use the TF-IDF algorithm to calculate the attributes of users and resources respectively, and estimate the user preference for the resources based on the two feature vectors. However, it is not comprehensive to calculate the rating score only using the records of user browsing on the resources, but more discriminative information should be taken into account.

Compared with existing approaches, our work considers from the perspective of resource provider, so as to generate the standard tag system to avoid tag sparsity. The actual user behaviors on the resources are returned as the feedbacks, which are more discriminative and informative to indicate the users' preference. Our recommendation algorithm will be based on these two components and follow collaborative filtering mechanism.

3 Resource Recommendation Based on Regular Tags

In this section, we will describe our proposed recommendation approach based on regular tags and user operations in details. We will formally introduce some basic concepts, and then interpret how to represent the users by tag-based information. After establishing the user preference model, we provide the recommendation algorithm in collaborative filtering mechanism.

3.1 Preliminaries

In our work, we consider from the perspective of resource provider, and design the recommendation model based on regular tags. The regular tags are those created by resource provider and assigned to each resource item to describe its characteristics. For example, a book item may be assigned the following regular tags: *science fiction*, *Chinese* and *space travel*. We employ regular tags because they are more accurate and strict than social tags, and based on which, the tag sparsity, redundancy and fuzziness can be effectively reduced in practical applications. Generally, in the recommendation system, we assume that the size of regular tag set is l, and the set can be denoted as $T = \{t_1, t_2, \ldots, t_l\}$. The user set contains m users and the item set contains n items, which can be represented as $U = \{u_1, u_2, \ldots, u_m\}$ and $I = \{i_1, i_2, \ldots, i_n\}$ respectively.

We also focus on the user feedbacks on the resources, which can be returned to the resource provider. Traditional user rating mechanism does not fully measure the user preference, because the rated items are usually no more than 1 % of the total number of items, so in large scale recommendation system, the user rating data will be extremely sparse, which will reduce the quality of the recommendation. Therefore, we collect the user behaviors after browsing the resources as the implicit user rating, e.g., reading, sharing or purchasing after browsing a book item. The different operations will reflect the user preference on the item. For example, a user shares item A with others after browsing it while does nothing after browsing item B, which indicates that compared with item B, the user prefers item A more. In the system, we assume f kinds of operation are defined, denoted as $O = \{o_1, o_2, \ldots, o_f\}$. For each operation o_i, we assign it a weight $w_i (1 \leq w_i \leq f)$ to represent its importance, so as to quantify the discrimination between users.

Based on the concepts above, we establish the relationship between tag, user, item and operation weight. Naturally, we can collect the information of regular tags describing each resource item, and the item-tag relationship is defined as

$$R = \begin{pmatrix} r_{11} & r_{12} & \cdots & r_{1l} \\ r_{21} & r_{22} & \cdots & r_{2l} \\ \vdots & \vdots & \ddots & \vdots \\ r_{n1} & r_{n2} & \cdots & r_{nl} \end{pmatrix}$$

where $r_{jk} = 1$ means tag t_k is employed to describe item i_j, and $r_{jk} = 0$ means tag t_k is not used. For each user, we can collect his operations on each resource item, and build the user-item relationship as

$$S = \begin{pmatrix} s_{11} & s_{12} & \cdots & s_{1n} \\ s_{21} & s_{22} & \cdots & s_{2n} \\ \vdots & \vdots & \ddots & \vdots \\ s_{m1} & s_{m2} & \cdots & s_{mn} \end{pmatrix}$$

where s_{jk} records the operation weight of user u_j on item i_k. Here if u_j makes no operation on i_k, $s_{jk} = 0$.

Since the operation behavior of the user on each item can express his preference, we can combine R and S to establish the relationship between users and tags, so the preference of a user on different tags can be estimated. Based on the weights assigned to each operation, we have the weighted user-tag relationship here:

$$G = \begin{pmatrix} g_{11} & g_{12} & \cdots & g_{1l} \\ g_{21} & g_{22} & \cdots & g_{2l} \\ \vdots & \vdots & \ddots & \vdots \\ g_{m1} & g_{m2} & \cdots & g_{ml} \end{pmatrix}$$

The g_{jk} denotes the weighted preference of user u_j on tag t_k, and it is calculated as:

$$g_{jk} = \sum_{e=1}^{n} s_{je} r_{ek}. \tag{1}$$

We can conclude that g_{jk} is the accumulated operation effect of u_j on tag t_k related resources. By such weighted user-tag relationship, the user preference on each tag can be more accurately estimated, so as to reflect the users' interests.

3.2   User Feature Representation

In order to make personalized recommendation, it is essential to find the effective representation to reflect the discriminative characteristics of different users, which is also important for discovering similar users.

The user features are derived from the property of items and the user operations showing different preferences. Similar as text processing, TF-IDF has been applied to express the tag feature vector for different users [2], but it only describes the information of tag frequency, which is not sufficiently discriminative to describe the user characteristics. For example, user A and B browse book C at the same time, but A purchases the book, showing that user A favors this book more than B. Considering this, our work takes into account three aspects of the users to form the user feature representation: the tag feature, the operation feature and time factor.

The Tag Feature. In this work, the tag feature is to employ the user's favorite tags to represent the user preference feature. The normalized TF-IDF is employed to calculate the tag feature vector, and for user u and tag t_k, we have:

$$F_{ut_k}^{tag} = \frac{v_{ut_k}}{\sum_j v_{ut_j}} \log \frac{m}{v_{t_k u}} \qquad (1 \le k \le l). \qquad (2)$$

Here v_{ut_k} is the counts that user u uses tag t_k, and $v_{t_k u}$ is the number of users that use tag t_k, so $F_{ut_k}^{tag}$ can reflect the preference of user u towards tag t_k.

The Operation Feature. The user operations on the resource items are important information to reflect the user preference on the items. In preliminary part, we have introduced the operation-weighted user-tag relationship, and the operation feature of the users will be based on this relationship.

In our work, the user's long-term average preference will be applied to generate the operation feature, so as to improve the discrimination of users towards different tags. For user u and tag t_k, the operation feature is calculated as

$$F_{ut_k}^{op} = e^{\frac{g_{uk}}{v_{ut_k}} - \lambda} \qquad (1 \le k \le l). \qquad (3)$$

Here $g_{uk}(1 \le j \le l)$ is the weighted user-tag preference calculated by Eq. (1), and v_{ut_k} is for normalization. λ represents the minimum value of operation weight of user u, so as to remove the operation bias of different users. Equation (3) reflects the user feature for different tags according to the actual operations.

The Time Factor. In addition, we focus on an interesting factor that reflects the user preference: the time factor. It is generally believed that the most recent collected resources can best reflect the user's interest, i.e., the tags used recently can best describe the user's preference. A user-interest model [4] based on the forgetting mechanism has been proposed based on the adaptive exponential decay function to deal with the time information of tags. In our work, we apply the adaptive time decay function and the idea of Ebbinghaus forgetting curve, to define the user feature based on time factor. The formula for user u on tag t_k is designed as:

$$F_{ut_k}^{time} = \beta + (1 - \beta)e^{-(d_{now} - d_{ut_k})} \qquad (1 \le k \le l). \qquad (4)$$

Here d_{now} represents the current time point, and d_{ut_k} means the last time when tag t_k marks the user. $\beta \in [0, 1]$ is used to adjust the influence of the time factor in the user's interest, and the influence of time factor is greater when β is smaller.

The Comprehensive Representation. According to the analysis above, we can now formally describe our user preference model. The user feature representation to describe his tag-based preference can be formulated by the tag feature, operation feature and time-based feature. For user u and tag t_k, we have:

$$F_{ut_k} = F_{ut_k}^{tag} \cdot F_{ut_k}^{op} \cdot F_{ut_k}^{time} \quad (1 \le k \le l). \tag{5}$$

So the feature vector of user u is: $F_u = \{F_{ut_1}, F_{ut_2}, \ldots, F_{ut_l}\}$.

3.3 User-Item Preference Analysis

The user preference for items is usually analyzed by the historical behavior of the user. Traditional collaborative filtering algorithm is based on user's rating to reflect the interest preferences of the user and to estimate the resource similarity, which ignores the characteristics of users and resources, and thus decreases the recommendation quality for new resource items significantly. We propose to analyze the user preference by the prediction score of the user himself and the similar users, based on the regular-tag characteristics of all items. The user-item preference can be divided into two categories: one is the preference of user for historical items, i.e., the browsed items; the other one is for new items.

Preference for Historical Items. The user preference for historical resources can be estimated by the user-item relationship S. For those browsed items, the user operations are utilized to weight the item browsing records. For user u_j on item i_k, the user-item preference can be estimated as:

$$P_{jk}^{hist} = \frac{s_{jk}}{\sum_i s_{ji}}. \tag{6}$$

Here the result is normalized to benefit further analysis, and s_{jk} is from user-item relationship matrix.

Preference for New Items. For new items, there are no existing user operation records. However, for each new item, the resource characteristics(tags) can be determined when the item is created, so the user preference can be estimated by comparing the user feature and the item-tag record. For user u_j on new item i_k, we design the normalized user preference for new item by the following equation:

$$P_{jk}^{new} = \frac{\sum_{i=1}^{l} r_{ki} \cdot F_{jt_i}}{\sum_{k=1}^{n} \sum_{i=1}^{l} r_{ki} \cdot F_{jt_i}}. \tag{7}$$

Here r_{ki} is from item-tag relationship matrix, and F_{jt_i} is the user feature value.

3.4   Top-K Recommendation Algorithm

Finally, we introduce the personalized recommendation algorithm, which recommend the new resource items to different users according to their personal preferences. Our framework follows the collaborative filtering mechanism. To recommend new items to a user, the algorithm will consider the preference of the target user and his similar users, and the preference scores will be ranked to provide the most favorite items for the target user.

Primely, for the target user, we need to find his similar users. Given the user feature representation by Eq. (5), we select the cosine similarity to calculate the similarity scores between the target user and other users, as follows:

$$sim(u_a, u_b) = \frac{F_{u_a} \cdot F_{u_b}}{\|F_{u_a}\| \|F_{u_b}\|}. \tag{8}$$

Formally, given a target user u_j, and a new item set for him, the estimation of the target user preference for each new item i_k can be divided into the following two parts:

1. **Target user preference**: Since we aim to recommend new items to the target user, his preference on the new item can be estimated by Eq. (7).
2. **Similar user preference**: We also take into account other users' interests as a reference. There are two situations for the referred user, that if this user has browsed i_k, his preference score can be calculated by Eq. (6), or else the score will be calculated by Eq. (7). The similarity between the target user and the referred user will be embedded as a weighting factor.

With the above two parts, we can design the preference score of u_j on item i_k as:

$$Score(u_j, i_k) = \alpha \cdot P_{jk}^{new} + (1 - \alpha) \frac{\sum\limits_{u_i \in U, i \neq j} sim(u_i, u_j) \times P_{ik}}{\sum\limits_{u_i \in U, i \neq j} sim(u_i, u_j)}. \tag{9}$$

Here $\alpha \in [0, 1]$ is a constant factor, which is used to express the significance of similar users on the recommendation result.

Based on Eq. (9), if we can calculate the preference scores of target user u_j on each new item, the preference result can be sorted in descending order, and the items with top K highest scores can be selected to recommend to the target user.

4   Experimental Evaluation

In this part, we will introduce the empirical study. The experiment setting will be introduced first, and then the effect of some important factors in our model will be reported. Finally we will present the comparison of our algorithm with some existing approaches.

4.1   Experiment Setting

Dataset. In our work, we evaluate the proposed recommendation algorithm on a practical resource publishing system, which is developed by WHUT Digital Communication Engineering Co., Ltd. The dataset contains a collection of 17735 tagged multimedia resource items, including books, articles, videos, etc. The system is maintaining 554 WeChat[1] public accounts, with a total of 674520 subscribers by the end of March 2016. The user data with operations on different resource items are collected through this platform.

Parameter Setting. According to the resource publishing on WeChat platform, the user behaviors on the resource items can be categorized into 6 operations. Different weights are assigned to each operation, and higher weight means higher degree of user preference on the item. The operations and the corresponding weights are listed in Table 1.

Table 1. The user operations and the weight assignment

User operation	Browse	Collect	Comment	Put into shopping cart	Purchase	Share with friends
Weight	1	2	3	4	5	6

Our user preference model contains two important parameters. β is used to control the influence of time factor in generating user feature representation. In our experiment, we set this parameter as $\beta = 0, 0.2, 0.5, 0.8, 1.0$. The other parameter is α, which is used to calculate the user preference score in recommendation. In order to verify the impact of reference from similar users on the recommendation results, the value of α is controlled as $\alpha = 0, 0.2, 0.5, 0.8, 1.0$.

In our experiment, in order to precisely evaluate our algorithm, we select 2000 tags with most number of use in the tag system. 2000 popular items associated with these tags are chosen together with 1000 active users. The dataset is divided into training and testing set at the ratio of $4 : 1$.

Evaluation Metrics. In this study, standard Precision and Recall rates are employed to evaluate the performance of the recommendation algorithm. Let $R(u)$ be the recommended item list calculated by the recommendation algorithm according to the user behaviors in the training set, and $T(u)$ is the actual behavior of users in the testing set. Precision is the ratio of the items in the list of recommendation hit the testing dataset, which is calculated as:

$$\text{Precision} = \frac{\sum_{u \in U} |R(u) \cap T(u)|}{\sum_{u \in U} |R(u)|} \tag{10}$$

---

[1] A popular instant messenger client in China, which integrates the functions of IM, social network, resource publishing and public services.

The Recall rate shows the proportion of recommended items in the users' actual items collection, which assesses the integrity of the user's interest, and is calculated as:

$$\text{Recall} = \frac{\sum_{u \in U} |R(u) \cap T(u)|}{\sum_{u \in U} |T(u)|} \tag{11}$$

In addition, we uses the F-measure to evaluate the quality of the recommendation model, which is calculated as:

$$\text{F-measure} = \frac{2 \times \text{Precision} \times \text{Recall}}{\text{Precision} + \text{Recall}} \tag{12}$$

4.2 Effect of β

We first test the effect of β, which adjusts the influence of time factor in user feature presentation. In this test, we set $\alpha = 0.5$, and verify the influence of β on the precision and recall rate, and the results are recorded in Figs. 1 and 2. It shows that the precision rate will decrease as more items are recommended while recall rate will significantly increase. If recommend a few items, the short-term interest statistics of the user are better to describe the user's need, while when recommend a large number of items, the long-term interest is more accurate to judge the user's preference.

4.3 Effect of α

α is the parameter to adjust the importance of similar users in recommendation process. In this test, we set $\beta = 0.5$ and change α value to study the effect of α on the recommendation, and the results are shown in Figs. 3 and 4. From the curves we can discover that the overall performance follows the same trend as that of β when the number of recommended items increases. In addition, small α value will result in high recommendation accuracy, and if $\alpha = 1$, the recommendation effect

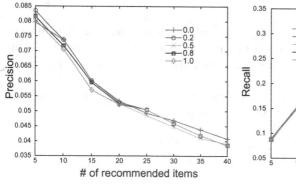

Fig. 1. Effect of parameter β on precision rate.

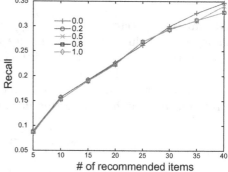

Fig. 2. Effect of parameter β on recall rate.

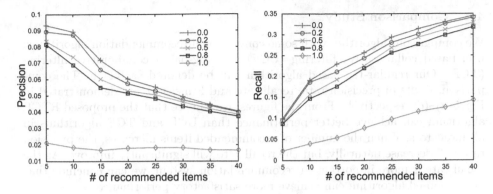

Fig. 3. Effect of parameter α on precision rate.

Fig. 4. Effect of parameter α on recall rate.

is relatively consistent. It means that, to a certain extent, the recommendation result is more accurate when the algorithm takes into account a large proportion of similar user information. The effect on recall rate behaves in the same way, which means the reference from similar users can introduce more resource items to cover the potential items selected by the target user.

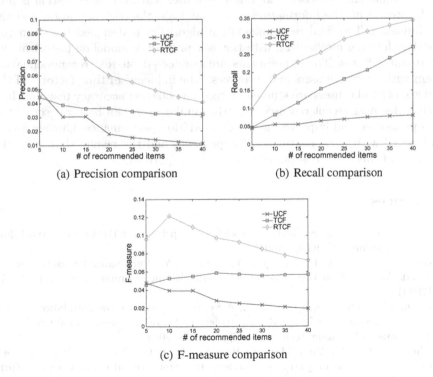

(a) Precision comparison

(b) Recall comparison

(c) F-measure comparison

Fig. 5. Comparison of different recommendation algorithms.

4.4 Comparison Study

We compare our algorithm with some conventional recommendation algorithms: user-based collaborative filtering (UCF) and tag-based collaborative filtering (TCF). Our regular-tag-based algorithm will be denoted as RTCF. The comparison results of precision rate, recall rate and F-measure are demonstrated in Fig. 5(a)–(c) respectively. From the figures, we can see that the proposed RTCF algorithm can achieve better performance than UCF and TCF algorithms in all indicators. When the number of recommended items increases, the precision rate will decrease naturally, but the recall rate will significantly improve. As the recall rate is more important in recommendation system, we can conclude that our proposed algorithm can achieve more satisfactory performance.

5 Conclusions

The conventional recommendation systems usually suffer from several problems, e.g., the item tag data and user rating data are extremely sparse, and the user feedbacks are not sufficiently utilized. This work considers from the perspective of resource provider, and proposes a recommendation algorithm based on the regular tags and user operation feedbacks. In order to precisely describe the user's personal characteristics, an innovative user feature representation is proposed, integrating the information of regular tags, the user operation and the time influence. The final recommendation algorithm is designed based on collaborative filtering mechanism, with the user preference model on historical and new resource items. The experiments are conducted on real recommendation system with extensive users and resources. The influence of time factor and the reference of similar users are studied by recommendation accuracy test, concluding that the most recent resources and the similar users can better describe the user's preference and improve the recommendation performance. Our algorithm is compared with some conventional approaches and the results show that the algorithm of this work is superior to the other algorithms.

References

1. Bennett, J., Lanning, S., Netflix, N.: The netflix prize. In: KDD Cup and Workshop in Conjunction with KDD (2009)
2. Cai, Q., Han, D.M., Li, H.S., Hu, Y.G., Chen, Y.: Personalized resource recommendation based on tags and collaborative filtering. Comput. Sci. **41**(1), 69–71 (2014)
3. Chen, J.M., Sun, Y.S., Chen, M.C.: A hybrid tag-based recommendation mechanism to support prior knowledge construction. In: IEEE International Conference on Advanced Learning Technologies, pp. 23–25 (2012)
4. Cheng, Y., Qiu, G., Bu, J., Liu, K., Han, Y., Wang, C., Chen, C.: Model bloggers' interests based on forgetting mechanism. In: International Conference on World Wide Web, pp. 1129–1130 (2008)

5. De Gemmis, M., Lops, P., Semeraro, G., Basile, P.: Integrating tags in a semantic content-based recommender. In: ACM Conference on Recommender Systems (2008)
6. Durao, F., Dolog, P.: A personalized tag-based recommendation in social web systems. In: International Workshop on Adaptation and Personalization for Web, pp. 40–49 (2012)
7. Firan, C.S., Nejdl, W., Paiu, R.: The benefit of using tag-based profiles. In: American Web Conference, pp. 32–41 (2007)
8. Jin, J., Chen, Q.: A trust-based top-k recommender system using social tagging network. In: International Conference on Fuzzy Systems and Knowledge Discovery, pp. 1270–1274 (2012)
9. Linden, G., Smith, B., York, J.: Amazon.com recommendations: item-to-item collaborative filtering. IEEE Internet Comput. **7**(1), 76–80 (2010)
10. Ma, T., Zhou, J., Tang, M., Tian, Y., Al-Dhelaan, A., Al-Rodhaan, M., Lee, S.: Social network and tag sources based augmenting collaborative recommender system. IEICE Trans. Inf. Syst. **98**(4), 902–910 (2015)
11. Mathes, A.: Folksonomies - cooperative classification and communication through shared matadata. Comput. Mediated Commun. **47**(10), 1–13 (2004)
12. Mishne, G.: Autotag: a collaborative approach to automated tag assignment for weblog posts. In: International Conference on World Wide Web (2006)
13. Pan, R., Xu, G., Dolog, P.: Improving recommendations in tag-based systems with spectral clustering of tag neighbors. In: Park, J.J., Chao, H.-C., Obaidat, M.S., Kim, J. (eds.) CSA 2011 and WCC 2011. LNEE, vol. 114, pp. 355–364. Springer, Netherlands (2012)
14. Sun, G., Liu, G., Zhao, L., Xu, J., Liu, A., Zhou, X.: A social trust path recommendation system in contextual online social networks. In: Chen, L., Jia, Y., Sellis, T., Liu, G. (eds.) APWeb 2014. LNCS, vol. 8709, pp. 652–656. Springer, Heidelberg (2014)
15. Yuan, Z.M., Huang, C., Sun, X.Y., Li, X.X., Xu, D.R.: A microblog recommendation algorithm based on social tagging and a temporal interest evolution model. J. Zhejiang Univ. Ser. C Comput. Electron. **16**(7), 532–540 (2015)

Academic Paper Recommendation Based on Community Detection in Citation-Collaboration Networks

Qisen Wang[1], Wenzhong Li[1,2,3](✉), Xiao Zhang[1], and Sanglu Lu[1,2,3]

[1] State Key Laboratory for Novel Software Technology,
Nanjing University, Nanjing, China
lwz@nju.edu.cn
[2] Sino-German Institutes of Social Computing, Nanjing University, Nanjing, China
[3] Collaborative Innovation Center of Novel Software Technology
and Industrialization, Nanjing, China

Abstract. Academic search engine plays an important role for science research activities. One of the most important issues of academic search is paper recommendation, which intends to recommend the most valuable literature in a domain area to the users. In this paper, we show that exploring the relationship of collaboration between authors and the citation between publications can reveal implicit relevance between papers. By studying the community structure of the citation-collaboration network, we propose two paper recommendation algorithms called Adaptive and Random Walk, which comprehensively consider several metrics such as textural similarity, author similarity, closeness, and influence for paper recommendation. We implement an academic paper recommendation system based on the dataset from Microsoft Academic Graph. Performance evaluation based on the assessments of 20 volunteers show that the proposed paper recommendation methods outperform the conventional search engine algorithm such as PageRank. The efficiency of the proposed algorithms are verified by evaluation.

Keywords: Paper recommendation · Social network · Citation-collaboration network · Community detection

1 Introduction

Searching for academic literature is of great important for researchers and engineers. As for the development of search engine theories and practices, scholar search engine such as Google Scholar [1], Microsoft Academic Search [2], and ArnetMiner [3], which specify the data source to science domain, are also built as more personalized means to search according to the content and influence of a publication. Most of the existing academic search engines rank the importance of a paper based on the relevance of keywords, the citation counts and influence index. However, such keyword-based search did not consider the collaboration

© Springer International Publishing Switzerland 2016
F. Li et al. (Eds.): APWeb 2016, Part II, LNCS 9932, pp. 124–136, 2016.
DOI: 10.1007/978-3-319-45817-5_10

between authors and the citation between papers, and sometimes cannot satisfy the need of academic paper recommendation, where the user hopes the search engine to recommend the most valuable literature in a domain area. For example, an author works on "natural language processing" may has coauthors who works on "information retrieval", which suggests that the areas of "natural language processing" and "information retrieval" maybe highly relevant. But a keyword based search without considering author collaborations cannot reveal such relevance. Another example is that the papers on "information diffusion" may frequently cite the papers on "epidemic model", which suggest that "epidemic model" could be quite valuable for "information diffusion", but a search engine without considering the citations between papers also fail to address such implicit correlations.

Therefore, to tackle the issues of academic paper recommendation, it is worthwhile to explore the relationship of collaboration and citation in literature. Given the large number of publications, such relationship can be described by a two-layer complex network called *citation-collaboration network*, where one layer is from the perspective of citation relations between publications, and the other is from the perspective of collaboration relations between authors. According to the homophily theory [4] of social networks, people are more likely to share their ideas with people from the same area, the same university, or at least speak the same language. Similar discipline also applys for the citation-collaboration network and forms the phenomenon so called "academic circles". Community detection algorithms [5,6] can be used to identify the "academic circles", which publications within the same community should be more relevant to each other, and publications from different communities are less relevant.

In this paper, we propose an academic paper recommendation system based on community detection in citation-collaboration networks. We first introduce a graph transform method to transform the complex network to a uniform directed graph with nodes representing papers and weights representing the closeness between them. Then we use a *weighted label propagation algorithm (WLPA)* to divide the citation-collaboration networks into smaller communities. After community detection, we can identify a group of relevant publications representing as a community and then perform paper recommendation algorithm in the same community. We propose two ranking algorithms called *Adaptive* and *Random Walk*, which comprehensively consider several metrics such as textural similarity, author similarity, closeness, and influence for paper recommendation. We implement an academic search engine based on the dataset from Microsoft Academic Graph [7]. Performance evaluation based on the volunteers' assessments shows that the proposed paper recommendation methods outperform the conventional search engine algorithm such as PageRank. The efficiency of the proposed algorithms are verified by the evaluation.

2 Related Work

2.1 Community Detection

Community detection is a classical problem aiming at splitting a graph into a given number of groups while minimizing the cost of edge cut [5,8]. In early work, the community detection algorithms require some other parameters except graph data, such as community sizes or numbers. Later, the GN algorithm [6] proposed by Newman and Girvan made the community detection more independent with a divisive way to split the whole graph into groups by the edges with biggest betweenness. Some other alternative methods emerged in the past decades, where the similarity between vertices were used as a vital measurement for community detection. An example is the hierarchical clustering proposed by Newman [9]. Rosvall and Bergstrom proposed a community detection method based on information theory [10]. Fu et al. [11] proposed a memetic algorithm for community detection in networks based on the genetic algorithm.

To reduce the time complexity and achieve efficient community detection in massive social networks, Raghavan, Albert, and Kumar proposed the label propagation algorithm (LPA), which has a relatively low time complexity of $O(n)$] [12]. However, LPA algorithm cannot be applied in directed weighted graphs, which is the focus of this paper.

2.2 Search Algorithms

Searching for web pages or academic papers in a network has been widely studied in the past. The most frequently adopted algorithm for search engine is the PageRank [13] algorithm from Google. The key idea is to initialize each node with a different value, and each nodes's value is given to its out links. After several rounds of iteration, each node's value will remain relatively stable. Therefore the importance of the nodes can be ranked according the value. Another well known search algorithm was the HITS algorithm proposed by Jon Kleinberg [14]. Taher Haveli-wala [15] proposed the Hilltop algorithm by putting the ideas of PageRank algorithm and HITS algorithm together. To the best of our knowledge, recommending academic paper based on community detection in citation-collaboration networks has not been studied in the past.

3 System Model and Problem Description

3.1 Citation-Collaboration Network

We introduce the citation-collaboration network to describe the citation among academic papers and the collaboration among authors. The citation-collaboration network is represented by a directed graph. A node in the graph represents a paper or an author. There are three different kinds of links in the graph representing different relationships: (1) citation relationship: if paper p_i is cited by paper p_j, there is a directed link from p_j to p_i; (2) collaboration relationship: if author a_i has a joint paper with author a_j, then there are bidirectional links between a_i and a_j; (3) author-paper relationship: if paper p_i has an author a_j, then there is a directed link from a_j to p_i. The citation-collaboration network is also a weighted graph, where the citation relationship and the author-paper relationship have weight 1, while the collaboration relationship has a weight equals to the number of joint papers of two authors.

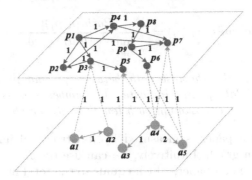

Fig. 1. A citation-collaboration network.

Figure 1 shows an example of a citation-collaboration network. It can be viewed as a two-layer complex networks: the citation network formed by citation between publications, and the collaboration network formed by joint works of authors or researchers. The cross-layer links indicate the author-paper mappings between the two layers.

3.2   Problem Description

Let $G = <V_1, V_2, E_1, W_1, E_2, W_2, E_3, W_3 | E_1 \subseteq V_1 \times V_1, E_2 \subseteq V_2 \times V_2, E_3 \subseteq V_2 \times V_1 >$ be a citation-collaboration graph, and $p_i, p_j \in V_1$ are two papers. We use the following metrics to measure the relevance of the two papers.

Definition 1 (Textural Similarity). *The textural similarity of two papers is calculated by the Jaccard Coefficient of their keywords, which is defined as*

$$TextSim(p_i, p_j) = \frac{|Key(p_i) \cap Key(p_j)|}{|Key(p_i) \cup Key(p_j)|},$$

where $Key(p_i)$ and $Key(p_j)$ are the set of keywords abstracted from the two papers.

Similarly, we define the author similarity of two papers as the Jaccard Coefficient of their authors.

Definition 2 (Author Similarity).

$$AuthorSim(p_i, p_j) = \frac{|Author(p_i) \cap Author(p_j)|}{|Author(p_i) \cup Author(p_j)|},$$

where $Author(p_i)$ and $Author(p_j)$ are the set of authors of the two papers.

Definition 3 (Closeness). *The closeness of two papers is defined as their normalized distance in the citation-collaboration graph. Specifically,*

$$Closeness(p_i, p_j) = \frac{|Path(p_i, p_j)|}{\max\limits_{\forall p_x, p_y \in V_1} |Path(p_x, p_y)|},$$

where $Path(p_i, p_j)$ is the shortest path between p_i and p_j in the citation-collaboration graph, $|Path(p_i, p_j)|$ indicates the number of hops in the path, and denominator indicates the maximum path length in the graph.

The influence of a paper is measured by its structural importance in the citation-collaboration graph. Intuitively, we can use the *Degree Centrality* [16] to define the influence of a paper, which equals to the total number of citations from other papers. The influence is indicated by the normalized degree centrality defined as follows.

Definition 4 (Influence). *The influence of a paper $p \in V_1$ is calculated by*

$$Influence(p) = \frac{\sum_{\forall q \in V_1} w_{qp}}{\max\limits_{\forall p' \in V_1} \sum_{\forall q \in V_1} w_{qp'}},$$

where w_{qp} is the indicator that $w_{qp} = 1$ if there exists a link from node q to node p in G and $w_{qp} = 0$ otherwise, and the denominator indicates the maximum citation count in the graph.

Given a paper $p \in V_1$, our objective is to find the most relevant and high influence papers. We rank the papers using a comprehensive function that considering the above metrics. For a paper $q \in V_1, q \neq p$, its ranking can me represented by

$$Rank(q, p) = f(TextSim(q, p), Closeness(q, p), Influence(q)), \qquad (1)$$

where $f()$ is a comprehensive function which will be discussed in our algorithms.

With the above notations, the academic paper recommendation problem can be described as follows. Given a citation-collaboration graph $G = \langle V_1, V_2, E_1, W_1, E_2, W_2, E_3, W_3 \rangle$ and a paper $p \in V_1$, find the top-K ranking papers in the graph. That is, we want to find a set S satisfies: $S \subset V_1, |S| = K$, and $\forall x \in S, y \in V_1 - S, Rank(x, p) \geq Rank(y, p)$.

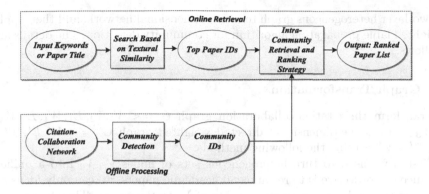

Fig. 2. The framework of paper recommendation system.

4 Methodology

To solve the academic paper recommendation problem, we need to search the high rank papers efficiently. A citation-collaboration graph may contains billion of nodes (papers and authors), thus exhausting search in the whole graph is highly inefficient. We propose a community-based strategy to solve the problem efficiently.

Figure 2 shows the framework of our solutions. The proposed method includes two stages: *offline processing* and *online retrieval*. In the offline processing stage, we apply community detection algorithm to divide the citation-collaboration graph into communities. Each community contains about thousands of papers. They are stored in a distributed way, and academic paper recommendation only consider the papers within the same community since they are more close in structure and more related due to their close citation relationships. Community detection can be run offline and can be updated daily or weekly when more papers are included in the system. The online retrieval process includes several steps. First, when a user input the keywords or the title of a paper, we run an initial search algorithm based on textural similarity and got the top-10 paper IDs in the system. Then for each paper ID, we locate its community, and run an intra-community retrieval algorithm and rank the papers according to their structural similarity and influence information. Finally, we sort the papers according to their ranks and output the top-K papers for recommendation.

The detailed algorithms are discussed in the following sections.

5 Community Detection

Community detection algorithms in complex networks have been widely studied in the past. However, since the citation-collaboration graph is a heterogeneous network with three different types of edges, the existing community detection strategy cannot be directly applied. To deal with the problem, we first transform the two-layer heterogeneous graph to a one-dimensional network, and then apply a weighted lable propagation algorithm for community detection. The details are as follows.

5.1   Graph Transformation

We transform the citation-collaboration graph $G = < V_1, V_2, E_1, W_1, E_2, W_2, E_3, W_3 >$ to a one-dimensional directed weighted graph $G' = < V', E', W' : E' \subseteq V' \times V' >$ using the following methods.

Firstly, we need to turn heterogeneous sets of nodes V_1, V_2 into a unified set. Since our objective is to recommend academic papers, we can only focus on the set of papers V_1 and map the weights of V_2 to the set V_1. Therefore we let $V' = V_1$ after graph transformation.

Secondly, we evaluate the relative importance of citation network and co-author network respectively. The weight of each edge in citation network is 1, while the edges in collaboration network have various weights depending on the number of co-publications. To deal with this issue, we introduce a factor γ to represent the relative importance when transforming the weights from the collaboration network to the citation network. Apparently, $0 < \gamma < 1$ is a turnable parameter and we let $\gamma = 0.5$ in our system.

We map the co-author relationship in the collaboration graph to the transformed graph as follows. If paper p_i and p_j have a common author, then there are bi-directional links between p_i and p_j. The weight on the link is calculated by considering both citation relationship and author similarity.

$$w'_{ij} = \begin{cases} 1 + \gamma * AuthorSim(p_i, p_j); & (p_i, p_j) \in E_1 \\ \gamma * AuthorSim(p_i, p_j); & otherwise; \end{cases}$$

After applying the above method, the citation-collaboration graph can be transformed to the new graph $G' = < V', E', W' >$, where V' is the set of papers, and E' and W' are the new set of edges and weights computed by the transforming method.

5.2   Weighted Label Propagation Algorithm

After transforming the two-level complex network into a single dimensional graph, we can apply a community detection algorithm to divide the graph to communities. To achieve community detection in an efficient way, we adopt the linear-time label propagation algorithm (LPA) which is introduced in [12]. However, the original LPA is designed for undirected unweighted graphs, which can not be applied directed in our system. To adapt the LPA to the directed weighted graph as in our case, we proposed a *weighted label propagation algorithm (WLPA)* shown in Algorithm 1.

In the algorithm, $Neighbor(v)$ is the neighbor set of v in G'. The algorithm contains the following steps.

(1) Initialization: assign each node a different community label.

Algorithm 1. Weighted Label Propagation Algorithm

Input: transformed network $G' = (V', E', W')$.
Output: Community labels for each node in G'.
1: initialize nodes with unique labels
2: **while** labels of nodes are not stable **do**
3:    **for all** $v \in V'$ **do**
4:       **for all** $u \in Neighbor(v)$ **do**
5:          $l_u = $ *the label of u*
6:          $sum(l_u) = sum(l_u) + w'_{vu}$
7:       **end for**
8:       update the label of v: $l_v = argmax_{l_i}\{sum(l_i) \; \forall i \in Neighbor(v)\}$
9:    **end for**
10: **end while**

---

(2) Iteration: In the iteration process, each node selects the most "adaptable" label in its neighbors and alter its own label to the same label. Unlike the original LPA, the WLPA only consider out-degree neighbors (which means the papers cited by the individual). Then the node consider the labels of its neighbors and the weights, and updates its label to the one with the maximum accumulative weights, which is

$$l_v = \underset{l_j}{argmax} \sum_{j \in Neighbor^{l_j}(v)} w'_{vj} \qquad (2)$$

where $Neighbor^{l_j}(v)$ represents the set of neighbors of v with label l_j, and the weight of edge between v and j is w'_{vj}.

Repeating this procedure until most of the nodes in graph remain stable labels and don't change their community labels in the next iterations. According to the analyze of LPA [12], the algorithm converges in constant rounds.

6   Online Retrieval and Ranking

When a user input the paper title or some keywords, we first use the textural similarity to perform initial search to obtain the top-K papers that are most relevant to the input keywords. In the system implementation, we use the Lucence engine [17], a search index engine, to automatically abstract keywords from the literature to form index, and to obtain the textural similarity values using the TF-IDF algorithm [18]. Lucence engine has been proved to be very efficient for text-based searching in large-scale distributed database] [17].

With the top-K papers from text searching, we can identify the corresponding community ID for each paper based on the community detection result. For each paper in a community, we evaluate its rank according to the network structure information. We introduce two ranking algorithms.

6.1   Adaptive Ranking Algorithm

The adaptive ranking algorithm works as follows. Firstly, given the top ith paper obtained from the textural similarity search, we can get its community ID, and then obtain the set of papers in the same community C_i. Secondly, we rank each paper in community C_i by comprehensively considering the metrics of textural similarity, influence and closeness. For an input paper p and each p_j in C_i,

$$Rank(p_j, p) = \lambda_1 * TextSim(p, p_j) + \lambda_2 * Influence(p_j) + \lambda_3 * Closeness(p, p_j),$$

where λ_1, λ_2 and λ_3 are turnable parameters to indicate the importance of each metric in the total rank.

Thirdly, we repeat the above process for all the top-K papers obtained from the textural similarity search, and sort the papers in $\bigcup_{i=1}^{K} C_i$ according to their ranking scores. Finally, we output the top-K papers with the highest ranking scores, which are recommended to the user.

6.2   Random Walk Algorithm

The other ranking algorithm we proposed is a random walk algorithm. Given a paper represented by a node in the graph, we make several rounds of random walk starting from the node with a fixed walking distance. If a node is reachable for multiple times, it means that it is important in the network structure and can be candidate for recommendation. To consider the closeness of two nodes, the probability of choosing the next waking node should be proportional to the weight on the links. The ranking of a paper can be computed by a comprehensive function of both the textural similarity and the times of visits in the random walks.

At each step of random walks, the walker randomly chooses one of its neighbors to move. The visiting sequence of this walker can be described as a Markov chain. The probability of walking from node p_i to node p_j is defined by

$$P(i, j) = \frac{w'_{ij}}{\sum_k w'_{ik}}. \tag{3}$$

The total length of walks is denoted by L. The longer walking distance, the less relevant of the papers. In our system, we set $L = 10$. The random walks repeat for several rounds R and we use $VisitTime(p_j)$ to denote the times that node p_j being visited. Intuitively, if two papers are close, they will be easy to reach each other, and the visit times should be high.

After random walks, we can compute the ranking of each node p_j being visited during walks starting from paper p.

$$Rank(p_j, p) = \lambda_1 * TextSim(p, p_j) + \lambda_4 * VisitTime(p_j), \tag{4}$$

where λ_1 and λ_4 are turnable parameters to indicate the importance of each metric in the total rank.

After sorting the result according to the ranking scores, the algorithm output the top-K papers for recommendation.

7   Performance Evaluation

We implement the proposed paper recommendation algorithms in the academic searching engine and evaluate their performance. The dataset we used is from the Microsoft Academic Graph [2], which contains the meta information of more than 120 million scientific publication records, citation relationships between those publications, as well as authors, institutions, journals, conference venues, and fields of study.

We compare the performance with a baseline algorithm based on the popular PageRank method [19]. Specifically, given an input paper p, we consider a modified PageRank algorithm to compute the ranking of paper p_j as follows.

$$Rank(p_j, p) = \lambda_1 * TextSim(p, p_j) + \lambda_5 * PageRank(p_j), \qquad (5)$$

where $PageRank(p_j)$ is the page rank value obtained from the conventional PageRank algorithm, and λ_1 and λ_5 are turnable parameters.

The default values of the system parameters are summarized in Table 1. The chosen of such values intends to provide fair weights to the impact factors.

Table 1. Default values of the system parameters.

System parameters	γ	λ_1	λ_2	λ_3	λ_4	λ_5	L	R
Default value	0.5	1.0	0.5	0.5	0.5	0.5	10	1000

To conduct the experiments, we invite 20 volunteers to help on evaluating the quality of academic paper recommendation results. We randomly choose 50 papers in the system and show the searching results to the volunteers. The ranking algorithms are anonymous and random to the volunteers. The volunteers are required to rank the quality of each paper' recommendation result with a score ranging from 1 to 5. The higher score means the better quality. We collect all the scores on the 50 papers and use them for performance evaluation.

Figure 3 compares the cumulative distribution function (CDF) of the scores of the three algorithms. As shown in the figure, for the page rank algorithm, about 70 % results are ranked 1, and there are less than 10 % results are ranked higher than score 3. The Adaptive algorithm and Random Walk algorithm perform much better than PageRank, where about 20 % results are ranked lower than 2, and about 50 % results are higher than 3. The Random Walk algorithm performs the best among all the strategies.

We also compare several performance metrics such as textural similarity, number of citations of the recommended paper, which results are shown in Figs. 4 and 5.

Figure 4 compares CDF of textural similarity of the three algorithms. As shown in the figure, PageRank has the lowest textural similarity, where about 90 % results have very small textural similarity lower than 0.05. The Random

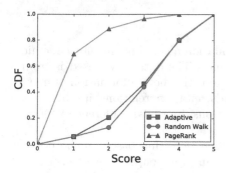
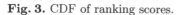

Fig. 3. CDF of ranking scores.

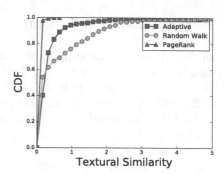

Fig. 4. CDF of textural similarities.

Walk algorithm have 7 % results lower than 0.05, and about 30 % results higher than 1. The Adaptive algorithm performs in between. Again, the Random Walk algorithm performs the best among the three.

The CDF of citation counts of the algorithms are shown in Fig. 5. The PageRank has the highest citation count, where 90 % results have citations higher than 2000. While for Adaptive and Random Walk, more than 90 % results have citations less than 1000. It means that PageRank tends to find the high citation papers. However, due to the low textural similarity of PageRank, such high cited papers maybe not relevant to the paper that a user want to search. Although PageRank emphasizes the influence of the paper, it fails to address the similarity of papers, which cannot achieve good performance in practice. The proposed algorithm considers textural similarity, citations and closeness in network structure, which performs better than PageRank in paper recommendation.

To show the time efficiency of the proposed algorithms, we compare the response time of the search engine using different algorithms. Figure 6 shows the CDF of response time of searching 50 papers. According to the figure, 80 % searching of Adaptive algorithm are replied within 200–400 ms; 80 % searching

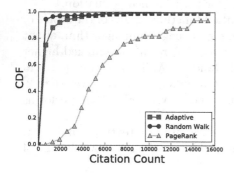

Fig. 5. CDF of citation counts.

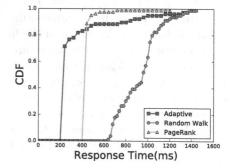

Fig. 6. CDF of response time.

of Random Walk are replied within 400–600 ms, and the PageRank algorithm is within 600–1200 ms most of the time. The proposed algorithms outperforms PageRank in response time and the Adaptive algorithm achieves the highest time efficiency in academic search.

8 Conclusion

In this paper, we focused on the issues of paper recommendation for academic search. We showed that by considering the relationships of author collaborations and paper citations can reveal the implicit correlation between publications. Therefore we proposed an academic paper recommendation system based on community detection in citation-collaboration networks. First, we introduced a graph transform method to transform the two-layer citation-collaboration network to a uniform directed graph with nodes representing papers and weights representing the closeness between them. Then, we introduced a weighted label propagation algorithm for community detection to find the most relevant cluster of papers. After dividing the graph into communities, we proposed two ranking algorithms called Adaptive and Random Walk to rank the papers within the same community. We implemented an academic paper recommendation system based on the dataset from Microsoft Academic Graph. The system performance were evaluated based on the volunteers' assessments, which show that the proposed paper recommendation algorithms outperform the conventional search engine algorithm such as PageRank. The efficiency of the proposed algorithms were verified by the evaluation.

Acknowledgements. This work was partially supported by the National Natural Science Foundation of China (Grant Nos. 61373128, 91218302, 61321491), the Key Project of Jiangsu Research Program Grant (No. BE2013116), the EU FP7 IRSES MobileCloud Project (Grant No. 612212).

References

1. Google scholar. http://scholar.google.com
2. Microsoft academic. https://academic.microsoft.com/
3. Aminer. https://aminer.org/
4. McPherson, M., Smith-Lovin, L., Cook, J.M.: Birds of a feather: homophily in social networks. Ann. Rev. Sociol. **27**(1), 415–444 (2001)
5. Filippo, R., Claudio, C.: Defining and identifying communities in networks. Proc. Nat. Acad. Sci. U.S.A **101**(9), 2658–2663 (2004)
6. Girvan, M., Newman, M.E.: Community structure in social and biological networks. Proc. Nat. Acad. Sci. U.S.A **99**(12), 7821 (2002)
7. Darrin, E., Bo-June (Paul), H.: Microsoft academic graph. http://research.microsoft.com/en-us/projects/mag/
8. Newman, M.E.: Fast algorithm for detecting community structure in networks. Phys. Rev. E Stat. Nonlinear Soft Matter Phys. **69**(6), 066 133 (2004)
9. Newman, M.E.J.: Detecting community structure in networks. Eur. Phys. J. B Condens. Matter Complex Syst. **38**(2), 321–330 (2004)

10. Rosvall, M., Bergstrom, C.T.: Maps of random walks on complex networks reveal community structure. Proc. Nat. Acad. Sci. U.S.A **105**, 1118 (2008)
11. Bao, F., Licheng, J., Maoguo, G., Haifeng, D.: Memetic algorithm for community detection in networks (2011)
12. Usha Nandini, R., Reka, A., Soundar R T, K.: Near linear time algorithm to detect community structures in large-scale networks. Phys. Rev. E **76**(3), 036106 (2007)
13. Pagerank: U.S. Patent 6 285 999
14. Jon, K.: Authoritative sources in a hyperlinked environment. J. ACM **46**(5), 604–632 (1999)
15. Haveliwala, T.H.: Topic-sensitive pagerank. In: Proceedings of the 11th International Conference on World Wide Web (WWW 2002), pp. 517–526. ACM, New York (2002)
16. Freeman, L.C.: Centrality in social networks conceptual clarification. Soc. Netw. **1**(3), 215–239 (1978)
17. Hatcher, E., Gospodnetic, O.: Lucene in action (in action series) (2004)
18. Hearst, M.A.: Tilebars: visualization of term distribution information in full text information access. In: Sigchi Conference on Human Factors in Computing Systems, pp. 631 – 637 (1995)
19. Page, L.: The pagerank citation ranking: Bringing order to the web. In: Stanford InfoLab, pp. 1–14 (1998)

Improving Recommendation Accuracy for Travelers by Exploiting POI Correlations

Kai Zhang, Dapeng Zhao, and Xiaoling Wang$^{(\boxtimes)}$

Shanghai Key Laboratory of Trustworthy Computing,
Institute for Data Science and Engineering,
East China Normal University, Shanghai, China
xlwang@sei.ecnu.edu.cn

Abstract. Personalized point-of-interest (POI) recommendation is a challenging task in location-based-service (LBS). Previous efforts on POI recommendation mainly focus on local users. According to user's activity areas, e.g., home and workplace, nearby locations have higher probability to be recommended. However, in many practical scenarios such as urban tourism, target users are usually out-of-town travelers. Their preferences are hard to model due to sparse distributed check-ins. In this paper, we manage to improve the location recommendation accuracy for travelers, via finding correlations between different POIs. For cross-city POIs, the influence of travel intent (I), e.g., business trip and family trip, is studied. For local POIs, we focus on their geographical neighbors (N). In addition, reviews (R) are introduced to bridge the gap between distant POIs and make recommendation explainable. Incorporating these three factors into the learning of latent space, a novel matrix factorization approach (INRMF) is proposed. Further experiments conducted on real dataset show our approach is competitive against state-of-art works.

Keywords: POI recommendation · Matrix factorization · Rating prediction · Urban tourism

1 Introduction

In recent years, significant rise of world-wide urban tourism has brought substantial benefits to the economy. Locations like restaurants, shops and hotels are called point-of-interests (POIs). One classical task of urban tourism is to offer personalized POI recommendations for travelers. Assume there are a set of user $U = \{u_1, u_2, ..., u_m\}$ and a set of POIs $I = \{i_1, i_2, ..., i_n\}$, where m and n denote the number of users and POIs respectively. In this paper, we study the recommendation problem as a rating predicting issue. Rating refers to users' preference on locations, higher score indicates more satisfaction. Rating predicting problem can be formally defined as: Given a specific user u, predict the rating $\hat{r}_{ui}$ he/she may give to an unvisited POI i.

POI recommendation is an application-driven issue, the accuracy of which depends a lot on the specific scenario and the corresponding data. Most of previous studies are based on users' historical check-ins on location-based social

© Springer International Publishing Switzerland 2016
F. Li et al. (Eds.): APWeb 2016, Part II, LNCS 9932, pp. 137–149, 2016.
DOI: 10.1007/978-3-319-45817-5_11

network (LBSNs), such as Foursquare and Gowalla. In this case, users are found to move within some certain regions. We call those regions *"activity areas"*, such as home and workplace. Based on this observation, many works recommend local locations near users' activity areas. However, in urban tourism scenario, users' check-ins distributes in many cities, making it too discrete to find any activity areas for a specific user. Different from prior works, target users are travelers in our work. We collect POI check-in records from Ctrip.com, a trip planning website. In Ctrip dataset, a check-in record is consisted of four parts: a rating score, a review, a corresponding travel intent and check-in time, formally defined as $d_{u,i}$<*rating, review, intent, time*>. While check-in history of traveler is sparse, POI's data is rich. In this paper, we try to learn the POI recommendation issue from POI perspective.

Geographical characteristic is a direct attribute possessed by POI. POIs are associated with real world locations, making distance an important factor. It is common to find POIs gathering together in some regions, like business areas, in order to attract more interests and popularity. From user perspective, they often co-check or co-visit POIs nearby and make similar ratings. From POI perspective, neighbors usually share similar surrounding environment and facility, leading to some common attributes and functions. From these observations, we have an intuition that POIs might be correlated with their neighbors.

Besides direct features, there are also many implicit feedbacks on POIs from users, such as travel intent and reviews. Travel intent refers to the purpose of a trip. For example, some go to New York city on business while some go there for sightseeing. During a trip, intent serves as a key for decision-making. E.g., efficiency is the first priority during a business trip, while family trip concerns more about enjoyment [6]. From another perspective, locations might be correlated. We use intent to classify users. And according to each POI's visited history, which kind of travelers frequently and rare check-in, serves as an implicit feedback indicating POI's attributes. For example, restaurants frequently visited by families may share something in common, on the other hand, distinct from those commonly visited by couples. From this point, not only local POIs, but also cross-city POIs are correlated. Similar results can also be found according to reviews. The benefit of reviews is that they give an explicit explanation why specific POI is recommended, making result persuasive.

Through preliminary explorations on the influence of travel intent and geographical neighbors, we find that both of these two factors have positive effects on rating prediction. Main contributions of this paper are summarized as follows:

- We study the POI recommendation issue for travelers from POI perspective, and find the positive effect of travel intent and geographical neighbors.
- We propose a novel matrix factorization approach, that integrates travel intent, geographical neighbors and review contents together.
- We conduct experiments on real dataset and show that our model outperforms other baseline approaches.

The remainder is organized as follows: Sect. 2 presents related works. In Sect. 3, we study the influence of travel intent and geographical neighbors; In

Sect. 4, we firstly present the basic matrix factorization model, then incorporate travel intent, geographical neighbors and review into the latent factor model, and propose a novel matrix factorization approach. Section 5 shows the experimental setting and details of results. In the last section, we summarize this paper and list some possible directions for improvement.

2 Related Work

POI recommendation has attracted significant attention in the past few years [14, 15]. The main issue of which is to recommend unvisited POIs to users they may be interested. Collaborative filtering is a well adopted technique in recommender system [13]. However, directly apply collaborative filtering approach to conduct POI recommendation receives limited performance.

Geographical characteristic then becomes a major consideration for recommending POIs. It is based on the observation that users tend to visit POIs near their home or office locations, as well as nearby locations of their favored POIs. Currently, geographical influence have been fused with traditional CF algorithms. For example, [10] models the check-in probability to the distance of the whole check-in history by power-law distribution, [1] models users' multicenter check-in behaviors via multi-center Gaussians.

Geographical characteristic is effective for accuracy improvement in POI recommendation. But the limitation is that the target users and POIs have to be local users and local locations. Many works then start to conduct POI recommendation for out-of-town users, i.e. travelers. Literature [11] introduces reviews to connect distant locations in different cities and build a probabilistic model. [2] adopt the idea of collaborative filtering and take the advantage of social network to find users with similar preference. Literatures [9,12] integrate several factors into a probabilistic generative model to overcome the issues of travel locality and interest drift.

In this paper, we focus on recommending POIs for travelers and introduce a novel factor travel intent. To our best knowledge, this is the first work that integrate travel intent, geographical neighbors and reviews into the POI recommendation scenario.

3 Observations

In this section, we conduct data analysis to find the influence of travel intent and geographical neighbors on POI's rating.

3.1 Exploration of Travel Intent

Check-in type, i.e. travel intent distribution in each POI's check-in records refers to an implicit feedback. We call this feedback *travel intent feedback*, formally

Table 1. List of notations

U	Set of users
I	Set of POIs
D_i	Set of check-in records of POI i
C_i	Set of POIs in the same intent function group as POI i
N_i	Set of geographical neighbors of POI i
d_{ui}	User u's check-in record in POI i
	one record includes $<rating, review, intent, time>$
r_{ui}	Observed ratings of user u to POI i
$\hat{r}_{ui}$	Predicted ratings of user u to POI i
μ	Mean of all observed ratings
b_u	Bias factors of user u
b_i	Bias factors of POI i
p_u	Latent factors of user u
q_i	Latent factors of POI i
q_z	Latent factors of aspect z

defined as Eq. 1. Intuitively, travel intent divide users into several groups, and travel intent feedback indicates the degree a POI is favored by each group.

$$v_{i,k} = \frac{|\{d \in D_i | d.intent = k\}|}{|D_i|} \quad (1)$$

where D_i is the check-in set of POI i; k is the identification of each travel intent.

We have the intuition that POIs with similar travel intent feedback may have some similar attributes or functions, thus receive similar ratings. To confirm this assumption, we first cluster POIs with similar travel intent feedback into several groups, by K-means clustering algorithm. Since POIs in each group might share close functions, we name the clustered groups as *intent function groups*. Next, we calculate the Pearson correlation coefficient between each POI's rating with group's average rating it belongs to. POI's mean rating in history is used to represent its rating.

The result is showed in Fig. 1. Horizontal axis represents the count of subsets all POIs grouped into. We compare the POI subsets clustered by travel intent feedback with randomly partitioned subsets. The number following "Intent" and "Random" is a filtering parameter, e.g., 100 means we only consider those POIs with more than 100 check-in times. From the result, we can find that POIs are strongly correlated with their corresponding intent function group. When the number of subsets reach seven, the correlation coefficient reach a relatively high level and then becomes stable. In the meantime, with the rise of filtering parameter, the correlation coefficient becomes stronger. Which means the more sufficient data is, the clearer positive influence travel intent shows.

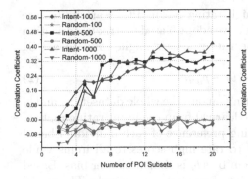

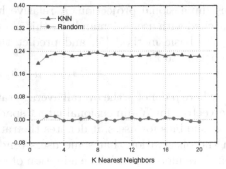

Fig. 1. Correlation between a POI's rating with average rating of the corresponding POI subset

Fig. 2. Correlation between a POI's rating with average rating of K nearest geographical neighbors

3.2   Exploration of Geographical Influence

Follow the exploration of travel intent in the last section, we also intend to study the influence of geographical neighbors. In detail, we calculate the Pearson correlation coefficient between each POI's rating with its K nearest neighbors' average rating.

The result is showed in Fig. 2. Horizontal axis represents the number of nearest neighbors considered, i.e. K. We compare the effect of geographical neighbors with randomly picked POIs. The exploration proves our assumption that POI is correlated with nearby neighbors. As the increase of K, correlation is relatively stable. Note that we set a maximal distance as two kilometers in this experiment. If the count of nearest neighbors is smaller than the requirement K within the distance threshold, we will lower K to include possible neighbors.

4   Our Approach: INRMF

In this section, we are intended to combine the prior knowledge into the bias matrix factorization, including travel intent (I), geographical characteristic (N) of POIs and reviews (R).

4.1   POI Recommendation by Matrix Factorization

Given a sparse rating matrix $\mathbf{R} \in \mathbb{R}^{m \times n}$, matrix factorization is an effective way to make item recommendation. Items in location recommendation scenarios are several point-of-interests (POIs). The primary idea of matrix factorization is to map users and POIs into a shared latent space with dimension $K \ll min(m, n)$. The latent space of a user indicates his/her demands on several aspects, while POI's latent space indicates its attributes on several aspects. A user's specific preference for a POI, i.e. rating, can then be modeled as the inner product of this dimensional projection. A relative high rating means POI's attributes are matched with user's demands. In this paper, we adopt biased matrix factorization as the basic method [5], and predict the rating a user might rank to a POI according to Eq. 2.

$$\hat{r}_{ui} = \mu + b_u + b_i + p_u^T q_i \tag{2}$$

where μ denotes the overall average rating among the whole dataset. p_u and q_i are vectors for K-dimensional latent space learned from the input rating matrix. b_u is the bias for users, it denotes the rating preference of each individual based on the observation that some users tend to rate high scores to all items while some are more critical. The adoption of item bias b_i is similar to user bias. Based on Eq. 2, the observed rating is factorized into four parts accordingly. The model parameters can be learned by minimizing the squared error function as showed in Eq. 3.

$$\min_{p*,q*,b*} \sum_{<u,i>\in S} (r_{ui} - \hat{r}_{ui})^2 + \lambda(\|p_u\|^2 + \|q_i\|^2$$
$$+ \|b_u\|^2 + \|b_i\|^2) \tag{3}$$

In this equation, parameter λ is a regularization used to avoid overfitting. $\|\cdot\|$ is the Frobenius norm.

4.2   Integration of Travel Intent

To recommend locations for travelers, travel intent serves a vital role. The exploration in Sect. 3.2 finds out that POIs in the same intent function group are correlated, i.e. they share similar attributes. Travel intent feedback is regardless of geographical distance, cross-city POIs can still be related. Here POI attributes can be quantified as several aspects, corresponding to the latent space learned by matrix factorization approach.

$$Sim_{i,j} = \frac{\sum_{k\in K} v_{i,k} \times v_{j,k}}{\sqrt{\sum_{k\in K} v_{i,k}^2} \times \sqrt{\sum_{k\in K} v_{j,k}^2}} \tag{4}$$

We incorporate travel intent feedback into the matrix factorization learning process from POI perspective. Let i be the target POI, q_i is the corresponding factorized latent factors, which represent i's attributes. We define C_i as a POI set, POIs of which are in the same intent function group as i. Since POIs in the same intent function group have many common attributes, we use the latent space of POIs in C_i to modify i's latent space. The influence of each is valued by the similarity metric defined in Eq. 4. The details of modification are shown in Eq. 5.

$$\hat{r}_{ui} = \mu + b_u + b_i + p_u^T((1-\alpha)q_i + \frac{\alpha}{|C_i|}\sum_{c\in C_i} sim(i,c)q_c) \tag{5}$$

where α is a weight parameter to control the influence of travel intent. $|C_i|$ serves as a normalization factor. Note that an alternative way for travel intent integration is to take all POIs' influence into consideration, which is very time-consuming. Adopt the intent function group that has already grouped highly related POIs together, is a better option.

4.3   Integration of Geographical Neighbors

Geographical characteristic distinguish POIs with common items in recommender system. Each POI is associated with a specific coordinate in the real world, and POIs nearby are called geographical neighbors. The exploration in Sect. 3.2 shows that geographical neighbors usually share similar POI functions, which has a positive effect on rating prediction. Geographical neighbors can be very effective for recommending local POIs.

Here we incorporate the influence of geographical neighbors into matrix factorization. Let $n \in N_i$ be the neighbors of the POI i. We use the latent space of POIs in N_i to modify i's original latent space. The corresponding predicted rating is then turned into Eq. 6.

$$\hat{r}_{ui} = \mu + b_u + b_i + p_u^{\mathrm{T}}((1 - \beta)q_i + \frac{\beta}{|N_i|} \sum_{n \in N_i} sig(i, n)q_n) \tag{6}$$

where β is a weight parameter that controls the geographical influence. $|N_i|$ serves as a normalization factor. $sig(i, j)$ measures the connection between one POI with its geographical neighbors, which is related with their distance, as shown in Eq. 7. Closer distance means tighter correlation.

$$sig(i, j) = \frac{2}{1 + exp(distance(i, j))} \tag{7}$$

4.4   Integration of Review Content

One problem of matrix factorization is that the latent factors are unexplainable, while review gives an explicit reason why users rate high/low scores. In the meantime, POI's attributes can also be indicated and correlated from review contents, regardless of their locations. For these reasons, reviews are adopted to improve the prediction accuracy and make the recommendation explainable.

We incorporate reviews into the latent factor model. More specifically, words are mapped into the same dimensional vector space as q_i. In another word, we use latent spaces of words in reviews to replace the original q_i. However, take all words into consideration is relatively time-consuming. It is wise to conduct aspect modeling before review integration, i.e. using a group of words to represent a specific aspect. [8] argues that frequency-based method shows better accuracy than topic model. In this paper, we use a combination of these two approaches. We firstly pick out top-K frequent nouns appear in the whole review corpus, representing important attributes. Then use word embedding technique, such as word2vec to generate a corresponding word vector for each word [3]. Based on word vectors, semantically approximate attributes can finally be clustered into the same aspect. Let $|Z|$ be the number of such aspects. The original latent factors of a POI is then decomposed into an integration of attribute space, as showed in Eq. 8.

$$q_i' = \frac{1}{\sum_{z \in Z} n_{iz}} \sum_{z \in Z} n_{iz} q_z \tag{8}$$

where z refers to an aspect, and n_{iz} denotes the count of words belong to aspect z in POI i's review documents.

4.5   INRMF Approach and Parameter Estimation

The final model INRMF is incorporated with aforementioned factors, i.e. travel intent (I), geographical neighbors (N) and reviews (R). The predicted rating is then modified as Eq. 9. To control the influence of intent function group and geographical neighbors, we set a constraint as $\alpha + \beta = 1$.

$$\hat{r}_{ui} = \mu + b_u + b_i + p_u^T((1 - \alpha - \beta)q_i'$$
$$+ \frac{\alpha}{|C_i|} \sum_{c \in C_i} sim(i, c)q_c' + \frac{\beta}{|N_i|} \sum_{n \in N_i} sig(i, n)q_n') \tag{9}$$

As for the corresponding error function, the regularization term of POI's latent factors are replaced by the new regularization term of aspects' latent space. We have the final error function following Eq. 10, which can be optimized by stochastic gradient descent (SGD) [4].

$$\min_{p*, q*, b*} \sum_{<u,i> \in S} (r_{ui} - \hat{r}_{ui})^2$$
$$+ \lambda(\|p_u\|^2 + \sum_{z \in Z} n_{iz} \|q_z\|^2 + \|b_u\|^2 + \|b_i\|^2) \tag{10}$$

5   Experiments

We evaluate the performance of INRMF in this section. All experiments are conducted on Windows platform with Intel Core i5-3470 Processor (3.20 GHz) and 8 GB memory.

5.1   Datasets

The dataset studied in this paper is crawled from Ctrip.com, a popular trip advisor website. The type of POI includes tourist spot, restaurant, shop and hotel. We collect check-in records in three different cities from July 2013 to July 2015. Each record contains POI's id, user's id, travel intent, review and corresponding time. The purpose of trip is pre-labeled into five groups, that is single trip, couple trip, group trip, family trip and business trip. Such data schema is also adopted by other famous travelling websites such as Booking and TripAdvisor. We also obtain some basic information for each POI such as longitude and latitude. Details of statistics are summarized in Table 1. POIs with less than 500 visit times has been filtered. Notice that the average check-ins of users is extreme sparse while POIs' average check-ins is rich enough. It is reasonable to conduct recommendation from POI perspective rather than user perspective (Table 2).

Table 2. Statistics of dataset

	Shanghai	Guangzhou	Shenzhen
#POIs	4,153	3,755	3,046
#users	499,474	290,618	234,214
#check-ins	2,470,658	1,134,802	832,609
Avg. #users per POI	594.91	302.21	273.35
Avg. #POIs per user	4.95	3.91	3.55

5.2 Performance Comparison

In the experiments, the compared approaches include:

- **GlobalAvg**: This method predicts unknown ratings as the average of all observed ratings.
- **UserKNN**: This is the user-based collaborative filtering, which predict ratings according to similar users.
- **ItemKNN**: This is the item-based collaborative filtering, which predict ratings according to similar items. Items refer to POIs in this paper.
- **SVD++**: SVD++ is a model-based collaborative filtering approach, which considers the implicit feedback of history data [4].
- **BiasedMF**: Biased MF is the baseline method in our paper, as illustrated in Sect. 3.1.
- **IRenMF**: This approach incorporates geographical information into the matrix factorization, and studies the geographical influence from two directions: neighbors and regions [7].

Our proposed methods incorporate three different prior information, including travel intent (I), geographical neighbors (N) and reviews (R).

Table 3. Comparison of all methods

Method	MAE	RMSE
GlobalAvg	0.69507	0.87444
UserKNN	0.70017	0.89753
ItemKNN	0.68239	0.90553
SVD++	0.67107	0.87030
BiasedMF	0.65069	0.84791
IRenMF	0.65092	0.84392
I-MF	0.64488	0.84324
IN-MF	0.64327	0.84239
INR-MF	0.64201	0.84215

- **I-MF:** This is our model that only incorporates travel intent.
- **IN-MF:** This approach incorporates both travel intent and geographical neighbors.
- **INR-MF:** This approach incorporates all of the three factors.

Since the target users are out-of-town travelers in this paper, we treat the data of three cities as a whole and evaluate the overall recommendation accuracy. 5-fold cross validation is applied, in each fold, 80 % of data are used for training and the remaining 20 % are used for test. The accuracy evaluation metrics adopted are Mean Absolute Error (MAE) and Root Mean Square Error (RMSE). Smaller MAE or RMSE means better rating prediction accuracy. For model-based collaborative filtering approaches, learning rate γ is set to 0.08 and regularization parameter λ is set to 0.02.

From experiment results showed in Table 3, we can find that all model-based collaborative filtering methods outperform memory-based ones. Item-based CF achieves better performance than User-based CF, because users' check-in data is much sparser than POIs'. Among all the baseline methods, BiasedMF achieves the best accuracy, both on MAE and RMSE. IRenMF integrates geographical characteristic and reduces the RMSE. Our proposed approaches named as I-MF, IN-MF, INR-MF all perform better than other approaches. Notice that the effect of reviews is lower than travel intent and geographical neighbors.

5.3 Parameter Tuning

The impact of travel intent and geographical neighbors are controlled by α and β respectively. Larger α and β means stronger influence. We conduct further experiments to select appropriate parameters that achieves the best prediction accuracy. Figure 3. shows different result while choosing various α. We can find that neither small nor large α performs well. Which indicates the attributes of POI itself and the influence of travel intent are both important. On the other

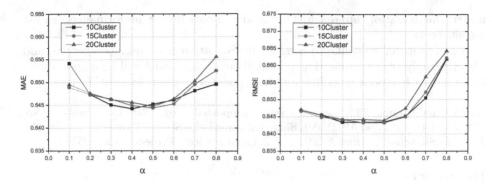

Fig. 3. Parameter tuning of travel intent

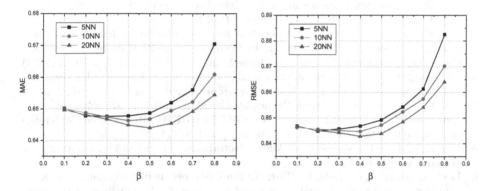

Fig. 4. Parameter tuning of geographical neighbors

hand, different number of clusters shows little affect. We finally set α to 0.4 and group POIs into 10 groups.

Figure 4 shows the effect of different β. We also want to find an appropriate K, i.e. number of geographical neighbors. The conclusion is similar with travel intent, neither small nor large β behaves well. Indicating the attributes of POI itself and the influence of geographical neighbors are both important. We set β and K as 0.4 and 20 respectively. Note that there is also a balance between travel intent and geographical neighbors. In this paper, we treat them with the equal importance, the final setting of α and β are 0.2 and 0.2.

6   Conclusions and Future Work

In this paper, we study the POI recommendation issue for travelers and focus on the POI perspective. We find that through travel intent and geographical neighbors, POIs can be correlated. Adopting the idea of latent factor model, we model the correlation as similar attributes, corresponding to the latent space. Incorporating travel intent, geographical neighbors and reviews, we propose a novel matrix factorization approach. Our model shows better performance against recent works.

In the future, we'd like to conduct investigation from user perspective and directly apply travel intent. The probabilistic graph model will be a good direction for user behavior modeling and incorporating various kinds of factors.

Acknowledgment. This work was supported by NSFC grants (No. 61472141U1501252 and 61021004), Shanghai Knowledge Service Platform Project (No. ZF1213) Shanghai Leading Academic Discipline Project (Project NumberB412).

References

1. Cheng, C., Yang, H., King, I., Lyu, M.R.: Fused matrix factorization with geographical and social influence in location-based social networks. In: Twenty-Sixth AAAI Conference on Artificial Intelligence (2012)
2. Ference, G., Ye, M., Lee, W.C.: Location recommendation for out-of-town users in location-based social networks. In: Proceedings of the 22nd ACM International Conference on Information and Knowledge Management, pp. 721–726. ACM (2013)
3. Goldberg, Y., Levy, O.: Word2vec explained: deriving Mikolov et al.'s negative-sampling word-embedding method. arXiv preprint (2014). arXiv:1402.3722
4. Koren, Y.: Factorization meets the neighborhood: a multifaceted collaborative filtering model. In: Proceedings of the 14th ACM SIGKDD International Conference on Knowledge Discovery and Data Mining, pp. 426–434. ACM (2008)
5. Koren, Y., Bell, R., Volinsky, C.: Matrix factorization techniques for recommender systems. Computer **8**, 30–37 (2009)
6. Levi, A., Mokryn, O., Diot, C., Taft, N.: Finding a needle in a haystack of reviews: cold start context-based hotel recommender system. In: Proceedings of the Sixth ACM Conference on Recommender Systems, pp. 115–122. ACM (2012)
7. Liu, Y., Wei, W., Sun, A., Miao, C.: Exploiting geographical neighborhood characteristics for location recommendation. In: Proceedings of the 23rd ACM International Conference on Conference on Information and Knowledge Management, pp. 739–748. ACM (2014)
8. Musat, C.C., Liang, Y., Faltings, B.: Recommendation using textual opinions. In: Proceedings of the Twenty-Third International Joint Conference on Artificial Intelligence, pp. 2684–2690. AAAI Press (2013)
9. Wang, W., Yin, H., Chen, L., Sun, Y., Sadiq, S., Zhou, X.: Geo-SAGE: a geographical sparse additive generative model for spatial item recommendation. In: Proceedings of the 21th ACM SIGKDD International Conference on Knowledge Discovery and Data Mining, pp. 1255–1264. ACM (2015)
10. Ye, M., Yin, P., Lee, W.C., Lee, D.L.: Exploiting geographical influence for collaborative point-of-interest recommendation. In: Proceedings of the 34th International ACM SIGIR Conference on Research and Development in Information Retrieval, pp. 325–334. ACM (2011)
11. Yin, H., Sun, Y., Cui, B., Hu, Z., Chen, L.: LCARS: a location-content-aware recommender system. In: Proceedings of the 19th ACM SIGKDD International Conference on Knowledge Discovery and Data Mining, pp. 221–229. ACM (2013)
12. Yin, H., Zhou, X., Shao, Y., Wang, H., Sadiq, S.: Joint modeling of user check-in behaviors for point-of-interest recommendation. In: Proceedings of the 24th ACM International on Conference on Information and Knowledge Management, pp. 1631–1640. ACM (2015)

13. Zhang, J.D., Chow, C.Y., Li, Y.: LORE: exploiting sequential influence for location recommendations. In: Proceedings of the 22nd ACM SIGSPATIAL International Conference on Advances in Geographic Information Systems, pp. 103–112. ACM (2014)

14. Zhang, J.D., Ghinita, G., Chow, C.Y.: Differentially private location recommendations in geosocial networks. In: 2014 IEEE 15th International Conference on Mobile Data Management (MDM), vol. 1, pp. 59–68. IEEE (2014)

15. Zhou, D., Wang, B., Rahimi, S.M., Wang, X.: A study of recommending locations on location-based social network by collaborative filtering. In: Kosseim, L., Inkpen, D. (eds.) Canadian AI 2012. LNCS, vol. 7310, pp. 255–266. Springer, Heidelberg (2012)

A Context-Aware Method for Top-k Recommendation in Smart TV

Peng Liu[1], Jun Ma[1(✉)], Yongjin Wang[2], Lintao Ma[2], and Shanshan Huang[2]

[1] School of Computer Science and Technology, Shandong University, Jinan, China
majun@sdu.edu.cn
[2] National Key Laboratory, Hisense Co., Ltd., Qingdao, China

Abstract. We discuss the video recommendation for smart TV, an increasingly popular media service that provides online videos by TV sets. We propose an effective video recommendation model for smart TV service (RSTV) based on the developed Latent Dirichlet allocation(LDA) to make personalized top-k video recommendation. In addition, we present proper solutions for some critical problems of the smart TV recommender system, such as sparsity problem and contextual computing. Our analysis is conducted using a real world dataset gathered from Hisense smart TV platform, JuHaoKan Video-on-Demand dataset(JHKVoD), which is an implicit watch-log dataset collecting sets of videos watched by each user with their corresponding timestamps. We fully portray our dataset in many respects, and provide details on the experimentation and evaluation framework. Result shows that RSTV performs better comparing to many other baselines. We analyse the influence of some of the parameters as well as the contextual granularity.

Keywords: Context-aware recommendation system · Smart TV

1 Introduction

Smart TV is an increasingly popular television service which provides various multimedia contents such as television programs, movies, news, musics and so on. Different from traditional one-way television services, smart TV, as a new-type and burgeoning media, greatly enriches the user interaction with the TV system by providing the interactive video-on-demand(VoD) service. So, recommendation system, as a booster providing appropriate suggestions of what these users might like, is becoming an indispensable part to the smart TV system.

Smart TV allows the collection of implicit user preferences, such as user's play-list and behavior log, for making a personalized recommendation. So video recommender system can automatically predict and recommend programs for user's personal tastes by analyzing other user patterns (Collaborative Filtering, CF) or clustering items by their content (Content-based Recommendations, CB). These two basic strategies are both well developed, and many other derived models are constructed on them [4].

© Springer International Publishing Switzerland 2016
F. Li et al. (Eds.): APWeb 2016, Part II, LNCS 9932, pp. 150–161, 2016.
DOI: 10.1007/978-3-319-45817-5_12

Despite the remarkable performance and the recommendation novelty, traditional CF recommendation systems suffer from the data sparsity problem. Because with the growth of the user (and item) scale, the overlap between two users choices become very few. Furthermore, smart TV system, unlike those online video sites such as *Youtube* or *FileTrust* [3], has hardly any social relations between users, which makes it impracticable to expand user data with their social trusts.

On the other hand, for context-aware recommender system, additional contextual information, basically temporal context in smart TV system, makes the data more sparse with much less associations among user feedbacks [5].

However, contextual information obviously benefits the smart TV recommendation system. It would enhance the efficiency of RS, if related items are recommended at higher rank and improper items are filtered out under specific period of time [7]. For example, if a user has an inclination for both drama and horror movies, but he only watch horror movies in the daytime. It would be extremely inappropriate to recommend the film *Silent Hill* to him at midnight.

In addition, context-aware approach provides a solution to another serious problem smart TV suffered from, shared-account problem [10]. In smart TV system, a TV client is generally used by not a single user but a group of people such as families. The compounded user interests recorded in each account cripple the traditional recommender system and bring forward the demand of taking into account additional contextual information, in order to separate the mixed user preferences apart. Under the basic assumption that different user in one account follow different time patterns, time-aware approach is an effective and state-of-art solution [10]. Therefore, contextual computing is of vital importance to the smart TV recommender system.

So, with the propose of solving the two problems mentioned above, we design a new context-aware recommendation model for smart TV, called RSTV. Our RSTV model inputs *JHKVoD* dataset, a randomly selected subset of 70 days watch-log dataset, gathered from *JuHaoKan* TV service of *Hisense Cloud Platform*, consisting of 3643 users' watching habits over 3849 movie items. Then, it builds up a developed topic model based on Latent Dirichlet Allocation [2] (LDA) to statistically learn the distribution of user interests from watching history. Recommendation are made based on associations between users past shown interests, and context-aware post-filtering method guarantees that user current requirements are satisfied.

This work is a cooperation research with National Key Laboratory, Hisense Co., Ltd.[1], and all the used data is provided by *Hisense Cloud Platform*[2] which is a large TV service provider keeping more than 18 million users in China.

The rest of this paper is organized as follows. Section 2 describes the video recommendation method. Section 3 presents the evaluation metrics and the corresponding experiments of RSTV and baselines. Section 4 makes conclusion and introduces our future work.

---

[1] http://www.hisense.com/.

[2] http://cloud.hisense.com/.

2   Model

RSTV estimates user interests with a developed topic model based on the video co-occurrence in users' watching lists. After deducing the user interest distributions, we enforce the user-based K-Nearest-Neighbor CF method. Those who have strong associations share their watching list according to their relevancy. Then we apply weight post-filtering to contextualize the top-N recommendation results. The recommended videos will be reordered by weighting its probability with the relevance of the specific context. The task we aim at is to recommend a k-size video list to user considering his current context, such that user will be more likely to watch the recommended videos.

2.1   Preliminaries

Table 1 shows the main notations of our model.

Table 1. Table of notations in RSTV

U, V, C	Set of user, video, context
b, B_u	A video-pair, video-pairs extracted from user u
K	Number of topics
α, β	Dirichlet priors
ϕ, θ	Topic-video distribution and user-topic distribution
$z_{u,b}$	A topic assignment on video-pair b for user u
$n_{z,u}$	Number of video-pairs of user u assigned to topic z
$n_{v,z}$	Number of times of video v assigned to topic z

The domains we consider consist of a set of users U, a set of videos V and a set of temporal contexts C. In RSTV, there are only two kinds of temporal context, $C = \langle C_w, C_d \rangle$. $C_w = \{Workdays, Weekends\}$ is a partition of a week, and $C_d = \{Day, Night\}$ is a partition of a day which uses 06:00 a.m. and 18:00 p.m. as dividing points.

Observations are available for user-context-item triads $\langle u, c, v \rangle$, where $u \in U$, $v \in V$, and $c = \langle c_w, c_d \rangle$, $c \in C$. For each user u, we collect his watching list $L_u = \{\langle u, c_1, v_1 \rangle, ..., \langle u, c_x, v_x \rangle\}$ without duplicate removal. Each watching list is regarded as a document, and each triad as a term. Therefore, semantic topic model in text domain such as LDA could be applied to analyze user latent interests in the recommendation system.

Let z be a topic(interest) and K be the number of topics(interests). We extract a list of video pairs B_u for each user from his watching list L_u, and $b \in B_u$ represents a video pair. Then we use all users' video pairs B as input to train the interest distribution for each user in our model, with pre-set parameters α and β representing the Dirichlet priors. $z_{u,b}$, $n_{z,u}$, $n_{v,z}$ respectively indicating the topic assignment on video-pair b for user u, the number of video-pairs assigned to topic z within B_u and the number of times of video v assigned to topic z.

2.2 Preprocess: Video-Pairs Extraction

Traditional topic model detects topic distribution based on the word co-occurrence in the documents, and topics are expressed as groups of correlated words. However, the length of user watching list(document) in our JHKVoD dataset is often short and most of them contain less than 50 videos in total (see Sect. 3.1), or even less if we consider the contextual dimensions. The data sparsity caused by lacking document-level video co-occurrence patterns will lead to bad performance if we directly enforce the sampling method on the documents.

Therefore, inspired by [11], we learn the topics by directly modeling the generation of video co-occurrence patterns instead of individual videos. Specifically, we obtain each user's watching list $L_u = \{\langle u, c_1, v_1 \rangle, ..., \langle u, c_x, v_x \rangle\}$ and then extract video-pairs as follows:

$$B_u = \{(v_i, v_j) | c_i = c_j, \langle u, c_i, v_i \rangle, \langle u, c_j, v_j \rangle \in L_u, i \neq j\} \qquad (1)$$

Each video-pair $b \in B$ contains two videos (v_i, v_j) from the same context.

So, In RSTV, topic assignment z will be sampled for each video-pair instead of each independent video. It explicitly models the video co-occurrence patterns and avoid the damage of sparsity problems.

2.3 RSTV

RSTV aims to discover the user interest, which could be represented as a mixture of topics. Each topic can be described as a multinomial distribution over videos, illustrating a latent factor of its relevant videos. Each user interest also can be described as a multinomial distribution over topics, indicating user bias on the corresponding factors. Each video-pair is respectively drawn from a specific topic. The probability of a video-pair drawing from a topic is decided by the probability that both videos in this video-pair are drawn from this topic.

RSTV captures the user's interest distribution(topic distribution) as θ_u and video distribution for each interest(topic) as ϕ_z. So the generative model can be described as follows:

(1) For each topic z, draw a topic-video distribution $\phi_z \sim Dir(\beta)$;
(2) For each user u, draw a user-topic distribution $\theta_u \sim Dir(\alpha)$;
(3) For each video-pair b in each user's video-pair list B_u:
    (i) draw a topic assignment $z_{u,b} \sim Multi(\theta_u)$;
    (ii) draw two videos $v_i, v_j \sim Multi(\phi_z), b = (v_i, v_j)$.

Figure 1 illustrates the graphical representation of the RSTV model.

Given a topic z, the probability of generating a video-pair $b = (v_i, v_j)$ can be written as:

$$P(b|z) = P(v_i|z)P(v_j|z) \qquad (2)$$

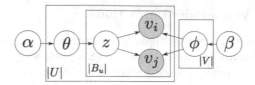

Fig. 1. Graphical representation of the developed topic model in RSTV.

and the joint probability of a video-pair b is represented as:

$$P(b) = \sum_z P(z)P(v_i|z)P(v_j|z)$$
$$= \sum_z \theta_z \phi_{z,v_i} \phi_{z,v_j}. \tag{3}$$

Contextual Computation. Although we extract the video-pairs according to the user context, RSTV does not directly model the process of context generation. Therefore, the user's topic distribution of a specific context cannot be directly achieved during the topic learning process.

To infer the user topics in a certain context, we consider that the contextual topic distribution is generally reflected by all the videos that appear in that context. So we use the expectation of these videos to calculate the contextual topic distribution:

$$P(z|c, u) = \sum_v P(z|v)P(v|c, u). \tag{4}$$

$P(z|v)$ can be calculated by Bayes' formula based on the parameters estimated in Model:

$$P(z|v) = \frac{P(z)P(v|z)}{\sum_{z'} P(z')P(v|z')}, \tag{5}$$

where $P(z) = \theta_z$ and $P(v|z) = \phi_{z,v}$. And we estimate the probability $P(v|c, u)$ as:

$$P(v|c, u) = \frac{n_{c,u}(v)}{\sum_c n_{c,u}(v)}, \tag{6}$$

where $n_{c,u}(v)$ is the frequency of user u choosing video v in the context c.

2.4   Parameter Estimation

We adopt Gibbs sampling methods to approximately infer the parameters θ, ϕ in our developed topic model. Gibbs sampling is a simple and widely used Markov chain Monte Carlo algorithm. It approaches the target distribution iteratively and asymptotically, thus it avoids falling into local optimum like variational inference. Besides, Gibbs sampling method demands only to maintain a few counters and state statistics, which is memory saving and very practicable to large dataset.

We adopt the Dirichlet distribution, which is a conjugate distribution to multinomial distribution. Therefore, with the Dirichlet conjugate priors $\boldsymbol{\alpha}$ and $\boldsymbol{\beta}$, we can simply integrate out the $\boldsymbol{\phi}$ and $\boldsymbol{\theta}$ without sampling them. We only have to sample the topic assignment $z_{u,b}$ for each video-pair $b \in \boldsymbol{B}_u$ from its conditional distribution given the remaining variables.

To implement the Gibbs sampling, we start at a random state for Markov chain. Then we calculate the conditional distribution $P(z|\boldsymbol{z}_{-b}, B_u, \alpha, \beta)$ as:

$$P(z|\boldsymbol{z}_{-b}, B_u, \alpha, \beta) \propto (n_{z,u} + \alpha) \cdot \frac{(n_{v_i,z} + \beta)(n_{v_j,z} + \beta)}{(\sum_v n_{v,z} + M\beta)^2}, \qquad (7)$$

where $n_{z,u}$ is the number of times of the video-pair b from B_u assigned to the topic z, and $n_{v,z}$ is the number of times of the video v assigned to the topic z. M is the total number of videos in the dataset. As is customary in LDA topic model, α and β are both symmetric Dirichlet priors. At Last, we easily estimate the topic distributions $\boldsymbol{\theta}$ and the video distributions $\boldsymbol{\phi}$ as:

$$\phi_{v,z} = \frac{n_{v,z} + \beta}{\sum_{v'} n_{v',z} + M\beta}, \qquad (8)$$

$$\theta_{z,u} = \frac{n_{z,u} + \alpha}{\sum_{z'} n_{z',u} + K\alpha}. \qquad (9)$$

The proposed Gibbs sampling algorithm is shown in Algorithm 1.

Algorithm 1. Gibbs sampling algorithm for the topic model in RSTV.

Input:
    the number of topics K
    hyper-parameters α and β
    video-pair set $B = \{B_u | u \in U\}$
    the number of iterations N_{iter}
Output:
    multinomial parameters ϕ and θ
1: initialize topic assignments z and the counters $n_{v,z}$, $n_{z,u}$
2: **for** $iter = 1$ to N_{iter} **do**
3:    **for** each user $u \in U$ **do**
4:       **for** each video-pair $b \in B_u$ **do**
5:          draw topic z from $P(z|z_{-b}, B, \alpha, \beta)$ in Eq. 7
6:          update $n_{z,u}$, $n_{v_i,z}$, $n_{v_j,z}$, while $b = (v_i, v_j)$
7:       **end for**
8:    **end for**
9: **end for**
10: compute the parameters ϕ in Eq. 8 and θ in Eq. 9
11: **return** ϕ and θ

2.5   Recommendation Strategy

Typically, recommender systems find the top-k recommendations for a user u by first computing the recommendation scores $s(u,v)$ for every candidate recommendation v and afterwards selecting the k recommendations v for which $s(u,i)$ is the highest.

Here, since our RSTV has already inferred out the distributions of user interests, we adopt the user-based collaborative filtering as our recommendation strategy. These user-based recommender systems are rooted in the intuition that users with similar watching habits tend to pick items from each other's preferred lists. Thus, for a target user u, this methods first find $KNN(u)$, the most similar users to u, by using a similarity measure $sim(u,u')$, where $u' \in KNN(u)$. Next, for each neighbor $u' \in KNN(u)$, it increases the recommendation score of every video watched by u' with the similarity score $sim(u,u')$. So the user-based recommendation score of a candidate recommendation v for user u is given by:

$$s(u,v) = \sum_{u' \in KNN(u)} sim(u,u') \cdot |I(u') \cap \{v\}|, \tag{10}$$

where $I(u)$ is the set of videos watched by u.

A typical choice for $sim(u,u')$ in Eq. 10 is the cosine similarity:

$$sim(u,u') = cosine(u,u') = \frac{\theta_u \cdot \theta_{u'}}{\|\theta_u\| \|\theta_{u'}\|}, \tag{11}$$

where θ_u is the topic distribution of user u estimated from Eq. 9.

In RSTV, we enforce the weight post-filtering approach to make contextual computation, which reorders the recommended videos by weighting the recommendation score $s(u,v)$ with the probability of video's relevance with the specific context. Therefore, the contextualized recommendation score computes as:

$$s_c(u,v) = s(u,v) \cdot w_c(u,v). \tag{12}$$

$w_c(u,v)$ is a weight factor indicating how likely user u will watch the video v in context c. It is defined as:

$$w_c(u,v) = 1 + \sum_z p(z|c,u)p(z|v), \tag{13}$$

where $p(z|c,u)$ is defined in Eq. 4 and $p(z|v)$ is:

$$p(z|v) = \frac{\phi_{v,z}}{\sum_{z'} \phi_{v,z'}}. \tag{14}$$

After completing weight all the recommendation scores $s_c(u,v)$ for each video given c and u, we rank them by descending order and recommend the top-k videos.

3   Experiment

3.1   Experimental Setup

Dataset. *JHKVoD* is a real-world watching-log dataset gathered from *JuHanKan* video-on-Demand service of Hisense smart TV system. All the data in *JHKVoD* was anonymized and collected in strict accordance with user privacy policies.

JHKVoD contains a randomly selected subset of 70 days watching-log from end July 2015 to September 2015. We focus on the movie data, and filter out inactive users whose records are less than 10 during 70 days. We randomly selected 1 % of all the user history data, which contains 109727 records from 3643 users demanding a set of 3849 videos and the sparsity of *JHKVoD* reaches 99.2 %. All the user and videos are denoted by their serial numbers.

JHKVoD dataset is a typical implicit feedback dataset. Each watching record consists of exact one user, one video and two timestamps indicating the start-time and the end-time. The duration of user watching the video thus can be captured. Figure 2 shows the video distributions for *JHKVoD* dataset under different context. We can see that user tends to watch more videos on weekends during a week (Fig. 2a) as well as afternoon and night during a day (Fig. 2b). Most of the users watch less than 50 videos in total (Fig. 2c).

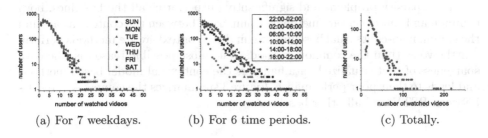

(a) For 7 weekdays.   (b) For 6 time periods.   (c) Totally.

Fig. 2. Video distributions for *JHKVoD* dataset

Note that there is no rating data available. However, in order to meet the demand of our baseline model, that rating data is needed as input for some of our baselines, we use the method mentioned in [4] to transform the time duration into the explicit rating.

We randomly split each dataset into five subsets, and use 5-fold cross validation to conduct training and testing on our model. We use the averaged evaluation score for each criterion as our final performance value.

Evaluation Metrics. For context-aware video recommendation task, we are given a context c for a user u and compute $s_c(u, v)$ for all $v \in V$ and rank them in decreasing order of $s_c(u, v)$. The evaluation metrics aim to access the positions of the right videos in the ranking list, noted as $S_{u,c}$, for each given user and context. In this study, we adopt four common metrics: *Recall@N, AUC, MAP, MRR.*

Baselines. For the common recommendation baselines, we introduce *PMF* [9], *BPFM* [8], *SVD++* [6] and *UserKNN(UCF)* methods. First three of them are well performed model-based CF methods, and the last one, *UCF*, is a widely used memory-based CF approach. For the context-aware recommendation baselines, we introduce *Tensor Factorization* [5](*TF*), *CAMF* [1], *CSLIM* [13] and *User-Splitting* [12](*Usplit-UCF*). *Usplit-UCF* is pre-filtering method, it uses general *UCF* method after splitting users according to the context.

Note that all the above baselines are computed using the open-source context-aware recommendation platform, CARSKit [14], with default settings.

3.2 Recommendation Results for RSTV

We conducted experiments on the JHKVoD dataset. For all experiments, the size of K-Nearest-Neighbor method was set to 10, and the number of recommendations was set to 10. All the baseline method were run under the default configuration of CARSKit platform. As for our model RSTV, the number of topics was set to 8.

We can see from the results that our proposed model, RSTV, overcomes the data sparsity problems and significantly outperforms all the baselines, both common and context-aware methods. Comparing between the context-aware and the common un-contextual baselines, we find the context-aware methods perform a little worse than the common methods. The most possible reason is, the data sparseness of contextualized data hampers the contextual model-based methods and leads to their poor performance. Figure 3 summarizes the comparison results between RSTV and all other baselines.

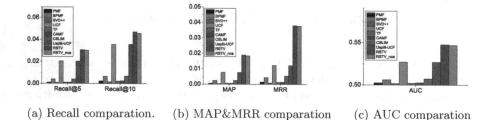

(a) Recall comparation.   (b) MAP&MRR comparation   (c) AUC comparation

Fig. 3. Evaluation results for RSTV and other baselines on *JHKVoD* dataset.

We also present the evaluation result for our RSTV model without the post-filtering, RSTV-nca, which is not context-aware. We can see that RSTV-nca recommendation result decreases a little bit in all evaluation criteria while removing the contextual information. It shows that contextual information contributes to the smart TV recommendation when using post-filtering method. These comparative results are consistent with the previous study of pre- and post-filtering approaches [7] that post-filtering method is better choice when common uncontextual approach beats the pre-filtering one.

Parameter Analysis: Number of Topic K. We vary the number of topic K from 3 to 100 to capture its influence on our RSTV performance. All the other parameters are fixed. Figure 4(a) plots the performance of RSTV. It shows that, when K is beneath 30, all the evaluation values go up along with the increase of parameter K, but the performance decreases gradually when K goes beyond 30. The most possible explanation seems to be that the proper value of parameter K leads to a better division of topics, and values beyond that can introduce noise into the pattern.

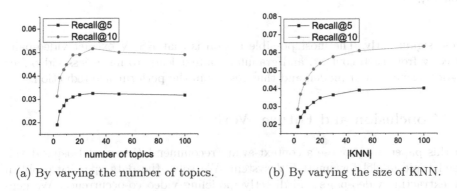

(a) By varying the number of topics.   (b) By varying the size of KNN.

Fig. 4. Parameter analysis: recall comparison

Parameter Analysis: Size of KNN. We vary the size of $KNN(u)$ to evaluate its effect on RSTV, and the number of topics is fixed to 8. Note that, the more $KNN(u)$ contains, the more flexibility the model has. Figure 4(b) illustrates its influence on RSTV. When $|KNN(u)|$ is under 20, performance of RSTV ascends sharply as its size goes up. And when $|KNN(u)|$ goes greater than 20, the growth of RSTV performance becomes slow.

Context Granularity Analysis. The granularity of contextual information is among the most important issues affecting the performance of context-aware recommender systems. Here we present the experiment results that conducted on four different granularity by varying the partitions of weekday and daytime context. We split the weekdays by either 7 parts for each day or 2 parts for Workdays and Weekends, and equally split the daytime by either 6 parts beginning at 02:00 a.m. or 2 parts for Day and Night. Figure 5 shows the Recall@10 and AUC results. It can be seen that for all the methods, their performance degraded as the contextual information goes granular, and the AUC score for RSTV drops

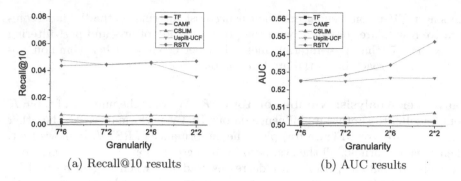

(a) Recall@10 results   (b) AUC results

Fig. 5. Results comparison by varying the granularity of contextual information. Term *7*6* for the granularity-coordinate means that a week is split into 7 parts and a day is equally split into 6 parts (beginning at 02:00 a.m.). The other terms are defined similarly.

most significantly. The most possible reason is that RSTV extract video-pairs directly from each context, and granular context leads to much less video-pair co-occurrence in our model, and thus results in the performance reduction.

4   Conclusion and Future Work

In this paper, we propose a context-aware recommendation model using developed topic model for smart TV system. We solve the data sparsity problem by extracting video-pairs and directly modeling video co-occurrence. We generate the user interest as a distribution over topics, and give the probabilistic representation of each context. Post-filtering method are used to reorder the recommendation list such that recommended videos may be more suitable for the current circumstance. We conduct our experiments on real smart TV dataset. Results show that our method significantly outperforms the other baselines in Recall, AUC, MAP and MRR. In the future, we plan to find a more precise method of probabilistically representing the context information. Besides, with the additional video information such as genre, we aim to further study the variation of user interests over time, and then figure out a way to automatically generate the user profiles.

Acknowledgements. This work was supported by National Key Laboratory, Hisense Co., Ltd., and Natural Science Foundation of China (61272240, 71402083).

References

1. Baltrunas, L., Ludwig, B., Ricci, F.: Matrix factorization techniques for context aware recommendation. In: Proceedings of the Fifth ACM Conference on Recommender Systems, pp. 301–304. ACM, New York (2011)

2. Blei, D.M., Ng, A.Y., Jordan, M.I.: Latent Dirichlet allocation. J. Mach. Learn. Res. **3**, 993–1022 (2003)
3. Golbeck, J.: Generating predictive movie recommendations from trust in social networks. In: Stølen, K., Winsborough, W.H., Martinelli, F., Massacci, F. (eds.) iTrust 2006. LNCS, vol. 3986, pp. 93–104. Springer, Heidelberg (2006)
4. Hu, Y., Koren, Y., Volinsky, C.: Collaborative filtering for implicit feedback datasets. In: 2008 Eighth IEEE International Conference on Data Mining, ICDM 2008, pp. 263–272 (2008)
5. Karatzoglou, A., Amatriain, X., Baltrunas, L., Oliver, N., Multiverse recommendation: n-dimensional tensor factorization for context-aware collaborative filtering. In: Proceedings of the Fourth ACM Conference on Recommender Systems, pp. 79–86. ACM (2010)
6. Koren, Y.: Factorization meets the neighborhood: a multifaceted collaborative filtering model. In: Proceedings of the 14th ACM SIGKDD International Conference on Knowledge Discovery and Data Mining, pp. 426–434. ACM (2008)
7. Panniello, U., Tuzhilin, A., Gorgoglione, M., Palmisano, C., Pedone, A.: Experimental comparison of pre- vs. post-filtering approaches in context-aware recommender systems. In: ACM Conference on Recommender Systems, pp. 265–268 (2009)
8. Salakhutdinov, R., Mnih, A.: Bayesian probabilistic matrix factorization using Markov chain Monte Carlo. In: Proceedings of the 25th International Conference on Machine Learning, pp. 880–887. ACM (2008)
9. Salakhutdinov, R., Mnih, A.: Probabilistic matrix factorization. In: Advances in Neural Information Processing Systems, vol. 20 (2008)
10. Verstrepen, K., Goethals, B.: Top-N recommendation for shared accounts. In: ACM Conference on Recommender Systems, pp. 59–66 (2015)
11. Yan, X., Guo, J., Lan, Y., Cheng, X.: A biterm topic model for short texts. In: Proceedings of the 22nd International Conference on World Wide Web, pp. 1445–1456 (2013)
12. Zheng, Y., Burke, R., Mobasher, B.: Splitting approaches for context-aware recommendation: an empirical study. In: Proceedings of the 29th Annual ACM Symposium on Applied Computing, pp. 274–279. ACM (2014)
13. Zheng, Y., Mobasher, B., Burke, R., CSLIM: contextual SLIM recommendation algorithms. In: Proceedings of the 8th ACM Conference on Recommender Systems, pp. 301–304. ACM (2014)
14. Zheng, Y., Mobasher, B., Burke, R., Carskit: a java-based context-aware recommendation engine. In: IEEE International Conference on Data Mining Workshop, pp. 1668–1671 (2015)

A Hybrid Method for POI Recommendation: Combining Check-In Count, Geographical Information and Reviews

Xiefeng Xu[1], Pengpeng Zhao[1(✉)], Guanfeng Liu[1], Caidong Gu[2], Jiajie Xu[1], Jian Wu[1], and Zhiming Cui[1]

[1] Department of Computer Science and Technology,
Soochow University, Suzhou, China
ppzhao@suda.edu.cn
[2] School of Computer Engineering, Suzhou Vocational University, Suzhou, China

Abstract. Due to the rapid development of mobile devices, global position systems (GPS), Web 2.0 and location-based social networks (LBSNs) have attracted millions of users to share their locations or experiences. Point of Interest (POI) recommendation plays an important role in exploring attractive locations. POI recommendation is associated with multi-dimensional factors, such as check-in counts, geographical influence, and review text. Although GeoMF can model geographical influence well by matrix factorization (MF), it ignores the impact of review text for POI recommendation. We propose a hybrid method to joint check-in counts, geographical information and reviews for POI recommendation. Specifically, we connect check-in counts and geographical information by incorporating geographical information into matrix factorization. In addition, we combine check-in counts with review text by aligning latent check-in counts in MF, and utilize hidden review topics obtained from LDA by a transformation. The results of our experiments on the real-world dataset show that our proposed model can improve the performance of recommendation.

Keywords: POI recommendation · Geographical information · Reviews

1 Introduction

With the rapid development of mobile devices and the Internet, social networking becomes more and more mature. At the same time, the popularity of global position systems (GPS) triggers the advent of location-based social networks (LBSNs), such as Foursquare, Gowalla, Facebook Places and so on. These location-based social networking services allow users to explore point of interests (POIs), post their physical locations by "check-ins" and share their experiences and reviews for POIs. POI recommendation makes a great value for both users and business owners of POIs. Hence, personalized POI recommendation has become a significant task.

© Springer International Publishing Switzerland 2016
F. Li et al. (Eds.): APWeb 2016, Part II, LNCS 9932, pp. 162–173, 2016.
DOI: 10.1007/978-3-319-45817-5_13

POI recommendation is different from traditional recommendation because geographical information plays an important role in POI recommendation. Toblers First Law of Geography says that "Everything is related to everything else, but near things are more related than distant thing" [15]. This indicates a spatial clustering phenomenon that users intuitively have a tendency to visit nearby POIs. Some researchers showed different ways on modeling the spatial clustering phenomenon [8,10,17,20]. GeoMF model [8] is a state-of-the-art method which combines geographical influence with matrix factorization. The author amplified users' latent factors with users' activity area vectors and expanded POIs' latent factors with POIs' influence area vectors. A user activity area vector shows the visiting area history of the user, and a POI's influence area vector indicates the influence level of the POI in different areas by its visiting history. However, GeoMF doesn't consider the influence of reviews of POIs written by users for recommendation.

There are some studies about how to exploit the reviews of users on items in traditional recommendation. Some articles [11,16] proposed topic matrix factorization (TopicMF) methods, which integrate users' ratings with review text for recommendation. McAuley et al. proposed a method called Hidden Factors and Hidden Topics (HFT) [11], which adopts a Latent Dirichlet Allocation (LDA) [2] model to fit review text and exploits a matrix factorization (MF) model to fit users' ratings. This method establishes a connection between the latent factors in MF model and the hidden topics obtained from LDA. However, it doesn't take geographical information into consideration for POI recommendation.

There is a tendency to integrate multi-dimensional information for POI recommendation [3,6,10]. In this paper, we combine three dimensional information, including check-in counts, geographical information and reviews, for POI recommendation. We connect check-in counts and geographical information by the method GeoMF mentioned above [8], incorporating geographical information into matrix factorization. While, we connect check-in counts and review text by the model HFT mentioned above [11], aligning the latent factors in MF model when fitting check-in counts, and aligning the stochastic vector obtained from LDA when fitting review text by a transformation. Our main contributions are as follows.

- We propose a hybrid method Topical-GeoMF to joint check-in counts, geographical information and reviews for POI recommendation.
- We evaluate our method with a real-world dataset. The extensive experimental results show that our method outperforms baselines in terms of Top-N recommendation.

The rest of the paper is organized as follows. We introduce the related work in Sect. 2. In Sect. 3, we list notations used in the following paper and provide the formal definition of the problem this paper focuses on. Some related methods contributing to our model are briefly described in Sect. 4. Our proposed method is explained in detail in Sect. 5. Our experimental results are shown in Sect. 6. Finally, we conclude our work in Sect. 7.

2   Related Work

The spatial clustering phenomenon in human mobility behavior is that users intuitively tend to visit nearby POIs including POIs close to users' home or the POIs they are favor of. In [17], the author proposed a power-law probabilistic model to capture geographical influence among POIs. Instead, the work [20] learned an individual distance distribution from the users' check-in history based on kernel density estimation. In [10], the author tried to estimate individual spatial distribution to model the spatial clustering phenomenon. There are also some methods [5,6,18] that incorporate geographical influence into topic model. In [18], the author proposed a model called LA-LDA, a location-aware probabilistic generative model that exploits location-based user ratings to model user profiles for POI recommendation. It covers three aspects of location based ratings, including spatial user ratings for nonspatial items, nonspatial user ratings for spatial items, and spatial user ratings for spatial items. To address these problems, the author proposed ULA-LDA, ILA-LDA and LA-LDA respectively.

There are some studies that take review text into consideration for recommendation. The model TopicMF integrates ratings with item contents or review text [11,16]. Reviews are associated with item attributes hidden in item latent factors that will justify a user's rating. TopicMF methods combine latent factors in ratings with latent topics in item reviews. In [16], the author proposed a model called collaborative topic regression (CTR), it exploits an additional latent variable that offsets the topic proportions when modeling user rating. In [11], the author proposed the Hidden Factors and Hidden Topics (HFT) model, which uses an objective that combines the accuracy of rating prediction (in terms of the mean squared error) with the likelihood of the review corpus (using a topic model). It can explain the variation present in ratings and reviews. The work [9] combined content-based filtering with collaborative filtering to model the reviews and ratings simultaneously.

Some methods have taken multi-dimensional factors into consideration for POI recommendation [3,4,6,10,21]. Among them, the work [10] is similar to ours. The author proposed a model Geographical-Topical Bayesian Non-negative Matrix Factorization(GT-BNMF), which incorporates the geographical influence and textual information. The author tried to estimate individual spatial distribution. However, geographical influence will be more close to the reality to learn from matrix factorization.

3   Problem Definition and Notations

We suppose that there are M users $U = \{u_1, u_2, ..., u_i, ..., u_M\}$ and N POIs $V = \{v_1, v_2, ..., v_j, ...v_N\}$. Let $R \in \mathbb{R}^{M \times N}$ be the rating matrix, where $R_{i,j}$ is the number of the check-ins that user i checked in POI j. The value 0 of $R_{i,j}$ denotes that user i never checked in POI j. The problem of POI recommendation is to recommend POIs to users. Model-based CF methods like PMF [14] are used to help our prediction. Due to One Class Collaborative Filtering (OOCF) problem in POI recommendation, some researchers' solution was to add a weighted matrix in matrix factorization (WMF).

In comparison with traditional recommendation, POI recommendation needs to consider the geographical factor. The vectors X,Y describe the geo-influence in POI recommendation. In [10], the author tried to estimate individual spatial distribution to model the spatial clustering phenomenon. Moreover, the model in [8] incorporated the geographical influence into matrix factorization.

Users will leave reviews if they visited a POI and then they intend to say something about it. The data $d_{i,j}$ is the review text of POI j written by user i, often along with a check-in behavior. In traditional recommendation, Topic MF methods like CTR [16], HFT [11] integrated review text for recommender and achieved good results.

GeoMF ignores the influence of review text to POI recommendation. Notations used in this paper are shown in Table 1.

Table 1. Notations

Symbols	Description
K	Number of latent dimensions or topics
$R_{i,j}$	Number of check-ins that user i checked in POI j
U_i	K-dimensional latent factors for user i
V_j	K-dimensional latent factors for POI j
$W_{i,j}$	Weight on the number of check-ins of POI j given by user i
X	Users' activity area matrix
Y	POIs' influence area matrix
$d_{i,j}$	Review of POI j by user i
$w_{d,n}$	The nth word of document d
$z_{d,n}$	The topic of the nth word in document d
θ_j	K-dimensional topic distribution of POI j
ϕ_k	Word distribution of topic k

4   Related Methods

In this section, we will briefly describe some related methods to learn our work better.

4.1   Two Basic Models: MF and LDA

MF. Due to the limitation in accuracy and scalability of Singular Value Decomposition (SVD) for factorizing the sparsity and frequency matrix, Koren proposed a matrix factorization (MF) method [7]. This method predicts rating $r_{i,j}$ for user i on item j according to

$$\hat{R}_{i,j} = U_i^T V_i \tag{1}$$

We factorize the rating matrix and it turns into two low rank matrices. Now, this matrix factorization (MF) method has been a state-of-the-art and standard recommendation method in dealing with rating information. The aim of MF is to find the latent user-specific and item-specific matrices $U \in \mathbb{R}^{K \times M}, V \in \mathbb{R}^{K \times N}$, where K is the number of latent factors ($K \ll M, N$). We learn the two matrices by the following function.

$$\min_{U,V} \sum_{R_{i,j} \neq 0} (R_{i,j} - \hat{R}_{i,j}) + \lambda(||U||_F^2 + ||V||_F^2) \tag{2}$$

LDA. Through rating information, we can discover the hidden factors by MF. Moreover, Latent Dirichlet Allocation (LDA) [2] is used to discover hidden topics from the collection of documents. In POI recommendation, review text of users hides a lot of useful information, such as what kind of restaurants does user i like and so on. Each document d has a topic distribution θ_d (a stochastic vector) which shows the proportion of each topic that document d covers. Similarly, each topic k has a word distribution ϕ_k which shows the proportion of each word belongs to topic k. The likelihood of the corpus M is as follow.

$$p(\theta, \phi, z) = \prod_{d=1}^{M} \prod_{n \in N_d} \theta_{z_{d,n}} \phi_{z_{d,n}, w_{d,n}} \tag{3}$$

4.2   GeoMF: Incorporating Geographical Information into MF

Due to the existing of geographical information in POI recommendation, some researchers put some ideas on how to describe it. In [17], the author proposed a power-law probabilistic model and the work [10] tried to estimate individual spatial distribution to model the spatial clustering phenomenon. Instead, the work [8] proposed that geographical information can be incorporated into matrix factorization.

In POI recommendation, check-in behaviors show the frequency of users' visiting POIs. Hence, there are only positive examples so that it's facing a well-known One Class Collaborative Filtering (OOCF) problem. In [8], the solution is to randomly sample some negative examples for each user and to allocate smaller weight to negative examples than positive examples. Based on the weighted matrix factorization, the objective function is as follow.

$$\min_{U,V} ||W \odot (R - U^T V)||_F^2 + \lambda(||U||_F^2 + ||V||_F^2) \tag{4}$$

where R is a rating matrix with the value of 0/1, which indicates that whether a user u has visited POI i.

In addition, matrices X, Y are proposed to describe geographical preferences. The whole activity area of all users is split into L subareas. Each user has an activity area vector x, and each value x_l in vector x indicates the possibility that a user may appear in the subarea l. Similarly, each POI has an influence areas vector y, each value y_l indicates the influence of a POI in subarea l. The objective function of this model is

$$\min_{U,V,X} ||W \odot (R - \hat{R})||_F^2 + \lambda_1(||U||_F^2 + ||V||_F^2) + \lambda_2||X||_1 \tag{5}$$

where $\hat{R} = U^T V + X^T Y$ is the predicted ratings. POI i's influence value at the subarea l is $y_l = \frac{1}{\sigma}K(\frac{d(l,i)}{\sigma})$, where $K()$ is standard normal distribution and σ is the standard deviation.

4.3   TopicMF:Incorporating Review Text into MF

Ratings provide the implicit feedback for recommender. Except it, review text of items written by users also provides some hidden information [1,9]. It will be effective when exploiting both of them for recommendation. Items preferences are hidden in ratings and review text. We can discover the preferences by matrix factorization (MF) [7] and Latent Dirichlet Allocation (LDA) [2] respectively. HFT method [11] emerged under this background.

This method integrates ratings with reviews by the transformation

$$\theta_{j,k} = \frac{exp(\kappa V_{j,k})}{\sum_K exp(\kappa V_{j,k})} \tag{6}$$

where the parameter κ is to control the 'peakiness' of the transformation. The value of $\Theta_{i,k}$ is always positive due to the exponent existing. The above function creates a connection between V_j and Θ_j and they can increase and narrow at the same time. It is reasonable because a user will describe the characteristics of a item by the review and it must be similar to the preferences of the item hidden in the rating.

The HFT method integrates ratings with review text by minimizing the following function

$$\sum_{R_{i,j}\neq 0} (R - U^T V)^2 - \lambda \sum_{d=1}^{M} \sum_{n\in N_d} log\theta_{z_{d,n}}\phi_{z_{d,n},w_{d,n}} \tag{7}$$

where the first part of this function is the error of the predicted ratings and the second part is the likelihood of reviews corpus. λ is the parameter that balances the contribution of these two aspects.

5   Our Proposed Model: Topical-GeoMF

Through the description of the previous section, we find that GeoMF incorporates geographical information into matrix factorization and achieved a better result than other similar methods in performance. But, it ignored the impact of review text of POIs by users in POI recommendation. Through reviews, users can show their favour of POIs they have visited. So, reviews written by users will help people a lot to choose which POI to visit. However, how to combine reviews becomes the key to the problem. We propose an effective model Topical-GeoMF by aligning latent factors and hidden topics to solve the problem. Our method combines check-in frequency, geographical information and review text by minimizing the following function.

$$f(\Theta, \Phi, X, z, \kappa) \triangleq ||W \odot (R - U^T V - X^T Y)||_F^2$$
$$- \mu \sum_{d=1}^{M} \sum_{n \in N_d} log\theta_{z_{d,n}} \phi_{z_{d,n}, w_{d,n}} \tag{8}$$
$$+ \lambda_1 ||x||_1 + \lambda_2 \Omega(\theta)$$

where $\Omega(\theta) = ||U||_F^2 + ||V||_F^2$ is a regularization term to avoid over-fitting problem, and $||x||_1$ is also the regularization term. The parameter μ balances the contribution of review text for the whole method. The first part of this equation is the error of predict times of check-ins. The biggest difference from traditional recommendation is that it incorporates geographical influence into MF. The expanding of matrices X, Y in above equation reflects the difference. The second part is the likelihood of the review corpus.

Parameter Estimation. We will simultaneously optimize the parameters related to check-in counts information $\Theta = \{U, V\}$, the parameters related to geographical influence X and the parameters related to reviews corpus $\Phi = \{\theta, \phi\}$. Our objective is to minimize the following function.

$$\underset{\Theta, \Phi, \kappa, X, z}{\operatorname{argmin}} \ f(\Theta, \Phi, \kappa, X, z) \tag{9}$$

We find that parameter Θ is related to parameter Φ by Eq. (4). So, we cannot optimize them independently. Θ and X can be learned by gradient descent and Φ by Gibbs sampling. Then, we learn these parameters through following two steps.

Step 1. This step covers two sub steps as follows

$$update \quad \Theta^{new}, \Phi^{new}, \kappa^{new} = \underset{\Theta, \Phi, \kappa}{\operatorname{argmin}} f(\Theta, \Phi, \kappa, z^{old}) \tag{10}$$

$$sample \ z_{d,n}^{new} \ with \ probability \ p(z_{d,n}^{new} = k) = \phi_{k, w_{d,n}}^{new} \tag{11}$$

In Eq. (10), topic assignment $z_{d,n}$ for each word in reviews corpus are fixed; then we update the parameters Θ, Φ and κ by gradient descent. We use L-BFGS here that is similar to the general gradient descent method [13], a quasi-newton method for non-linear optimization problems with many variables [12] because it's very convenient in use. It can help us compute gradients easily when facing so many parameters.

In Eq. (11), parameters associated with θ and ϕ are fixed, then we sample the topic assignment $z_{d,n}$ by iterating through all reviews corpus d and all words' positions n. As with LDA, we initiate topic assignment $z_{d,n}$ a random value between 1 and k and update $z_{d,n}$ with probability proportion to $\theta_{d,n}\phi_{k,w_{d,n}}$. The only point differs from LDA is that topic proportion θ is not sampled from a Dirichlet prior, instead through the former sub step.

Step 2. In this step, we will learn users' activity area matrix X. We fix the parameters Θ, Φ, κ, and the objective function respect to X is similar to a non-negative weighted least square problem. Gradient descent algorithm is used to update parameters. It's difficult to alter all the parameters at the same time. Hence, we learn each user's activity area vector independently.

The users' activity areas X is initiated to a zero matrix. The gradient of the objective function $L(x_u)$ respect to x is

$$\nabla L(x_u) = Y^T W^u (Y x_u - (r_u - V_{U_u})) + \lambda_1 \tag{12}$$

Then, we can update x_u as follows

$$x_u^{(t+1)} = U_+(x_u^{(t)} - \alpha \nabla L(x_u)) \tag{13}$$

where α is the learning rate and $P_+(x_l)$ projects a vector $x \in \mathbb{R}^L$ onto its non-negative orthant $\mathbb{R}^L_+$.

$$P_+(x_l) = \begin{cases} x_l, & x_l > 0 \\ 0, & otherwise \end{cases}, l \in \{1, ..., L\} \tag{14}$$

6   Experiments

We conduct several experiments to evaluate the performance of our proposed method. We evaluate the method on the real-world dataset for POI recommendation. The goal of our experiments is to answer three questions:

- How does our method compare with the state-of-the-art method GeoMF for POI recommendation?
- How does the effect of changing the proportion of train sets?
- How does parameter μ affect the prediction of recommendation?

Table 2. Description of our dataset foursquare

Statistics	Counts
Number of users	5439
Number of POIs	48103
Number of reviews	112317
Number of the whole check-ins	122117
Number of words in reviews	1535454
Density of check-ins	0.00043

6.1   Dataset and Metrics

We evaluate the proposed model on a LBSNs dataset Foursquare. Table 2 shows the description of our dataset. We select users who have been to at least 10 locations. Finally, our dataset contains 5439 users and 48103 POIs. In this dataset, each user can make 22.45 check-ins on average and each POI can make 2.54 check-ins on average.

For each user, we randomly select $x\%$ of his visited locations as training dataset and the remaining $(1-x)\%$ as testing dataset. Our method is measured by two metrics Recall and Precision in the Top-N POI recommendation. The metric Recall means the percentage of how many visited locations emerge in the top N recommended POIs. The metric Precision means the percentage of how many Top N recommended POIs will emerge in visited locations. For each user, the indication of two metrics is as follows.

$$Recall@N = \frac{number\ of\ POIs\ that\ user\ u\ visited\ in\ Top\ N}{total\ number\ of\ POIs\ that\ user\ u\ have\ visited} \tag{15}$$

$$Prescision@N = \frac{number\ of\ POIs\ in\ Top\ N\ that\ user\ u\ have\ visited}{N} \tag{16}$$

We will obtain the average metric of all the M users. The final results of methods are obtained by averaging on the metrics of three times independently.

6.2   Comparison with Baselines

In this section, we compare our proposed model with the following baselines:

PMF. Probabilistic Matrix Factorization (PMF) method models the user preference matrix as a product of two lower-rank matrices of user and POIs and this is a basic model.

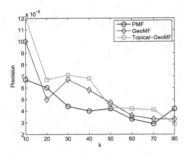
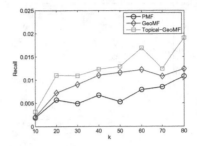

Fig. 1. Comparison of Topical-GeoMF with baselines in metrics precision and recall.

GeoMF. This framework is a component of our proposed model and we have introduces it in Sect. 4.2. It incorporated the geographical information into matrix factorization. However, it doesn't consider the influence of review text on POIs written by users.

The experimental results of Topical-GeoMF and the two baselines are shown in Fig. 1.

We select 70 % of data as training dataset when comparing with the two baselines. And we also fix the parameter $\mu = 0.01$ and $\lambda_1 = \lambda_2 = 0.1$. From the results, we observe that our method generally outperforms two baselines. When K is 80, our model doesn't perform well because it should not be reasonable when there are too many topics in users' reviews.

Then, we modify the size of training dataset into 80 % and 60 % respectively to evaluate our method. The results are as described in Fig. 2. Our method performs better that the percentage of training dataset is higher.

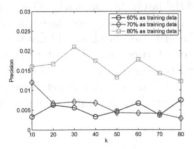
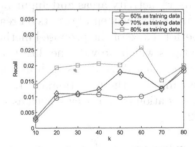

Fig. 2. Evaluation of our proposed method Topical-GeoMF by varying percentage of training dataset.

6.3 Sensitivity to Parameters: K and μ

In our model, there are two important hyperparameters: the number of latent factors K and μ. The parameter μ controls the contribution of review text. First, we vary the number of latent factors $K = \{10, 20, 30, 40, 50, 60, 70, 80\}$ with 80 %,

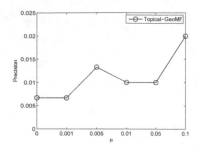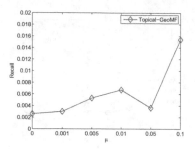

Fig. 3. The performance of our proposed method Topical-GeoMF by varying μ.

70 %, 60 % as the training dataset respectively. The results in Fig. 2 show that our model is relatively stable to K. So we consider $K = 10$ as the default value.

Then, we study how the reviews associated hyperparameter μ affect the whole performance of Topical-GeoMF. As shown in Fig. 3, we have observed that Topical-GeoMF is relatively stable when the parameter μ small. So, we select $\mu = 0.01$ as default in our experiments.

7 Conclusion

Our proposed model Topical-GeoMF is a hybrid method that combines check-in count, geographical information and review text by a mixture of MF [7] and LDA [2] for POI recommendation. On the one hand, we incorporate reviews into matrix factorization by aligning hidden factors and topics. On the other hand, we incorporate geographical influence into MF by augmenting users' latent factors with users' activity areas and augmenting POIs' latent factors with POIs' influence areas. Experimental results on real-world LBSNs data Foursquare validated the performance of the proposed method. The proposed method still has several limitations which provide the directions for our future work. It's interesting to integrate other dimensions such as social and temporal influence and it should further improve the performance [19,21]. In addition, we can extract some useful information from reviews, such as sentiment, geographical information and so on [22].

Acknowledgments. This work was partially supported by Chinese NSFC project (61472263, 61402312, 61402311, 61472268).

References

1. Bao, Y., Fang, H., Zhang, J.: Topicmf: simultaneously exploiting ratings and reviews for recommendation. In: AAAI, pp. 2–8 (2014)
2. Blei, D.M., Ng, A.Y., Jordan, M.I.: Latent Dirichlet allocation. J. Mach. Learn. Res. **3**, 993–1022 (2003)

3. Fu, Y., Liu, B., Ge, Y., Yao, Z., Xiong, H.: User preference learning with multiple information fusion for restaurant recommendation. In: SDM, pp. 470–478. SIAM (2014)
4. Gao, H., Tang, J., Hu, X., Liu, H.: Content-aware point of interest recommendation on location-based social networks. In: AAAI, pp. 1721–1727 (2015)
5. Hu, B., Ester, M.: Spatial topic modeling in online social media for location recommendation. In: RecSys 2013, pp. 25–32. ACM (2013)
6. Hu, B., Jamali, M., Ester, M.: Spatio-temporal topic modeling in mobile social media for location recommendation. In: ICDM 2013, pp. 1073–1078. IEEE (2013)
7. Koren, Y., Bell, R., Volinsky, C.: Matrix factorization techniques for recommender systems. Computer **8**, 30–37 (2009)
8. Lian, D., Zhao, C., Xie, X., Sun, G., Chen, E., Rui, Y.: GeoMF: joint geographical modeling and matrix factorization for point-of-interest recommendation. In: SIGKDD 2014, pp. 831–840. ACM (2014)
9. Ling, G., Lyu, M.R., King, I.: Ratings meet reviews, a combined approach to recommend. In: RecSys 2014, pp. 105–112. ACM (2014)
10. Liu, B., Fu, Y., Yao, Z., Xiong, H.: Learning geographical preferences for point-of-interest recommendation. In: SIGKDD 2013, pp. 1043–1051. ACM (2013)
11. McAuley, J., Leskovec, J.: Hidden factors, hidden topics: understanding rating dimensions with review text. In: RecSys 2013, pp. 165–172. ACM (2013)
12. Nocedal, J.: Updating quasi-newton matrices with limited storage. Math. Comput. **35**(151), 773–782 (1980)
13. Ricci, F., Rokach, L., Shapira, B.: Introduction to recommender systems handbook. In: Ricci, F., Rokach, L., Shapira, B., Kantor, P.B. (eds.) Recommender Systems Handbook, pp. 1–35. Springer, New York (2011)
14. Salakhutdinov, R., Mnih, A.: Probabilistic matrix factorization. Citeseer (2011)
15. Tobler, W.R.: A computer movie simulating urban growth in the detroit region. Econ. Geogr. **46**, 234–240 (1970)
16. Wang, C., Blei, D.M.: Collaborative topic modeling for recommending scientific articles. In: SIGKDD 2011, pp. 448–456. ACM (2011)
17. Ye, M., Yin, P., Lee, W.-C., Lee, D.-L.: Exploiting geographical influence for collaborative point-of-interest recommendation. In: SIGIR 2011, pp. 325–334. ACM (2011)
18. Yin, H., Cui, B., Chen, L., Zhiting, H., Zhang, C.: Modeling location-based user rating profiles for personalized recommendation. TKDD **19** (2015)
19. Yuan, Q., Cong, G., Sun, A.: Graph-based point-of-interest recommendation with geographical and temporal influences. In: CIKM 2014, pp. 659–668. ACM (2014)
20. Zhang, J.-D., Chow, C.-Y.: iGSLR: personalized geo-social location recommendation: a kernel density estimation approach. In: SIGSPATIAL 2013, pp. 334–343. ACM (2013)
21. Zhang, J.-D., Chow, C.-Y.: GeoSoCa: exploiting geographical, social and categorical correlations for point-of-interest recommendations. In: SIGIR 2015, pp. 443–452. ACM (2015)
22. Zhao, K., Cong, G., Yuan, Q., Zhu, K.Q.: SAR: a sentiment-aspect-region model for user preference analysis in geo-tagged reviews. In: ICDE 2015, pp. 675–686. IEEE (2015)

Improving Temporal Recommendation Accuracy and Diversity via Long and Short-Term Preference Transfer and Fusion Models

Bei Zhang[1,2] and Yong Feng[1,2(✉)]

[1] College of Computer Science, Chongqing University,
Chongqing 400030, China
fengyong@cqu.edu.cn
[2] Key Laboratory of Dependable Service Computing in Cyber Physical Society,
Ministry of Education, Chongqing University, Chongqing 400030, China

Abstract. Temporal factor plays an important role in products and services recommended process. It is necessary to combine temporal factors with effective methods to improve recommendation performance. In this paper, we present a novel approach to improve personalized recommendation performance with changing user preferences based on temporal dataset. In the approach, we take consideration of different influence of long and short-term user preferences and construct a preference transfer model based on our enhanced Hidden Markov Model. Then we accomplish preference fusion by adopting our Long and Short Term Graph, a graph model modified from Session-based Temporal Graph, to recommend unknown items. Finally, the experimental results show that our approach achieves important improvements compared to some existing approaches in performance.

Keywords: HMM · Long and short term · Graph model · Preference transfer · Preference fusion

1 Introduction

Recent years, there have been a flouring of artificial intelligence and mass data analysis, which aims to seek precise information in different domains. Recommendation system emerged as a more intelligent tool than search engine, needing less precise descriptive information. And there have been several recommendation algorithms proposed which can be classified to several types based on data source. Collaborative filter recommendation system is first RS which is proposed and is extensively applied in real life. It makes recommendations by exploring user-item interaction information to find correlations between users or items [1–4], which just needs a little information and achieves a relatively effective result. Another method, content-based approach, ranks candidate items based on how well they match the topic interest of users as their preference [5, 6]. Besides, hybrid methods take advantage of both information available from previous user-item interactions and content of users or items to recommend items [7, 8].

© Springer International Publishing Switzerland 2016
F. Li et al. (Eds.): APWeb 2016, Part II, LNCS 9932, pp. 174–185, 2016.
DOI: 10.1007/978-3-319-45817-5_14

However, there is still some weakness on these traditional methods. Collaborative filter method which overwhelmingly reliant on history data cannot make appropriate recommendation [6, 7]. And the method based on content has a great demand for the degree of information and lacks compatibility to incomplete characteristic data. And, there is still a serious problem about "cold start" and sparseness of data [6–8]. Content-based algorithm try to solve this question but it need more related information which in some way historical data cannot provide [7]. Besides, the traditional methods only concentrate on the static and global characteristic of user or item [7]. They consider users or items as something static and unchanging. In fact, the changes of users' preference occupy a very important position in recommendation and the way users select items often has a sequential manner.

In this paper, we firstly study preference transfer with long and short-term preference, in which transfer is modeled by a modified recommendation system based on Hidden Markov Model. Then we design a dynamical model which classifies time period into long term and short term and short terms belongs to corresponding long term. In last, we construct a graph model to fuse long and short-term preference appropriately, enhancing recommender to improve performance with achieve balance between long and short term. We systematically compare our approach with other algorithms on real dataset. The result demonstrates that our approach is effective for temporal dataset and shows better performance compared to other related algorithms.

2 Related Works

In this section, we introduce several approaches focusing on utilizing temporal data and briefly review some relevant works about time period.

Early there have been some new algorithms proposed to compute the time weight for each item by increasing weights to history data. Besides, the collaborative filtering (CF) have been transformed into a novel algorithm to solve univariate time series problem through transforming the history data to encode time order information [2]. Also in CF, Lathia [9] formalized a method which automatically assign and update each user neighborhood sizes other than setting global parameters.

Recommendation based on a Markov chain model is another temporal recommendation method by predicting users' next selection based on their previous selection [10–12]. The m-th selection by users are determined only by the last m-1 recent observations from the user. And the m is an important and troublesome variable, if m is a small value, the recommender can generate a proper result because of ignorance of a large part of the information in user's profile. But if m equals a large value it would increase the complexity to learn the parameters of model [10, 11].

Another alternative is the approach based on Hidden Markov Model (HMM), where the states are latent variables and the items selected by users are viewed as observed variables [10, 13, 14]. Each state has a probability distribution over the set of items and the states transmit according to transition probability matrix. And Aghdam [11] proposed a novel method to improve the schema of HMM by add a new latent layer on the HMM, forming hierarchical Hidden Markov model (HHMM). The novel model views the distribution of current state changes as result of changes of users' history behaviors.

Besides, some researchers discovered that size of time period of dataset has an important influence on preference of dynamic recommendation system [15]. Xiang [15] Used graph to explicitly model users' long and short-term preferences. The approach have noticed the importance of the role time period play in recommendation system, but they still do not integrate long and short-term preference adequately.

3   Content

In this section, we present long and short-term preference transfer model based on HMM and propose LSTG model which focusing on fusion of long and short-term preference. In our approach, the first part gathers the users' changeable preference in two types of terms with monitoring variations in users' preference and dynamically adapting to changes. The second part is in charge of fusion of long and short-term preference from first part and finds unknown items for users to mitigate sparseness of data problem in recommendation system.

3.1   Long and Short-Term Preference Transfer

At first, we introduce a model of changeable preference based on Hidden Markov Model to simulate real scene in users' lives. In HMM, latent states represent users' preference and the observable variables represent corresponding items [11, 14].

In this model, the observable variable for each user is a sequence of types of items or items self. In one period, the observable variable consists of type IDs of items or IDs of items which users frequently browsed or select in some period. The latent states can be defined as the preference of one user, which are susceptible to influence from its environment over time.

Intuitively, the size of time period sometimes has influence on users' preference and transfer between latent states because users show different preference in time period of different length in common sense. Time dimension can viewed as a local effect variable and it cannot be compared among users while long time period results in not grasping detail changes of users and vice versa.

We classify time period into two type: **Long Term** and **Short Term**, where the division of time period can lead to discrepant latent states distribution and different observable variable. The construction can bring about more consideration of temporal change in our model to balance the long and short-term effect in users' choices, which avoids weakness in usage of temporal historical data. Hence, we can model a novel preference transfer model distinguished by length of time period.

The long-term transfer model can detect overall preference state change in a long time period. On the contrary, the short-term transfer model can discover temporary detailed change in a short cycle. Combination of long and short-term transfer model can overcome instability in the process of preference transfer. As illustrated in Fig. 1, long-term transfer model links to short-term transfer model with long and

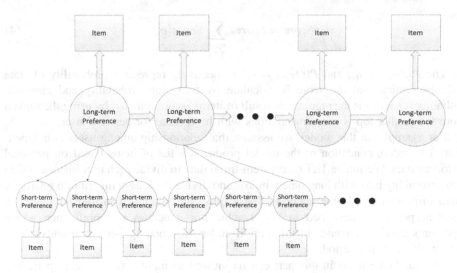

Fig. 1. A simple illustration of Long- and Short-term model

short-term preference which is latent states in transfer model based on HMM. In our model, long and short-term preference generate different items with respective feature in different periods as described in Fig. 1.

Using Baum-Welch algorithm to train parameter of HMM [17], the model calculates a maximum value of $p(O|\lambda)$ according to the sequences of items $O = \{o_1, o_2, \ldots o_T\}$ and sequence of preference $Q = \{q_1, q_2, \ldots q_T\}$ (Table 1).

Table 1. Description of parameter in HMM

N	Number of latent states
M	Number of observable variables
A	The transition probabilities among latent states
B	The observation probability matrix between states and observable variables
π	The initial latent states probabilities distribution

According to parameter achieved in train process, we can predict new popular items in next period and calculate the regularized score of candidate items.

$$P(S_{T+1} = q_i) = P(S_{T+1} = q_i | S_T = q_j)P(S_T = q_j)$$
$$= A * P(S_T = q_i) \tag{1}$$

$$P(O_{T+1} = o_i) = P(O_{T+1} = o_i | S_{T+1} = q_i)P(S_{T+1} = q_i)$$
$$= B * P(S_{T+1} = q_i) \tag{2}$$

$$Score = B * A^n * P(S_{T+1} = q_i) \tag{3}$$

$$nScore = Score / \sum_{o_i \in O} Score(o_i) \tag{4}$$

The $P(S_{T+1} = q_i)$ and $P(O_{T+1} = o_i)$ respectively represents probability of state and observation and the score is calculate by transition probability and emission probability. $nScore$ is the regularized result of items' score. Finally, by using the ranked score we predict top-N recommendations with high score for personal user.

For example, in this model, we assume that relationship among users is independent. The recommendation of the model produces a list of items based on personal historical data. We utilize UIT (users-item-time) data to divide each user historical data into several big bins with long time window and divide the long-term bin into relatively small bins with small time window. These bins can be applied to transfer model in different period as described in Fig. 1. One bin generates one preference which represents user's preference in this period and corresponding item represents user's favorite item in this period.

The candidate items in this part can be viewed as input source of next part. The long-term candidate items represent long-term preference and short-term candidate items represent short-term preference.

3.2   Long and Short-Term Preference Fusion

In the section, we introduce a novel graph model LSGT (Long and Short Term Graph) based on STG (Session-based Temporal Graph) [15]. The novel graph model take more consideration on capture on implicit user preference over time compared to STG. Through this graph model, we can realize fusion of the two preference with recommending top-k items.

For handing recommendation, graph model uses a bipartite graph <user, item> to handle raw data [15]. In STG, it transforms this form into <user, item> and <session, item> by diving the temporal data into bins and binding the bins with the corresponding user [15]. Different from STG, LSTG pays more attention to user self, which detect users' preference from individual historical data. We divide the personal overall time of user by two step into long and short term. We extend one user node with several long-term nodes and the short-term nodes are from long-term node.

In LSTG, we create two node types: long term node and short term node, to enable new links between items. The novel graph model is composed of bipartite graph $G(ULT, SLT, I, E, w)$. The long-term node is named as ULT, the $ULT(1,2)$ denotes user U_1 in long term LT_2 in Fig. 2. And the short-term node is named as SLT, the $SLT(2,2,1)$ denotes user U_2 in short term ST_1 from long term LT_2 in Fig. 2. And I is the set of item nodes which denote items users select, $w : E \rightarrow R$ denotes a non-negative weight function for edges. As shown in Fig. 2, term node is linked to items, representing user selected some items in one period. LSTG transforms recommendation problem into calculating the overall scores of the shortest paths between the corresponding nodes pair.

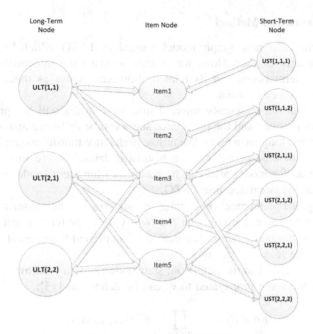

Fig. 2. An example of LSTG

Here, we illustrate the propagation process of long and short-term preference in novel graph model. In some way, we can view these nodes as voter and they are able to select most possible items for users. Note that the term node is combined with one time period and one user. In fact, we can choose different size of time period to define term node, such as one day or one week to short term and one month or one quarter to long term etc. It almost depends on the influence of the size of period and the size of number of overall items.

As described in Fig. 2, it is an example of simple LSTG with two users' long-term nodes and their short-term nodes separately. In LSTG, first long-term node $ULT(1,1)$ injects preference and spreads to all items selected by user U_1 in long term LT_1 [15]. And short-term node can inject preference in similar way.

$$w(v_i, v_j) = \begin{cases} 1 & v_i \in L \cup S, v_j \in I \\ \eta_l & v_i \in I, v_j \in L \\ \eta_s & v_i \in I, v_j \in S \end{cases} \qquad (5)$$

As described in Eq. (5), the weight of edge is defined as $w(v_i, v_j)$, which is determined by the type of nodes. In Eq. (5), we assume that the preference spread with full amount from term nodes but part amount from items nodes, modeling the influence of long term and short term more precisely.

3.3   Recommendation Method

We has constructed a new graph model named as LSTG which fuses long and short-term preference of user. Now, we prepare to make recommendation at time t, considering long-term nodes and its related short-term nodes as these nodes to be injected with preference of users.

We assume that the preference injected into these nodes will be spread to items selected by users in time t and the preference score can be delivered among item node, short-term node and long-term node. We utilize preference transfer model in part one to predict the items in time period t based historical dataset before time t. The items gathered from transfer model also can be made into long-term node and short-term nodes, which can be integrated into LSTG.

In some way, the preferences propagated among nodes is the consensus of the long and short-term preference, harmonizing the two types of preference influence. In the addition, the two start nodes denote the item sets in different time period in the current time t, which come from transfer model.

If $P\{v_0, v_1, v_2, \ldots, v_n\}$ is the path from start node v_0 to an unknown item node v_n, the final preference score propagated to v_n can be defined as [15]:

$$Rate(P) = \prod_{v_k \in P, 0 \leq k \leq n} \psi(v_k, v_{k+1})\alpha(v_0) \tag{6}$$

where $\alpha(v_0)$ is an indicator of the influence of different nodes, which is defined as Eq. (7) [15]:

$$\alpha(v_0) = \begin{cases} \beta & v_0 = v_L \\ 1 - \beta & v_0 = v_S \end{cases} \tag{7}$$

$\Psi(v_k, v_{k+1})$ is the propagation function that measures the degree of the preferences propagation from v_k to v_{k+1}. It's defined as Eq. (8):

$$\psi(v_k, v_{k+1}) = \frac{w(v_k, v_{k+1})\mu(v_k, v_{k+1})}{\sum_{v_i \in out(v_k)} w(v_k, v_i)\mu(v_k, v_i)} \tag{8}$$

$w(v_i, v_j)$ represents weight of edge between v_i and v_j in Eq. (5). And $\mu(v_i, v_j)$ denotes the number of direct link between v_i and v_j, which represents the frequency user choose this item in the session. $out(v_k) = \{v_i \in V : e(v_k, v_j) \in E\}$ is node set of v_k's all adjacent nodes.

We define $PR(u, i)$ which represents score users give to items as Eq. (9) [15]:

$$PR(u, i) = \sum_{P \in path(u,i)} Rate(P) \tag{9}$$

where $Rate(P)$ represents weight of path defined in Eq. (6). P represents some path belonging to the nodes' path set $path(u, i)$ which contains all paths pointing to item node v_i of user u.

In the end, we rank all items ordered by $PR(u, i)$ in Graph and keep top-k items for recommendation to user.

4 The Experiment

In order to compare the performance of the proposed recommendation algorithm with that of some traditional or temporal algorithms, we choose real-world rating datasets and evaluate them from various metrics.

4.1 Dataset

We used dataset of Last.fm, an Internet based personalized radio station and music recommender system. Because more frequent users' records can be achieved from music website and we can get frequent preference transfer in this condition. We can achieve user dataset as followed:

The dataset is organized in formation tuples <user, artist, song, timestamp>, which represents the whole listening habits (till May, 5th 2009) for nearly 1,000 users. And the dataset contains nearly 20,000,000 lines with nearly 1000 users and 107,000 artists, across almost 40 months.

In experiment, we compare our approach with STG, TimeSVD++. And we also select popular items and items last time user listened to compare. The approach are evaluated on this dataset on the task of predicting the artists a user will listen to in a particular time period. Artists listened by users contains previous artists users have listened and new artists they never listened.

4.2 Performance Metrics

In our experiment, the result from novel dynamical system just predict the artists a user will listen to in a particular time period which we set as a long term. As a result, the relevant evaluation such as precision and recall will show a low value to us. Hence, we add some new metrics to evaluate our approach and others.

Precision [14] is the fraction of recommendation set that is correct and Recall [14] is the fraction of the correct items that is recommended. If more items are recommended the precision will decrease, but recall will increase. The F-value [14] balances the precision and recall to summarize the two numbers.

$$Percision = \frac{|TopN \cap Listened|}{|TopN|} \tag{10}$$

$$Recall = \frac{|TopN \cap Listened|}{|Listened|} \tag{11}$$

We label precision as P and label recall as R, the F-value as F.

$$\frac{1}{F} = \frac{1}{2}\left(\frac{1}{P} + \frac{1}{R}\right) \tag{12}$$

Besides, we introduce hit ratio to evaluate the degree recommendation lists hit items users select. $hit(u)$ represents whether users' recommendation list hits. If items in recommendation list intersect with items user select we view it as a hit. The Hit Ratio is average value of all test users' hit value.

$$hit(u) = \begin{cases} 0 & \text{if recommend list hits litems user selects} \\ 1 & \text{otherwise} \end{cases} \tag{13}$$

$$HitRatio = \frac{\sum\limits_{u \in TestSet} hit(u)}{|TestSet|} \tag{14}$$

Diversity illustrates degree of richness about users' selection recommended by recommendation system. If recommendation list contains more types of items, diversity will present a relatively higher value and vice versa. Considering the importance of the diversity in users' decision we introduce metrics diversity to evaluate performance from different angles [19].

$$Diversity = \sum_{u \in TestSet} \sum_{i,j \in candidates} dist(i,j) \tag{15}$$

$dist(i,j)$ represents the distance between item i and item j. *candidates* represents the items list recommended to user u.

4.3   Evaluation

In the experiment, the approach we propose has by far better performance than other approaches in precision, recall and F-value as illustrated in Table 2. In fact, the approaches based on graph model (Ours and STG) show better performance than other methods. In details, we can see that recall and F-value displays an upward trend in all methods. And we can find that the F-value of the graph model including our approach and STG grows faster than others along with increase of list size. Hence, we can select approaches based on graph model if we predict items in short term to pursue better performance.

In Fig. 3, we find that hit ratio of approach we propose outperforms other methods most of time. But with increasing of the size of recommendation list the TimeSVD++ begins to transcend our approach. In fact, it's normal phenomenon because TimeSVD ++ is train by the historical dataset includes global information before prediction time but our approach just utilize dataset with one-year time span. So when size of recommendation list increases probability recommendation list hits items user selects will increases with more information, too. However, size of recommendation should be

Table 2. Comparative evaluation results in different Top-K

Recommendation metrics

(a)Top-5 recommendation

	Ours	STG	TimeSVD++	Most popular	Last listen
P	**0.1325**	0.095	0.0479	0.057	0.0314
R	**0.047**	0.0295	0.0148	0.0171	0.0065
F	**0.0694**	0.045	0.0227	0.0263	0.0108

(b)Top-10 recommendation

	Ours	STG	TimeSVD ++	Most popular	Last listen
P	**0.1229**	0.0888	0.0471	0.0508	0.0256
R	**0.078**	0.0502	0.0315	0.0283	0.0125
F	**0.0955**	0.0642	0.0377	0.0364	0.0168

(c)Top-20 recommendation

	Ours	STG	TimeSVD ++	Most popular	Last listen
P	**0.1133**	0.0864	0.0498	0.045	0.0169
R	**0.1295**	0.0901	0.0639	0.0477	0.015
F	**0.1209**	0.0882	0.056	0.0463	0.0159

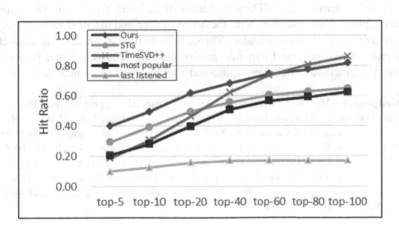

Fig. 3. Hit ratio trend among different Top-K

controlled in the range less than a limited value because excessive scale of list will result in low recall, which causes users' poor experience.

We evaluate diversity of all approaches with average distance between items in recommendation list. As illustrated in Fig. 4, there is low disparity among approaches, but the approach we propose show a relatively higher value but not highest value. And the diversity increases along with growth of size of recommendation list.

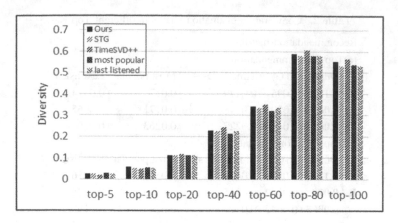

Fig. 4. Comparison of diversity among different Top-K

5 Conclusion

In this paper, we study users' preference transfer and fusion with long and short-term preference and presents a novel approach to improve performance in recommendation. In usage of HMM we model process of user preference transfer. And we achieve top-k recommendation items from LSTG with fusion of different preference. Our approach efficiently captures long and short-term factors over time and discover influence of long and short term in preference transfer. Through experiments on real-life dataset, we demonstrate that the proposed dynamic model shows a relatively better performance than other base methods, including traditional methods and sequential methods.

Acknowledgement. The authors acknowledge the financial support of the Frontier and Application Foundation Research Program of CQ CSTC (No. cstc2014jcyjA40037), National High-tech R&D Program of China (No. 2015AA015308), the Chongqing Graduate Student Research Innovation Project (No. CYS16032), Fundamental Research Funds for the Central Universities (No. CDJZR188801), Degree and Graduate Education Research Program (No. 2015Y0414), and the Project of Chongqing Higher Education Reform (No. 153003). The author is grateful to the anonymous referee for a careful checking of the details and for helpful comments that improved this paper.

References

1. Sarwar, B., Karypis, G., Konstan, J.: Item-based collaborative filtering recommendation algorithms. In: Proceedings of the 10th International Conference on World Wide Web, Hong Kong, pp. 285–295 (2001)
2. Koren, Y.: Collaborative filtering with temporal dynamics. In: ACM SIGKDD International Conference on Knowledge Discovery and Data Mining, Paris, pp. 89–295 (2009)
3. Peng, J., Zeng, D.D., Zhao, H.: Collaborative filtering in social tagging systems based on joint item-tag recommendations. In: ACM International Conference on Information and Knowledge Management, Toronto, pp. 809–818 (2010)

4. Liu, Q., Chen, E., Xiong, H.: Enhancing collaborative filtering by user interest expansion via personalized ranking. IEEE Trans. Syst. Man Cybern. Part B Cybern. A Publ. IEEE Syst. Man Cybern. Soc. **42**, 218–233 (2012)
5. Popescul, A., Ungar, L.H., Pennock, D.M.: Probabilistic Models for Unified Collaborative and Content-Based Recommendation in Sparse-Data Environments. ePrint arXiv, 437-444 (2013)
6. Ninaus, G., Reinfrank, F., Stettinger, M.: Content-based recommendation techniques for requirements engineering. In: 22nd IEEE International Conference on Requirements Engineering, pp. 1439–1446 (2014)
7. Shih, Y.Y., Liu, D.R.: Hybrid recommendation approaches: collaborative filtering via valuable content information. In: Proceedings of the 38th Annual Hawaii International Conference on IEEE, Hawaii, p. 271b (2005)
8. Rojsattarat, E., Soonthornphisaj, N.: Hybrid Recommendation: Combining Content-Based Prediction and Collaborative Filtering. Intelligent Data Engineering and Automated Learning, pp. 337–344 (2003)
9. Lathia, N., Hailes, S., Capra, L.: Temporal collaborative filtering with adaptive neighbourhoods. In: Proceedings of the 32nd International ACM SIGIR Conference on Research and Development in Information Retrieval, pp. 796–797. ACM (2009)
10. Adomavicius, G., Tuzhilin, A.: Context-Aware Recommender Systems Recommender Systems Handbook, pp. 217–253. Springer, US (2011)
11. Hosseinzadeh, A.M., Hariri, N., Mobasher, B.: Adapting recommendations to contextual changes using hierarchical hidden Markov models. In: Proceedings of the 9th ACM Conference on Recommender Systems, pp. 241–244. ACM (2015)
12. Javari, A., Jalili, M.: A probabilistic model to resolve diversity-accuracy challenge of recommendation systems. Knowl. Inf. Syst. **44**(3), 1–19 (2015)
13. Stefanidis, K., Pitoura, E., Vassiliadis, P.: Managing contextual preferences. J. Inf. Syst. **36**(8), 1158–1180 (2011)
14. Sahoo, N., Singh, P.V., Mukhopadhyay, T.: A hidden Markov model for collaborative filtering. J. Ssrn Electron. J. **36**(4), 1329–1356 (2010)
15. Xiang, L., Yuan, Q., Zhao, S.: Temporal recommendation on graphs via long-and short-term preference fusion. In: Proceedings of the 16th ACM SIGKDD International Conference on Knowledge Discovery and Data Mining, pp. 723–732 (2010)

An Approach for Cross-Community Content Recommendation: A Case Study on Docker

Yang Yong[1], Li Ying[1,2(✉)], Tang Hongyan[1], Jia Tong[1],
and Shao Wenlong[3]

[1] School of Software and Microelectronics, Peking University, Beijing, China
{yang.yong,li.ying,hytang,jia.tong}@pku.edu.cn
[2] National Engineering Center of Software Engineering, Peking University,
Beijing, China
[3] VMware, Inc., Beijing, China
wshao@vmware.com

Abstract. With the boom of open source software, open source communities are formed and involved in software development, deployment and application with unprecedented level. However, the rapid expansion of open source communities results in a lot of redundant contents within the community, and most importantly, among communities since they overlap each other with shared issues. On the one hand, redundant contents that are expressed in informal free texts highly increase the size of contents, which makes people suffering from finding what they exactly need from communities; on the other hand, these communities are mutually complementary that the knowledge sharing across communities can be very beneficial to users. It is crucial to recommend content for users' need through retrieving knowledge across communities. Current studies mainly focus on acquiring knowledge from one specific community to treat communities as isolated islands, and few of them have tackle the problem of content recommendation across multiple communities. In this paper, we firstly analyze five popular open source communities, and then propose an approach of cross-community content recommendation based on LDA topic model, integrating and distilling information from multiple communities to make knowledge acquisition easier and more efficient. Taking Docker as the case study, extensive experiments show that after performing a cross-community recommendation, more than 34 % overall unanswered questions find matched answers when similarity threshold β is set to 0.85. When setting β to 0.6, almost 90 % unanswered question can be answered with existing community content. It effectively leverages various communities to recommend valuable content to users.

Keywords: Cross-community · LDA · Recomendation · Information aggregation

The work is supported by Shenzhen Municipal Science and Technology Program (Grant No. JSGG201 4051616 2852628), and VMware UR project.

F. Li et al. (Eds.): APWeb 2016, Part II, LNCS 9932, pp. 186–197, 2016.
DOI: 10.1007/978-3-319-45817-5_15

1 Introduction

The recent years have seen a prosperity of open source software and a successful open source software can not work without active developer and user communities, such as Github and Stack Overflow. In communities, users share their knowledge under the same subject, and the value of information can be improved because of concentration. However, with the rapid growth of both the content volume and the number of communities, information becomes redundant within and among communities. Valuable content is distributed in isolated communities. Users can hardly find what they exactly want and then post content that actually has existed in communities, which further exacerbates the content redundancy. For example, without considering community overlapping, more than 12 % of issues that related with Docker on Github are redundant. Although the appearance of search engines alleviates this problem to some extent, it is still a tough job for people to pick useful information out in a crowd of retrieval results from various communities separately. Therefore, there is an urgent need to aggregate and distill information from different communities to enhance the information value for users. Several recommendation systems [5, 8] for community have been proposed to recommend preferable contents and communities to users. But they are limited to recommend content in the same community [8] and can't take full advantage of rich contents of various communities.

Although community is a self-explanatory concept without a widely-accepted, unified, rigorous definition, most of open source software communities can be simplified into a question-answer or subject-discussion model. A user raises a question or subject while other users offer answers or discussions closely related. In this way, we will not distinguish between question and subject, as well as between answer and discussion. There are two barriers for people to find preferable content across communities.

- Community contents are redundant seriously due to similar questions being posted onto the community. Although many open source communities will recommend users to search similar questions before raising a new question, users usually fail to get what they want due to expression differences or lexical gaps between the queried question and existing questions. The repeated questions are posted again and lead to a distribution of valuable information. In Sect. 3, we will reveal the content redundancy in real communities.
- Currently communities are treated as information islands and users are unlikely to participate in all relevant communities. It happens often that no one answers the question in one community but the question has been actually answered in another community. All of these will affect the user experience.

In this paper, we tackle these problems by proposing a cross-community recommendation approach based on topic model with LDA (Latent Dirichlet Allocation) [12], named CCRTM, which combines and makes full use of rich content of different communities. Besides, we implement a prototype of CCRTM and conduct a case study of Docker content recommendation with experiments on real community datasets to validate the approach. What makes CCRTM different with other community recommendation methods is that it can recommend preferable and valuable content to users by aggregating information from different communities and taking advantage of their content evaluation metrics. In summary, we make the following contributions in this paper:

- We analyze 5 popular Docker communities including Docker-dev and Docker-user of Google Groups, Docker Forums, Stack Overflow and Github Issues, and reveal the question redundancy within and among these communities.
- We provide a topic model based cross-community recommendation approach to effectively help users acquire knowledge across various communities.
- We quantitatively evaluate *effectiveness* and *richness* of the proposed approach with extensive experiments.

The rest of this paper is organized as follows. Section 2 presents some related works. Section 3 defines the redundancy of communities and shows the redundancy situation of real communities. CCRTM is described in detail in Sect. 4. In Sect. 5, new metrics are introduced and experiments to evaluate CCRTM are conducted. Section 6 draws some conclusions and gives a picture of future work.

2 Related Works

Rich information is buried in communities and many researchers have attached great importance on discovering useful knowledge from community data. Cheng et al. [1] leverage community-contributed data, specifically the freely available community photos, for personalized travel recommendation. Twitter community has been used to recommend real-time topical news [4]. Kamahara et al. [5] develop a multimedia TV recommendation system to recommend unexpected content based on the communities on the network.

And various models and algorithms have been used to explore communities. Duan et al. [2] model networks with dynamic weighted directed graphs (DWDG) and put forward *Stream-Group* to discover the community structure. A local search based genetic algorithm [3] and a hierarchical latent Gaussian mixture model [6] are proposed to detect communities from complex networks.

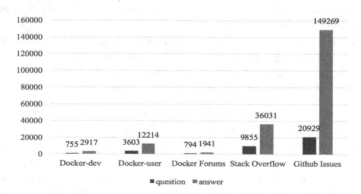

Fig. 1. Basic information of five communities

Topic models have been widely used in community research area. Zongchen et al. [9] use topic model to improve the question retrieval results in QA (question-answer) community by alleviating the lexical gap problem. Li et al. [10] develop a social community tagging system based on LDA topic model. Yin et al. [11] integrate topic model with community discovery to analyze latent community topics.

Content-based, collaborative, hybrid recommendation approaches have been applied into communities to recommend both communities and community content for users. A collaborative personalized recommendation system is presented based on ARM or LDA model [7] to recommend interesting communities and users. And a question recommendation [8] system based on Probabilistic Latent Semantic Analysis (PLSA) is proposed to recommend question to users for QA communities. But it is limited to recommend content in the same community and can not make full use of other communities' data.

3   Redundancy of Community

3.1   Datasets

Docker is a famous open source container engine and it has flourishing communities. We crawled data from five English Docker communities, which are, Docker-dev and Docker-user of Google Groups, Docker Forums, Stack Overflow and Github Issues, from January 1, 2013 to March 6, 2016. Finally, a total number of 238,308 items were collected, which are composed of 35,936 questions and 202,372 answers. In this paper, an item means a question or answer. The scale of different communities differs a lot and the detailed data description is shown in Fig. 1.

3.2   Definition of Redundancy

For the sake of simplification, we only use the question redundancy to illustrate the situation of duplicated content of open source software communities, since all answers are closely related to corresponding questions. Then we provide a community redundancy definition and redundancy calculating algorithm based on similarity measures.

Algorithm 1 Algorithm of redundancy calculating

Input: question set S, similarity threshold θ
Output: redundancy of set S
1:   create redundant question set rqSet
2:   **for** every question Q_i in S, $Q_i \notin$ rqSet
3:      **for** every question Q_j in S, $Q_i \neq Q_j$
4:         **if** $sim(Q_i, Q_j) > \theta$ **then**
5:            **add** Q_j to rqSet
6:   redundancy $\alpha = length(\text{rqSet})/ length(S)$
7:   **return** redundancy α

Redundancy: For a question set S and a similarity threshold θ, if $sim(Q_i,Q_j) > \theta$, $Q_i,Q_j \in S$, then Q_i can be called a redundant question. The proportion of redundant questions α for set S is the redundancy of S. When S refers to a question set of a community, α means the redundancy of the community. When S represents questions of all communities, α is the global redundancy.

The redundancy calculating algorithm is described in Algorithm 1. The *sim* function is a method to measure the similarity between questions. Different implementations, such as variations based on Levenshtein Distance [15] and Cosine Similarity [16], can be used.

3.3   Redundancy Results

Cosine similarity is adopted for the *sim* method. We use LDA topic model and TF-IDF (term frequency–inverse document frequency) [17] separately to model the question and conduct a comparison between the results of two models.

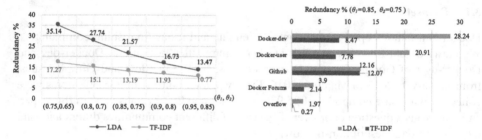

Fig. 2. Global redundancy   **Fig. 3.** Community redundancy

In LDA topic model, a topic is defined to be a distribution over a fixed vocabulary of words and a document can be treated as a mixture of topics. Each document exhibits these topics with different proportions. A document means a question here. After using VSM (Vector Space Model) to represent questions, we assign the topic number $k = 200$ to train a LDA model from all questions we collected. And then, we can find a match for every question in the corpus of the model, which is a mixture of 200 topics learnt from all questions.

TF-IDF is a commonly used weight factor in information retrieval area. It can indicate the importance of word to a document in all documents. Combining with the Vector Space Model, a question can be denoted as a vector in which each dimension corresponds to the TF-IDF value of a separate word. And the number of dimension of a question depends on how many unique word exists in all questions.

Compared to LDA, TF-IDF, which ignores the semantic relations of words, is stricter when used to measure similarity of questions. Therefore, when performing a comparison, we set a threshold θ_1 for LDA and a lower threshold θ_2 for TF-IDF model. We plot the global redundancy with different threshold for two models in Fig. 2. And Fig. 3 shows each community redundancy within their own contents with a threshold $\theta_1 = 0.85$ and threshold $\theta_2 = 0.75$.

From Figs. 2 and 3, we can see that redundant questions are common within and among communities, not to mention the answers. With threshold $\theta_1 = 0.85$ and $\theta_2 = 0.75$, overall redundancy reaches 21.57 % with LDA and 13.19 % with TF-IDF. For Docker-dev community, 28.24 % questions are redundant within its own contents. With the redundancy, valuable answers are distributed under similar questions in various communities and some questions are left unanswered in some communities. It impedes users to acquire knowledge by taking full advantages of all information available in communities. Therefore, it is necessary to propose a method to aggregate and recommend valuable content across communities to users.

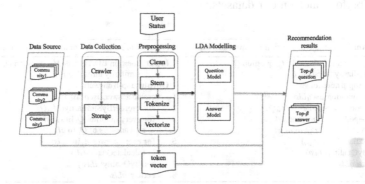

Fig. 4. Overview of cross-community recommendation

4 Methodology

4.1 An Overview of Cross-Community Recommendation

Figure 4 shows an overview our CCRTM approach for cross-community recommendation. In data collection stage, communities are specified and necessary information including questions, answers and their quality data are crawled and stored. It's optional whether to reserve the initial questions and answers data after training LDA models from them. A question model and an answer model are trained from the items after the preprocessing stage. Then, according to user status, related questions and high-quality answers from different communities will be recommended to users.

4.2 Preprocessing

In this stage, a question or answer will go through cleaning, tokenization, stemming and vectorization. Whether to perform lemmatization and whether to use TF-IDF value as the weight of a word are optional. Empty or incomplete items are dropped and HTML tags are filtered in cleaning stage and other items are split into tokens. Lancaster stemming algorithm based on Paice' work [14] are used to process word tokens. Finally, a term will be denoted as a word vector.

4.3   LDA Modeling

We assign a topic number k, which is the key parameter for LDA, to train two models from the processed community data, a question model and an answer model. There is no best principle or standard for selection of k, though some advices are provided to leverage HDP (Hierarchical Dirichlet Process) [13] or calculate the perplexity of the test corpus [12]. However, according to our experimental experience, neither of them work well on our datasets. To conduct the evaluation, we set $k = 200$ as the default topic number and change the k value to build new LDA models according to changes of community data. The experimental result, as shown in Figs. 9 and 10, illustrates how we choose the topic number k for making CCRTM have a stable richness and effectiveness performance on our datasets.

Algorithm 2 Algorithm of question recommendation	**Algorithm 3** Algorithm of answer recommendation
Input: question LDA model M_q, question Q, similarity threshold β	**Input**: question LDA model M_q, question Q, similarity threshold β
Output: top-β questions	**Output**: top-β answers
1. create question list *qList*	1. create answer list *aList*
2. *sList*=orderBySimilarity(M_q,Q)	2. *sList*=orderBySimilarity(M_q,Q)
3. *qList*=getSimilarQuestions(sList, β)	3. **for** every question Q_i in *sList*:
4. **for** every question Q_i in *aList*	4. add answers of Q_i to *aList*
5. calculateQuality(Q_i)	5. **for** every answer A_j in *aList*
6. sortByQuality(*qList*)	6. calculateQuality(A_i)
7. **return** *qList*	7. sortByQuality(*aList*)
	8. **return** *aList*

4.4   Recommendation

According to users' status which can be one of *searching, viewing, answering* a question, a series of closely related questions and answers can be recommended to users, based on the topic similarity and quality of content. All questions whose similarity to users' question exceed similarity threshold β are recommended to users. We call it a top-β recommendation. The detailed procedures are shown in Algorithm 2. Recommending answers by a question is based on the top-β question recommendation and recommending answers by an answer is similar to steps described in Algorithm 2. Steps of the former are presented in Algorithm 3. Similarity of questions and answers can be produced by the LDA models with the cosine similarity.

The recommendation results are sorted by their quality score. The quality evaluation method of question and answer varies from community to community. In CCRTM, to make a unified item quality metric, we use ratio of the key features' value and average feature value of its community to measure item quality. For example, key feature of a question in Github Issue is the answer number, then the quality of a question can be calculated by N_i / N, in which N_i is the answer number of Q_i and N is the average answer number of a question in Github Issue. Similarly, the quality of an answer can be estimated according to voting, reference or favor number. In this way, the quality of an item in its original community can be used to rank the cross-community recommendation results to show valuable contents to community users.

5 Evaluation

In this section, we introduce two new metrics and conduct experiments to evaluate proposed cross-community recommendation approach. Then we show how the topic number k and similarity threshold β affect the evaluation result and how we choose the default topic number on our datasets. Finally, we show the experiment results and provide our analysis.

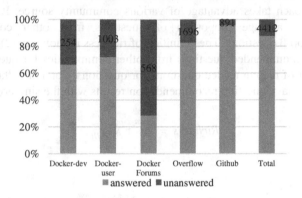

Fig. 5. Unanswered questions in each community

5.1 Experiment Setup

We implement a prototype of CCRTM with python and design our experiments with the Docker community datasets mentioned in Sect. 3. We adopt the genism [18] implementation of LDA whose parameter estimation is based on Hoffman's work [19] and assign a default topic number $k = 200$ to train models.

5.2 Evaluation Metrics

Few of studies have been working with cross-community recommendation through aggregating information from multiple communities. The commonly used metrics in recommender system, such as *precise* and *recall*, are not proper to evaluate cross-community recommendation results, because they can't reflect how cross-community recommendation differs and to what extent it solves island problems in community. Currently there is no public labeled dataset of different open source software communities. Therefore, as the first attempt, in this paper, we introduce two metrics to evaluate the performance of cross-community recommendation approach as following:

Effectiveness: It measures that to what extent a cross-community recommendation approach solves unanswered questions. There are quite a lot of questions unanswered in each community, which is shown in Fig. 5. In Docker Forums, the proportion of unanswered question takes up 71.54 %. Globally, more than 14 % questions haven't been answered. Thus a metric to evaluate how many unanswered questions are solved by the cross-community recommendation is necessary. Given the unanswered questions set $BQ_{a=0}$ before recommendation and the unanswered questions set $AQ_{a=0}$ after a recommendation process with similarity threshold β, the effectiveness can be calculated as the ratio of $AQ_{a=0}$ and $BQ_{a=0}$ in (1).

$$Effectiveness = \frac{AQ_{a=0}}{BQ_{a=0}}, BQ_{a=0} \neq 0 \tag{1}$$

Richness: Richness is a metric to measure whether the cross-community recommendation approach takes advantage of various community source. It can be represented with the average proportion of questions from other communities in recommended top-β questions. The definition of richness is shown in (2), where OQ_i is the number of recommended questions from other communities for question Q_i, N_i is the total number of recommended questions for question Q_i and m is the total number of questions that have question recommendation results with the similarity threshold β.

$$Richness = \sum_{i=1}^{m} \frac{OQ_i}{N_i} \Big/ m \tag{2}$$

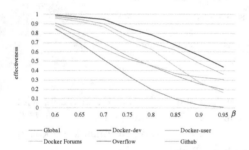 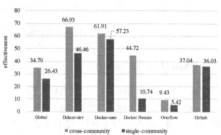

Fig. 6. Effectiveness for each community

Fig. 7. Effectiveness for cross-community and single-community recommendation

5.3   Results

Effectiveness result: To evaluate the effectiveness of CCRTM, we simulate scenarios that users are viewing unanswered questions in a community. Then a list of related questions is recommended to users according to the unanswered question. After performing a question recommendation for each unanswered question, we get the effectiveness of CCRTM for overall five communities and single community. In order to evaluate the impact of similarity threshold on the effectiveness, we conduct several experiments with various threshold β. Figure 6 shows the effectiveness of the recommendation results for overall five communities and each community with a series of similarity threshold β. More than 25 % unanswered questions find matched answers overall when similarity threshold β is set to 0.9 and the effectiveness for Docker-dev reaches 56.69 %. When setting β to 0.6, almost 90 % answered question can be answered with existing community content.

To figure out the advantage of cross-community recommendation compared to single-community recommendation, a comparative experiment with $\beta = 0.85$ is conducted to assess the improvement of effectiveness through cross-community recommendation. According to the comparison result in Fig. 7, another 8 % unanswered questions find matched answers from other communities after using CCRTM. In Docker Forums, a cross-community recommendation achieves a 33.98 % improvement for effectiveness, which means 33.98 % unanswered questions in Docker Forums can find useful answers in other four communities.

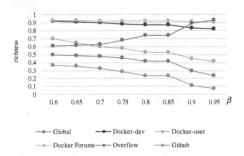

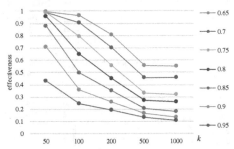

Fig. 8. Richness for each community

Fig. 9. Effectiveness for all communities with different topic number

Richness result: Similar to the evaluation process of effectiveness, scenarios that users are searching or viewing a question in a community are simulated. CCRTM recommends related questions for each question in overall five communities and then the richness of CCRTM is calculated. The similarity threshold β ranges from 0.6 to 0.95. Figure 8 shows that with the similarity threshold β in a certain range, the richness of recommendation result for each community is stable. And the richness result differs widely from community to community. Generally speaking, for a question, more than 40 % recommendation results are recommended from other communities with $\beta = 0.85$. And when it comes to Docker-dev and Docker Forums, recommended questions from other communities take up more than 80 %. The richness result for Github is a bit low because it has rich content as shown in Fig. 1 and its redundancy is high as shown in Fig. 3.

Topic number k and similarity threshold β: In order to observe whether effectiveness and richness of CCRTM are sensitive to the selection of k and β. We assign different k values in a wide range to train LDA models from the same Docker community datasets. Then these models are used with different β values for cross-community recommendation.

The effectiveness for overall communities with various k and β is showed in Fig. 9 and the richness results for each community are plotted in Fig. 10. From the global point of view, with the same LDA model, the lower β is, the higher effectiveness and richness are. As shown in Fig. 9, with a specific β, a topic number k between 200 and 500 makes the effectiveness become stable no matter how the k grows. As for the richness shown in Fig. 10(f), when k exceeds a value between 200 and 500, the richness become unstable and unpredictable. Based on the observation, we set 200 as the default topic number to gain a stable performance without considering the recommendation accuracy. When it comes to single community, the richness result differs a lot according to Fig. 10(a)–(e) and so does the effectiveness for single community.

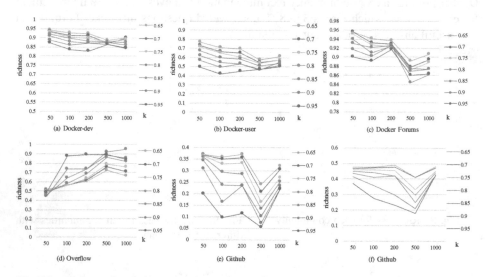

Fig. 10. Richness for each community and overall communities with different topic number

6 Conclusion and Discussion

We propose the cross-community content recommendation approach CCRTM by integrating information from different communities. The approach can leverage rich content from various communities to recommend what users exactly need in community. After conducting experiments on the Docker community datasets, we show how CCRTM solve content redundancy within and among communities and how it helps users make full use of cross-community data. Efficient knowledge acquisition from communities will improve user experience, and promote the development of open source software.

As the first attempt, our proposal has opened new opportunities for community recommendation, though it still has some limitations. So far it is basically a content-based recommendation method which mostly focuses on text contents and their similarity. In our future work, we would like to apply collaborative filtering to CCRTM by mining user relation among multiple communities. Another limitation is that accuracy of cross-community recommendation has not been evaluated yet due to the limitation of available datasets. We will explore how to evaluate the accuracy of cross-community recommendation. Finally, how to select topic number k and similarity threshold β will be studied in the future with more experiments for different datasets. Besides, we are planning to make CCRTM as a service to serve open source software community users and continuously improve its performance.

References

1. Cheng, A.J., Chen, Y.Y., Huang, Y.T., et al.: Personalized travel recommendation by mining people attributes from community-contributed photos. In: Proceedings of the 19th ACM International Conference on Multimedia, pp. 83–92. ACM (2011)
2. Duan, D., Li, Y., Jin, Y., et al.: Community mining on dynamic weighted directed graphs. In: Proceedings of the 1st ACM International Workshop on Complex Networks Meet Information and Knowledge management, pp. 11–18. ACM, MLA (2009)
3. Jin, D., He, D., Liu, D., et al.: Genetic algorithm with local search for community mining in complex networks. In: 2010 22nd IEEE International Conference on Tools with Artificial Intelligence (ICTAI), vol. 1, pp. 105–112. IEEE (2010). MLA
4. Phelan, O., McCarthy, K., Smyth, B.: Using Twitter to recommend real-time topical news. In: Proceedings of the Third ACM Conference on Recommender Systems, pp. 385–388. ACM (2009)
5. Kamahara, J., Asakawa, T., Shimojo, S., et al.: A community-based recommendation system to reveal unexpected interests. In: Proceedings of the 11th International Multimedia Modelling Conference, MMM 2005, pp. 433–438. IEEE (2005)
6. Zhang, H., Giles, C.L., Foley, H.C., et al.: Probabilistic community discovery using hierarchical latent Gaussian mixture model. In: AAAI, vol. 7, pp. 663–668 (2007)
7. Chen, W.Y., Chu, J.C., Luan, J., et al.: Collaborative filtering for orkut communities: discovery of user latent behavior. In: Proceedings of the 18th International Conference on World Wide Web, pp. 681–690. ACM, MLA (2009)
8. Qu, M., Qiu, G., He, X., et al.: Probabilistic question recommendation for question answering communities. In: Proceedings of the 18th International Conference on World Wide Web, pp. 1229–1230. ACM (2009)
9. Ji, Z., Xu, F., Wang, B., et al.: Question-answer topic model for question retrieval in community question answering. In: Proceedings of the 21st ACM International Conference on Information and Knowledge Management, pp. 2471–2474. ACM (2012)
10. Li, D., He, B., Ding, Y., et al.: Community-based topic modeling for social tagging. In: Proceedings of the 19th ACM International Conference on Information and Knowledge Management, pp. 1565–1568. ACM (2010)
11. Yin, Z., Cao, L., Gu, Q., et al.: Latent community topic analysis: integration of community discovery with topic modeling. ACM Trans. Intell. Syst. Technol. (TIST) 3(4), 63 (2012)
12. Blei, D.M., Ng, A.Y., Jordan, M.I.: Latent dirichlet allocation. J. Mach. Learn. Res. 3, 993–1022 (2003)
13. Teh, Y.W., Jordan, M.I., Beal, M.J., et al.: Hierarchical Dirichlet processes. J. Am. Stat. Assoc. (2012)
14. Chris, D.P.: Another stemmer. ACM SIGIR Forum 24(3), 56–61 (1990)
15. Levenshtein Distance. http://en.wikipedia.orgjwikillevenshtein_distance
16. Cosine Similarity. https://en.wikipedia.org/wiki/Cosine_similarity
17. Salton, G., Buckley, C.: Term-weighting approaches in automatic text retrieval. Inf. Process. Manag. 24(5), 513–523 (1988)
18. Gensim. http://radimrehurek.com/gensim/
19. Hoffman, M., Bach, F.R., Blei, D.M.: Online learning for latent dirichlet allocation. In: Advances in Neural Information Processing Systems, pp. 856–864 (2010)

Research Full Paper: Data Quality and Privacy

Scalable Private Blocking Technique
for Privacy-Preserving Record Linkage

Shumin Han[✉], Derong Shen, Tiezheng Nie, Yue Kou, and Ge Yu

College of Information Science and Engineering, Northeastern University, Shenyang, China
hanshumin_summer@yeah.net,
{shenderong,nietiezheng,kouyue,yuge}@ise.neu.edu.cn

Abstract. Record linkage is the process of matching records from multiple databases that refer to the same entities and it has become an increasingly important subject in many application areas, including business, government, and health. When we collect data which is about people from these areas, integrating such data across organizations can raise privacy concerns. To prevent privacy breaches, ideally records should be linked in a private way such that no information other than the matching result is leaked in the process, this technique is called privacy-preserving record linkage (PPRL). Scalability is one of the main challenges in PPRL, therefore, many private blocking techniques have been developed for PPRL. They are aimed at reducing the number of record pairs to be compared in the matching process by removing obvious non-matching pairs without compromising privacy. However, they vary widely in their ability to balance competing goals of accuracy, efficiency and security. In this paper, we propose a novel private blocking approach for PPRL based on dynamic k-anonymous blocking and Paillier cryptosystem. In dynamic k-anonymous blocking, our approach dynamically generates blocks satisfying k-anonymity and more accurate values to represent the blocks with varying k. We also propose a novel similarity measure method which performs on the numerical attributes and combines with Paillier cryptosystem to measure the similarity of two blocks in security, which provides strong privacy guarantees that none information reveals. Experiments conducted on a public dataset of voter registration records validate that our approach is scalable to large databases and keeps a high quality of blocking. We compare our method with other techniques and demonstrate the increases in security and accuracy.

Keywords: Record linkage · Private blocking · k-anonymity · Paillier cryptosystem · Scalability

1 Introduction

Nowadays, large amounts of data from several domains like businesses, government agencies and research projects has been generated, collected and stored. Matching records that relate to the same entities from several databases has been recognized to be of increasing importance in many application domains. However, when data about individuals or sensitive attributes is to be integrated across organizations, privacy has to be

© Springer International Publishing Switzerland 2016
F. Li et al. (Eds.): APWeb 2016, Part II, LNCS 9932, pp. 201–213, 2016.
DOI: 10.1007/978-3-319-45817-5_16

considered. Therefore, we need to protect these data from unauthorized disclosure. For example, in a decentralized healthcare system, where the personal medical records are distributed among several hospitals, it is critical to integrate the information of a patient without disclosing his/her sensitive attributes. Thus, making sure that privacy of individuals is maintained whenever databases are linked across organizations is vital.

Privacy-Preserving Record Linkage (PPRL) [1] is the process of identifying records from multiple data sources that refer to the same individuals, without revealing other information besides the matched records. PPRL has been widely used in many fields. For example, Microsoft has acquired Yahoo, by applying record linkage technique on their client databases, we cannot only obtain common clients between them but also acquire the potential new clients from Yahoo, which has significant business value for Microsoft. However, the client databases are confidential, exposing client data to other companies would cause heavy loss. Therefore comparing client databases without data disclosure excepting matched records is crucial.

Considering the growing large volumes of available data, developing PPRL which are scalable to large databases is necessary. Therefore, blocking techniques are developed. Blocking techniques are used to divide records into mutually exclusive blocks and only the records within the same block can be linked. A naive pair-wise comparison of two databases in record linkage has a quadratic complexity in their sizes. Thus, blocking techniques [2] reduce a large number of potential comparisons by removing as many record sets as possible that correspond to non-matches.

Private blocking [3] aims to generate candidate record pairs which are remained to perform PPRL without revealing any sensitive information that can be used to infer individual records and their attribute values. The k-anonymity and Paillier cryptosystem are two main privacy techniques which are applied on private blocking. Although many previous private blocking techniques have used these two privacy techniques [3, 4], there still exist some drawbacks to be solved. In [3], the Two-Party Private Blocking (TPPB) method avoids the use of a third party and cryptographic techniques, and instead, trades off privacy for blocking quality. In [4], Inan et al. suggest creating forming generalized hierarchies (FGH) for reducing the cost of PPRL. However, the forming hierarchies may cause the blocks over-generalization and reduce the accuracy of blocking. We propose a novel private blocking technique based on dynamic k-anonymous blocking and Paillier cryptosystem which can deal with the problems above. Our approach accurately creates blocks without revealing any private information and takes less time than previous approaches which apply cryptographic techniques.

The contributions of this paper are: (1) we propose a novel dynamic k-anonymous blocking algorithm which generates k-anonymous blocks and more accurate values to represent the blocks with varying k, the values are called representative values (RVs) in the following text. (2) We apply a cryptographic technique Paillier cryptosystem on the RVs of each block without revealing any information, which provides stronger privacy than previous approaches. And we propose a novel measure method which performs on the numerical attributes and combines with Paillier cryptosystem to measure the similarity of two blocks in security. (3) Experimental evaluation conducted on a real-world dataset shows our method has an advantage of keeping a high accuracy even k becoming very large. This advantage is meaningful because it is acknowledged that blocks become more secure with the increasing k. We compare our method with other techniques and demonstrate the increases in security and accuracy.

The remainder of this paper is organized as follows. In the following section we mention some previous works related to ours. In Sect. 3 we introduce definitions and background. In Sect. 4 we describe our approach. In Sect. 5 we analyze the privacy of our approach. In Sect. 6 we show its experimental evaluation. Finally we summarize our findings in Sect. 7.

2   Related Work

Due to the growing size of databases, various private blocking methods have been developed in recent years. Most methods rely on the use of a third party. Al-Lawati et al. [5] proposed a secure three-party blocking protocol in 2005 which achieves high performance PPRL by using secure hash encoding for computing the TF-IDF distance measure in a secure fashion. Inan et al. [4] proposed a hybrid approach that combines generalization and cryptographic techniques to solve the PPRL problem in 2008. An approach to PPRL was proposed by Karakasidis et al. [6] in 2011 that a secure blocking based on phonetic encoding algorithms. The records that have similar (sounding) values are divided into the same block. In 2012 a k-anonymous private blocking approach based on a reference table was proposed by Karakasidis et al. [7] for three-party PPRL techniques. Durham [8] proposed a framework for PPRL using Bloom filters in 2012. Recently, Karakasidis [9] proposed a novel privacy preserving blocking technique based on the use of reference sets and Multi-Sampling Transitive Closure for Encrypted Fields (MS-TCEF). As to the two-party techniques, Inan et al. [10] in 2010 presented an approach for PPRL based on differential privacy. The approach combines differential privacy and cryptographic methods to solve the PPRL problem in a two-party protocol. A two-party approach based on the use of Bloom filters for approximate private matching was developed by Vatsalan et al. [11] in 2012. Vatsalan [3] proposed an efficient Two-party private blocking based on privacy techniques k-anonymous clustering and public reference values.

The methods in [3, 4] are closest to our approach. However the approach in [3] uses public reference values as the RVs, although the attributes values of records are not revealed, to a certain degree, public reference values also expose some information about corresponding block. And when k becomes very large, the public reference values cannot sufficiently represent the blocks causing the quality of blocking reduces heavily. The approach in [4] uses forming generalized hierarchies to generate k-anonymous blocks, which may make the RVs over-generalization and reduces the accuracy of generating candidate pairs. We create blocks using dynamic k-anonymous blocking instead of forming hierarchies, which generates the RVs more accurately and flexibly. Applying Paillier cryptosystem provides a stronger guarantee of privacy, which takes less time than previous approaches that apply cryptographic techniques.

3 Preliminaries

3.1 Problem Formulation

We assume two databases D_A and D_B are to be matched, potentially each record from D_A needs to be compared with each record from D_B, resulting in a maximum number of $|D_A| \times |D_B|$ comparisons between two databases. Private blocking contributes to removing obvious non-matching pairs and generating candidate record pairs without revealing any information about the originating plaintexts, which reduces the complexity of comparisons. Considering the privacy, the process of private blocking is different from the traditional blocking. In private blocking, the records of one database should not be exposed to other parties. Further details involved in private blocking are outlined as follows [12]:

Blocking Key Selection. The blocking key is the criteria by which the records are partitioned.

Block Partitioning. Once a blocking key has been selected, this blocking key is as an input to partition each database *respectively* by the same principle where the output is a set of blocks and their RVs.

Candidate Blocks Generation. Given the blocks of each database, through measuring the similarity between the RVs, we can decide whether the records in two blocks compare, then the candidate record pairs would be generated.

3.2 K-anonymity

We now give the definitions of k-anonymity [13].

- Explicit Identifier is a set of attributes, such as name and social security number (SSN), containing information that explicitly identifies record owners;
- Quasi Identifier (QI) is a set of attributes that could potentially identify record owners;
- Sensitive Attributes consists of sensitive person-specific information such as disease, salary, and disability status;
- Non-Sensitive Attributes contains all attributes that do not fall into the previous three categories.

To prevent record linkage through QI, Samarati and Sweeney proposed [13] the notion of k-anonymity:

k-anonymity: If one record in table T has some value QI, at least $k - 1$ other records also have the value QI. Table T is k-anonymity with respect to the QI.

In other words, the minimum group size on QI is at least k. In a k-anonymous table, each record is indistinguishable from at least $k - 1$ other records with respect to QI. Consequently, the probability of linking a victim to a specific record through QI is at most $1/k$. Consider a table T contains no sensitive attributes (such as the voter list). An attacker could possibly use the QI in T to link to the sensitive information in an external source. A k-anonymous T can still effectively prevent this type of record linkage without revealing the sensitive information. In this paper, the RVs are QI.

3.3 Paillier Cryptosystem

The Paillier cryptosystem [14], named and invented by Pascal Paillier in 1999, is a probabilistic asymmetric algorithm for public-private key cryptosystem. The scheme is an additive homomorphic cryptosystem, this means that, given only the public key and the encryption of m_1 and m_2, one can compute the encryption of $m_1 + m_2$. More formally, let $Enc_{k_{pub}}$ and $Dec_{k_{priv}}$ be the Paillier encryption and decryption functions with keys k_{pub} and k_{priv}, m_1 and m_2 be messages, $c(m_1)$ and $c(m_2)$ be ciphertexts such that $c(m_1) = Enc_{k_{pub}} c(m_1)$, $c(m_2) = Enc_{k_{pub}} c(m_2)$. So Homomorphic addition can be expressed by operators "$\cdot$" and "$+$" as follow:

$$Dec_{k_{pri}}(c(m_1) \cdot c(m_2)) = m_1 + m_2. \tag{1}$$

4 Proposed Solution

Our proposed solution conducts private blocking by dynamic k-anonymous blocking and Paillier cryptosystem. It is composed of three parts: Data Preparation, Local k-anonymous Blocks Construction and Candidate Blocks Generation. The framework is described in Fig. 1.

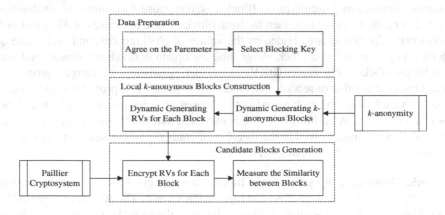

Fig. 1. The framework of our approach

4.1 Data Preparation

In data preparation, we agree on the parameters used in our approach and select one or more attributes as blocking keys.

Agree on the Parameter. We assume three participants in our method *Alice*, *Bob* and *Charlie*. *Alice* and *Bob* are the owners of databases D_A and D_B who participate in the protocol to perform private blocking on their databases. *Charlie* is used to generate candidate blocks or in other words decide whether to compare the records in two blocks. *Alice* and *Bob* agree on the parameter k that the minimum number of elements in a block.

Select Blocking Key. Blocking key is used to partition the records into blocks. Selecting an appropriate blocking key is necessary. To protect the privacy of blocks, our approach generates blocks satisfied k-anonymity, in other words each block contains at least k records. The method in [3] also uses k-anonymity and select Given name and Surname as blocking keys. However, when k becomes large, the RVs in method [3] cannot sufficiently represent the blocks causing the quality of blocking reduces heavily. To avoid the deficiency above, our approach selects the numerical attributes such as age, zip code or salary as the blocking key. The numerical attributes represent the blocks more accurately and flexibly with varying k. They also take less time than other attributes.

4.2 Local k-anonymous Blocks Construction

The local blocks construction phase partitions the records into blocks by blocking key. To construct blocks on distinct data sources without leaking any private information, our approach utilizes k-anonymity and Paillier cryptosystem privacy techniques. We generate k-anonymous blocks and obtain the RVs of each block using dynamic k-anonymous blocking algorithm.

Dynamic Generating k-anonymous Blocks. We suppose A_N (numerical attribute) is selected to be the blocking key, then we form blocks on the databases of *Alice* and *Bob* respectively. The blocks are divided by the values of blocking key, and each value of blocking key construct one block. After this, we obtain equivalence classes and sort them by the blocking key values (BKVs). Considering privacy, we merge equivalence classes until the number of records in a block being at least k. It provides k-anonymous privacy characteristics, as each record in the database can be seen as similar to at least $k - 1$ other records. Algorithm 1 (which is executed independently by *Alice* and *Bob*) shows the main steps involved in the merging of equivalence classes to create k-anonymous blocks (Algorithm 1, lines 4–7).

Dynamic Generating RVs for Each Block. We assume L is a block satisfied k-anonymity, and x, y are the smallest and biggest BKVs in L. The RVs are composed by $[x, y]$. Then, the BKVs of each record in block L is replaced by $[x, y]$, more specifically each record in block L has at least $k - 1$ records with the same BKVs. Therefore, the block L is k-anonymity respecting to $[x, y]$ and $[x, y]$ is the RVs of the block L.

Comparing the approach in [4], which uses forming generalized hierarchies may lead to the RVs over-generalization and reduce the accuracy of generating candidate blocks, our approach dynamically adjusts the RVs with the change of k and has a good influence on keeping high accuracy even k becoming very large. Algorithm 1 shows the main steps involved in dynamic generating the RVs of each block (Algorithm 1, line 8).

Algorithm 1: Dynamic k-anonymity Blocking

Input:
- **E:** Equivalence classes divided and sorted by A_N $\{E_1, E_2, E_3, \ldots, E_n\}$
- Minimum number of elements in a block k
Output:
- L_A: Set of k-anonymous blocks $\{ L_{A1}, L_{A2}, L_{A3}, \ldots, L_{Am} \}$

- $V[L_{Am}]$: RVs of L_{Am}

1:    $i=1; j=1; L_{Aj}=\emptyset;$

2:    **while** $i \leq n$ **do:**
3:        $Kset = \emptyset$
4:            **while** $\left| L_{Aj} \right| \leq k$ **do:**
5:                $L_{Aj} = L_{Aj} \cup E_i$
6:                $Kset.\mathrm{add}(\mathrm{E}_i . A_N)$
7:                $i++$
8:        $V[L_{Aj}] = [Kset[0], Kset[size-1]]$
9:        $j++$

4.3   Candidate Blocks Generation

After generating k-anonymous blocks and corresponding RVs, we need to decide candidate blocks to eliminate record pairs that are expected to be non-matches. To protect the privacy of RVs and generate candidate blocks, Algorithm 2 shows the process that encrypts the RVs with Paillier and performs a novel measure method on the encrypted RVs to measure the similarity between blocks. And Fig. 2 shows the process of generating candidate blocks in privacy, from which we know that our approach is absolute security with none information revealing.

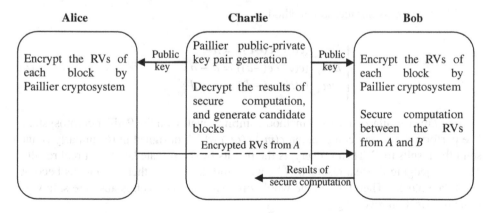

Fig. 2. The process of generating candidate blocks in privacy

Encrypt RVs for Each Block. To measure the similarity between blocks, the RVs of blocks should be released by at least one data owner. Before releasing, the RVs in both A and B are encrypted by Paillier to guarantee privacy.

Charlie generates Paillier public-private key and send the public key to A and B. Then, A and B respectively encrypt their RVs with the public key (Algorithm 2, lines 3–5). We assume that the RVs of block L_A (from A) is $[a, b]$ and the RVs of block L_B (from B) is $[c, d]$. The RVs are encrypted as follow:

$$c(-a) = Enc_{k_{pub}}(-a); \quad c(b) = Enc_{k_{pub}}(b) \tag{2}$$

$$c(-c) = Enc_{k_{pub}}(-c); \quad c(d) = Enc_{k_{pub}}(d); \quad c(-d) = Enc_{k_{pub}}(-d) \tag{3}$$

Measure the Similarity Between Blocks. After getting encrypted RVs in A and B, we pass the encrypted RVs in A to part B. In part B who lacks the private key, Bob cannot infer the plaintexts of records in A.

As to the party B, Bob has gained the encrypted RVs from A, then he uses the encrypted RVs of two blocks from A and B to decide whether two blocks match. We design a novel similarity measure method which combines with Paillier cryptosystem to measure the similarity between blocks (Algorithm 2, lines 7–16). The novel similarity measure method is expressed as follow:

$$\begin{cases} b < c \text{ or } d < a, & L_A \text{ and } L_B \text{ non - match} \\ b < d, & L_A \text{ and } L_B \text{ match, but } L_A \text{ does} \\ & \text{not match with other blocks in } B \\ \text{otherwise} & L_A \text{ and } L_B \text{ match} \end{cases} \tag{4}$$

According to the Homomorphic addition in Paillier cryptosystem:

$$Dec_{k_{pri}}(c(m_1) \cdot c(m_2)) = m_1 + m_2. \tag{5}$$

We can express our measure method as:

$$\begin{cases} Dec_{k_{pri}}(c(b) \cdot c(-c)) = b - c \\ Dec_{k_{pri}}(c(d) \cdot c(-a)) = d - a \\ Dec_{k_{pri}}(c(b) \cdot c(-d)) = b - d \end{cases} \tag{6}$$

Our novel similarity measure method combines well with the Paillier cryptosystem. We perform the secure computation $c(m_1) \cdot c(m_2)$ which designed in (6) in party B and send the results to C. Then C decrypts the results by the private key to get real results. Through judging the real results by (4), we could decide whether two blocks become candidate blocks. Therefore, in the whole process, our approach is absolute safe with none information revealing.

The last step PPRL conducts on each candidate record pairs individually by using a private matching technique, which should not reveal any information regarding the sensitive attributes and non-matches (this step is outside of our approach).

Algorithm2: Generating Candidate Blocks

Input:
- $V(L_A)$: RVs of each block in A $\{[a_1, b_1], [a_2, b_2],...,[a_n, b_n]\}$
- $V(L_B)$: RVs of each block in B $\{[c_1, d_1], [c_2, d_2],...,[c_m, d_m]\}$

Output:
- Candidate blocks match or non-match

1:    **for** $i=1$; $i \leq n$; $i++$ **do**
2:      **for** $j=1$; $j \leq m$; $j++$ **do**
3:        $c(-a_i) = Enc_{k_{pub}}(-a_i)$; $c(b_i) = Enc_{k_{pub}}(b_i)$;
4:        $c(-c_j) = Enc_{k_{pub}}(-c_j)$; $c(d_j) = Enc_{k_{pub}}(d_j)$;
5:        $c(-d_j) = Enc_{k_{pub}}(-d_j)$;
6:        send $c(-a_i)$ and $c(b_i)$ to B
7:        $S_1 = c(b_i) \cdot c(-c_j)$; $S_2 = c(d_j) \cdot c(-a_i)$;
8:        $S_3 = c(b_i) \cdot c(-d_j)$;
9:        send S_1, S_2, S_3 to C
10:       **if** $Dec_{k_{priv}}(s_1) < 0$ or $Dec_{k_{priv}}(s_2) < 0$ **then**
11:         **return** non-match;
12:       **else if** $Dec_{k_{priv}}(s_3) < 0$ **then**
13:         **return** match;
14:         **break**;
15:       **else**
16:         **return** match;

---

5   Privacy Analysis

In this section we will discuss the privacy guarantees offered by our approach. We assume *Alice*, *Bob* and *Charlie* will follow the protocol honestly, but may try to infer private information based on messages they receive during the process without collusion [15]. Next we summarize the information that our approach discloses to each of the participants.

Alice: This party does not receive any messages regarding *Bob*'s database.
Bob: This party receives encrypted RVs of blocks from A. With the protection of Paillier cryptosystem, B cannot infer the real values from A.
Charlie: This party does not receive any messages regarding the RVs of blocks in A or B but receives the encrypted results of secure computation from B. After decrypting the encrypted results with private key, the real results only show final results without revealing the specific information from A and B. For example, C only knows the result of $b-c$ and does not know the respective value of b and c.

6   Experiments

To perform the experimental analysis, we selected a publicly available dataset of real personal identifiers, derived from the North Carolina voter registration list (NCVR). The database NCVR contains 375,314 records. We selected attribute Age as the blocking key. For blocking evaluation, we need to generate two different sizes of datasets which are 10,000 and 100,000. Therefore, we respectively sampled 10,000 and 100,000 number of records randomly drawing from NCVR for *Alice*. Then we generated datasets for *B* composed of 10,000 and 100,000 records as well. Of these records, 8000 (80000) were randomly selected from NCVR (excluding those in *A*), while 2000 (20000) were randomly selected from *A*. The goal was to privately identify the 2000 (20000) matching records between *A* and *B*. Our experiments also perform on datasets of different sizes, we sampled 0.1 %, 1 %, 10 % and 100 % of records in the full database twice each for *A* and *B*. All tests were conducted on a computer server with a 64-bit, 8.0G of RAM Intel Core (3.30 GHz) CPU.

6.1   Evaluation Measures

We use the following measures to evaluate the performance of private blocking techniques in terms of complexity and quality of blocking. Complexity is evaluated by the total time required for blocking. We utilize reduction ratio (*RR*) and pair completeness (*PC*) as evaluation metrics for private blocking approaches [15]. Specifically, suppose c is the number of candidate record pairs produced by the private blocking, c_m is the number of true matches among c candidate pairs, $n = |D_A| \cdot |D_B|$ is the number of all possible pairs and n_m is the number of true matches among all pairs. Then, *RR* and *PC* are defined as follows:

$$RR = 1 - c/n \quad PC = c_m/n_m \tag{7}$$

6.2   Performance Evaluation

We compare our approach with previous two approaches TPPB [3] and FGH [4]. The approach TPPB generates candidate blocks satisfying k-anonymity and uses public reference values as the RVs of blocks. Since each block consists of at least k records, only when revealing one reference value from each block can guarantee k-anonymity privacy. If several reference values are released by a block, the k-anonymity privacy would not be guaranteed. As to FGH, it generates k-anonymous blocks by forming generalized hierarchies.

We set the parameters of two approaches according to the settings provided by the authors [3, 4]. We compared three private blocking techniques on two different sizes of datasets which are 10,000 and 100,000 to measure the change of *RR*, *PC* and *blocking time* against k. The changing trends of *RR*, *PC* and *blocking time* against k are similar in two datasets. We also measure the *blocking time* with different dataset sizes for the three approaches. Then, we discuss the results of our experiments.

RR with Varying k. Figure 3 shows the RR with varying k in three approaches. Our approach and FGH keep a high RR with the increasing k. When k increases to 1000, RR is still above 0.86 in the smaller dataset. Towards TPPB, at first RR reduces when k is less than 200. Then, with k becoming bigger, RR increases and at last RR almost closes to 1. It can be explained that when k becomes larger, in TPPB, representing a block by only one reference value is not sufficient to represent all the values in block, which might lead to the number of candidate blocks reduces and the RR increases.

Fig. 3. RR with different values for k (a) Dataset Size = 10,000 (b) Dataset Size = 100,000

PC with Varying k. Because of the reason above, some true candidate blocks being missed with the increasing k, therefore the PC reduces heavily in TPPB as shown in Fig. 4. In FGH, PC also reduces heavily with the reason that the bigger the k the higher level in the VGHs the records are generalized which may cause over-generalization. With regard to our approach, PC is always 1 on both datasets. This owns to our good similarity measure method.

Fig. 4. PC with different values for k (a) Dataset Size = 10,000 (b) Dataset Size = 100,000

Blocking Time with Varying k. To the aspect of *blocking time* in Fig. 5, the *blocking time* reduces with k in three approaches because the number of resulting blocks (n/k) becomes less as k gets bigger. As shown in Fig. 5, the blocking time of our approach is more than the other two approaches. It is because that our approach applies Paillier cryptosystem.

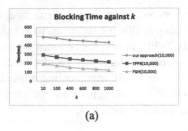
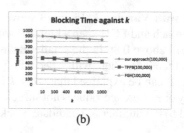

(a)   (b)

Fig. 5. *Blocking Time* with different values for *k* (a) Dataset size = 10,000 (b) Dataset size = 100,000

***Blocking Time* with Varying Database Sizes.** In Fig. 6, we compare the *blocking time* for three approaches with different dataset sizes. Our approach takes a little more time than the others with different dataset sizes. All the three approaches do not consider the communication cost. Through inferring, we can get the knowledge that all encrypted RVs are totally transmitted at most 500 times in our approach, which far less than the communication cost of previous approaches applying cryptographic techniques.

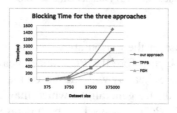

Fig. 6. *Blocking Time* with different dataset sizes for the three approaches

Hence, we conclude that our approach performs better in accuracy and privacy with a little loss of efficiency.

7 Conclusion

We present a novel scalable private blocking technique which is more accurate and secure than previous approaches. Dynamic *k*-anonymity blocking guarantees that each block has at least *k* records and meanwhile generates more accurate RVs with varying *k*. We also propose a novel similarity measure method which combines with Paillier cryptosystem and guarantees absolute security without revealing any information. As experiments show, our approach exhibits high performance both in accuracy and security with a little loss of *blocking time*. A limitation in our approach is the application of Three-Party Private Blocking. In future work, we plan to extend our approach to Multi-Party Private Blocking that is applicable for several datasets.

Acknowledgment. This work is supported by the National Basic Research 973 Program of China under Grant No. 2012CB316201, the National Natural Science Foundation of China under Grant No. 61472070.

References

1. Vatsalan, D., Christen, P., Verykios, V.S.: A taxonomy of privacy-preserving record linkage techniques. Inf. Syst. **38**(6), 946–969 (2013)
2. Christen, P.: A survey of indexing techniques for scalable record linkage and deduplication. IEEE Trans. Knowl. Data Eng. **24**, 1537–1555 (2011)
3. Vatsalan, D., Christen, P., Verykios, V.S.: Efficient two-party private blocking based on sorted nearest neighborhood clustering. In: ACM CIKM (2013)
4. Inan, A., Kantarcioglu, M., Bertino, E., Scannapieco, M.: A hybrid approach to private record linkage. In: ICDE, pp. 496–505 (2008)
5. Al-Lawati, A., Lee, D., McDaniel, P.: Blocking-aware private record linkage. In: IQIS, pp. 59–68 (2005)
6. Karakasidis, A., Verykios, V.S.: Secure blocking + secure matching = secure record linkage. J. Comput. Sci. Eng. **5**, 223–235 (2011)
7. Karakasidis, A., Verykios, V.S.: Reference table based k-anonymous private blocking. In: 27th Annual ACM Symposium on Applied Computing, Trento (2012)
8. Durham, E.: A framework for accurate, efficient private record linkage. Ph.D. Thesis, Vanderbilt University (2012)
9. Karakasidis, A., Verykios, V.S.: Scalable blocking for privacy preserving record linkage. In: ACM KDD, Sydney (2015)
10. Inan, A., Kantarcioglu, M., Ghinita, G., Bertino, E.: Private record matching using differential privacy. In: EDBT, Lausanne, Switzerland, pp. 123–134 (2010)
11. Vatsalan, D., Christen, P.: An iterative two-party protocol for scalable privacy-preserving record linkage. In: Aus DM, CRPIT, Sydney, Australia, vol. 134 (2012)
12. Durham, E.A.: A framework for accurate, efficient private record linkage. Ph.D. thesis, Graduate School of Vanderbilt University, Nashville (2012)
13. Sweeney, L.: *k*-anonymity: a model for protecting privacy. Int. J. Uncertainty Fuzziness Knowl. Based Syst 10, 557–570 (2002)
14. Paillier, P.: Public-key cryptosystems based on composite degree residuosity classes. In: Stern, J. (ed.) EUROCRYPT 1999. LNCS, vol. 1592, pp. 223–238. Springer, Heidelberg (1999)
15. Kuzu, M., Inan, A.: Efficient privacy-aware record integration. In: ACM EDBT (2013)

Repair Singleton IDs on the Fly

Xingcan Cui[1], Xiaohui Yu[1,2(✉)], and De Guo[1]

[1] School of Computer Science and Technology,
Shandong University, Jinan, China
{xccui,guode}@mail.sdu.edu.cn, xyu@sdu.edu.cn
[2] School of Information Technology, York University,
Toronto, ON M3J 1P3, Canada

Abstract. Tracking moving entities at predefined locations plays an essential role in many surveillance related applications. Occasionally, the IDs of those entities are incorrectly recorded due to various reasons such as errors in recognition. Such errors need to be repaired on the fly as those IDs are often involved in some time-sensitive query processing or data analysis tasks. In this paper, we address a specific case where the errors result in singleton IDs, i.e., IDs that appear only once during a specific period of time and thus could be safely presumed to be erroneous. The repair of the IDs is based on constraints posed by the data itself (e.g., constraints posed by the road network). We present a tracking tree structure to index the candidate repairs for each singleton ID, which enables repairing of the IDs on the fly. We implement a distributed repair system on the Apache Storm platform. Experiments on both real and synthetic datasets demonstrate the effectiveness and efficiency of our singleton detection and repair approach.

Keywords: Data quality · Location sequence constraint · Stream process

1 Introduction

Tracking moving entities in predefined locations plays an essential role in many surveillance related applications. For example, in a cargo container management system, the ID printed on each container's body is captured [15]; or in a traffic surveillance system [3], surveillance cameras installed at a predefined set of locations automatically capture and recognize vehicle license plates. In these applications, the key operation is to identify the ID for each entity. However, in reality, the recognition rate is influenced by many factors, (e.g., lighting and weather conditions). In these scenarios, errors in the captured/recognized IDs are not uncommon. Though new techniques are always being developed to improve the recognition rate, errors cannot be completely eliminated in real-world applications [2].

We are concerned with detecting and repairing the erroneous IDs by exploring the correlations between tracked records. Specifically, in this paper, we tackle

© Springer International Publishing Switzerland 2016
F. Li et al. (Eds.): APWeb 2016, Part II, LNCS 9932, pp. 214–226, 2016.
DOI: 10.1007/978-3-319-45817-5_17

the case of singleton IDs, i.e., IDs that appear only once over a period of time. Since in many applications, an ID is normally captured more than once over a sufficiently long period (e.g., a vehicle is likely to be captured more than once over an hour by the surveillance cameras if the locations are dense), there is a high probability that it is an erroneous one if it does not appear again over a given time window. Since each entity is usually captured in multiple locations, even if an ID is misidentified in one location, we could try to find the true ID from other locations by consulting the moving sequences of locations they make.

The detection and repair of IDs errors can benefit from the fact that in many applications, there are inherent constraints on how the entities could move (and thus on the sequence of locations for the entity IDs). For example, if there exist three capture locations A, B, C on the same road (appearing in that order), all vehicles passing through location A and C should also pass location B, since B falls between A and C. Similarly, vehicles that pass through location B would also have passed through location A and may pass through location C in the near future. As more capture locations are added, the location sequences may become complex, but they can always be enumerated. We call that the "location sequence constraint" for entities.

In addition, we observe that the records with erroneous ID often show similarities or connections with their corresponding true records. If the IDs are strings, the edit distance between an erroneous ID and the true ID tends to be short compared with the average distance between a random pair of IDs. We can make use of this fact to help identify the true ID.

Since in most cases the records are continuously collected and analyzed, the quality of data will directly affect some time-sensitive tasks, (e.g., fake or duplicated plate detection and travel time analysis in a traffic surveillance system [3]), we aim to detect and repair the ID errors on the fly.

In this paper, we mainly make the following contributions.

- We present a new approach to for ID repair in the case of singleton IDs based on location sequence constraints. This approach is nontrivial since we cannot easily pinpoint the location sequence that each entity follows.
- We develop a tracking tree data structure to index prefixes of location sequences that support the repair of IDs on the fly. We present the structure of the tree and describe how the tree is maintained.
- We implement an ID repair system on a distributed stream processing platform, Apache Storm [1], to support the ID repair over large volumes of data.
- We show through experiments on both real and synthetic datasets the effectiveness and efficiency of our approach.

2   Problem and Solution

2.1   Definitions

The key terms used in this paper are defined below.

Definition 1. *Tracking record. A tracking record is defined as a triplet like* $\langle \mathbf{k}, \mathbf{loc}, \mathbf{t} \rangle$, *where* $\mathbf{k}$ *represents the recognized ID that may be erroneous,* $\mathbf{loc}$ *represents the location where the record is captured and* $\mathbf{t}$ *represents the capture time.*

Definition 2. *Trajectory. A trajectory* $\mathbf{T}_k$ *is defined as a sequence of tracking records with identical ID k and are chronologically ordered, like* $\mathbf{T}_k = \langle k, loc_1, t_1 \rangle \rightarrow \langle k, loc_2, t_2 \rangle \rightarrow \cdots \rightarrow \langle k, loc_n, t_n \rangle$, *where* $t_1 < t_2 < \cdots < t_n$.

Definition 3. *Location sequence. A location sequence is defined as a finite sequence of locations with length* ≥ 2, *like* $loc_1 \rightarrow loc_2 \rightarrow \cdots \rightarrow loc_l$. *In reality, there should always be a set of location sequences that each entity should follow. We call them* **feasible location sequences**.

Definition 4. *Singleton trajectory and singleton ID. If the length of a trajectory is always 1 (containing only one tracking record) in a pre-defined time window* $\mathbf{tw}$, *we call it a singleton trajectory and the corresponding ID a singleton ID.*

According to our assumptions, singleton IDs are considered to be errors since they do not show up again in other locations, and thus violate the location sequence constraints. To repair them, we try to find correct IDs from other trajectories and rewrite the singleton IDs to them. By ID rewriting, two or more trajectories can be merged to compose a new trajectory as if they belong to the same entity. Notably, each entity's true trajectory must follow one of the feasible location sequences. Thus the location sequence for a newly composed trajectory should always be a prefix of one of them.

Definition 5. *Candidate repair and target ID. Given a set of feasible location sequences* $\mathcal{S}$, *suppose a trajectory* $\mathbf{T}_r$, *as well as one or more singleton trajectories* $\mathbf{T}_1, \mathbf{T}_2, \cdots, \mathbf{T}_p$ *can be merged (by writing* $ID_1, ID_2, \cdots, ID_p$ *to r) to compose a new trajectory* $\mathbf{T}'_r$ *and the location sequence for* $\mathbf{T}'_r$ *is the prefix of one feasible location sequence in* $\mathcal{S}$, *we call the set of trajectories a candidate repair and the ID r the target ID.*

Actually, we take the target IDs as the correct ones that other errors should be repaired to. For each candidate repair, there could be multiple singleton trajectories, while only one with the target ID. This can be easily decided when there is only one trajectory that is not a singleton (must contain the target ID). However, when all trajectories in a candidate repair are singletons, we must evaluate the confidence of correction for each trajectory and choose the most likely target ID.

Definition 6. *Confidence function. Given a candidate repair* $CR = \{T_1, T_2, \cdots, T_p\}$ *with all trajectories are singletons, we define the confidence function to evaluate the probability of correctness for trajectory with* $ID = k$ *as:*

$$conf(CR, k) = \frac{1}{p-1} \sum_{i=1}^{p} \frac{dist(k, ID_i)}{max(|k|, |ID_i|)},$$

where $dist(k, ID_i)$ *represents the edit distance for string k and* ID_i.

The confidence function we introduced actually reveals the average edit distance from an ID to all other IDs in a candidate repair. In reality, more factors could be added to the function, e.g., the inherent recognition accuracy of devices, the rarity of ID strings, etc. In this paper, we simply adopt the edit distances for ID pairs. Moreover, as each singleton trajectory may be contained in multiple candidate repairs, the confidence function cannot only be used to decide the target ID in a single candidate repair, but also can be used to select the most likely target ID for all candidate repairs that contain the same singleton ID.

2.2   Problem Statement and Solution

For a continuously collected stream of tracking records, we concentrate on the problem of detecting singleton IDs, as well as on choosing the most likely candidate repairs and target IDs for them.

The repair approach we developed can be separated into two processes: singleton detection (detect ID error) and target selection (repair ID error).

In keeping with goal – repair singleton ID on the fly, we accomplish the two processes on a stream data model. The singleton detection process is trivial, since we can easily group the incoming tracking records by their IDs and construct trajectories, separately. Once an ID was asserted to be a singleton, we find all its candidate repairs immediately, decide the target IDs and then choose the most likely one. This process will be accomplished with the assistant of a data structure called tracking tree.

3   The Tracking Tree

To continuously construct and index the candidate repairs for each trajectory, we developed a data structure called tracking tree.

As shown in Fig. 1, a tracking tree is actually a prefix-tree for a feasible location sequence ($A \rightarrow B \rightarrow C \rightarrow D \rightarrow E$ in this figure). Each node in this

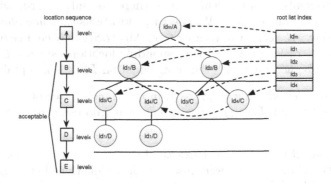

Fig. 1. The structure of a tracking tree for location sequence $A \rightarrow B \rightarrow C \rightarrow D \rightarrow E$ and main ID id_m

tree represents a tracking record (may be duplicated) whose level is consistent with the position of its location in the feasible location sequence. We call the root's ID *main ID* and each tracking tree can be decided by the main ID and a corresponding feasible location sequence. There are two extra definitions used in the tracking tree:

Definition 7. *dPath. We define dPath as a path from the root of the tree to an arbitrary leaf node.*

Note that, any location sequence derived from a *dPath* must be a prefix of the corresponding feasible location sequence and thus trajectories in a *dPath* can compose a candidate repair. In other words, a tracking tree actually represents trajectory combinations for multiple candidate repairs.

Definition 8. Dominant ID. *For any dPath of a tracking tree, if the node with ID k exists, we say ID k dominants this dPath.*

As each *dPath* is started with the root, thus the main ID dominant all *dPath*es in a tracking tree.

When encountering a tracking record with ID never seen before, we first create tracking trees for this ID (as the main ID) and all feasible location sequences headed by its location attribute. Then we try to update existing tracking trees with the new record to construct prefixes of feasible location sequences. Specifically, the update operation is done by insert and delete operations. We will introduce them separately.

In the rest of this paper, we use the notation LS to represent the corresponding location sequence of a tracking tree, $LS[i]$ to represent the ith location attribute of the location sequence and id_m to represent the main ID.

3.1 The Insert Operation

Checking Acceptability. Before inserting a tracking record as nodes to a tracking tree, we should first check if the record is acceptable. Due to the location sequence constraints, at a particular time, there are only records with specific locations that may be accepted. We call these locations the "*acceptable location set*" (denoted by $ALSet$) for a tracking tree. An $ALSet$ may change during data collecting. For a tracking tree with corresponding location sequence LS, location $LS[0]$ can only be accepted with the root. Location $LS[i]$ is acceptable if:

1. there exist nodes with $loc = LS[i-1]$ in the tracking tree or
2. nodes with location $LS[j](j \geq i)$ and $ID = id_m$ do not exist in the tracking tree.

These rules restrict a level by level growth of the tracking tree. Also, once nodes with main ID and $loc = LS[j]$ were inserted, other locations before $LS[j]$ in LS $(LS[j-1], LS[j-2], \cdots)$ can never be accepted again. In Fig. 1, the $ALSet$ should contain all locations except location A. With the help of a first-in, first-out queue, we can update the $ALSet$ by a level traversal. Note that, once a record with $ID = id_k$ is not accepted by a tracking tree, then all other records with the same ID cannot be accepted since the combination is done on trajectories rather than isolated records.

Nodes Insertion. The basic idea to insert a new coming tracking record with location $LS[i]$ is to replicate and append it as children to existing nodes with location $LS[i-1]$. During node insertion, conflicts may appear. We use a set $CSet$ to store nodes that conflict with the new record and should be removed from the tracking tree. For a tracking record $r_k = \langle id_m, LS[i], t_1 \rangle$ and an existing node with $node_k = \langle id_n, LS[i-1], t_0 \rangle$, the specific insertion rules are as follows:

1. If $node_k$ has no children, we can directly insert r_k as a child to $node_k$.
2. If $node_k$ has children, and id_m do not dominate $dPath$es containing $node_k$, we can insert r_k as a new child of $node_k$.
3. If $node_k$ has children, and id_m dominate $dPath$es containing $node_k$, we should remove the children to the $CSet$ first and then insert r_k as the only child of $node_k$.
4. If $node_k$ has an only child with id_e, and both id_m and id_e dominate the $dPath$es containing $node_k$, we should just remove the node with id_e to the $CSet$.

For a tracking record with ID k, if there already exist nodes with the same ID in the tree, the new record must be inserted as successors of them. In other words, different records from the same trajectory should always be in the same $dPath$. As shown in Fig. 1, we can index roots as $rList$ for each ID (trajectory) and speed up positioning the parent nodes for new records to append to.

3.2   The Delete Operation

During data collection, some IDs may proved to be unsuitable for a tracking tree. Nodes corresponding to these IDs should be deleted. There are four situations to do that. They are:

1. As illustrated before, when inserting tracking records, existing nodes that show confliction will be added to $CSet$ and then should be removed from subtrees rooted by the inserted nodes' roots.
2. If a node has been deleted, all its successor nodes should also be deleted.
3. Once a trajectory breaks on a level of the tree (all existing nodes from a level have been deleted), other nodes from the same trajectory should also be deleted.
4. If a new coming record cannot be inserted to a tracking tree, all existing nodes with the same ID should be deleted and this tree will never accept records from the same trajectory.

It is not hard to see that the deletion of some nodes may cause other nodes to be not suitable for a tracking tree and so should also be deleted. For example, suppose a $dPath$ in a tracking tree is $id_1 \rightarrow id_2 \rightarrow id_3 \rightarrow id_1 \rightarrow id_2$. If there comes a tracking record with $ID = id_3$ and the location is the same with the last node ($ID = id_2$), then nodes with $ID = id_2$ will be not suitable and should be deleted. After that, nodes with $ID = id_3$ will also turn out to be not suitable and eventually the whole $dPath$ will be deleted. The delete operation is actually a chain process. Thus we need two cycle procedures to accomplish it: *node remove* and *recursive delete*. We first call the node remove procedure to remove nodes whose ID show confliction in the node insertion procedure (contained in $CSet$). Then we call the recursive delete procedure to find out nodes that become not suitable after the remove procedure (see delete situation 3) and call the remove procedure again until there are no new nodes that should be deleted.

To check if a trajectory breaks, we can store the nodes' count on each level. Each delete operation will be conducted on subtrees rooted by the newly inserted ID's roots, since confliction caused by node insertion may only affect them. Once the *addedList* outputted an insert operation is empty, it means no new nodes were added. All existing nodes with that ID, including the roots, should be deleted.

After a series of insert and delete operations, some tracking trees may prove to be fake, if the trajectory corresponding to the main ID cannot be accepted. The choice of a wrong feasible location sequence for that trajectory may be the cause. Moreover, to avoid indefinitely expanding, we set an *expiration time* for roots of tracking trees. Expired tracking trees will also be taken as fake that do not accept tracking records any more.

4   Performing Repair in a Distributed Fashion

Though our proposed method is more suitable for regions with restricted feasible location sequences, in reality, the tracking records may come from multiple isolated regions and should be processed concurrently. Thus we implement the system on a distributed stream-processing platform - the Apache Storm (Other distributed processing platforms with similar architectures could also be used).

In Storm, a job is logically abstracted to a directed graph - the *topology*, which consists of the entrance unit - *spout* and a series of processing units - *bolts*. Data stream containing infinite records (or other data unit) are processed and transferred from one bolt instance to another without accessing the disk.

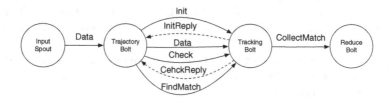

Fig. 2. The storm topology

Table 1. Seven streams in the topology

Name	Source	Destination	Grouping method
Data	InputSpout	TrajectoryBolt	Fields grouping by ID
	TrajectoryBolt	TrackingBolt	All grouping
Init	TrajectoryBolt	TrackingBolt	Fields grouping by ID
InitReply	TrackingBolt	TrajectoryBolt	Direct grouping
Check	TrajectoryBolt	TrackingBolt	All grouping
CheckReply	TrackingBolt	TrajectoryBolt	Direct grouping
FindMatch	TrajectoryBolt	TrackingBolt	All grouping
CollectMatch	TrackingBolt	ReduceBolt	Fields grouping by ID

As shown in Fig. 2, we accomplish the trajectory composing and singleton detection process in the *TrajectoryBolt*. When encountering a tracking record with new ID, it will send requests (via the *Init* stream) to the downstream *TrackingBolt*, where the tracking trees are located in, to initialize tracking trees rooted by the record. Then, the *TrajectoryBolt* will send "acceptable checking" requests by the *Check* stream to gain a list of tracking trees that may accept records from this trajectory.

The main challenges here lies on how to distribute the tracking trees and how to reduce the cost of meaningless update operations. In our system, tracking trees with the same main ID are grouped to construct star structures (named tracking tree star) which can be distributed to different *TrackingBolt* instances by their ID. Also, we maintain indices from trajectories to tracking tree stars that indicate which tracking trees can accept a trajectory. As illustrated before, tracking records from a trajectory can not be accepted by a tracking tree if any other records from the same trajectory can not be accepted. Due to that, the number of indices will monotonically decrease during data collection which can effectively eliminate unnecessary update operations.

Once a trajectory (ID) is judged as a singleton, the *TrajectoryBolt* will send requests to all downstream tracking trees to generate candidate repairs via the *FindMatch* stream. The last *ReduceBolt* is responsible for collecting these candidate repairs and choosing the most likely one, referring to the confidence function. All these singleton IDs will be repaired.

There are seven streams defined in the topology. The source, destination and grouping method of them are shown in Table 1.

5 Experiments

Based on the repair system on Apache Storm, we conduct experiments to evaluate the effectiveness and efficiency of our proposed method.

5.1 Experiment Settings

We use two kinds of dataset: a real dataset from the traffic surveillance system of a provincial capital city in China and a series of synthetic datasets.

The real dataset which contains 2045 tracking records of 699 trajectories were collected from a restricted region from 8:00 a.m. to 9:00 a.m. on a particular day. Figure 3 shows the distribution of surveillance locations in the region. There are five feasible location sequences: $A \to B \to C \to D$, $A \to C \to D$, $A \to B \to C$, $A \to C$ and $A \to B$. Due to error recognition, some license plates were misidentified. We manually label the plate numbers by examining the original photos taken by each camera – that gives us a labeled dataset that can be used as ground truth. Also, to explore the performance of our system, we generate a series of synthetic datasets. All of them share the same features with the real dataset in time interval and edit distance distribution for the error/correct ID pairs.

The system is implemented in Java language and is deployed on a storm cluster consisting of three desktop PCs with 2.5 GHz Intel i5 CPU and 8 GB of memory. The error detection and error repair processes are considered separately. We adopt the recall, accuracy and f-measure, which are often used in information retrieval, as the metrics for effectiveness evaluation. Using $error$ to represent the set of records with ID error, $detected$ to represent the set of records that have been labeled as singletons (error) and repaired, and $correct$ to represent the number of records that have been correctly repaired, in the error detection process, $recall_d = \frac{|error \cap detected|}{error}$, $precision_d = \frac{|error \cap detected|}{detected}$ and $f\text{-}measure_d = \frac{2 \cdot (recall_d \cdot precision_d)}{recall_d + precision_d}$. In the error repair process, $recall_r = \frac{correct}{error}$, $precision_r = \frac{correct}{detected}$, and $f\text{-}measure_r = \frac{2 \cdot (recall_r \cdot precision_r)}{recall_r + precision_r}$. According to a sample of correctly recognized trajectories, we set the time window for singleton judgment to 200 s and the expiration time to 20 min by default.

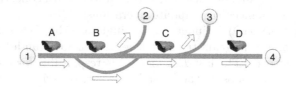

Fig. 3. Road network of the region

5.2 Experiments on Real Dataset

The results of the error detection and error repair processes are shown in Fig. 4(a) and (b). From Fig. 4(a), we can conclude that singleton IDs can properly represent errors. However, in two situations the detection method may fail.

1. If the ID for an entity is all misidentified except at one location, it may cause the only correct ID to be a singleton, especially when the feasible location sequence is 2.
2. Missing records may also cause a correct ID to be a singleton since it will prolong the interval of two adjacent records or even make it infinite.

To verify the effectiveness of the location sequence constraints, we implement another repair approach, which only take the edit distance of two ID strings into account. Specifically, when an ID is judged as a singleton, we choose another ID whose edit distance is the shortest with the singleton as its target. The comparison of the two approaches is shown in Fig. 4(b). The f-measure of our repair approach and the approach only considering edit distance is 63 % to 35 %, which shows that the location sequence constraints can effectively improve the repair rate. However, the two situations that may cause singleton ID detection failure will also affect the repair process. Besides, the confidence function we present in this paper only consider the edit distance of ID strings, which may lead to a mismatch when the edit distance between an erroneous ID and its true one is relatively long.

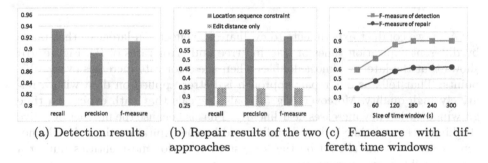

(a) Detection results   (b) Repair results of the two (c) F-measure with dif-
                           approaches   feretn time windows

Fig. 4. Results of the detection and repair processes on real dataset

Figure 4(c) shows the f-measure results with different sizes of time window. We can see from this figure that, with time window varying from 30 s to 180 s, both the quality of detection and repair increase. However, after 180 s the f-measures tend to be stable. That is because the time window is already long enough to cover the consecutive records for the same entity. A well set time window should be the longest interval for most entities moving from one location to the next.

5.3 Experiments on Synthetic Datasets

To evaluate scalability of our system, we generated a series of synthetic datasets and emit tracking records from *InputSpout* at maximum speed. As bottlenecks may only occur in the tracking bolt, we just explore the throughput by adopting different parallelisms for it. All the results shown in Fig. 5(a) are obtained when the processing rate tends to be stable. From this figure we can see that, when there is only one tracking bolt, the system can process about 250 tracking records in a second. With the parallelism growing, the throughput also increases near-linearly.

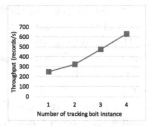
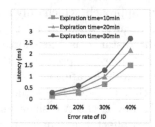

(a) Throughputs with different number of tracking bolt instance

(b) Latencies with different expiration times and error rates

Fig. 5. Throughput and latency results on synthetic datasets

Also, we conduct experiments to evaluate the effects on latencies that made by different expiration times of tracking tree and error rates of ID. Figure 5(b) shows that the latency almost doubles when the error rate increases every 10 % points. That tells us our repair approach is better applied on data with a relatively high quality. Moreover, Fig. 5(b) also reveals that with expiration time growing, the latency increases near linearly. That is because the number of active tracking trees at a time is directly influenced by the expiration time. A well chosen expiration time should be the longest interval for most entities during a whole feasible location sequence.

6   Related Work

Entity identification is of great interest in recent years. In the scenarios of natural language text, it can be used to disambiguate named entity and enhance readability [11]. With the greater prevalence of social media, there is a fast-growing wealth of data of named entity. For instance, Twitter has seen tremendous growth in recent years and the richness of data has attracted great interest from both private and public organizations. Named entity recognition (NER) is used for helping them collect and understand users' opinions about particular organizations [10]. Moreover, entity identification can be adopted in some specific scenarios, e.g., Chinese named entity recognition [6]. There is also some research on skewed class distributions for entity identification [8].

Our work can be also regarded as record linkage. One of the most popular technologies of this kind is private record linkage [9,16]. Besides, it can be used over Web Databases to integrate data from multiple web sources. Some research are on social record linkage has studied study the problem across different social media platforms [12]. There are also some interactive tools to deal with record linkage [4,7].

Data repair is an essential approach to guaranteeing the quality of data. Errors may arise because of the imperfection of the scanning technologies used to collect data. Data repair is included in the data cleaning process, thus many works about data cleaning involve data repair [13]. Also, there are some interactive system for data repair [17]. Automatically identifying and repairing data in a dependable manner is an important challenge. One of the solution is to formulate a constraint [14], as we used in this paper. In many scenarios, record matching and data repairing are both involved in the process of data cleaning. One category of related work is concerned with finding an algorithm that can both match records and repair data [5].

7 Conclusion

In this paper, we present a novel method to repair erroneous IDs on the fly. Singleton and location sequence constraints based approaches are adopted to settle the problem. We introduce the structure and some basic operations of the tracking tree, which could help to index the candidate repairs in real time. The effectiveness and efficiency of our method are verified by conducting experiments on the distributed repair system we built. In future work, consideration could be given to adding more factors to the confidence function, and that would improve the precision of choosing the target ID.

Acknowledgement. This work was supported in part by the National Basic Research 973 Program of China under Grant No. 2015CB352502, the National Natural Science Foundation of China under Grant Nos. 61272092 and 61572289, the Natural Science Foundation of Shandong Province of China under Grant Nos. ZR2012FZ004 and ZR2015FM002, the Science and Technology Development Program of Shandong Province of China under Grant No. 2014GGE27178, and the NSERC Discovery Grants.

References

1. Apache storm. http://storm.apache.org/
2. Chang, S., Chen, L., Chung, Y., Chen, S.: Automatic license plate recognition. IEEE Trans. Intell. Transp. Syst. **5**(1), 42–53 (2004)
3. Cui, X., Dong, Z., Lin, L., Song, R., Yu, X.: Grandland traffic data processing platform. In: 2014 IEEE International Congress on Big Data, Anchorage, AK, USA, 27 June–2 July 2014, pp. 766–767 (2014)
4. Elfeky, M.G., Elmagarmid, A.K., Verykios, V.S.: TAILOR: a record linkage tool box. In: Proceedings of the 18th International Conference on Data Engineering, San Jose, CA, USA, 26 February–1 March 2002, pp. 17–28 (2002)

5. Fan, W., Li, J., Ma, S., Tang, N., Yu, W.: Interaction between record matching and data repairing. In: Proceedings of the ACM SIGMOD International Conference on Management of Data, SIGMOD 2011, Athens, Greece, 12–16 June 2011, pp. 469–480 (2011)
6. Fu, G., Luke, K.: Chinese named entity recognition using lexicalized hmms. SIGKDD Explor. **7**(1), 19–25 (2005)
7. Galhardas, H., Florescu, D., Shasha, D., Simon, E., Saita, C.: Declarative data cleaning: language, model, and algorithms. In: Proceedings of 27th International Conference on Very Large Data Bases, VLDB 2001, Roma, Italy, pp. 371–380, 11–14 September 2001
8. Gliozzo, A.M., Giuliano, C., Rinaldi, R.: Instance filtering for entity recognition. SIGKDD Explor. **7**(1), 11–18 (2005)
9. Inan, A., Kantarcioglu, M., Bertino, E., Scannapieco, M.: A hybrid approach to private record linkage. In: Proceedings of the 24th International Conference on Data Engineering, ICDE 7–12, 2008, Cancún, México, pp. 496–505, April 2008
10. Li, C., Weng, J., He, Q., Yao, Y., Datta, A., Sun, A., Lee, B.: Twiner: named entity recognition in targeted twitter stream. In: The 35th International ACM SIGIR Conference on Research and Development in Information Retrieval, SIGIR 2012, Portland, OR, USA, pp. 721–730, 12–16 August 2012
11. Li, Y., Wang, C., Han, F., Han, J., Roth, D., Yan, X.: Mining evidences for named entity disambiguation. In: The 19th ACM SIGKDD International Conference on Knowledge Discovery and Data Mining, KDD 2013, Chicago, IL, USA, pp. 1070–1078, 11–14 August 2013
12. Liu, S., Wang, S., Zhu, F., Zhang, J., Krishnan, R.: HYDRA: large-scale social identity linkage via heterogeneous behavior modeling. In: International Conference on Management of Data, SIGMOD 2014, Snowbird, UT, USA, pp. 51–62, 22–27 June 2014
13. Raman, V., Hellerstein, J.M.: Potter's wheel: an interactive data cleaning system. In: Proceedings of 27th International Conference on Very Large Data Bases, VLDB 2001, Roma, Italy, pp. 381–390, 11–14 September 2001
14. Wang, J., Tang, N.: Towards dependable data repairing with fixing rules. In: International Conference on Management of Data, SIGMOD 2014, Snowbird, UT, USA, pp. 457–468, 22–27 June 2014
15. Wu, W., Liu, Z., Chen, M., Yang, X., He, X.: An automated vision system for container-code recognition. Expert Syst. Appl. **39**(3), 2842–2855 (2012)
16. Yakout, M., Atallah, M.J., Elmagarmid, A.K.: Efficient private record linkage. In: Proceedings of the 25th International Conference on Data Engineering, ICDE 2009, Shanghai, China, pp. 1283–1286, 29 March–2 April 2009
17. Yakout, M., Elmagarmid, A.K., Neville, J., Ouzzani, M.: GDR: a system for guided data repair. In: Proceedings of the ACM SIGMOD International Conference on Management of Data, SIGMOD 2010, Indianapolis, Indiana, USA, pp. 1223–1226, 6–10 June 2010

Modeling for Noisy Labels of Crowd Workers

Qian Yan[1], Hao Huang[1(✉)], Yunjun Gao[2], Chen Ying[1], Qingyang Hu[2],
Tieyun Qian[1], and Qinming He[2]

[1] State Key Laboratory of Software Engineering, Wuhan University, Wuhan, China
{qy,haohuang,yingchen,qty}@whu.edu.cn
[2] College of Computer Science, Zhejiang University, Hangzhou, China
{gaoyj,huqingyang,hqm}@zju.edu.cn

Abstract. Crowdsourcing services can collect a large amount of labeled data at a low cost. Nonetheless, due to some influence factors such as the unqualified crowd workers and the controversiality of instances to be labeled, the collected labels often contain noisy data, i.e., they sometimes are randomly given, incorrect, or missing. Although approaches have been proposed to infer these influence factors to help better model the labeling results, the inferences are not guaranteed to reflect the true effects of the influence factors on the uncertainty and errors in the labels. In this paper, we propose to conduct probability fitting over the noisy labeled data with Bernoulli Mixture Model. Workers with similar behaviors correspond to a same Bernoulli component in the mixture model. The effects of influence factors are fused in the Bernoulli parameter of each Bernoulli component, which directly reflects the uncertainty of labels, and can help identify labeling errors, predict real labels, and reveal the behavior patterns of crowd workers. Experiments on both benchmark and real datasets verify the efficacy of our model.

1 Introduction

By distributing labeling tasks on a crowdsourcing platform, such as Amazon Mechanical Turk, labels can be collected at a relatively low cost [5]. Nevertheless, as crowd workers on such platforms are usually not strictly qualified, they may be unreliable and give low quality responses [7], resulting in noisy labeling results in which some labels are randomly given, incorrect, or missing. Therefore, there is a strong demand of modeling such noisy labels.

Since the uncertainty and errors in collected labels are mainly caused by unreliable crowd workers, the idea of inferring workers' behavior patterns or labeling abilities to properly model the noisy labels has been widely proposed [13,20]. Another important source of the uncertainty and errors in collected labels is the controversiality of instances to be labeled, which is often inferred simultaneously [15,18]. Besides, the sparsity of labeled data caused by missing labels also lowers the quality of crowdsourcing data. Efforts have been made to cope with such sparse data or to recover missing labels [6].

The inferences mentioned above tend to be helpful under certain situations though, they often have the following three flaws. (1) Firstly, for an isolated

© Springer International Publishing Switzerland 2016
F. Li et al. (Eds.): APWeb 2016, Part II, LNCS 9932, pp. 227–238, 2016.
DOI: 10.1007/978-3-319-45817-5_18

task, these inferences would overly complicate the modeling work. To quantify the above two influence factors and their effects on the uncertainty and errors in the labels, one needs to predict more model parameters. Nonetheless, as a rule, this kind of prediction is a non-convex optimization problem, in which the parameters are often updated by an EM-like algorithm, such that prediction results usually depend on the selection of initial parameters and fall into local optimum. (2) Secondly, the uncertainty and errors in labels are caused by the joint action of known and hidden influence factors. Hence, even if the inferences for the known ones can be always done precisely, they are not guaranteed to reflect the true effects of all influence factors. (3) Thirdly, among these inferring approaches, a mainstream assumption is that each crowd worker has a fixed labeling ability level for all instances, and each instance has a fixed controversiality level for all crowd workers [18]. This assumption is a simplification for parameter estimation, but fails to consider that crowd workers with different expertise and purposes, and controversial instances often cause higher uncertainty. To relax this assumption, some other work [19] tried to estimate the variable expertise of crowd workers by using parametric models, which introduce more parameters to estimation and further complicate the modeling work.

To avoid the above flaws, we propose to directly model the noisy labels by conducting probability fitting over them with a Bernoulli Mixture Model, in which each group of crowd workers with a similar behavior pattern corresponds to a Bernoulli component. The model no longer cares how the influence factors work jointly to affect the labels. Instead, the effects of influence factors are fused in the Bernoulli parameter of each Bernoulli component. We also present a model parameter estimation approach, which enables the model to handle the situations that partial labels are missing. Experiments on both benchmark and real datasets demonstrate that our model has a reasonably better performance than the existing state-of-the-art approaches on label prediction, and it is also able to identify crowd workers with different behavior patterns.

The remaining sections are organized as follows. We present the framework of our model in Sect. 2, and introduce the model parameter estimation in Sect. 3, following which report the experimental results in Sect. 4 and review the related work in Sect. 5 before concluding the paper in Sect. 6.

2   Model Framework

2.1   Problem Statement

We first formalize the crowdsourcing labeling problem before presenting our method. For a task $\mathcal{X} = \{x_i\}_{i=1}^{N}$ containing N instances to be labeled, there are L crowd workers to label them. Each crowd worker ℓ ($\ell \in \{1, \ldots, L\}$) labels an arbitrary nonempty subset $\mathcal{X}^{\ell}$ of $\mathcal{X}$, i.e., $\mathcal{X}^{\ell} \subseteq \mathcal{X}$, and there will be L label sets $\mathcal{Y} = \{\{y_{\ell i}\}_{i=1}^{N}\}_{\ell=1}^{L}$, in which there would be some missing labels $y_{\ell i} = \varnothing$. In this paper, we focus on binary labeling problems (as the extension to multi-labeling cases is conceptually straightforward), namely, if label $y_{\ell i}$ is not missing (i.e., $y_{\ell i} \neq \varnothing$), then $y_{\ell i} \in \{0, 1\}$. Since each instance is labeled by different crowd workers, contradictory labels may exist for a same instance.

To model $\mathcal{Y}$, many of the existing work assume the existence of a hidden true label set $\mathcal{Y}^* = \{y_i^*\}_{i=1}^N$, and often infer it together with the behavior patterns or labeling abilities of workers and the controversiality of the instances. Nevertheless, these inferences involve more parameters to be estimated, and make it more complicated to model the problem and search the optimum.

Aiming at a simple yet more flexible and functional model, we try to directly understand and simulate the generation process of $\mathcal{Y}$. Towards this, we consider a crowd worker as a unit, and model the probability distribution of all the labels $\{y_{\ell 1}, \ldots, y_{\ell N}\}$ of each crowd worker ℓ. We denote the distribution as $p(\boldsymbol{y}|\boldsymbol{\theta})$ and assume that every $\boldsymbol{y}_\ell = [y_{\ell 1}, \ldots, y_{\ell N}]^T$ is drawn from this distribution. Given this, all we need to do is choosing an appropriate model form for $p(\boldsymbol{y}|\boldsymbol{\theta})$ and inferring the corresponding model parameter $\boldsymbol{\theta}$ based on observed labels in $\mathcal{Y}$.

2.2 Bernoulli Mixture Model

Model Form. We choose N-dimensional Bernoulli Mixture Model to fit the probability distribution of variable $\boldsymbol{y}$. The basic reason is two-fold. (1) Firstly, the value of each label is subject to Bernoulli distribution, i.e., it is either 1 or 0. (2) Secondly, the variable $\boldsymbol{y}$ is N-dimensional since it reflects the labeling results of a crowd worker on the given N instances. Without loss of generality, we assume that there are K Bernoulli components in the Bernoulli Mixture Model. Formally, the model can be described as follows.

$$p(\boldsymbol{y}|\boldsymbol{\mu}, \boldsymbol{\pi}) = \sum\nolimits_{k=1}^{K} \pi_k \prod\nolimits_{i=1}^{N} \mu_{ki}^{y_i}(1 - \mu_{ki})^{(1-y_i)}$$

where $\boldsymbol{\mu} = \{\boldsymbol{\mu}_1, \ldots, \boldsymbol{\mu}_K\}$, $\boldsymbol{\mu}_k$ refers to the Bernoulli parameter of the kth Bernoulli component, $\boldsymbol{\pi} = \{\pi_1, \ldots, \pi_K\}$, and π_k is the coefficient (or weight) of the kth Bernoulli component.

Parameter Meanings. In our model, the labeling result $\boldsymbol{y}_\ell$ of each crowd worker ℓ is regarded to be generated by a Bernoulli component. If a group of crowd workers behave similarly (e.g., spammers give labels randomly, malicious workers give wrong labels deliberately), their labels tend to be generated by the same Bernoulli component. Moreover, if the members of this group occupy the majority of crowd workers, the coefficient π_k of the corresponding Bernoulli component (say the kth one) will be larger than that of the other ones.

In our model, we do not assume how influence factors work jointly to affect the uncertainty and errors in given labels $\mathcal{Y}$. Instead, we simulate the generation process of $\mathcal{Y}$ directly, and fuse the effect of influence factors into the Bernoulli parameter $\boldsymbol{\mu}_k$. The ith element μ_{ki} of $\boldsymbol{\mu}_k$ indicates the probability that the ith instance's label is 1 and is generated by the kth Bernoulli component.

Functionality. Our model has the following main functionalities.

- Label prediction. In practice, as the majority of crowd workers usually try to give right labels, their corresponding Bernoulli component (say the kth one) will be the main Bernoulli component which has a greater final π_k. Meanwhile, if the final μ_{ki} is greater than a threshold (in this paper, we set the threshold as 0.5), we can consider the ith instance's real label as 1.
- Reflecting the uncertainty of labels. If the final μ_{ki} is close to 0.5, the uncertainty of ith label generated by the kth Bernoulli component is high. Furthermore, if the kth Bernoulli component is the main Bernoulli component at the same time, it means the controversiality of the ith instance is high.
- Categorizing crowd workers. Denote a latent variable $z = \{z_1, \ldots, z_L\}$, in which z_ℓ ($\ell \in \{1, \ldots, L\}$) is a K-dimensional vector with one element equals to 1 and the others equal to 0. Equation $z_{\ell k} = 1$ ($k \in \{1, \ldots, K\}$) indicates that the crowd worker ℓ's labeling result y_ℓ is generated by the kth Bernoulli component. Then, the expectation $h_{\ell k}$ of $z_{\ell k}$ reflects crowd worker ℓ's membership in the kth Bernoulli component. According to Bayes theorem, $h_{\ell k}$ can be calculated as

$$h_{\ell k} = E[z_{\ell k}] = \frac{\pi_k p(y_\ell | \mu_k)}{\sum_{j=1}^{K} \pi_j p(y_\ell | \mu_j)} \tag{1}$$

where $p(y_\ell | \mu_k) = \prod_{i=1}^{N} \mu_{ki}^{y_{\ell i}} (1 - \mu_{ki})^{(1 - y_{\ell i})}$. Furthermore, the properties of crowd workers grouped by the kth Bernoulli component can be reflected by the value of μ_k. For example, if each element in μ_k is around 0.5, this crowd worker group would be spammers since they give random labels and have an about 50 % accuracy; if the elements in μ_k are complementary to that of the main Bernoulli component, this crowd worker group would be malicious workers since they give opposite labels on purpose.

Advantages. Compared with the existing work, our model is more flexible and practical. (1) Firstly, in our model, the influence factors (e.g., instance controversiality, workers' labeling ability) are not assumed to be set at a fixed level. When they change with different instances and crowd workers, our model still works. (2) Secondly, our model does not require the existence of true labels. (3) Thirdly, if the options of crowd workers diverge from each other, our model can utilize different Bernoulli components to model the label results and extract the different options.

3   Model Parameter Estimation

Given the model, what to do next is to estimate the model parameters μ and π based on the label set $\mathcal{Y}$. Usually, parameter estimations for mixture probability distribution are carried out by EM steps. For example, estimating $h_{\ell k}$ via Eq. (1) is a standard E (Expectation) step. With the estimation on $h_{\ell k}$, parameters μ and π could be updated via a standard M (Maximization) step, i.e., maximizing the likelihood function. Formally, the updating could be as follow.

$$\mu_k = \frac{1}{L_k} \sum_{\ell=1}^{L} h_{\ell k} y_\ell, \quad \pi_k = \frac{L_k}{L} \tag{2}$$

where $L_k = \sum_{\ell=1}^{L} h_{\ell k}$ can be regarded as the number of crowd workers whose labels are generated by the kth Bernoulli component.

However, the above standard EM steps can not be applied on the estimation of our model parameters, since they require the integrity of labels. But, in practice, there are many missing labels in the labeling results of crowd workers.

3.1 Missing Label Problem

To deal with the missing labels, there are two common methods. (1) A straightforward method to cope with the missing label problem is regarding $\boldsymbol{y}_\ell^m$ as another latent variable, using current model parameters to estimate the expectation of the missing labels, and then combining this expectation and observed labels to re-estimate model parameters. Nonetheless, this method may not properly recover the missing labels. Instead, it may bring bias for the missing labels. For example, assuming that we are using Gaussian distribution to fit a 2-dimensional dataset $\{x_i, y_i\}_{i=1}^{N}$, partial values of y_i are missing, and the missing values are greater than a threshold $y' < \max\{y_i\}_{i=1}^{N}$, if we only use the observed data to model the dataset distribution, then the mean of y will be less than the true value. Using this mean as the expectation of missing values of y to recover the complete dataset brings bias. (2) Another common method is to utilize the dependency between x_i and y_i and predict the missing data via linear or non-linear regression. But, this method may concentrate the missing values of y on regression hyperplane, and brings another kind of bias.

In brief, simply using observed labels to recover missing labels often makes the recovery result be biased towards the observed labels.

3.2 Parameter Estimation with Incomplete Labels

To decrease the bias brought by recovering missing labels, a reasonable approach to the parameter estimation for our model is that for each item containing latent variable in the likelihood function, we regard it as a whole to estimate its expectation. Similar idea can be found in incomplete data modeling [4].

Let's refer back to Eq. (2), for the observed part $\boldsymbol{y}_\ell^o$ in $\boldsymbol{y}_\ell$, we only need to estimate one latent variable, i.e., the expectation $h_{\ell k}$ of $z_{\ell k}$ which can be denoted as $E[z_{\ell k}|\boldsymbol{y}_\ell^o, \boldsymbol{\pi}, \boldsymbol{\mu}]$; for the missing part $\boldsymbol{y}_\ell^m$ in $\boldsymbol{y}_\ell$, since both $\boldsymbol{y}_\ell^m$ and $h_{\ell k}$ are latent variables, we estimate the overall expectation of the product of them, which can be denoted as $E[z_{\ell k}\boldsymbol{y}_\ell^m|\boldsymbol{y}_\ell^o, \boldsymbol{\pi}, \boldsymbol{\mu}]$. Since dimensions in a Bernoulli distribution are independent to each other, the value of $E[z_{\ell k}\boldsymbol{y}_\ell^m|\boldsymbol{y}_\ell^o, \boldsymbol{\pi}, \boldsymbol{\mu}]$ is equal to $h_{\ell k}\boldsymbol{\mu}_k^m$.

Our parameter estimation method is outlined in Algorithm 1. It takes as inputs the number K of Bernoulli components which is defined by users and the label set $\{\boldsymbol{y}_\ell\}_{\ell=1}^{L}$ collected from crowd workers. The algorithm is carried out by four phases, i.e., (1) the parameter initialization phase, in which users select appropriate initial values for $\boldsymbol{\mu}$ based on their priori knowledge or according to their purposes, e.g., purpose of identifying spammers or malicious workers from accurate crowd workers (a more detailed discussion is given in Sect. 4.1); (2) the expectation phase, in which the algorithm respectively estimates the

---

Algorithm 1. Improved EM Algorithm for Updating Bernoulli Mixture Model on Incomplete Crowdsourcing Data

---

Input : The number K of Bernoulli components, and the label set $\{\boldsymbol{y}_\ell\}_{\ell=1}^L$ returned by L crowd workers in which each $\boldsymbol{y}_\ell$ includes the observed part $\boldsymbol{y}_\ell^o$ and the missing part $\boldsymbol{y}_\ell^m$.

Output: Parameters $\boldsymbol{\mu}$ and $\boldsymbol{\pi}$, and the probability $\{\{h_{\ell k}\}_{k=1}^K\}_{\ell=1}^L$ of the labels of a crowd worker ℓ ($\ell \in \{1, \ldots, L\}$) being generated by the kth ($k \in \{1, \ldots, K\}$) Bernoulli component.

1 **Initialization:** Select appropriate initial values for $\boldsymbol{\mu} = \{\boldsymbol{\mu}_1, \ldots, \boldsymbol{\mu}_K\}$;

2 **Expectation:** For each $\ell = 1, \ldots, L$, substitute the observed part $\boldsymbol{y}_\ell^o$ of $\boldsymbol{y}_\ell$ into Eq.(1) to calculate the expectation $E[z_{\ell k}|\boldsymbol{y}_\ell^o, \boldsymbol{\pi}, \boldsymbol{\mu}]$; meanwhile, for the missing part $\boldsymbol{y}_\ell^m$, calculate the expectation $E[z_{\ell k}\boldsymbol{y}_\ell^m|\boldsymbol{y}_\ell^o, \boldsymbol{\pi}, \boldsymbol{\mu}] = h_{\ell k}\boldsymbol{\mu}_k^m$;

3 **Maximization:** Use $E[z_{\ell k}\boldsymbol{y}_\ell^m|\boldsymbol{y}_\ell^o, \boldsymbol{\pi}, \boldsymbol{\mu}]$ to replace the corresponding part $h_{\ell k}\boldsymbol{y}_\ell^m$ of $h_{\ell k}\boldsymbol{y}_\ell$ in Eq. (2), use $E[z_{\ell k}|\boldsymbol{y}_\ell^o, \boldsymbol{\pi}, \boldsymbol{\mu}]$ to replace $h_{\ell k}$ in Eq. (2), and then update $\boldsymbol{\mu}$ and $\boldsymbol{\pi}$;

4 **Convergence Checking:** Compare the difference between the current and old values of $\boldsymbol{\mu}$ and $\boldsymbol{\pi}$, if the difference is not small enough, return to the Expectation step;

---

expectation $E[z_{\ell k}|\boldsymbol{y}_\ell^o, \boldsymbol{\pi}, \boldsymbol{\mu}]$ for the observed part of each crowd worker's labels, and the expectation $E[z_{\ell k}\boldsymbol{y}_\ell^m|\boldsymbol{y}_\ell^o, \boldsymbol{\pi}, \boldsymbol{\mu}]$ for the missing part; (3) the maximization phase, in which model parameters $\boldsymbol{\mu}$ and $\boldsymbol{\pi}$ are updated; and (4) the convergence checking phase, in which the algorithm checks whether our EM steps converge.

4   Experimental Evaluation

We conduct experiments on benchmark and real crowdsourcing datasets to verify the efficacy of our model in the following two aspects, i.e., (1) it can categorize crowd workers by their behavior patterns, (2) it can accurately predict real labels.

4.1   UCI Benchmark Data

In this section, we test our model on Waveformnoise dataset from UCI Machine Learning Repository [3]. The dataset consists of 5000 instances, and each instance has a true label which is treated as the ground truth for label prediction.

We simulate the labels of crowd workers for the instances in the dataset. The total number of crowd workers varies from 10 to 100. In the simulation, the majority of the crowd workers give relatively accurate labels, and we set their proportion α as a random value between 0.7 and 0.8. The rest of crowd workers are low-quality crowd workers, i.e., spammers and malicious workers. We use Hammer-Spammer Model and One-Coin Model that are commonly used in the existing work [10–13] to generate these crowd workers.

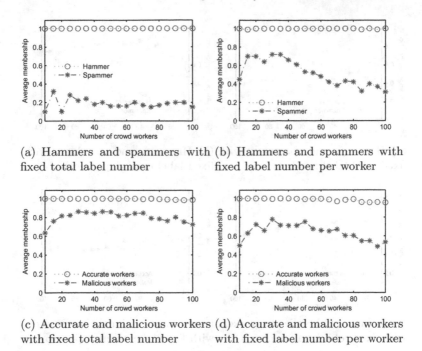

(a) Hammers and spammers with fixed total label number

(b) Hammers and spammers with fixed label number per worker

(c) Accurate and malicious workers with fixed total label number

(d) Accurate and malicious workers with fixed label number per worker

Fig. 1. Workers' membership in main Bernoulli component (with proper initial μ).

- Hammer-Spammer Model. Each worker is either a "hammer" with a probability α or a "spammer" with a probability $1 - \alpha$. A hammer is a worker who always gives accurate labels, while a spammer gives labels randomly.
- One-Coin Model. The ability of worker ℓ is parameterized by a single parameter p_ℓ, which reflects the probability of correctness when worker ℓ labels an instance. For accurate workers, we set $p_\ell = 0.8$; for low-quality workers we set $p_\ell = 0.2$. The latter are also called malicious workers in crowdsourcing as their accuracy is less than 50%, the accuracy of random case.

Crowd Worker Categorization. In this experiment, we verify the capability of our model on categorizing crowd workers with different behaviour patterns. To this end, we generate crowd workers with either Hammer-Spammer Model or One-Coin Model, and set $K = 2$ to categorize crowd workers into two groups since each model generates two types of crowd workers. For the number of labels, (1) firstly, we fix the total label number at 2500, and vary the number of labels given by each worker from 250 to 25 by increasing workers; (2) secondly, we fix the number of labels given by each worker at 50 and increase workers.

When we have a priori knowledge on the behaviour patterns of crowd workers, we can initialize the model parameter μ properly to achieve a faster convergence and a reasonably better result. For example, (1) if we know that there are both hammers and spammers, we can set the initial μ_{1i} ($i \in \{1, \ldots, N\}$) to the mean

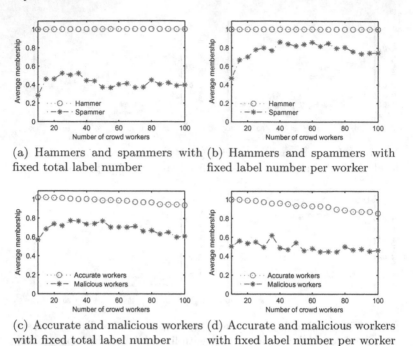

(a) Hammers and spammers with fixed total label number

(b) Hammers and spammers with fixed label number per worker

(c) Accurate and malicious workers with fixed total label number

(d) Accurate and malicious workers with fixed label number per worker

Fig. 2. Workers' membership in main Bernoulli component (with improper initial μ)

of $y_{\ell i}$ ($\ell \in \{1, \ldots, L\}$), and set each initial μ_{2i} ($i \in \{1, \ldots, N\}$) to 0.5; (2) if we know that there are both accurate and malicious workers, we use the same initialization for μ_{1i}, and perform a random initialization for μ_{2i}. Then, for each type of crowd workers, we calculate the average expectation of their membership (i.e., the average value of their $h_{\ell k}$) in the main Bernoulli component. We record the average expectations in each execution and report their mean values of 10 times of run. Figure 1 illustrates the experimental results, from which we can observe that the average membership of hammers and accurate workers in main Bernoulli component is significantly higher than that of spammers and malicious workers, indicating that our model accurately categorizes these crowd workers.

To evaluate our model under the situation that there is no priori knowledge on the behaviour patterns of crowd workers, we exchange the initial setting on μ_{2i} for spammers and malicious workers (setting on μ_{1i} is the same), i.e., (1) for the hammer-spammer case, we set each μ_{2i} ($i \in \{1, \ldots, N\}$) to a random initial value, (2) for the case of accurate and malicious workers, we set each initial μ_{2i} ($i \in \{1, \ldots, N\}$) to 0.5. Figure 2 shows corresponding results, from which we can observe that (1) our model can effectively categorize hammers and spammers by their average membership in main Bernoulli component, and (2) although average membership of accurate workers in main Bernoulli component decreases a little with the growth of the number of workers, it is still significantly higher

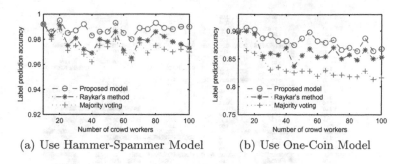

(a) Use Hammer-Spammer Model   (b) Use One-Coin Model

Fig. 3. The accuracy of label prediction (with fixed total label number).

than the average membership of the malicious workers, indicating that without priori knowledge, our model can also reasonably categorize crowd workers.

Label Prediction. In this experiment, we verify the effectiveness of our model on label prediction. We adopt the majority voting method [16] and the model provided by Raykar et al. [13] as the baselines for accuracy comparison. Figure 3 illustrates the comparison result, which shows that our model has a reasonably better accuracy than baselines. The reason behind is that in our model, spammers or malicious workers are assigned to non-main Bernoulli component, leaving hammers and accurate workers in main Bernoulli component.

4.2 Affective Text Analysis Data

In this section, we evaluate the efficacy of our proposed model on the Affective Text Analysis Data [14]. There were 38 workers to label the emotions of 100 news headlines. For each headline, 10 workers rated a score (within $[0, 100]$) for each of six emotions, namely, anger, disgust, fear, joy, sadness, and surprise. Most workers labeled 20 to 40 headlines so that most labels are missing.

We normalize the scores to binary values (0 or 1) by using a threshold 30. We set $K = 2$, set initial μ_{1i} ($i \in \{1, \ldots, N\}$) to the mean of $y_{\ell i}$ ($\ell \in \{1, \ldots, L\}$), and set each initial μ_{2i} ($i \in \{1, \ldots, N\}$) to 0.5. Figure 4 shows each worker's final membership in main Bernoulli component. The workers with low membership in main Bernoulli component have been highlighted in red. From the figure, we can have the following observations. (1) The divergence between the workers in the task of rating surprise emotion is the largest, while that in the task of rating fear emotion is the smallest. This observation means that crowd workers have different feeling strength for the surprise emotion, while the identification of fear emotion is easier and clearer for different workers. (2) The 14th and 33th workers have low membership in main Bernoulli component for all tasks, indicating that they are low-quality workers (very likely spammers as their final μ_{2i} is close to 0.5). Some workers have low membership in main Bernoulli component for half of the tasks. The reliability of their labels is also relatively lower than others.

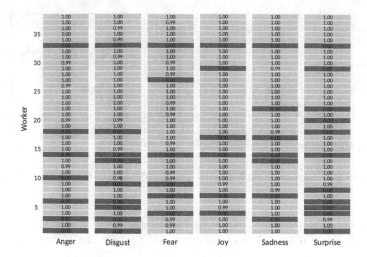

Fig. 4. Workers' membership in main Bernoulli component in affective text analysis.

5   Related Work

Collecting labels via crowdsourcing services has been proven to be effective in many applications such as natural language processing and medical data processing [1]. The collected labels are usually duplicated, i.e., different crowd workers provide multiple, possibly contradictory labels for one instance. This repeated-labeling scheme makes a good balance between labeling cost and data quality, and attracts attentions to the research subject of modeling data from multiple, unreliable sources.

For labeling tasks in crowdsourcing, the most common target is label prediction, i.e., obtaining reliable labels for instances. To this end, the mainstream approaches assumes the existence of a ground truth label for each instance, and tries to predict the true labels based on crowd labels. Karger et al. [10] proposed an algorithm which transfers messages between instances and workers iteratively; Liu et al. [12] generalized this algorithm by introducing priors on workers' labeling abilities and inferring the respective generative model with variational methods of graph models. Whitehill et al. [18] proposed different worker abilities as well as instance controversiality, which are inferred together with true labels via a probabilistic model. Furthermore, some latest techniques such as noise correction [21] and imbalanced learning [22] are used to improve label quality, especially under the circumstance that examples have a few noisy labels.

A closely related subject is learning a classifier from the labels when instances can be represented in a vector space. This work can be done simply by first inferring true labels with label prediction techniques mentioned above, and then learning a classifier by traditional classification approaches. More sophistic schemes include learning directly from crowd labels while inferring hidden worker abilities simultaneously [13], treating workers as personal classifiers which is assumed to be related to the final classifier [8,9], and modeling worker abilities as functions of the instance space, of which parameters are inferred together with the final classifier [20]. These work modeled worker abilities in different ways, but did not involve instance controversiality explicitly as they treated instance space as a whole. A flaw of such work is that vector representation for instances can not always be obtained easily for many real-world tasks.

Although most of the existing work aims at predicting a single reliable label for each instance [23], a few ones try to address the problem from other aspects. Wang and Zhou [16] presented a general theoretical framework to help identify (or eliminate) low-quality crowd workers from non-low-quality crowd workers with high accuracy. Welinder et al. [17] proposed the concept of "School of Thoughts" which allows workers to label differently and extracts different groups of opinions; Tian and Zhu [15] extended this concept and estimated worker abilities and instance controversiality by clustering workers responses. Ertekin et al. [2] studied the problem of approximating the crowd which estimates the crowd's majority opinion by querying only a subset of the workers.

As we can see, existing related work on modeling crowdsourcing data is abundant and diversified. Nevertheless, most existing work only covered a certain aspect. In contrast, our model predicts the data generation of the crowd labels themselves instead of such certain aspects, and is functional enough to include behavior patterns and different opinions of crowd workers while keeping the model itself simple and flexible.

6   Conclusion

In this paper, we have proposed a Bernoulli Mixture Model to model the noisy labels of crowd workers, and presented an improved EM algorithm for model parameter estimation. This model directly simulates the generation process of the labels without any extra assumption or inferences on influence factors such as the labeling ability of workers and the controversiality of instances. Experimental results on both benchmark and real crowdsourcing datasets have verified the effectiveness of our model on label prediction and categorizing crowd workers with different behavior patterns.

Acknowledgements. This work was supported in part by NSFC Grants (61502347, 61522208, 61572376, 61472359, and 61379033), the Fundamental Research Funds for the Central Universities (2015XZZX005-07, 2015XZZX004-18, and 2042015kf0038), the Research Funds for Introduced Talents of Wuhan University, and the International Academic Cooperation Training Program of Wuhan University.

References

1. Difallah, D.E., Catasta, M., Demartini, G., Ipeirotis, P.G., Cudré-Mauroux, P.: The dynamics of micro-task crowdsourcing: the case of Amazon MTurk. In: WWW 2015, pp. 238–247 (2015)

2. Ertekin, S., Rudin, C., Hirsh, H.: Approximating the crowd. Data Min. Knowl. Disc. **28**(5), 1189–1221 (2014)
3. Frank, A., Asuncion, A.: UCI machine learning repository (2010). http://archive. ics.uci.edu/ml/
4. Ghahramani, Z., Jordan, M.I.: Supervised learning from incomplete data via an EM approach. In: NIPS 1993, pp. 120–127 (1994)
5. Guo, X., Wang, H., Yangqiu, S., Gao, H.: Brief survey of crowdsourcing for data mining. Expert Syst. Appl. **41**(17), 7987–7994 (2014)
6. Hu, Q., Chiew, K., Huang, H., He, Q.: Recovering missing labels of crowdsourcing workers. In: SDM 2014, pp. 857–865 (2014)
7. Hu, Q., He, Q., Huang, H., Chiew, K., Liu, Z.: Learning from crowds under experts' supervision. In: Tseng, V.S., Ho, T.B., Zhou, Z.-H., Chen, A.L.P., Kao, H.-Y. (eds.) PAKDD 2014, Part I. LNCS, vol. 8443, pp. 200–211. Springer, Heidelberg (2014)
8. Kajino, H., Tsuboi, Y., Kashima, H.: A convex formulation for learning from crowds. In: AAAI 2012, pp. 73–79 (2012)
9. Kajino, H., Tsuboi, Y., Kashima, H.: Clustering crowds. In: AAAI 2013, pp. 1120–1127 (2013)
10. Karger, D.R., Oh, S., Shah, D.: Iterative learning for reliable crowdsourcing systems. In: NIPS 2011, pp. 1953–1961 (2011)
11. Karger, D.R., Oh, S., Shah, D.: Efficient crowdsourcing for multi-class labeling. ACM SIGMETRICS Perform. Eval. Rev. **41**(1), 81–92 (2013)
12. Liu, Q., Peng, J., Ihler, A.T.: Variational inference for crowdsourcing. In: NIPS 2012, pp. 692–700 (2012)
13. Raykar, V.C., Yu, S., Zhao, L.H., Valadez, G.H., Florin, C., Bogoni, L., Moy, L.: Learning from crowds. J. Mach. Learn. Res. **11**(4), 1297–1322 (2010)
14. Snow, R., O'Connor, B., Jurafsky, D., Ny, A.Y.: Cheap and fast–but is it good? evaluating non-expert annotations for natural language tasks. In: EMNLP 2008, pp. 254–263 (2008)
15. Tian, Y., Zhu, J.: Learning from crowds in the presence of schools of thought. In: KDD 2012, pp. 226–234 (2012)
16. Wang, W., Zhou, Z.H.: Crowdsourcing label quality: a theoretical analysis. Sci. Chin. Inf. Sci. **58**(11), 1–12 (2015)
17. Welinder, P., Branson, S., Belongie, S., Perona, P.: The multidimensional wisdom of crowds. In: NIPS 2010, pp. 2424–2432 (2010)
18. Whitehill, J., Wu, T.-F., Bergsma, J., Movellan, J.R., Ruvolo, P.L.: Whose vote should count more: optimal integration of labels from labelers of unknown expertise. In: NIPS 2009, pp. 2035–2043 (2009)
19. Yan, Y., Rosales, R., Fung, G., Dy, J.G.: Modeling multiple annotator expertise in the semi-supervised learning scenario. In: UAI 2010, pp. 674–682 (2010)
20. Yan, Y., Rosales, R., Fung, G., Schmidt, M., Hermosillo, G., Bogoni, L., Moy, L., Dy, J.G.: Modeling annotator expertise: learning when everybody knows a bit of something. In: ASITATS 2010, pp. 932–939 (2010)
21. Zhang, J., Sheng, V.S., Wu, J., Fu, X., Wu, X.: Improving label quality in crowdsourcing using noise correction. In: CIKM 2015, pp. 1931–1934 (2015)
22. Zhang, J., Wu, X., Sheng, V.S.: Imbalanced multiple noisy labeling. IEEE Trans. Knowl. Data Eng. **27**(2), 489–503 (2015)
23. Zhao, Y., Zhu, Q.: Evaluation on crowdsourcing research: current status and future direction. Inf. Syst. Front. **16**(3), 417–434 (2014)

Incomplete Data Classification Based on Multiple Views

Ming Sun, Hongzhi Wang$^{(\boxtimes)}$, Fanshan Meng, Jianzhong Li, and Hong Gao

Departmemt of Computer Science and Technology,
Harbin Institute of Technology, Harbin, China
{mingsun,wangzh,lijzh,honggao}@hit.edu.cn, fanshan.mfs@gmail.com

Abstract. Missing values have negative impacts on big data analysis. However, in absence of extra knowledge, exact imputation can hardly be conducted for many data sets. Therefore, we have to tolerate missing values and perform data mining on incomplete data sets directly. To achieve high quality data mining on incomplete data, we propose a classification approach based on multiple views. We use various complete views of the data set to generate the base classifiers and combine the results of base classifiers. Since the amount of base classifiers will affect the effectiveness and efficiency of the classification, we aim to find proper view sets. We prove that the view set selection problem is an NP-hard problem and develop an approximation algorithm with approximate ratio $ln|S| + 1$ where S is the feature set of original data set. Extensive experimental results demonstrate the efficiency and effectiveness of the proposed approaches.

1 Introduction

Classification is an important kind of techniques of data mining. Clearly, missing values will affect the quality of classification result. Taking medical data sets as an example, patient A's blood pressure value is lost, and we cannot predict what it is exactly in absence of extra knowledge. If we fill in a value which is in the blood pressure normal range but the true data is beyond the scope, the model being trained will be affected by the wrong data and emerge fatal mistakes, and vice versa. Someone's life will be lost regrettably due to the mistakes. Thus, incompleteness has to be handled for classification.

Current solutions can be divided into two types. One solution [1–6] trains the model from complete data subset and imputes missing values based on that model. The other deletes samples with missing values, and then performs classification directly. In absence of extra knowledge, the cost of the former solution is high, and it can hardly get exact values, which may lead to a low accuracy of classification and inefficiency. The efficiency of the latter one indeed wins, but the accuracy is disturbed by the samples with missing values.

Motivated by this, in this paper, we attempt to conduct classification on incomplete data directly to acquire relatively high-quality results. To achieve this goal, we need to utilize the complete share in data sufficiently. We use

© Springer International Publishing Switzerland 2016
F. Li et al. (Eds.): APWeb 2016, Part II, LNCS 9932, pp. 239–250, 2016.
DOI: 10.1007/978-3-319-45817-5_19

Table 1. A dirty data set D

	s_1	s_2	s_3	s_4	s_5	s_6	T
r_1	√	×	√	√	×	√	1
r_2	√	√	√	√	×	√	1
r_3	√	×	√	√	√	√	0
r_4	×	×	√	√	√	√	1
r_5	×	√	√	×	√	×	0
r_6	√	√	√	×	√	√	0
r_7	×	√	×	√	√	√	1

Table 2. Part of extracted views

V_i	S_i	R_i
V_1	$S_1 = \{s_1, s_3, s_4, T\}$	$R_1 = \{r_1, r_2, r_3\}$
V_2	$S_2 = \{s_2, s_4, s_5, T\}$	$R_2 = \{r_7\}$
V_3	$S_3 = \{s_1, s_2, s_5, T\}$	$R_3 = \{r_6\}$
V_4	$S_4 = \{s_2, s_5, s_6, T\}$	$R_4 = \{r_6, r_7\}$
V_5	$S_5 = \{s_3, s_4, T\}$	$R_5 = \{r_1, r_2, r_3, r_4\}$
V_6	$S_6 = \{s_2, s_6, T\}$	$R_6 = \{r_2, r_6, r_7\}$
...	...	

an example to illustrate this point. Consider an incomplete data set D with the feature set $S(s_1, s_2, s_3, s_4, s_5, s_6, T)$, where T is the column of type and the sample set $R(r_1, r_2, r_3, r_4, r_5, r_6, r_7)$, as shown in Table 1, where "√" denotes the data is available while "×" denotes it is missing. Table 2 shows the part of views from D. V_i denotes the ith view queried, S_i is the feature set of the ith view, and R_i is the sample set. Each view is a complete share of original data. The challenge is that the complete share of data may not be sufficient enough for classification due to the difference in incomplete values. Fortunately, the data have relevance. That is, some part of data could be implied from other part.

With the consideration of keeping relevance among data and not trying useless imputation, we propose a novel method which considers different complete share of the incomplete data. A base classifier is trained from each view. With these base classifiers, we adopt the voting mechanism to generate the final classification results. Clearly, the amount of views affects the efficiency as well as effectiveness of the algorithms. To achieve high performance, we do not have to use all the views, but only to find a combination of views which can cover all the features of dirty data set and a great majority of samples.

The contributions of this paper could be summarized as three points.

1. We take different views of incomplete data into account and regard them as base classifiers, which mitigates decline of accuracy caused by missing data.
2. To keep all features in the data set during training and ensure the efficiency, we propose view selection problem which we prove to be NP-hard and develop an efficient weighted greedy algorithm called VS with approximate ratio, $ln|S| + 1$, where S is the feature set of original data.
3. Experimental results show that our approach performs better and spend less time than the imputation-based solution.

The remainder of this paper is organized as follows. We introduce our framework in the next section which can acquire a relatively high-quality classification result from an incomplete data set directly. Section 3 introduces our view selection phase. Experimental results are reported in Sect. 4 and conclusions are presented in Sect. 5.

2 The Framework

In this section, we introduce the framework of our approach. Our method has three phases, preprocessing, view selection and combination. We acquire views satisfying the rules in preprocessing phase, which prepares for view selection phase. After selecting the views used to train base classifiers, we combine their classification results to obtain the final result. To describe the approach clearly, we use $\mathcal{F}$ to represent the set of views satisfying the rules which will be described concretely in Sect. 2.1, and use ℓ to represent the combination of views selected to be the base classifiers.

2.1 Preprocessing

In this phase, we give some rules to limit the amount of views for selection which mainly reduce the calculation burden on view selection phase.

Rule 1: $\forall\ V_i \in \mathcal{F}, |\ R_i\ | \geq \theta\ |\ R\ |, \theta \in [0.1, \delta]$.

With this rule, we restrict that the sample coverage of each view should exceed a threshold θ which avoids the extreme case that one view indeed covers many features but only cover few samples. We also restrict the value of θ should be less than δ which agrees with common sense.

Rule 2: $\forall\ V_i \in \mathcal{F}, |\ S_i\ | > 2$.

This rule ensures that the views only containing two features (including the column T) are abandoned because it is ridiculous to classify with a single feature.

Rule 3: If $\exists\ V_i$ and V_j, where $R_i = R_j$ and $S_i \subseteq S_j$, then delete V_j from $\mathcal{F}$.

Once there exists two views satisfying the rule above, the view with more features will be deleted. From the point of samples, they both have the same performance. The only distinctive between them is that one or more extra features are considered redundant.

Rule 1 makes this phase similar to the mining of frequent item set because the set of threshold θ is same as the support count in Apriori [7], while Rule 3 change the process into the mining of frequent closed item sets. As the combination of these three rules, we choose the algorithm in [8] to filter the views for $\mathcal{F}$.

2.2 View Selection

Since the first phase have filtered the views for $\mathcal{F}$, in this phase, our goal is to find the combination ℓ which contain the fewest views but can generally express the incomplete data set with current available data from $\mathcal{F}$.

We will formally define view selection problem which is proven to be NP-hard in Sect. 3. Furthermore, we develop an approximate algorithm whose approximate ratio is $ln|S| + 1$.

2.3   Combination

In this phase, each view selected into ℓ will be trained to be a base classifier and we adopt the voting mechanism to generate the final classification results. We can use any mining technique to train the base classifiers.

Given an instance X_q to be classified, because of the voting mechanism, the final result $F(X_q)$ depends on the result $f_i(X_q)$ of all the base classifiers. We regard the result of each classifier equally, that is,

$$F(X_q) \leftarrow \arg\max_{V_i \in \ell} f_i(X_q).$$

3   View Selection

In this section, we focus on view selection problem in the second phase. We formally define the view selection problem as follows:

$$\min \ |\ell|, \ \ell \subseteq \mathcal{F}$$

$$s.t. \begin{cases} \bigcup_{V_i \in \ell} S_i = S, \\ |\bigcup_{V_i \in \ell} R_i| \geq \delta \, |R| \quad \delta \in [0.5, 1]. \end{cases}$$

In this definition, on condition that we attempt to select the fewest views to cover all the features and a great majority of samples, each view abundantly keeps the relationship among different features and guarantees a good generalization ability.

Theorem 1. View selection problem is NP-hard.

Proof (Sketch). To prove that view selection problem is NP-hard, we show that set-covering problem [9] could be reduced to view selection problem. In other words, we need to show how to reduce any instance of set-covering problem to an instance of view selection problem in polynomial time. When $\delta = 1$ and $\forall \ V_i \in \mathcal{F}, |R_i| = \delta \, |R|$, we can use induction to express any instance of set covering as that of view selection. Thus, if the view selection problem has a solution, the set covering problem has a solution. Since set-covering is NP-hard, view selection problem is NP-hard.   □

Our problem is similar to the set-covering problem in two dimensions. Unlike the set-covering problem, we cannot directly use the greedy strategy. Concretely speaking, when we want to use the greedy strategy to pick a view, we may indeed choose the one that covers the greatest number of remaining uncovered elements of one dimension, but we cannot guarantee that how many uncovered elements of the other dimension we have covered. We must consider the coverage of features and the coverage of samples simultaneously. Moreover, we do not need to cover all the elements of samples while we must cover all the elements of features. With regard to our problem, we decide to change one dimension into the weight of the other dimension which will affect the selection of view.

Algorithm 1. View Selection $(S, R, \mathcal{F})$

1: $U = S, T = R$
2: $\ell = \varnothing, M = \varnothing$
3: **while** $U \neq \varnothing$ **do**
4:    **if** $\mid M \mid \geq \delta \mid R \mid$ **then**
5:       **for all** i such that $V_i \in \mathcal{F}$ **do**
6:          $\omega_i = 1$
7:       **end for**
8:    **else**
9:       **for all** i such that $V_i \in \mathcal{F}$ **do**
10:          $\omega_i = \mid R_i \backslash M \mid$
11:       **end for**
12:    **end if**
13:    select an $V_i \in \mathcal{F}$ that maximizes $\omega_i \times \mid S_i \cap U \mid$
14:    $U = U \setminus S_i, T = T \backslash R_i$
15:    $\ell = \ell \cup \{V_i\}, M = M \cup R_i$
16: **end while**
17: return ℓ

The algorithm works as follows. The set U and T contains, at each stage, the set of remaining uncovered features and samples. The set ℓ is used to contain which views have been chosen while the set M is to record which samples have been covered so that we can compute the weight of every view. Line 4–12 update the weight of each subset. When the second constraint that the samples should be covered at least $\delta \mid R \mid$ is satisfied, ω_i is 1 which means the value of weight will not affect the selection of view any more. Otherwise, ω_i is $\mid R_i \backslash M \mid$. Line 13 is the greedy decision-making step, choosing a view V_i whose product between ω_i and the amount of features will covered the first time is the largest. After V_i is selected, line 14 removes its features from U and its samples from T, and line 15 places V_i into ℓ and add R_i, i.e. the samples of V_i into M. When the algorithm terminates, the set ℓ contains a subset of $\mathcal{F}$ that covers S.

Time Complexity Analysis. We can easily verify VS to run in time polynomial in $\mid S \mid$, $\mid R \mid$ and $\mid \mathcal{F} \mid$. Since the number of iterations of the loop on lines 3–16 is bounded from above by $\min(\mid S \mid, \mid \mathcal{F} \mid)$, and we can implement the loop body to run in time $O(\mid R \parallel \mathcal{F} \mid + \mid S \parallel \mathcal{F} \mid)$. For most data sets, the amount of samples is over that of features a lot ($\mid R \mid \gg \mid S \mid$), so $O(\mid R \parallel \mathcal{F} \mid + \mid S \parallel \mathcal{F} \mid) = O(\mid R \parallel \mathcal{F} \mid)$. Even a straightforward implementation of our algorithm runs in time $O(\mid R \parallel \mathcal{F} \mid \min(\mid S \mid, \mid \mathcal{F} \mid))$.

Approximate Ratio Bound Analysis. We now show that the weighted greedy algorithm returns a combination of views with the cost not too much larger than an optimal combination. For convenience, in this paper we denote the dth harmonic number $H_d = \sum_{i=1}^{d} 1/i$ by $H(d)$. As a boundary condition, we define $H(0) = 0$.

Theorem 2. VS algorithm is polynomial-time $\rho(n)$-approximation algorithm, where $\rho(n) = \ln \mid S \mid + 1$.

Proof. We have already shown that *VS* runs in a polynomial time in time complexity analysis. We define the cost of the algorithm is the sum of the reciprocal of the weight when the view is selected in that our problem attempt to find the minimal views satisfying two constraints. We try to select a view with a larger weight even though it may not cover the most uncovered features till the second constraint is satisfied. To show that *VS* is a $\rho(n)$-approximation algorithm, we denote $V(i)$ as the ith view selected by *VS*. The algorithm incurs a cost of $\frac{1}{\omega(i)}$ when it adds $V(i)$ to ℓ where $\omega(i)$ is the weight of $V(i)$ when $V(i)$ is selected. We spread this cost of selecting $V(i)$ evenly among the features covered for the first time by $S(i)$. Let c_x denote the cost allocated by to feature x, for each $x \in S$. Each feature is assigned a cost only once, when it is covered for the first time. If x is covered for the first time by V(i), then $c_x = \frac{1}{\omega(i)|S(i)-(S(1) \cup S(2) \cup \cdots \cup S(i-1))|}$.

The cost of ℓ is the sum of the cost of every feature in S after the implementation of *VS*, so we have

$$cost(\ell) = \sum_{x \in S} c_x \tag{1}$$

Each element $x \in S$ is in at least one set in the optimal cover ℓ^*, and so we have

$$\sum_{V_j \in \ell^*} \sum_{x \in S_j} c_x \geq \sum_{x \in S} c_x \tag{2}$$

Combining Eq. (1) and inequality (2), we have that

$$cost(\ell) \leq \sum_{V_j \in \ell^*} \sum_{x \in S_j} c_x \tag{3}$$

Now, we estimate the equation $\sum_{x \in S_j} c_x$. Consider any j such that $V_j \in \mathcal{F}$ and $i = 1, 2, \ldots, |\ell|$, and let $u_i = |S_j - (S(1) \cup S(2) \cup \cdots \cup S(i-1))|$ be the number of features in S_j that remain uncovered after the algorithm has selected sets $V(1), V(2), \ldots, V(i-1)$. We define $u_0 = | S_j |$ to be the number of features of S_j, which are all uncovered. Let k be the least index such that $u_k = 0$, so that every feature in S_j is covered by at least one of sets $S(1), S(2), \ldots, S(k)$ and some features in S_j is uncovered by $S(1) \cup S(2) \cup \cdots \cup S(k-1)$. Then, $u_{i-1} \geq u_i$, and $u_{i-1} - u_i$ features of S_j are covered for the first time by S_j, for $i = 1, 2, \ldots, k$. Thus, $\sum_{x \in S_j} c_x = \sum_{i=1}^{k} (u_{i-1} - u_i) \cdot \frac{1}{\omega(i)|S(i)-(S(1) \cup S(2) \cup \cdots \cup S(i-1))|}$. Observe that $|S(i) - (S(1) \cup S(2) \cup \cdots \cup S(i-1))| \geq |S_j - (S(1) \cup S(2) \cup \cdots \cup S(i-1))| = u_{i-1}$, because the greedy choice of $S(i)$ guarantees that S_j cannot have a bigger product between the weight and the amount of new features covered than $S(i)$ does (otherwise, the algorithm would have chosen S_j). Consequently, we obtain

$$\sum_{x \in S_j} c_x \le \sum_{i=1}^{k}(u_{i-1} - u_i) \cdot \frac{1}{\omega(i) \cdot u_{i-1}} \le \sum_{i=1}^{k}(u_{i-1} - u_i) \cdot \frac{1}{\omega_j \cdot u_{i-1}}$$

$$= \frac{1}{\omega_j}\sum_{i=1}^{k}(u_{i-1} - u_i) \cdot \frac{1}{u_{i-1}}$$

$$= \frac{1}{\omega_j}\sum_{i=1}^{k}\sum_{q=u_i+1}^{u_{i-1}} \frac{1}{u_{i-1}}$$

$$\le \frac{1}{\omega_j}\sum_{i=1}^{k}\sum_{q=u_i+1}^{u_{i-1}} \frac{1}{q} = \frac{1}{\omega_j}\sum_{i=1}^{k}\left(\sum_{q=1}^{u_{i-1}} \frac{1}{q} - \sum_{q=1}^{u_i} \frac{1}{q}\right) \qquad (4)$$

$$= \frac{1}{\omega_j}\sum_{i=1}^{k}(H(u_{i-1}) - H(u_i)) = \frac{1}{\omega_j}[H(u_0) - H(u_k)]$$

$$= \frac{1}{\omega_j}[H(u_0) - H(0)] = \frac{1}{\omega_j}H(u_0)$$

$$= \frac{1}{\omega_j}H(|S_j|) \le \frac{1}{\omega_j}H(\max|S_q|: V_q \in \mathcal{F})$$

From the inequality (3) and (4), we have

$$cost(\ell) \le \sum_{V_j \in \ell^*}\sum_{x \in S_j} c_x \le \sum_{V_j \in \ell^*} \frac{1}{\omega_j}H(\max|S_q|: V_q \in \mathcal{F})$$

$$= H(\max|S_q|: V_q \in \mathcal{F}) \sum_{V_j \in \ell^*} \frac{1}{\omega_j}$$

$$= H(\max|S_q|: V_q \in \mathcal{F}) \sum_{V_j \in \ell^*} cost(\ell^*)$$

For harmonic progression $H(n)$, there is $\ln(n+1) \le H(n) = \sum_{k=1}^{n} 1/k \le \ln n + 1$. Since $|S_q| \le |S|$, $cost(\ell) \le H(|S|)cost(\ell^*) \le (\ln|S|+1)cost(\ell^*)$, thus proving the theorem. $\qquad\square$

4   Experimental Results

In this section, we conduct extensive experiments to verify the effectiveness and efficiency of the proposed framework on real data sets. The basic information of data sets from UCI machine learning repository[1] we used is shown in Table 3. The amount of features and samples of both data set is suitable for our approach. Besides, Heart Disease Data Set hardly has missing values so that we need to randomly inject missing values, while Chronic Kidney Disease Data Set itself has some real missing values. We generate the incomplete data sets by randomly

---

[1] http://archive.ics.uci.edu/ml/.

Table 3. Data sets used in experiments (from UCI)

Data set	Features	Samples
Heart disease	14	300
Chronic kidney disease	25	400

removing some values evenly to control the ratio of missing values. We generate the test data set from the original data set with three parameters, the number of extracted tuples (#tuple), the number of extracted features (#feature) and the missing ratios (#missing). The default values of these parameters are 300, 13 and 0.2. The default values of θ and δ are 0.5 and 0.8, respectively. Moreover, we choose KNN [10] to be the classification technique of basic classifiers and comparative methods where k is set to be 3. The reason why we choose KNN is its insensitivity to the amount of features in each base classifiers, unlike neural networks. Experimentally, we have discovered the accuracy we acquired is the best when $k = 3$. We use run time and accuracy to measure the efficiency and effectiveness, respectively. Accuracy is defined as $Accuracy = \frac{|D_{right}|}{|D|}$ where D represents the original data set and D_{right} represents the set containing the tuples which is classified correctly.

4.1 Comparison to Other Methods

First of all, we compared our approach with another two approaches. The former one classification on the data set which is imputed based on Bayesian network [11]. The latter one performs classification on the incomplete data set directly.

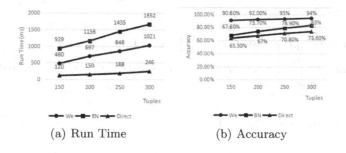

(a) Run Time   (b) Accuracy

Fig. 1. Scalability in tuples

Varying #tuple. To test the scalability, we use Heart Disease Data Set and vary #tuple from 150 to 300. The experimental results are shown in Fig. 1.

The experiment shows that with the increase of sample amount, the accuracy increases while it costs more time. More samples means a better generalization of original data set. Besides, we can observe that our framework acquires higher accuracy than another two methods but run time only outperforms BN [11]. DirectKNN is disturbed by the samples with missing values. The fewer amount of samples indeed reduce the run time but sacrifices the accuracy. The results of BN [11] are also interpretable. It first spends much to use the correlation between features to construct Bayesian networks and then uses it to predict the possible value, whose cost is large. Our framework just finds a suitable combination of views which regards each sample as a entirety, but not considering the inner relationship of features in each sample. In terms of accuracy, BN [11] utilizes the correlation between features to impute the missing values which usually appear in some certain features. However, when the correlation is weak and the missing value appears evenly, the classification accuracy decreases reasonably.

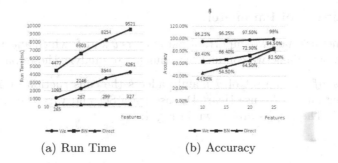

(a) Run Time   (b) Accuracy

Fig. 2. Scalability in features

Varying #feature. To test the scalability, we use Chronic Kidney Disease Data Set and vary #feature from 10 to 25. The experimental results are shown in Fig. 2.

The experiment shows that with the increase of sample feature, the accuracy is improved significantly while it costs more time, since more features and relationships are kept to express the original data set. The comparison of run time and accuracy between these three methods is similar as the experimental results of varying #tuple for the same reasons. Besides, we observe that the impact of feature amount on run time is more significant on that of the number of samples. Because more features lead to larger computation cost especially in the phase of filtering views for $\mathcal{F}$.

Varying #missing. To test the impacts of missing ratio, we use Heart Disease Data Set and vary #missing from 0.1 to 0.4. The results are shown in Fig. 3.

From Fig. 3, it is observed that the increase of ratio will reduce run time and accuracy. In terms of run time, more missing values decrease the scale of $\mathcal{F}$ and Bayesian networks to great extent. With regard to accuracy, its decrease with the increase of ratio agrees with common sense since the larger missing ratio means the fewer amount of available data.

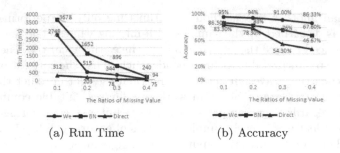

(a) Run Time   (b) Accuracy

Fig. 3. Impacts on missing ratio

4.2   Impacts of Parameters

Our framework has two important parameters θ and δ, which affect the run time and accuracy. Thus, we conduct experiments to analyze the impact of them.

Impacts of θ. The set of θ mainly affects the construction of $\mathcal{F}$ in the pre-processing phase. We use Heart Disease Data Set and vary the value of θ from 0.1 to 0.7. The results are shown in Fig. 4.

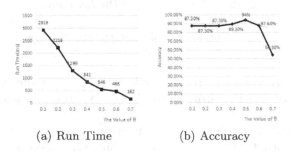

(a) Run Time   (b) Accuracy

Fig. 4. Impacts of θ

In terms of run time, the increase of θ heightens the standard to filtering views for $\mathcal{F}$, then fewer views meet the standard, which leads to the decrease of run time. Considering accuracy, it is observed that when $\theta = 0.5$, the accuracy is the highest. The reason is that when θ is less than 0.5, some useless views whose amount of sample is few added to $\mathcal{F}$ disturb the view selection because of the low ability to generalize. When θ is more than 0.5, some views with more samples but fewer features are selected preferentially which cannot represent the data set more complicatedly and some views with suitable samples and features are ignored. That is why in most experiments, we keep the value of θ 0.5.

Impacts of δ. The set of δ mainly affects the implementation of VS in the view selection phase. We use Heart Disease Data Set and vary the value of δ from 0.5 to 1. The results are shown in Fig. 5.

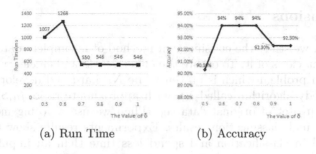

(a) Run Time   (b) Accuracy

Fig. 5. Impacts of δ

With regard to run time, the increase of δ extends the time for calculation of the weight of each view in $\mathcal{F}$ which causes the increase of run time. For accuracy, when δ is 0.5, too few samples are not covered so the low ability to generalize leads to the low accuracy. When δ is 0.9 or 1, some views with fewer features but more samples are selected preferentially which can not represent the data set more complicatedly and some views with suitable samples and features are ignored. However, for this data set, when δ is from 0.6 to 0.8, we get the same accuracy which means *VS* may select the same combination on the condition that the view selected finally covers at least 20 % uncovered samples. Setting $\delta = 0.8$ is a balance action for both feature coverage and sample coverage.

We should take attention to the fact that the most suitable value of θ and δ in this part varies due to different data sets.

4.3 Effectiveness of View Selection Algorithm

To test the effectiveness of our *VS* algorithm. we use Heart Disease Data Set and the experimental results are shown in Table 4. For comparison, we randomly generate 10 groups of views in $\mathcal{F}$ with the same number of the views selected with our approach, train the learners with each group and report the run time and accuracy. We also compare with the optimal brute-force searching algorithm.

The approximate answer obtained from *VS* whose accuracy is still high spends far less time than that of the optimal solution. In addition, either the maximal accuracy, the average one or the minimal one, the random selection of views cannot beat *VS*. Both facts show the effectiveness of our *VS* algorithm.

Table 4. Comparison

	OPT	*max*	*min*	*avg*	*OPT**
Time (ms)	1521	—	—	—	4223
Accuracy	94 %	87.6 %	84.3 %	86 %	—

5   Conclusions

In this paper, we study the classification method of incomplete data which considers different views of it. To ensure the effectiveness and efficiency, we propose view selection problem which is proven to be NP-hard and develop an efficient weighted greedy algorithm called *VS* with approximate ratio, $ln|S| + 1$, where S is the feature set of original data set. Then we use a voting mechanism to generate the final classification results. Experiments results show that our *VS* performs well in classification and spend less time than an imputation-based solution. Our future work includes taking inconsistent data into consideration and performing more experiments and comparisons.

Acknowledgement. This paper was partially supported by National Sci-Tech Support Plan 2015BAH10F01 and NSFC grant U1509216,61472099,61133002 and the Scientific Research Foundation for the Returned Overseas Chinese Scholars of Heilongjiang Provience LC2016026.

References

1. Troyanskaya, O., Cantor, M., Sherlock, G., Brown, T., Hastie, T., Tibshirani, R., Botstein, D., Altman, R.B.: Missing value estimation methods for DNA microarrays. Bioinformatics **17**(6), 520–525 (2001)
2. Oba, S., Sato, M.A., Takemasa, I., Monden, M., Matsubara, K.I., Ishii, S.: A Bayesian missing value estimation method for gene expression profile data. Bioinformatics **19**(16), 2088–2096 (2003)
3. Zhu, X., Zhang, S., Jin, Z., Zhang, Z., Xu, Z.: Missing value estimation for mixed-attribute data sets. IEEE Trans. Knowl. Data Eng. **23**(1), 110–121 (2011)
4. Setiawan, N.A., Venkatachalam, P.A., Hani, A.F.M.: Missing attribute value prediction based on artificial neural network and rough set theory. In: International Conference on BioMedical Engineering and Informatics, BMEI 2008. IEEE (2008)
5. Abdella, M., Marwala, T.: The use of genetic algorithms and neural networks to approximate missing data in database. In: IEEE 3rd International Conference on Computational Cybernetics, ICCC 2005. IEEE (2005)
6. Hagan, M.T., Demuth, H.B., Beale, M.H., De Jesús, O.: Neural Network Design. PWS publishing company, Boston (1996)
7. Agrawal, R., Srikant, R., et al.: Fast algorithms for mining association rules. In: Proceedings of 20th International Conference very large Data Bases, VLDB (1994)
8. Pei, J., Han, J., Mao, R., et al.: Closet: an efficient algorithm for mining frequent closed itemsets. In: ACM SIGMOD Workshop on Research Issues in Data Mining and Knowledge Discovery (2000)
9. Christofides, N.: Graph Theory–An Algorithmic Approach. Academic Press Inc., New York (1975)
10. Cover, T.M., Hart, P.E.: Nearest neighbor pattern classification. IEEE Trans. Inf. Theor. **13**(1), 21–27 (1967)
11. Jin, L.: Research on missing value imputation of incomplete data. Harbin Institute of Technology (2013)

Fuzzy Keywords Query

Shanshan Han, Hongzhi Wang($^{\boxtimes}$), Hong Gao, Jianzhong Li,
and Shenbin Huang

Harbin Institute of Technology, Harbin 150001, Heilongjiang, China
shanshanHan0731@126.com,
{wangzh,honggao,lijzh,huangshenbin}@hit.edu.cn

Abstract. Considering advantages and disadvantages of search engines
and SQL query of keywords search, we propose a novel method of constructing fuzzy ontology model combining ontology with fuzzy sets. We
introduce membership functions and the fuzzy operator, then expand
semantic keywords based on the semantic synonym dictionary. To process
query effectively, we design an optimized storage structure based on B*
tree. Some techniques, such as index mechanism, are used in our system. Based on these techniques, we propose a top-k query which uses
dynamic filtering to accelerate processes. We experimentally demonstrate
the accuracy and efficiency of the methods.

1 Introduction

The world is faced with fuzzy query, along with vague expressions, informal languages, abbreviated forms, and so on. For example, when search "tall middle-
aged women". Here, "tall", "middle-aged" are fuzzy terms. Such problem motivates to fuzzy query. Synonyms is another common problem in query. For example, "Database Management System" is abbreviated to "DBMS". Such kind of
problems requires a synonym retrieval. These two major retrieval methods, used
by relational database and engines, have advantages to accuracy, while they are
recognized to fall short of dealing with fuzzy or synonyms data effectively.

SQL [1] is a basic and declarative programming language in relational databases. However, users must master SQL and the data model before using it.
What's more, SQL can't deal with fuzzy or synonyms keywords query very
well. As for search engine [2], while retrieving data, it can only match keywords mechanically. That is, it can only scrap web pages by index, but can't
understand the meaning of keywords. When dealing with fuzzy queries, results
are imprecise.

To combine benefits and eliminate defects of the both, we propose a more
versatile method to perform keywords query based on the relational database.
We introduce the membership function in fuzzy sets and fuzzy operators to
build fuzzy ontology. By characterizing the distance between tuples, we expand
keywords query. We also describe semantic similarities by combining semantic
distance with co-occurrence frequency. In order not to deviate from query intention, we propose timely results to meet users' needs.

© Springer International Publishing Switzerland 2016
F. Li et al. (Eds.): APWeb 2016, Part II, LNCS 9932, pp. 251–262, 2016.
DOI: 10.1007/978-3-319-45817-5_20

In the subsequent parts of this paper, we show the overview of our system. Next, we propose a fuzzy query algorithm based on fuzzy ontology, and demonstrate its correctness. Besides, we provide a detail description of the query process and the optimization algorithm, which contains improved score function and the top-k query. Finally, we demonstrate the performance and accuracy of our method.

The contribution of this paper are summarized as follows:

1. A novel query form is proposed to involve fuzzy semantic constraints in the query based on the ontology, which overcomes the gap between queries and data while support fuzzy semantic matching of query processes and data;
2. A system is proposed to process such query efficiently. We design data storage, query processing algorithm as well as query optimization strategies;
3. Extensive experiments are performed to test the proposed system. From the results, the system could process the queries efficiently and effectively.

2   Background

Even though some methods [3–10] have been proposed for fuzzy search over relations or string set, these work focus on the efficiency of similarity search. However, without knowledge, the similarity search is based on similarity functions and cannot handle the similarity in semantics. As a comparison, we propose a knowledge-based approach and could distinguish the similarity and differences in semantics, especially for the words with similar semantics but very difference appearance. Besides, many similarity measures based on structured knowledge have been proposed, such as WordNet, domain ontology. They are regarded as the foundation of similarity calculation. Some of the methods are based on features, which estimate similarity between concepts of same and different attributes of concepts in the ontology description model. Methods based on edge count, which measure similarity by the number of semantic connections through dividing two concepts in ontology. However, Method based on feature is more suitable for image matching, while similarity measures based on attributes of concept need to parse each ontology, and organize them into a concept tree according to the is-a relationship. By dividing this tree from up to down, we can get feature properties different between each concept after each division. Yet, the whole process is complicated. And in the aspect of nature language, such as synonyms and polysemy, it's not so specific.

In this paper, we combine membership function in fuzzy sets with fuzzy operators to build fuzzy ontology. With the convenience of ontology, such as better hierarchy, stronger association between logical implication, and accessible correlation storage of data, we can abstract reality to fuzzy sets. Through compared with existing data models and depending on objective laws of study, we can obtain membership functions to characterize semantic similarities.

3 Overview of the System

The system starts from analyzing query keywords and converting fuzzy words to exact keywords by relationships between words. We put forward strategies of semantic keywords query in relational databases. In our system, fuzzy data management is divided into five modules: model construction, data storage, keywords organization, query process, and results generation.

Model construction module constructs fuzzy ontology models. To strengthen compact between independent layers and improve maintainability and reusability, the structure of knowledge is divided, which provides a concept layer to interact between users and the system and infers logical connection between users' needs. What's more, it filters useless information and classify information.

Data storage module determines the way data stored. We design novel caching mechanism to effect query processes. If the number of nodes in index is small, the whole index can be contained in the main memory to save time.

Keywords organization module parses keywords. It expands keywords according to fuzzy ontology models and sorts them by similarity in semantics. Different from accurate queries on relational databases, query may be fuzzy or semantic. Thus, we propose semantic analysis techniques. Based on such techniques, we design a weight-based score function which describes the similarity of semantics and the top-k query defined on such score function.

The query process module performs the following process: First, the system grabs and analyzes the property of keywords to find relevant content based on full-text index. Similar to BANKS [11], query results are defined as a join tree containing at least one node, and the root node of the tree is connected to keywords information. In this way, it can find the shortest path from the root to leaf nodes by Dijkstra algorithm. Such operations reduce query frequency. Results are expressed as SQL to operate multi-table join and then returned to users. This module accesses data through the data storage module.

The results generation module provides results returned by the query process module. Previous systems return final results to users in unification. When users are not satisfied, the system repeat the last work. To avoid this waste of time we propose a novel strategy. When users are not satisfied, the system changes previous query errors timely, then replace keywords with other related keywords automatically. Processes don't stop until users are satisfied.

The data flows of these components are shown in Fig. 1. These five modules are all interconnected to complete the process tasks together efficiently.

4 Related Techniques

4.1 Fuzzy Query

Fuzzy Sets [12]. Suppose set A maps set X to $[0,1]$, A: $X \rightarrow [0,1]$, $x \rightarrow A(x)$. Then we say A is a fuzzy set on X. A(x) is called the membership function of fuzzy set A, or A(x) is membership degree of fuzzy set A on x.

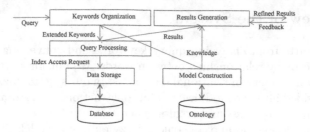

Fig. 1. Five modules of system: model construction, data storage, keywords organization, query processing, and results generation. Arrows indicate data flows.

Membership Function. The membership function can be expressed as similarity, importance and so on. Value in membership function is between 0 and 1.0 indicates that the element completely does not belong to the set, while 1 indicates that the element completely belongs to the set. Elements in range from 0 to 1 (excluding 0 and 1) indicate the share of elements in the set. The larger the value, the more relevance is shared between the words.

The membership function is a quantitative description of the fuzzy concept. In practice, we can study the characteristics of objects, and sum up in comparison to overall characteristics of applications. Besides, we can compare six kinds of classical fuzzy distribution (rectangular, trapezoidal and semi trapezoidal, parabolic, normal distribution, Cauchy distribution, ridge distribution), and choose the closest one. In practice, fuzzy distribution function is often used to approximate membership functions.

The most common fuzzy distribution functions are as follows: (C is an arbitrary point in universe U, and a, b and k are parameters that are greater than 0).

1. Partial small decreasing function of large left and small right.

$$u_x = \begin{cases} 1 & x \leq C \\ [1 + a(x-c)^b]^{-1} & x > C \end{cases}$$

2. Partial small rising function of large right and small left.

$$u_x = \begin{cases} 1 & x < C \\ [1 + a(x-c)^b]^{-1} & x \geq C \end{cases}$$

3. Symmetric intermediate normal convex function.

$$u_x = e^{-k(x-c)^2}$$

Words of fuzzy concepts, such as "about", "almost", are fuzzy operators. Thus fuzzy membership function with fuzzy operators can be described as follows:

$$U_{(y,x)} = \begin{cases} e^{-(y-x)^2} & |y - x| < \theta \\ 0 & |y - x| \geq \theta \end{cases}$$

where y is the keyword, x is a fuzzy operator, and θ is the threshold value.

4.2   Semantic Analysis

The purpose of semantic analysis is to expand query. That is to say, use existing technology to add related words to the query, and make up for original query. Moreover, it can improve recall and precision.

Definition 1 (Query Expansion). *Suppose $c \in W$, $p \in P$, the intersection between W and P is not empty, so we say c and p is extensible. We say that there is similarity between them, and the value is between 0 and 1.*

For given words c and p, their co-occurrence frequency appeared in domain is defined as: $DomainRel(c,p) = \frac{Freq(c,p)}{Freq(c)+Freq(p)-Freq(c,p)}$.

In literature [13], the semantic distance between two concepts is the sum of n edges weights in shortest paths: $Dist(c,p) = \sum_{i=1}^{n} weight_i$.

We use semantic distance to depict the similarity between concepts. The Max stands for a large number and it is defined by users in practice. The word a is chosen based on the algorithm and a is computed to adjust the result. The conversion formula between them is described as follows:

$$Sim(c,p) = 1 - \sqrt{\frac{a-1}{a}} * \beta * Dist(c,p) \quad a \geq 2$$

$$\beta = \begin{cases} DomainRel(c,p)^{-1} & if\ DomainRel(c,p) \neq 0 \\ Max & else \end{cases}$$

In this paper, we do not simply rely on the semantic distance or co-occurrence frequency to characterize the semantic similarity, but combine both together. When combined with the co-occurrence frequency, they can make the search results more optimized.

4.3   Expansion Technique

Expansion technique is described as follows: we regard keywords typed by users as the initial keywords, then, sort the query results by matching degree, and select the text closer to users' intention. Next, search the word with the highest frequency. Finally, system chooses words of which not only in high correlation but also in high co-occurrence frequency as expansion words, and then notes such expansion words as query keywords to search. In order to interact with users frequently and make queries constantly meet users' needs, we propose query expansion based on feedback. We still view initial keywords typed by users as the first keywords to query, and then return results. If users are satisfied with the results, system stops searching. Otherwise, system returns the top ranked expansion keywords to users until they're satisfied. Users choose similar contents from these, and then system queries these relevant words by SQL.

4.4   Fuzzy Ontology Model

Ontology, generally accepted as "a definition of derivation rules, which is adequately integrated by related terms and relations on basis of constructing related relationship between words" [14], refers to concepts of knowledge in field and unified description of relationships between concepts. In literature [15], models based on ontology can be divided into two stages: (a) using natural languages and figures to describe the domain model to form an original model; (b) using languages of knowledge to encode domain model, which leads to easier communication with users and the forming of descriptive languages to be compiled by software. When constructing fuzzy ontology, it's mainly to store relations between semantic words, analyze characteristics, and store corresponding fuzzy functions in the first stage. Ontology was often used to deal with semantic distances in previous studies. Here we combine fuzzy sets with ontology to construct fuzzy ontology models, which stores basic concepts and knowledge as well as links between concepts.

4.5 Data Storage Structure

Using index can speed up query. When targeted contents are in index, we just access results in index instead of querying all data. Today, most databases use B or B+ tree index to optimize query. B+ tree, the outer search tree, can undertake sequential and random search while maintaining a dynamic balance. However, once the division is no longer able to continue, the consequent division results in the significant dropping of space utilization. So the space utilization ratio of B+ tree is only 50 %. B* tree, adding a pointer to brothers in none-root node and none-leaf node in B+ tree, defines that the total number of keywords in none-leaf node is at least $(2/3)*M$, which means the lowest utilization is $2/3$.

When faced with large data, we use B* tree index. It improves space utilization and the query speed. Meanwhile, when insert records, we need I/O operations. In the worst case, such operations are proportional to the height of the tree. Each time when an item is added to leaf node, system needs disk seek to retrieve block containing the leaf node. When another item wants to be added to a block and the buffer is full, past blocks may be eliminated in some replacement strategies such as LRU to insert the item into the buffer. Dispatching blocks away will waste time on disk seeking. So, when modify and update existing data, we propose to use bulk load and create temporary files for relational indexes, then sort files according to established index. Finally, insert data to item index.

5   Query Process

In this section we talk about the query process techniques of fuzzy queries.

5.1 Online Query Optimization

Translating keywords searching into SQL query is called online query. To improve online query, it's crucial to reduce the sum of data stored in memory, especially the sum of disk accesses. To save time, we establish the index. We set buffer block in memory, and take appropriate replacement strategy to update the cache contents. Just when we need to replace this content, we write it back to the disk. In this way, we could not only speed up the query process, but also reduce the time of transferring data between the disk and the memory.

5.2 Top-k Query

Score Function. In this paper, we propose a weight-based score similar to BANKS. Top-k results are selected according to the score. We assume that different locations of keywords have different weights denoted by a_i, and i denotes the location of the query keywords. The membership degree of the first category is denoted by b_i, which expresses the membership degree of the ith query keywords and query results. The multiplication of the semantic similarity and the co-occurrence frequency is denoted by c_i.

Define $t[f_i]$ as the projection of tuple t on attribute a_i, and K represents for the keyword set: $K_q = k_1, k_2, k_3, \ldots, k_q$. Score($t[f_i]$, K) represents for the relevance of query results and the keywords query. The score function is as follows.

Algorithm 1. Weight-based score function.
Input: keywords set K, results set A.
Output: the overall similarity degree of keywords and results.

```
Program WeightBasedScoreFunction()
   for each result A[i] in the query result set A
      Score(t[fi],Kq) =weight * membershipDegree;     //ai*bi
         for each tuple in result A                   //calculate ci
            if it is satisfied with Fuzzy Query
               correlation=membership degree;
            else
               correlation=semanticSimilarity*co-occurFrequency;
      Score(t[fi], Kq)*=correlation;
   output the Score(t[fi],Kq) of each query result A[i];
end.
```

Dynamic Filtering. In the top-k algorithm, we propose the thought of dynamic filtering. First, store part of the data, and filter the data of lower score values to make the number of the rest data approach to k. Then store part of the rest data, deal with the data left and newly inserted, filter out the low score values to make the number of the rest data that tend to be k in the dynamic pool. Repeat this process until no data left. Filtering out intermediate data that may not be engaged in the final results makes the retention of top-k easier and time-saving.

5.3 Query Processing Algorithm

We propose a top-k query algorithm based on dynamic filtering to filter out those not in top-k through a real-time filtering. It performs queries that may consist top-k tuples, and finally returns top-k results. Filtering is executed in aspect of candidate tuples and query results. The middle filter layer filters unreliable data so that the subsequent process gets much simpler to perform. The second filter layer is to make results closer to users' demand.

This algorithm is mainly for massive data. Data is read for one time. It's unpractical to bring entire data into memory, thus, we define a sample pool of a fixed volume with a capacity of C. After selecting currently needed tuples, we push tuples with functioned scores into the sample pool. After the number of tuples in the pool reaches C, we delete tuples in the first round in this way: (1) If the number with high scores is more than k, then we mark this area (which have high-k value) as Q_1, and record the minimum score in Q_1 as W_1. After that, we delete other areas in the sample pool besides Q_1. (2) Select currently needed tuples. If the number of high scores is more than k, we mark this area as Q_2, and record the minimum score containing in Q_2 as W_2 (now $W_2 \geq W_1$). Then delete other areas in sample pool besides Q_2. (3) Select currently needed tuples, then execute similar processes. If the score of the current tuple is less than W_2, delete this tuple directly, else, push this tuple into the sample pool. Execute such process until all data having been selected. At this point the number of tuples in the sample pool should be no larger than C. The rest tuples are performed in the sample pool, getting query results, with each score values calculated in the query result set. Finally, based on the dynamic filtering algorithm, system sorts the rest of relatively small query results which have been filtered, then return top-k.

This algorithm can be achieved by the Minimum heap. Based on such dynamic filtering technique, intermediate filter results don't produce the top-k as soon as possible. It makes later sorts much easier to implement. Compared with previous algorithms, it has a better optimization considering space and time complexity. We can not only use this method to deal with intermediate results, but also use it to implement the final results.

Algorithm 2. dynamic filtering top-k query algorithm.
Input: Original data.
Output: Tuple or query results whose number tend to k.

```
Program DynamicFiltering ()
    MinHeap.maxsize=C, k=0;
    d=select the minimum rank, initial value is 0;
    while(!test[num].IsFull()&&k<MAX)
        if(each sample in results value>=d)
            insert sample into MinHeap;
        q=MinHeap.CurrentSize;
    while(q>=k)
        Remove the sample from MinHeap;
    d=get min from MinHeap;
    show the rest of MinHeap;
end.
```

6   Results and Discussion

In this section, we evaluate our keyword query approach from the evaluation criteria of precision ratio and recall ratio. The experimental environment is Code-Blocks on Windows. In order to fully evaluate the performance and efficiency of the algorithm, we use real data (taken from the website EBay) and artificial data (produced by data generator) as experimental data sets. We measure the performance of the algorithm by several factors, such as the size of data, the number of retrieved keywords, the change of k. Without special explanation, the default unit of time is second while the default unit of data size is MB.

6.1   Accuracy of the Algorithm

In experiment, the selection of the rest data is not arbitrary. We choose words having links with keywords as far as possible, which provide more comparable measurement accuracy so that can it fully show the accuracy of the algorithm.

We use ratios to measure the ratio of tuples related to keywords semantically in the data set. We have top-10 and top-3 query, with the ratios of contained keywords vary from 10 % to 80 % and 3 % to 9 %. Results are shown in Table 1.

Table 1. The degrees of relevance in top-10 and top-3 query

	Top-10								Top-3						
Data (%)	80	70	60	50	40	30	20	10	9	8	7	6	5	4	3
Result (%)	100	100	100	100	100	90	90	80	66.8	66.7	66.7	66.7	66.7	66.6	33.3

Experiments show that, for the top-10 query, results are ideal (having reached 100 %) when the ratio is no less than 30 %. Due to relatively strict experimental data sets, when the ratio of keywords contained is less than 30 %, results drop slowly. As for the top-3 query, when the ratio is no less than 4 %, the results have a constant proportion.

Because the choice of experimental data is not arbitrary as in real world, and still needs to meet certain conditions strictly, it leads the entire experimental results to the low side. But thanks to strict data, we can reflect the accuracy of the algorithm very well.

6.2   The Efficiency of Algorithms

The Influence of the Data Size. The experiment adopts large amounts of data to detect time of algorithm execution. Taking data sources from the website EBay, each tuple contains seven attributes. Results are shown in Fig. 2.

Through experimental tests, we can see that with the increase of data number, run time increases slowly. When the data number is up to 10M, the run time of join results for four keywords is kept at about 10 s. After increasing the

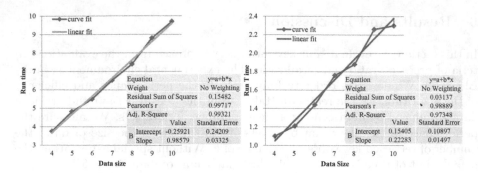

Fig. 2. The influence of data size. The left shows results when use 4 keywords to acquire top-2, and the right shows when use a single keyword to have top-10 query. The size of the dynamic pool is 20.

number of keywords, run time will be controlled at m times compared to the run time of a single keyword query (m is multiples). For the whole algorithm, when the size of the dynamic pool is fixed, as the data size increases, the increase of time is relatively slow. Hence the algorithm has a relatively high efficiency for big data.

The Influence of Dynamic Filtering Algorithm. Here we test the influence of the size of the dynamic pool on execution efficiency. We choose 10M and 4M data, and type four keywords. Results are shown in Fig. 3.

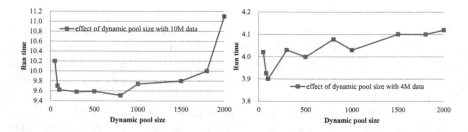

Fig. 3. Use four keywords to test the influence of the size of dynamic pools. The dynamic pool sizes vary in 50, 80, 100, 300, 500, 800, 1000, 1500, 1800 and 2000. The left shows results when the size of data is 10M, while the right is 4M.

From the experiment we see that the algorithm performance doesn't increase as the sample pool capacity increases. For the universal data, the sample pool which is too big or too small will affect run time, thus selecting appropriate sample pools for different sizes of data is crucial. By testing on different sizes of data (10M and 4M), we find that in the same size of the sample pool, different sizes of data perform differently. For data which has larger capacity, the

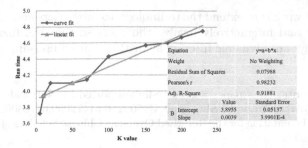

Fig. 4. Use 4M data and 4 keywords to test the influence of k. The capacity of sample pool is fixed with 3000 and the intermediate value of the dynamic filter is a prime number 7.

most suitable sample pool capacity should be generally greater. Both have the property of abandoning head and tail, retaining middle in common.

The Influence of the Value of k. In experiment, by fixing the capacity of the sample pool and the number of keywords, we test the influence of k changed on execution efficiency. The number of keywords is fixed with 4. The capacity of the sample pool is fixed with 3000, and the volume of data is 4M. Intermediate value of the dynamic filter is a prime number 7. Results are shown in Fig. 4.

We find that with the increase of k, run time increases gradually, but more slowly, thus we can infer the performance of the dynamic filter algorithm is more superior.

6.3   Summary of Experimental Results

From experiments we see that for queries of massive data, the algorithm has high execution efficiency, and can also save more time. Execution efficiency is related to the change of k value, the size of the dynamic pool, the number of keywords, intermediate value of the dynamic filtering, and so on. For example, a too big or too small sample pool affects run time., thus we should select appropriate pool for different sizes of data to achieve the best efficiency.

7   Conclusions

In this paper, we propose a novel method to deal with fuzzy keywords query. We combine the ontology with fuzzy sets to construct fuzzy ontology, and use some existing technologies, such as index, to speed execution efficiency. Moreover, the dynamic filtering algorithm gets the final results easier and faster by continuously simplifying the intermediate process. Finally, as shown in our experiments, the dynamic filtering algorithm has high execution efficiency under the influence of different factors.

Our current work is to extend the technology to meet users need. Future work is to efficiently and interactively involve the users so that the fuzzy keywords query technology and dynamic filtering algorithm can be user-approved in real-time.

Acknowledgments. This paper was partially supported by National Sci-Tech Support Plan 2015BAH10F01 and NSFC grant U1509216,61472099,61133002 and the Scientific Research Foundation for the Returned Overseas Chinese Scholars of Heilongjiang Province LC2016026.

References

1. International Standards Organization: Database Language SQLPart 2: Foundation (SQL/Foundation). International Standards Organization (1999)
2. Yin, J., Chen, Y., Zhang, G.: Research and development of search engine technology. Comput. Eng. **31**, 54–57 (2005)
3. Lu, J., Lin, C., Wang, W., et al.: String similarity measures and joins with synonyms. In: SIGMOD International Conference on Management of Data, pp. 373–384 (2013)
4. Wang, J., Li, G., Feng, J.: Extending string similarity join to tolerant fuzzy token matching. ACM Trans. Database Syst. **39**, 95–97 (2014)
5. Deng, D., Li, G., Feng, J.: A pivotal prefix based filtering algorithm for string similarity search. In: ACM Sigmod International Conference on Management of Data, pp. 673–684. ACM (2014)
6. Li, G., Deng, D., Feng, J., et al.: Top-k string similarity search with edit-distance constraints. In: 2013 IEEE 29th International Conference on Data Engineering (ICDE), pp. 925–936. IEEE (2013)
7. Fan, J., Li, G., Zhou, L., et al.: Seal: spatio-textual similarity search. VLDB **5**(9), 824–835 (2012)
8. Zeng, Z., Bao, Z., Dobbie, G., Lee, M.L., Ling, T.W.: Semantic path ranking scheme for relational keyword queries. In: Decker, H., Lhotská, L., Link, S., Spies, M., Wagner, R.R. (eds.) DEXA 2014, Part II. LNCS, vol. 8645, pp. 97–105. Springer, Heidelberg (2014)
9. Zeng, Z., Bao, Z., Lee, M.L., Ling, T.W.: A semantic approach to keyword search over relational databases. In: Ng, W., Storey, V.C., Trujillo, J.C. (eds.) ER 2013. LNCS, vol. 8217, pp. 241–254. Springer, Heidelberg (2013)
10. Feng, J., Li, G., Wang, J.: Finding top-k answers in keyword search over relational databases using tuple units. IEEE Trans. Knowl. Data Eng. **23**, 1781–1794 (2011)
11. Hulgeri, A., Nakhe, C.: Keyword searching and browsing in databases using BANKS. In: ICDE, pp. 431–440. IEEE Computer Society (2012)
12. Jiao, Y., Lei, C.: Application of the fuzzy theory in information retrieval. J. China Soc. Forentific Techn. Inf. (2000)
13. Zuo, W., Wang, Y., Gao, J., Zhao, J., Shao, H.: Semantic query optimization based on ontology. J. Comput. Res. Devel. (2009). ISSN 1000–1239/CN 11–1777/TP
14. Lu, Y.: Storage of fuzzy ontologies based on relational databases. Comput. Sci. **38**(6), 217–222,245 (2011)
15. Zhu, H., Liang, Y., Tian, Q., et al.: A method for ontology module selection. Key Eng. Mater. **440**(2), 577–583 (2010)

EPLA: Efficient Personal Location Anonymity

Dapeng Zhao[1], Kai Zhang[1], Yuanyuan Jin[1], Xiaoling Wang[1(✉)],
Patrick C.K. Hung[2], and Wendi Ji[1]

[1] Shanghai Key Laboratory of Trustworthy Computing,
Institute for Data Science and Engineering,
East China Normal University, Shanghai, China
xlwang@sei.ecnu.edu.cn
[2] Faculty of Business and Information Technology, University of Ontario
Institute of Technology (UOIT), Oshawa, ON, Canada

Abstract. A lot of researchers utilize side-information, such as map which is likely to be exploited by some attackers, to protect users' location privacy in location-based service (LBS). However, current technologies universally model the side-information for all users. We argue that the side-information is personal for every user. In this paper, we propose an efficient method, namely EPLA, to protect the users' privacy using visit probability. We selected the dummy locations to achieve k-anonymity according to personal visit probability for users' queries. AKDE greatly reduces the computational complexity compared with KDE approach. We conduct comprehensive experimental study on the realistic Gowalla data sets and the experimental results show that EPLA obtains fine privacy performance and efficiency.

Keywords: LBS · Privacy · Anonymity · KDE · Cloaking region

1 Introduction

In recent years, Location-based services (LBSs) are becoming more and more popular in our lives. While users benefit from the convenience of LBSs, they face problems caused by the privacy disclosure.

In a lot of applications, such as Meituan (www.meituan.com), LBS service providers aren't completely trusted. They appeal to users to login by electronic coupons and discount, and then, they obtain the login information. Since LBS service providers have obtained login information, users' are more fearful that their location information is collected by service providers. Once LBS service providers get users' location information, they can precisely analyze the users and get their privacy information.

To protect users' location privacy, a lot of scholars have proposed large amount of techniques including spatial cloaking technique [1,2], pseudonyms technique [3,4] and so on. Spatial cloaking technique is a very popular approach and it reduce the spatial resolution to obscure the real locations. Moreover, most of the current methods assumed that adversary don't consider side-information, such as the location Semantics [5]. Therefore, some unlikely locations, such as

© Springer International Publishing Switzerland 2016
F. Li et al. (Eds.): APWeb 2016, Part II, LNCS 9932, pp. 263–275, 2016.
DOI: 10.1007/978-3-319-45817-5_21

lakes and mountains, are included to hide users' real locations. As we know, the adversary can easily filter out the unlikely locations. Even though, a few literatures [4] make use of the side-information, researchers model them as universal model for all users. If location is the same one, the query probability is the same in [4] for every user. As we know, movement trajectories for everyone are personal. For instance, most of users travel around their residences and workplaces. So, the query probability of the same location is different for everyone. From the analysis, current methods are difficult to ensure the desired privacy degree.

To overcome the drawback of above methods, we present EPLA, an efficient personal location anonymity, to protect users' location privacy. Differentiate from the current approaches, EPLA takes the users' visited locations into account and selects dummy locations based on visit probabilities that are possibility visiting all locations. Since the visit probability is personal, we respectively model visit probability for each user in this paper. EPLA is a two-step method. In the first phase, we divided the space into cells and make sure the dummy locations candidate set P. And then, Approximate Kernel Density Estimation (AKDE) is utilized to compute the personal visit probability of each location p_i in candidate dummy locations set P based on the sampling user's visited locations. Computational complexity of the Kernel Density Estimation (KDE) [6] method is decreased from $O(|P|n^3)$ to $O(|P|n)$ (where $|P|$ is the number of elements in set P and n is the number of sampling user's visited locations). In the second phase, we achieve k-anonymity via maximizing location information entropy and the area of cloaking region (CR).

The contributions made in this research are three-fold::

1. We proposed a new method, namely Approximate Kernel Density Estimation (AKDE) to compute the personal visit probability. AKDE greatly reduces the computational complexity compared with Kernel Density Estimation (KDE) from $O(|P|n^3)$ to $O(|P|n)$.
2. We analyze the error between AKDE and KDE, and then, proof the error upper bound is $\frac{2}{(p!)^{1/2}}(\frac{3}{2q})^p$.
3. We conduct extensive experiments to evaluate the proposed method.

2   Related Work

Spatial cloaking technique [8] is a very popular method. Wang et al. [9] proposed a new model which can solve Location-aware Location Privacy Protection(L2P2) problem. ICliqueCloak [2] was proposed against location-dependent attacks.

Mix-zone [10] is one representative of Pseudonyms techniques, which enables users only to change their pseudonyms inside a special region where users do not report the exact locations. Guo et al. [11] combines a geometric transformation algorithm with a dynamic pseudonyms-changing mechanism and user-controlled personalized dummy generation to achieve strong trajectory privacy preservation. [12] exploited dummy locations to achieve anonymity.

Cryptography technique [13] is also one of the main approaches. Ghinita et al. [13] proposed a novel framework to support private location dependent queries, based on Private Information Retrieval PIR. Considering the insecure wireless net environment, [14] presented a k-anonymity algorithm using encryption for location privacy protection that can improve security of LBS system by encrypting the information transmitted by wireless. [15] designs a suite of novel fine-grained Privacy-preserving Location Query Protocol (PLQP) which allows different levels of location query on encrypted location information.

Differential privacy technique [16] is a new technique to protect location privacy. Andrés et al. [17] presented a mechanism for achieving geo-indistinguishability by adding controlled random noise to the users' location with differential privacy.

3 Preliminaries

In our method, we adopt a client/server framework and pay close attention to location privacy when users dispatch snapshot queries in LBS.

Our method firstly divides the space into $n \times n$ cells and select the $(n-1) \times (n-1)$ corners set P' of inner cells. The user location of a user R is indicated as a corner p_R which is closest to him. The candidate location anonymity set is $P = P' - p_R$. As shown in the Fig. 1, the space is divided into 4×4 cells and the set P' is $\{p_1, p_2, \cdots, p_9\}$. The red location R is a real user's location and it is close to p_5. Therefore, the candidate location anonymity set is $P = P' - \{p_5\}$.

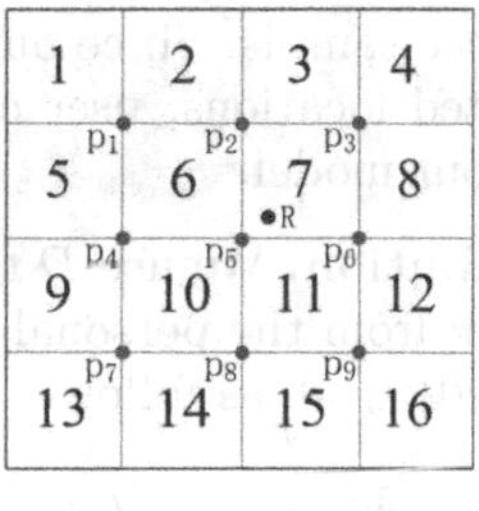

Fig. 1. Candidate locations (Color figure online)

3.1 The Personal Visit Preference

To know the personal user's visit preference, we conducted an analysis on the China Telecommunications data set which is collected in Shanghai. We randomly choose two different users from data. As Fig. 2 depicts, Their visit preference are different. The visited places of U_1 are centralized which the visited places of U_2 is dispersive. We can know the distribution over the distances between every pair of a user's visited locations is also personal and it can reflect the user personal Preference.

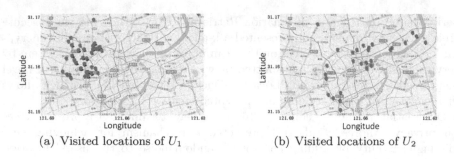

(a) Visited locations of U_1 (b) Visited locations of U_2

Fig. 2. Distributions of personal visited locations

3.2 Kernel Density Estimation of Distance Distribution

The analysis of Sect. 3.1 inspires us to use the distance distribution to research the personal visit preference. As we know, the form of the distance distribution is uncertain and we don't get the parameters of the probability density function. In this paper, we use KDE to model the personal distance distribution because KDE can be used to model arbitrary distributions and don't assume the form of the probability density function. There two steps in our method: sampling distances, and estimating distance distribution.

Sampling Distances. Firstly, we randomly sample some locations from the user's visited locations. Then, we compute the Euclidean distance of every pair of sampling locations as distance sample. Since our method achieve anonymity in client and only use his visited locations, user can input some of his visited locations when he firstly uses our model.

Estimating Distance Distribution. We use D to denote the distance sample for a certain user, which is stem from the personal distance distribution density function f. $\hat{f}$ is KDE of f based on D, as follows:

$$\hat{f}(d) = \frac{1}{|D|\sigma} \sum_{d' \in D} K \left(\frac{d - d'}{\sigma} \right) \tag{1}$$

where $K(.)$ is the kernel function and σ is the smoothing parameter, called the bandwidth. In our method, it is the normal kernel $K(x) = \frac{1}{\sqrt{2\pi}} e^{-\frac{x^2}{2}}$ and the bandwidth $\sigma = \left(\frac{4\hat{\sigma}^5}{3|D|} \right)^{1/5} \approx 1.06\hat{\sigma}|D|^{-1/5}$ [6], where $\hat{\sigma}$ is the standard deviation of distance sample D.

3.3 Fast Gauss Transform

A Gaussian $e^{-(d_i-d')^2/2\sigma^2}$ can be transformed to Hermite polynomials centered at x_0 by the Hermite expansion. And the fast Gauss Transform [18] is given by:

$$e^{-(d_i-d')^2/2\sigma^2} = \sum_{s=0}^{\infty} \frac{1}{s!} \left(\frac{d'-d_0}{\sqrt{2}\sigma}\right)^s h_s\left(\frac{d_i-d_0}{\sqrt{2}\sigma}\right)$$

$$= \sum_{s=0}^{p-1} \frac{1}{s!} \left(\frac{d'-d_0}{\sqrt{2}\sigma}\right)^s h_s\left(\frac{d_i-d_0}{\sqrt{2}\sigma}\right) + \varepsilon(p) \tag{2}$$

where $h_s(x) = (-1)^s \frac{d^s}{dx^s}(e^{-x^2})$ and $\varepsilon(p)$ is the error when we truncate the infinite series after p terms. When d' is close to d_0, a small p is enough to guarantee that $\varepsilon(p)$ is negligible.

3.4   Metrics for Location Privacy

In this paper, entropy [4] and the area of CR [2] are used to measure the level of privacy and the location information entropy is a very popular metric. The entropy H is defined as:

$$H(x) = -\sum_{i=1}^{k} p_i \cdot \log_2 p_i \tag{3}$$

where p_i is the probability which the location i is the user's real location.

The area of CR is also an important metric. The higher privacy requirement demand the bigger area of CR. Since it is difficult to compute the area of polygon, we make use of an approximate method to substitute for it. Intuitively, the sum of the distances between pairs of locations in anonymity set P^* can be used to substitute for it, which is $\sum_{i \neq j} d(P_i^*, P_j^*)$, where $d(P_i^*, P_j^*)$ denotes the distance between location P_i^* and P_j^* in anonymity set P^*.

4   Personal Visit Probability

In this section, we firstly calculate the personal visit probability by KDE, and then, an approximate method, namely AKDE, is proposed. Finally, we analyze the error between AKDE and KDE.

4.1   Exact Personal Visit Probability (EPVP)

In order to exactly compute the personal visit probability, We use KDE to estimate the personal distance distribution for every user U and compute the visit probability of any cell p_j in candidate anonymity set P according to personal distance distribution. The user's sampling visited locations is $L = \{l_1, l_2, \cdots, l_n\}$. We can compute the Euclidean distance between every location l_i in L and p_j, as follows:

$$d_i = \text{distance}(l_i, p_j), \forall l_i \in L \tag{4}$$

we can use Eq. (1) to compute a probability for each d_i as follows:

$$\hat{f}(d_i) = \frac{1}{|D|\sigma} \sum_{d' \in D} K\left(\frac{d_i - d'}{\sigma}\right) \tag{5}$$

The probability which U visits a location p_j can be computed as follows:

$$p(p_j) = \frac{1}{n} \sum_{i=1}^{n} \hat{f}(d_i) = \frac{1}{\sqrt{2\pi}n\sigma|D|} \sum_{i=1}^{n} \sum_{d' \in D} e^{-\frac{(d_i - d')^2}{2\sigma^2}} \tag{6}$$

Eventually, we derive visit probability of any p_j in candidate anonymity set P. As is show in Algorithm 1, the computational complexities of line 3 and line 4 are both $O(n^2)$. The computational complexity line 8 is $O(n^2)$. Since we should calculate all p_j in P, the total computational complexity from line 8 to line 11 is $O(n^3)$. Therefore, the total computational complexity of from line 5 to line 14 is $O(|P|n^3)$. The total complexity of Algorithm 1 is $O(n^2) + O(n^2) + O(|P|n^3) = O(|P|n^3)$.

Algorithm 1. EPVP (Exact Personal Visit Probability)

---

1 Input: sampling user's visited locations $L = \{l_1, l_2, \cdots, l_n\}$.
2 Output: $p(p_j)$ which denotes the user's visit probability of any p_j in candidate anonymity set P.
3 Calculate the distance sample D;
4 Compute the bandwidth σ using Equation (3);
5 **for** *each p_j in P* **do**
6 $\quad$ $z \leftarrow 0$;//Initializing auxiliary variable z
7 $\quad$ Calculate the distance d_i between l_i and p_j using Equation (7);
8 $\quad$ **for** $i : 1 \rightarrow n$ **do**
9 $\quad\quad$ $\hat{f}(d_i) \leftarrow 0$;
10 $\quad\quad$ **for** *each d' in D* **do**
11 $\quad\quad\quad$ $\hat{f}(d_i) \leftarrow \hat{f}(d_i) + K\left(\frac{d_i - d'}{\sigma}\right)$;
12 $\quad$ **for** $i : 1 \rightarrow n$ **do**
13 $\quad\quad$ $z \leftarrow z + \hat{f}(d_i)$;
14 $\quad$ $p(p_j) \leftarrow \frac{1}{n}z$;

4.2 Approximate Personal Visit Probability

As we know, the computational complexity $O(|P|n^3)$ of EPVP grows rapidly with n increasing. In this part, we design an approximate method, namely APVP, to compute personal visit probability through the fast Gauss transform and three-sigma rule of Gaussian distribution. And we finally reduce the complexity of EPVP to $O(|P|n)$.

For Eq. 6, the personal visit probability $p(p_j)$ that a user visits the location p_j can be approximately calculated using Eq. 2. To reduce the error, we should not only select center, such as the mean of d'. And with the number of centers increasing, computational complexity increases.

Nearly all values of d' in the distance sample D lie within interval $[\bar{d} - 3\hat{\sigma}, \bar{d} - 3\hat{\sigma}]$ according to three-sigma rule of Gaussian distribution. Where $\bar{d}$ is the mean of the sample D and $\hat{\sigma}$ is the standard deviation of the sample D. In order to facilitate the calculation, we evenly divide the interval $[\bar{d} - 3\hat{\sigma}, \bar{d} - 3\hat{\sigma}]$ into $2q$

small intervals set $I = \{I_1, I_2, \cdots, I_{2q}\} = \{[\bar{d} - 3\hat{\sigma}, \bar{d} - 3\hat{\sigma} + \frac{3\hat{\sigma}}{q}], [\bar{d} - 3\hat{\sigma} + \frac{3\hat{\sigma}}{q}, \bar{d} - 3\hat{\sigma} + \frac{6\hat{\sigma}}{q}], \cdots, [\bar{d} + 3\hat{\sigma} - \frac{3\hat{\sigma}}{q}, \bar{d} + 3\hat{\sigma}]\}$ and the distance sample D is divided into 2q small distance sample set $\{D_1, D_2, \cdots, D_{2q}\}$ according to I. Where, $q = 3k$ and $k \geq 1$. Each small distance sample D_i is approximately shifted respectively by the fast Gauss transform and we select the middle value $\mu_i = \bar{d} - 3\hat{\sigma} + \frac{3i\hat{\sigma}}{2q}$ of I_i as transforming center of D_i. Based on the partition D_i and three-sigma rule of Gaussian distribution, we have

$$
\sum_{d' \in D} e^{-\frac{(d_i - d')^2}{2\sigma^2}} \approx \sum_{r=1}^{2q} \sum_{d' \in D_r} \sum_{s=0}^{p-1} \frac{1}{s!} \left(\frac{d' - \mu_r}{\sqrt{2}\sigma}\right)^s h_s\left(\frac{d_i - \mu_r}{\sqrt{2}\sigma}\right)
$$

$$
= \sum_{r=1}^{2q} \sum_{s=0}^{p-1} \frac{1}{s!} \sum_{d' \in D_r} \left(\frac{d' - \mu_r}{\sqrt{2}\sigma}\right)^s h_s\left(\frac{d_i - \mu_r}{\sqrt{2}\sigma}\right) \tag{7}
$$

$$
= \sum_{r=1}^{2q} \sum_{s=0}^{p-1} A(s,r) h_s\left(\frac{d_i - \mu_r}{\sqrt{2}\sigma}\right)
$$

where $A(s,r) = \frac{1}{s!} \sum_{d' \in D_r} \left(\frac{d' - \mu_r}{\sqrt{2}\sigma}\right)^s$. We can transform the Eq. 6 according to Eq. 8, as follows:

$$
p(p_j) = \frac{1}{\sqrt{2\pi}n\sigma|D|} \sum_{i=1}^{n} \sum_{r=1}^{2q} \sum_{s=0}^{p-1} A(s,r) h_s\left(\frac{d_i - \mu_r}{\sqrt{2}\sigma}\right) \tag{8}
$$

As is show in Algorithm 2, the computational complexities of line 3 and line 4 are both $O(n^2)$. The complexities of line 5 and line6 are also both $O(n^2)$. Since the p and q are constant parameters and $|p| \gg n$, the computational complexity of from line 7 to line 13 is $O(pq|P|n) = O(|P|n)$. Therefore, the total complexity of Algorithm 2 is $O(n^2) + O(n^2) + O(n^2) + O(n^2) + O(|P|n) = O(|P|n)$.

Algorithm 2. APVP (Approximate Personal Visit Probability)

---

1 Input: sampling user's visited locations $L = \{l_1, l_2, \cdots, l_n\}$.
2 Output: $p(p_j)$ which denotes the user's visit probability of any p_j in candidate anonymity set P.
3 Calculate the distance sample D;
4 Calculate the bandwidth σ based on Equation (3);
5 Group the distance sample D into small set D_i and obtain the center μ_i;
6 Calculate A(s,r) using Equation (11) and obtain a two-dimension array;
7 **for** each p_j in P **do**
8     $z \leftarrow 0$;//Initializing auxiliary variable z
9   Calculate the distance d_i between l_i and p_j using Equation (7);
10   **for** $i : 1 \rightarrow n$ **do**
11   **for** $r : 1 \rightarrow 2q$ **do**
12   **for** $s : 0 \rightarrow p - 1$ **do**
13              $z \leftarrow z + A(s,r) h_s(\frac{d_i - \mu_r}{\sqrt{2}\sigma})$;

4.3 Error Analysis

Theorem 1 *(The upper bound of the error ε). By truncating the infinite series after p terms in Eq. 4 and dividing interval $[\bar{d}-3\hat{\sigma}, \bar{d}-3\hat{\sigma}]$ into 2q small intervals, Algorithm 2 guarantees that the upper bound of the error ε satisfies $\frac{2}{(p!)^{1/2}}(\frac{3}{2q})^p$.*

Proof. At first, according to Cramers inequality [21] $h_s(x) \leq 2^{s/2}(s!)^{1/2}e^{-x^2/2}$ and $e^{-x^2/2} \leq 1$, so $h_s(x) \leq 2^{s/2}(s!)^{1/2}$

Hence,

$$
\begin{aligned}
|\varepsilon(p)| &\leq \sum_{s=p}^{\infty} \frac{1}{s!} |\frac{d'-d_0}{\sqrt{2}\sigma}|^s |h_s(\frac{d_i-d_0}{\sqrt{2}\sigma})| \\
&\leq \sum_{s=p}^{\infty} \frac{1}{s!} |\frac{d'-d_0}{\sqrt{2}\sigma}|^s 2^{s/2}(s!)^{1/2} \\
&= \sum_{s=p}^{\infty} \frac{1}{(s!)^{1/2}} |\frac{d'-d_0}{\sigma}|^s \\
&\leq \frac{1}{(p!)^{1/2}} \sum_{s=p}^{\infty} |\frac{d'-d_0}{\sigma}|^s
\end{aligned}
\tag{9}
$$

As depicted in Sect. 4.2, every d' in distance sample D is assigned to a small distance set D_i whose transforming center is μ_i. And we have $|d'-d_0| = |d'-\mu_i| \leq \frac{3\sigma}{2q}$. Therefore,

$$
|\varepsilon(p)| \leq \frac{1}{(p!)^{1/2}} \sum_{s=p}^{\infty} (\frac{3}{2q})^s
\tag{10}
$$

Moreover, $q \geq 3$, Accordingly

$$
|\varepsilon(p)| \leq \frac{1}{(p!)^{1/2}}(\frac{3}{2q})^p \frac{2q}{2q-3} \leq \frac{2}{(p!)^{1/2}}(\frac{3}{2q})^p
\tag{11}
$$

As depicted in Theorem 1, the upper bound of the error decreases faster than the exponential decay with the p and q increasing, as shown in Fig. 3.

5 Anonymity Set Selection (ASS)

In the process of dummy location selection, we consider two factor which are location information entropy and the area of CR to achieve anonymity. Therefore, the process of dummy selection can be formulated as MCDM model. Let $P^* = \{P_1^*, P_2^*, \cdots, P_k^*\}$ denote the location anonymity set in our scheme. The MCDM model can be described as:

$$
Max\{-\sum_{i=1}^{k} p(P_i^*) \cdot \log p(P_i^*), \sum_{k \neq j} d(P_i^*, P_j^*)\}.
\tag{12}
$$

where $P_i^*, P_j^* \in P^*$, $p(P_i^*)$ and $p(P_j^*)$ denote the personal visit probabilities of P_i^* and P_j^* respectively.

For MCDM model, It is hard to find a location set to meet all requirements simultaneously. So, we use heuristic solution [4] to select the proper dummy location set. The process has two steps as follows: (1) maximizing location information entropy; (2) maximizing area, and Algorithm 3 depicts the process of dummy location selection.

Maximizing Location Information Entropy: A user whose location need to protect first input real location R. So, we can find the closest corner p_R to R as Sect. 3.1 depicts. Secondly, our algorithm should get the user's personal visit probabilities which are computed by Algorithm 2 and then sorts all locations in P based on the personal visit probabilities. Then, our algorithm selects $4k$ candidates locations set which contains the $2k$ locations before p_R and the $2k$ locations after p_R to ensure that they are as similar as possible to R. After that, our algorithm select m location set $S = \{S_1, S_2, \ldots, S_m\}$ from $4k$ candidates locations, each set containing the real location and $2k - 1$ dummy location are contained. The $j^{th}(j \in [1, m])$ set can be denoted as $S_j = \{S_{j1}, S_{j2}, \ldots, S_{ji}, \ldots, S_{j2k}\}$. According to the personal visit probabilities, we should normalize visit probabilities. We denote them using $P_{j1}, P_{j2}, \ldots, P_{ji}, \ldots, P_{j2k}$ and $P_{ji} = \frac{p(S_{ji})}{\sum_{i=1}^{2k} p(S_{ji})}$, where $p(S_{ji})$ is the personal visit probability of the location S_{ji} and $\sum_{i=1}^{2k} P_{ji} = 1$. So, the entropy H_j of anonymity set S_j can be derived based on Eq. 3. Finally, the location set $S' = \{S_1', S_2', \cdots, S_j', \cdots, S_{2k}'\}$ whose entropy is maximum in S is selected.

Algorithm 3. ASS (Anonymity Set Selection)

---

1 Input: $p(P_i)$ denoting the user's visit probability of P_i in P; R denoting the real location; m denoting the number of sets.
2 Output: P^* the dummy anonymity set
3 Choosing $4k$ dummy candidates including $2k$ locations before and $2k$ locations after the user's real locations R;
4 Constructing m location sets $S = \{S_1, S_2, \cdots, S_m\}$, each S_j contains R and $2k - 1$ dummy locations randomly selected from $4k$ dummy candidates;
5 **for** *each S_j* **do**
6 $\quad$⌊ Calculating the entropy H_j via Equation (11);
7 $S' = argmax H_j$;
8 Initializing $P^* = \{R\}$ and $S' = S' - R$;
9 **for** $i = 0 : k - 1$ **do**
10 $\quad$ **if** p^* *in S' and* $\sum_{P_j^* \in P^*} d(p^*, P_j^*)$ *is greatest* **then**
11 $\quad\quad$⌊ $P^* = P^* + p^*$;
12 $\quad\quad$ $S' = S' - p^*$;

13 Return P^*;

Maximizing Area: In this process, we use greedy algorithm to get the final location anonymity set $P^* = \{P_1^*, P_2^*, \cdots, P_k^*\}$ from S' based on $\sum_{i \neq j} d(S_i', S_j')$. Firstly, a set $P^* = \emptyset$ is constructed. The user's real location is added into P^*

and removed from S'. Then, the next location p^* is selected to be added into P^* and to be removed from S', when $\sum_{P_j^* \in P^*} d(p^*, P_j^*)$ is greatest. As lines 9–12 in Algorithm 3 depicts, ASS repeats this step $k - 1$ times. Finally, we get the anonymity set P^*.

6   Performance Evaluation

6.1   Experiment Setup

In the experiments, we use the publicly available Gowalla [19] data set to instead of the users' locations sample. The statistics of the data sets are shown in Table 1. We evaluate the privacy degree of our scheme by comparing three algorithms, namely dummy [12], DLS [4] and enhanced-DLS [4]. The dummy method is taken as baseline in this paper.

Moreover, we select 40 km × 40 km American region, and then, it is divided into 80 × 80 sells in our experiment. So the candidate locations anonymity set P includes $79 \times 79 - 1 = 6240$ locations. The personal visit probability is calculated by AKDE.

Table 1. Gowalla data set

Number of users	196,591
Number of POIs	1,280,969
Number of check-ins	6,442,890

6.2   Evaluation of Privacy Degree

In the process, location information entropy and the sum of distance are used as metrics. Anonymity cost is also an important trait in LBS and we measure it using online time. Moreover, we compare the time costs of APVP and EPVP, and then, analyze the relationship between the time cost of APVP and the constant parameters (namely p and q).

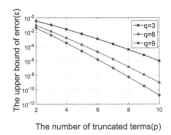

Fig. 3. Anonymity times vs. k

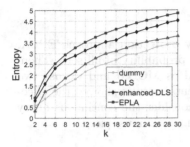

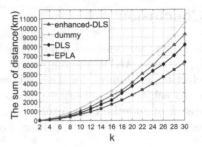

Fig. 4. ε vs. k

Fig. 5. Entropy vs. k

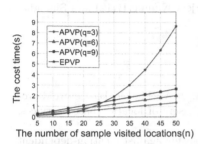

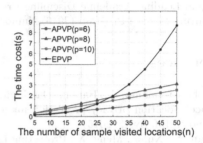

Fig. 6. The sum of distances vs. k

Fig. 7. The time cost vs. n

Entropy vs. k. Larger information entropy implies that it more difficult to ensure the user's real location. As Fig. 4 depicts, the location information entropies of all methods increase following k. The entropy of dummy is terrible than other. Our scheme *EPLA* is optimal and provides better privacy level than *DLS* and enhanced-*DLS*.

Area of CR vs. k. The area of CR is focused on in location privacy domain. As we know, when the area is too small, attacker can know user's real location. So, we should consider the area to evaluate the privacy level. As Fig. 5 depicts, the area of dummy method is biggest. EPLA method is slightly short of the other methods. From above analysis, we know every user's zone of action is limited by a lot of factors and most of visit location is close to our residence and workplace.

The Time Cost vs. n. As we know, the different p and q will lead to the different time cost of APVP when n is the same. We select different p and q to test the time cost of APVP. In the Fig. 6, we fix p = 8 and we fix q = 8 in the Fig. 7. And the cost of APVP linearly grows with n increasing while the EPVP looks like exponential growth. Moreover, when p and n is fixed, the time cost is increasing with P increasing. And when q and n are fixed, the time cost is also increasing with q increasing. As Fig. 6 depict, the computational complexity of EPVP is far outweigh the computational complexities of APVP.

7   Conclusion

In this paper, we designed a new location privacy protection approach EPLA. Firstly, we used AKDE to compute personal visit probability of all sells. Then, the selection of dummy location set was modeled as MCDM and we use two factors to achieve the selection process. Experimental results showed the effectiveness of our method. In future work, we will consider the continuous query problem and try to solve the privacy protection in continuous query.

Acknowledgment. This work was supported by NSFC grants (No. 61532021, 61472141 and 61021004), Shanghai Knowledge Service Platform Project (No. ZF1213), Shanghai Leading Academic Discipline Project (Project NumberB412) and Shanghai Agriculture Science Program (2016) Number 2-1.

References

1. Ashouri-Talouki, M., Baraani-Dastjerdi, A., Seluk, A.A.: The cloaked-centroid protocol: location privacy protection for a group of users of location-based services. Knowl. Inf. Syst. (2015)
2. Pan, X., Xu, J., Meng, X.: Protecting location privacy against location-dependent attacks in mobile services. TKDE **24**, 1506–1519 (2012)
3. Lu, H., Jensen, C.S., Yiu, M.L.: Pad: privacy-area aware, dummy-based location privacy in mobile services. In: Proceedings of the Seventh ACM International Workshop on Data Engineering for Wireless and Mobile Access. ACM (2008)
4. Niu, B., Li, Q., Zhu, X., et al.: Achieving k-anonymity in privacy-aware location-based services. In: 2014 IEEE Proceedings of INFOCOM. IEEE (2014)
5. Lee, B., Oh, J., Yu, H., et al.: Protecting location privacy using location semantics. In: Proceedings of the 17th ACM SIGKDD International Conference on Knowledge Discovery and Data Mining. ACM (2011)
6. Silverman, B.W.: Density Estimation for Statistics and Data Analysis. CRC, London (1986)
7. Zhang, J.D., Chow, C.Y.: iGSLR: personalized geo-social location recommendation: a kernel density estimation approach. In: Proceedings of the 21st ACM SIGSPATIAL International Conference on Advances in Geographic Information Systems (2013)
8. Jia, J., Zhang, F.: K-anonymity algorithm using encryption for location privacy protection. Int. J. Multimedia Ubiquit. Eng. **10**, 155–166 (2015)
9. Gruteser, M., Grunwald, D., Wang, Y., Xu, D., He, X., et al.: L2P2: location-aware location privacy protection for location-based services. In: INFOCOM (2012)
10. Palanisamy, B., Liu, L.: Attack-resilient mix-zones over road networks: architecture and algorithms. IEEE Trans. Mob. Comput. **14**, 495–508 (2015)
11. Guo, M., Pissinou, N., Iyengar, S.S.: Pseudonym-based anonymity zone generation for mobile service with strong adversary model. In: Consumer Communications and Networking Conference (CCNC) (2015)
12. Kido, H., Yanagisawa, Y., Satoh, T.: An anonymous communication technique using dummies for location-based services. In: Proceedings of International Conference on Pervasive Services, ICPS 2005. IEEE (2005)

13. Ghinita, G., Kalnis, P., Khoshgozaran, A., et al.: Private queries in location based services: anonymizers are not necessary. In: Proceedings of the 2008 ACM SIGMOD International Conference on Management of Data. ACM (2008)

14. Jia, J.: K-anonymity algorithm using encryption for location privacy protection. Int. J. Multimedia Ubiquit. Eng. (2015)

15. Li, X.Y., Jung, T.: Search me if you can: privacy-preserving location query service. In: 2013 IEEE Proceedings of INFOCOM. IEEE (2013)

16. Clifton, C., Tassa, T.: On syntactic anonymity and differential privacy. In: 2013 IEEE 29th International Conference on Data Engineering Workshops. IEEE (2013)

17. Andrés, M.E., Bordenabe, N.E., Chatzikokolakis, K., et al.: Geo-indistinguishability: differential privacy for location-based systems. In: Proceedings of the 2013 ACM SIGSAC Conference on Computer Communications Security (2013)

18. Greengard, L., Strain, J.: The fast Gauss transform. SIAM J. Sci. Stat. Comput. **12**, 79–94 (1991)

19. Cho, E, Myers, S.A., Leskovec, J.: Friendship and mobility: user movement in location-based social networks. In: Proceedings of the 17th ACM SIGKDD International Conference on Knowledge Discovery and Data Mining. ACM (2011)

20. Pukelsheim, F.: The three sigma rule. Am. Stat. **48**, 88–91 (1994)

21. Indritz, J.: An inequality for Hermite polynomials. Proc. Am. Math. Soc. **12**, 981–983 (1961)

A Secure and Robust Covert Channel
Based on Secret Sharing Scheme

Xiaorong Lu[1]([✉]), Yang Wang[2], Liusheng Huang[1], Wei Yang[1], and Yao Shen[1]

[1] School of Computer Science and Technology,
University of Science and Technology of China, Hefei 230027, China
{ldayy,shenyao}@mail.ustc.edu.cn, {lshuang,qubit}@ustc.edu.cn
[2] Suzhou Institute for Advanced Study,
University of Science and Technology of China, Suzhou 215123, China
angyan@ustc.edu.cn

Abstract. Network covert channel (referred to as network steganography) is a covert communication technique that uses the redundancies of network protocols to transfer secret information. While encryption only protects communication from being decoded by unauthorised parties, a covert channel aims to hide the very existence of the secret communication. More recently focus has shifted towards network protocols based covert channels because the huge amount of data and vast number of different protocols in the Internet seems ideal as a high-bandwidth vehicle for covert communication. However, few approaches which can embed secret information with both great security and robustness ensured have been worked out by so far. In this paper, we propose a novel packet length based covert channel exploiting the secret sharing scheme in order to overcome the drawbacks of existing schemes. A comprehensive set of corresponding experiment results and security analysis show that the proposed covert channel is provably secure and with great robustness than that of the existing algorithms.

Keywords: Covert channel · Network steganography · Packet length · Network security · Information hiding · Network protocols

1 Introduction

Steganography is the art or practice of concealing a covert message within another message. The first work about covert channel was introduced by Lampson [7] to avoid information leakage in computer systems. The advantage of steganography over cryptography alone is that the intended secret message does not attract attention to itself as an object of scrutiny. Plainly visible encrypted messages-no matter how unbreakable-will arouse interest, and may in themselves be incriminating in countries where encryption is illegal. Thus, whereas cryptography is the practice of protecting the contents of a message alone, steganography is concerned with concealing the fact that a secret message is being sent, as well as concealing the contents of the message.

F. Li et al. (Eds.): APWeb 2016, Part II, LNCS 9932, pp. 276–288, 2016.
DOI: 10.1007/978-3-319-45817-5_22

Generally network covert channels are classified into two broad categories: network storage covert channel [8,9] and network timing channel [10]. In storage covert channel, a process directly or indirectly writes to a particular storage location whereas other process reads from that particular storage location. On the other hand timing covert channel focuses on conveying the message by the arrival pattern of packets rather than the contents of message.

Another notable branch of network covert channel is based on packet length. In general the packet length is adjusted intentionally to transfer the secret data across the network. This technique was first introduced by Padlipsky [12] in 1978 while another similar scheme was proposed by Girling [13] in 1987. Yao [18] proposed a packet length based scheme in 2008 and Liping two papers [15,16] improved this technique by imitating the normal traffic stream of packets. Recently A. S. Nair [14] designed a kind of covert channel by modifying the length of UDP packets.

While these covert channels mentioned above are secure when they have been proposed, they all are vulnerable in case of current network condition. They require packets to reach the receiver ordered and no packet is lost regardless of packet loss and all these requirements are ensured by a so-called upper layer application and the covert channel itself doesn't care about these cases.

In this paper, we propose a novel packet length based covert channel exploiting the secret sharing scheme in order to overcome the drawbacks of existing schemes. The main contributions of this paper are three-fold:

- Our scheme doesn't send the secret message itself to the receiver. It send shadows of secret message to the network in this case that an eavesdropper can not get the secret message. But an eavesdropper can get the secret message from the schemes mentioned above and the remaining thing is that he takes time to analyze the secret message from the packets he gets from the network. Our scheme doesn't need the packets to reach the receiver in the order that the sender sends them. Actually when our scheme starts to work, the packets order will be rearranged in a random way.
- Our scheme allows that network can introduce packet losses. In this case some packets will be lost because of the bad network condition. Our scheme will work well when some packets are lost while existing schemes will definitely fail without an upper layer application caring.

The rest of the paper is organized as follows. In Sect. 2 we review related work. The network model and necessary background are presented in Sect. 3. Section 4 introduces the proposed secret sharing based covert channel and communication protocol. A comprehensive set of corresponding experiments are performed in Sect. 5 which show that the proposed covert channel follows the normal traffic statistical features to fail most of the detection schemes proposed before and thus ensures more security and robustness than that of the existing algorithms. Finally, Sect. 6 concludes this paper.

2   Related Work

As mentioned in the Sect. 1, there are several packet length based methods in literature. In order to get familiar with the trend of the packet length based steganography quickly, we only review some notable work of this field. The concept of packet length based covert channel was first proposed by Padlipsky et al. in [12]. In their paper, the authors have designed a covert channel by modulating the length of link layer frames for secret message transmission. The secret data file is divided into several bytes and each byte secret message is represented by a certain frame length. The approach exploits 256 different frame lengths to describe the bytes. This scheme is not successful in imitating the network traffic flow. Girling [13] also have proposed another link layer staganographic scheme. However, it is proved that the packet length distribution has been changed hugely by these schemes and they can be detectable easily.

Yao [18] proposed a model named LAWB based on the packet length. In his method, the sender shares a secret matrix of which all the cells represent unique length with the receiver. The sender selects a cell randomly and gets the corresponding length of the message to be send, and the receiver uses the same way to find the random number from the matrix to retrieve the secret message.

Recently Anand [14] designed another covert channel using the UDP packet length. In [14], the secret message is divided into 4-bit binary string and thus there are 16 kind of 4-bit binary string. The scheme takes 1 to 500 to construct a matrix where each number represented a length is added to the corresponding (x mod 16) row in the matrix. Whenever a string needed to be sent, the 4-bit binary string is converted into its decimal form to find out the corresponding length to be sent to the receiver. By the way, the authors also has presented detecting schemes for length based covert channel [6,17] which used the statistical techniques. In their two articles, the authors found that data embedding reduces the smoothness of the Packet Length Vector and thus the presence of data hiding based on packet length can be detected by the analysis of packet length frequency and other techniques they proposed.

However, there would be certain applications that send the packets of random sizes without any particular pattern whereas steganographers could exploit packet length to design more secure covert channel than that of the existing works. Moreover, the payload of UDP packets can be modulated because the length of the messages has a random distribution and this situation is suitable for the design of covert channel. In this paper, we are motivated by this situation to design a more secure and robust covert channel based on the secret sharing scheme and our scheme is proved to be less detectable and fault-tolerant as it can fails the detecting scheme proposed in [6,17].

3   Preliminaries, Models and Assumptions

3.1   Secret Sharing Scheme

In cryptography, a secret sharing scheme is a method for distributing a secret among a group of participants, each of which is allocated a share of the secret.

The secret can be reconstructed only when a sufficient number, of possibly different types, of shares are combined together; individual shares are of no use on their own. This scheme was invented independently by Shamir [1] and Blakley [2] in 1979. This subsection we illustrate a simplified version of secret sharing scheme. Methodical discussion about this scheme can be found in [3–5].

A decisive mathematical definition of secret sharing scheme is described as:

The goal is to divide secret S (e.g., a safe combination) into n pieces of data $D_1,\ldots,D_n$ in such a way that:

- Knowledge of any k or more D_i pieces makes S easily computable.
- Knowledge of any $k-1$ or fewer D_i pieces leaves S completely undetermined (in the sense that all its possible values are equally likely).

This scheme is called *(k, n)* threshold scheme. Each piece D_i is called a shadow or a share of the secret S. If $k = n$ then all participants are required to reconstruct the secret. Suppose we want to use a *(k, n)* threshold scheme to share our secret S, without loss of generality assumed to be an element in a finite field F of size P where $0 < k \leq n < P$; $S < P$ and P is a prime number.

Choose at random $k-1$ positive integers $a_1,\cdots,a_{k-1}$ with $a_i < P$, and let $a_0 = S$. Build the polynomial:

$$f(x) = a_0 + a_1x + a_2x^2 + a_3x^3 + \cdots + a_{k-1}x^{k-1} \bmod P \qquad (1)$$

Let us construct any n points out of it, for instance set $i = 1,\cdots,n$ to retrieve $(i, f(i))$. Every participant is given a point (an integer input to the polynomial, and the corresponding integer output). Given any subset of k of these pairs, we can find the coefficients of the polynomial using interpolation. The secret is the constant term a_0. Secret sharing schemes are ideal for storing information that is highly sensitive and highly important. In this paper we exploit this advantage to design a highly secure and robust covert channel.

3.2 Model for Network Covert Channel

Network Model for Covert Channel. In general, a traditional model for network based covert channel is shown in Fig. 1. Alice is the secret message sender and Bob is the secret message receiver while Warden is the administrator who can monitor all the network packets. When transmission starts, Alice will send overt packets to Bob with the secret message represented by the length of the packets to be send. By the way, Bob also sends messages to Alice in order that it appears as a normal chat between them as the model is a chat application scenario.

Data Transmission Model for Covert Channel. A pervasive data transmission model for covert channel is showed in Fig. 2. With this model, each network packet will carry a piece of the secret message itself. Alice directly embeds the secret information itself into the packet, and then the packet will be sent to Bob

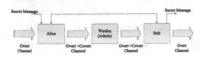

Fig. 1. Model of traditional packet length based covert channel

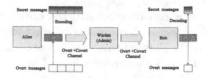

Fig. 2. A pervasive data transmission model for covert channel

through the overt channel. When Bob receives the packet, he can retrieve the secret message easily from the packet. However there is a lurking peril that if the embedding method is too direct that Warden can also get the secret message from the packets by some statistical analysis tools. And what's worse, if any packet is lost then the secret message will never be retrieved by Bob although he receives the remain packets. In this paper, we design our new covert channel using the secret sharing scheme to avoid the problems mentioned above. Therefore the new data transmission model is illustrated in Fig. 3. Under this model, the original secret message will be transformed into several pieces which are called shadows of the secret message. Then Alice embeds the shadows into packets and sends the packets to Bob. By a reversible transformation process, Bob can retrieve the secret message using the shadows instead of the secret message itself. It is not surprising that Warden can get nothing from the packet stream, because he just gets some shadows of the secret message. Even if some packets are lost through the transmission between Alice and Bob, Bob can still retrieve the secret message by using a *(k, n)* threshold scheme with $k < n$.

3.3 Additional Assumptions

In this paper, we aim to design a highly secure and robust covert channel. In the following we give some reasonable assumptions about the security and robustness.

- **Security**: We assume that Alice and Bob both will never reveal the information about the *(k, n)* threshold scheme. In other word, Warden can not get the k, n, P from Alice and Bob.
- **Robustness**: In our scheme, we allow that the network will make a "mistake" sometime. That is to say, some packets sent by Alice will never be received by Bob and the remain packets will not reach Bob in the same order which Alice sends them. Under this assumption, most existing covert channel will never work well as it is designed. We will show later that our scheme can be effective under this condition.

4   Secret Sharing Based Secure Covert Channel

In this section, we present our novel covert channel using secret sharing scheme. Since we design our channel to be a packet length based covert channel, the

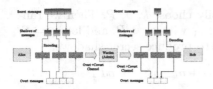
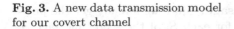
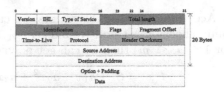

Fig. 3. A new data transmission model for our covert channel

Fig. 4. Relevant fields to be modified in the IP header

original packet length must be changed for decoding the shadows of secret message. Based on this idea, we alter the relevant fields in the IP header as shown in Fig. 4. The details will be described in the following. The scheme is divided into two parts: encoding process and decoding process and we start with the first part.

4.1 Encoding Process

As we use secret sharing scheme to design our staganographic method, the encoding process consists of three parts: designing *(k, n)* threshold scheme, constructing shadows of secret message, embedding shadows into network packets. Each part of the encoding process will be described respectively.

Designing *(k, n)* Threshold Scheme. This part we want to choose a appropriate *(k, n)* threshold, i.e. choose three numbers: k, n, P. According to Sect. 2, there are some constraints when choosing these numbers:

$$0 < k \leq n < P \tag{2}$$

$$P \ is \ a \ prime \ number \tag{3}$$

Except the mathematical constraints, some other constraints should be stated according to the TCP/IP protocols. We wish that each packet will carry a shadow from Alice to Bob. If the packet is too big that it will be divided into several fragments, this will cause the mistaken decoding by Bob. Therefore, the packet length must be smaller than the maximum length defined by the IP protocol which is 1500. We denote the packet length as p_{len}. On the other hand, each shadow of the secret must be at least as large as the secret itself. This result is based on information theory, but can be understood intuitively. As we use the shadow value to modify the packet length, thus the shadow value and the secret should be no more than 1500. Let $S = S_1 S_2 \cdots S_m$ be a $(m*w)$-bits binary secret message to be sent by Alice to Bob. The remain constraints are as follows:

$$P < 1500 \tag{4}$$

$$p_{len} < P \ for \ each \ packet \tag{5}$$

$$S_i < P, \ 1 \leq i \leq m \tag{6}$$

According to these constraints, we can easily choose k, n, P. Then we randomly choose $k-1$ positive integers $a_1, \cdots, a_{k-1}$ with $a_i < P$, and let $a_0 = S_i$ (for $1 \leq i \leq m$) then we can get polynomial to produce secret shadows:

$$f(x) = S_i + a_1 x + a_2 x^2 + a_3 x^3 + \cdots + a_{k-1} x^{k-1} \ mod \ P \qquad (7)$$

Constructing Shadows of Secret Message. Since we already get the polynomial (7), we can calculate the shadows for each S_i of the secret message S. As introduced in Sect. 2, we denote the shadow as D_i. Therefore, for each S_i (for $1 \leq i \leq m$), there are n shadows, i.e. $D_{i,j}$(for $1 \leq j \leq n$). We set x from 1 to n, for the whole secret message S, the shadow matrix with size of $m * n$ is therefore:

$$D_{m,n} = \begin{bmatrix} D_{1,1} & D_{1,2} & \cdots & D_{1,n-1} & D_{1,n} \\ D_{2,1} & D_{2,2} & \cdots & D_{2,n-1} & D_{2,n} \\ \vdots & \vdots & \vdots & \vdots & \vdots \\ D_{m-1,1} & D_{m-1,2} & \cdots & D_{m-1,n-1} & D_{m-1,n} \\ D_{m,1} & D_{m,2} & \cdots & D_{m,n-1} & D_{m,n} \end{bmatrix} \qquad (8)$$

The each row(D_i) of matrix (8) represents n shadows of each S_i. Even more, note that each $D_{i,j}$ represents the $f(x)$ when $x = j$.

Embedding Shadows into Network Packets. According to Sect. 2, each shadow consists of two number: x and $f(x)$. When we embed a shadow into a network packet, we must embed both the x and $f(x)$. As described at the beginning of this section, we attempt to modify the IP header to embed the shadows. In the following, we will introduce the method for embedding x and $f(x)$ respectively.

Firstly, we embed x into the identification field of IP header. According to the TCP/IP protocols documents, identification field is primarily used for uniquely identifying the group of fragments of a single IP datagram. When a single IP has not any fragments, it can be uniquely identifying by the identification field. Practically the value of identification field is set in ascending order by the computer. That's to say, the value of identification field of the IP datagram sent first will be smaller than that of the IP datagram sent after, unless the value get the maximum value then it will be reset to the minimum value by the computer. As x is 1 to n in ascending order, we choose identification field to embed x. Note that we just need to embed t ($k < t \leq n$) shadows into network packets because at least k shadows can reconstruct the secret. And we set $t > k$ in order to ensure that Bob can reconstruct the secret even if some packets are lost during the transmission but there are still at least k shadows can successfully reach Bob. For each S_i, randomly select t $D_{i,j}$ from its matrix $D_{m,n}$ in the ascending order of j, that is $(j_1, j_2, \cdots, j_{t-1}, j_t)$. Denote the modified value as $Idnew$ and give a random positive number y, here is the formula to modify identification field to embed x.

$$Idnew_{i,j_r} = (y \ll 12) + i * n + j_r$$
$$for \ 1 \leq i \leq m, \ 1 \leq r \leq t \qquad (9)$$

where $(y \ll 12)$ is a arithmetic left shift of y by 12. Actually we set the most left four bits of $Idnew$ with a random number of y and the remain twelve bits with the binary value of $(i*n+j_r)$ as the identification field value is a 16 bits binary number. Because the right bits in a identification field vary more frequently than the left bits, we can change the sending order of the shadows-carrying packets on a massive scale in order that the eavesdropper can not get anything about secret message with a sequenced analysis of packets stream. On the other hand, we use formula 9 to help Bob classify the shadows from different S_i. Under the operation of $(i*n+j_r)$, for the shadows from the same S_i, the difference between any two value of $D_{i,j}$ is smaller than n under rule 9. If we do not use this formula, then there maybe exists two value of $D_{i,j}$ and $D_{l,r}$ from S_i and S_l but their difference is smaller than n. Thus Bob will make a mistake when he reconstruct the secret. We use formula 9 to avoid this mistake. Actually with this formula, the difference between any two $idnew$ from the same S_i is always smaller than n while that from different S_i is definitely larger than n. Thus we can embed x correctly into IP field.

Secondly, we modify the total length field to embed $f(x)$. Under the constraints (7), (8) and the polynomial (8), we know that each $D_{i,j}$ satisfies $D_{i,j} < P$. Thus we just set the value of total length field as $D_{i,j}$. Denote the modified value of total length field as $Plenew$, that is:

$$Plenew_{i,j} = D_{i,j}, \ for \ Idnew_{i,j} \tag{10}$$

Protocol 1. Secret Message Encoding Protocol(SMEP)

Require: S=$S_1 S_2 \cdots S_m$, $(k,\ n)$ threshold
1: For each subgroup S_i, find the corresponding $f(x)$ and calculate its shadows $D_{i,1}$, $D_{i,2}, \cdots, D_{i,n}$. Thus all the S_i's shadows form the matrix $D_{m,n}$.
2: For each element $D_{i,j}$ in matrix $D_{m,n}$, randomly select a integer y, computer the value $I_{i,j}$ of $(y \ll 12) + i*n+j$. And all these values make up a matrix $I_{m,n}$ in accordance with the matrix $D_{m,n}$. Each $I_{i,j}$ corresponds to $D_{i,j}$ on the basis of the index i and j.
3: Sort all the elements in $I_{m,n}$ to be an ascending sequence $I^*_{m,n}$ in term of the value $I_{i,j}$. Meanwhile the sequence $D^*_{m,n}$ will be formed based on $I^*_{m,n}$. The order of $D^*_{m,n}$ is determined by $I^*_{m,n}$ and each $D^*_{i,j}$ corresponds to the $I^*_{i,j}$ according to the i and j.
4: This implementation iterates over the $D^*_{m,n}$ and acts delete operation according to δ_{max} and t. For each position λ of maxima or minima value of $D^*_{m,n}$, calculate δ using formula (13). If $\delta > \delta_{max}$ and the shadows number of the subgroup which λ belongs to is more than t, then delete the element with position λ in $D^*_{m,n}$ and the corresponding element in $I^*_{m,n}$. This procedure will be continuing unless any condition of λ and t is not satisfied.
5: Sequentially choose the value in $I^*_{m,n}$ and $D^*_{m,n}$, each pair of $I^*_{i,j}$ and $D^*_{i,j}$ can structure a message packet with length $D^*_{i,j}$ and identification field value $I^*_{m,n}$. Then send the packet to Bob.
6: Fetch and transfer the rest secret messages.

Since we get these value pairs, we can structure the packets with new identification value and new total length and then send them to Bob. But there maybe exists a latent danger that Warden can find the existence of the covert channel. Because we just modify the total length field with our designed $D_{i,j}$, that will cause that the packet stream will be abnormal with the normal packet stream. In order to avoid this abnormality, we want to find a proper quantization about the difference between the normal traffic flow and the secret traffic flow. Sur A's work [6] has a definition of the distortion metric and according to the definition, the distortion metric of stego packet stream is usually greater than that of normal stream. And we define the maximum distortion metric of the packets from normal packet stream for our paper inspired by [6]. The maximum distortion metric is defined as follows:

$$\delta_{max} = \max_{n}\{2v(\lambda) - v(\lambda - 1) - v(\lambda + 1)\} \tag{11}$$

where n is the number of packets in the packets stream. $v(.)$ is the length value and λ is the position of maxima or minima. Before sending secret messages, we calculate the δ_{max} from a prepared normal packets stream. We assume that the secret messages is partitioned into several sub-groups with w-bits in each. Let $S = S_1 S_2 \cdots S_m$ be a (m*w)-bits binary secret message to be sent by Alice to Bob. And Alice and Bob share the value of k, n, P in advance. The secret message encoding protocol (SMEP) is shown in Protocol 1.

4.2 Decoding Process

As the Encoding protocol is shared by both Alice and Bob, Bob can reconstruct the matrix $D_{m,n}$ effortlessly. When Bob receives a message, he needs to find the packet length, also say len_{recv} which represents a piece of shadow. And its position in matrix $D_{m,n}$ will be got from the identification field value of the packet named as I_{recv}. The secret message decoding protocol(SMDP) is shown in Protocol 2.

Protocol 2. Secret Message Decoding Protocol(SMDP)

1: For each packet received, set I_{recv} as the identification field value and len_{recv} as the total length field value.

2: Calculate $i = (I_{recv} \& 0x0111)\% n$ and $j = (I_{recv} \& 0x0111)/n$. Then set $D_{i,j} = len_{recv}$. Repeat step 1 and step 2 until there is no new packet received by Bob.

3: For i from 1 to m, reconstruct the secret message using the shadows D_i with a system of n-variables linear congruence equations.

---

5   Evaluation Study

Our experiment study is conducted based on a UDP chat application which is developed using Java. Experiment is done by a comparison of the normal packet

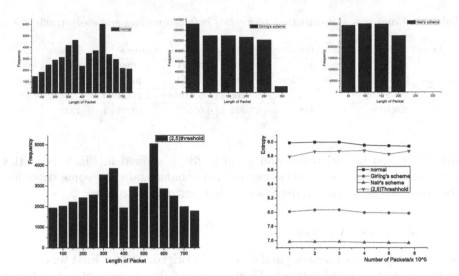

Fig. 5. Packet length frequency of normal traffic and stego traffic

stream with that of our scheme and other two schemes proposed in [13,14] as
these schemes are based on packet length, especially the scheme in [14] is based
on UDP packet. We have collected about 1.7 million packets from a gateway
of our institute which all the packets are UDP chatting messages and from 197
local hosts in five days. Then we divide these packets into 12 groups according
to the time and calculate the statistical parameters. Each subgroup with sample
packets is served as a reference of normal packets stream and also as the data
to be communicated between the clients. When we conduct our scheme, we use
one sample to calculate δ_{max} and also to be a normal stream to compare with
our stego stream.

5.1 Traffic Distribution Comparison

Firstly we choose the k, n, and p for our experiments. When choosing these
parameters, there are two rules to follow: one is that the k and n should not
be too large because this is a secret communication that the duration should be
short otherwise larger k and n means that more packets would be transmitted
and the duration will be longer; the second one is that k should be large enough
that the algorithm should work well in a relatively bad network condition. We
set the threshold (2,5) when P=1499 and $t = 3$. Firstly we compare the packet
length of normal traffic and the three stego traffic included our schem's and
the results are showed in Fig. 5. By comparing the results of packet stream, it
is shown that our proposed scheme greatly imitates the normal network traffic
flow. Besides the comparison of packet stream, we also compare the different
statistical data between normal traffic and stego traffic from one sample data
as is showed in Table 1. Also a comparison of entropy of different size of packet

Table 1. Comparison of statistical properties between normal and stego traffic flow

Scheme	Correlation	Mean	Deviation	Variance	Skewness	Kurtosis	Min	Median	Max
Normal	1	723.4	415.3	172494.4	0.07711	−1.11369	1	691	1472
Girling	−0.01658	108.8	74.5	5552.2	0.18214	−1.20346	0	103	255
Nair	−0.01153	93.8	55.7	3106.1	−0.01958	−1.21321	0	96	191
(2,5)	0.00012	734.6	418.3	174821.1	0.07052	−1.10985	1	732	1469

stream between normal traffic and stego traffic is showed in Fig. 5. Statistics shows that our proposed scheme can be more undetectable by some detection schemes based on statistical detection such as entropy detection.

5.2 Robustness Analysis

Here we compare our threshold in different network conditions. First we conduct our experiment in a real network. Then we repeat the experiment in a network which introduces packet losses. We simulate the network with packet losses by the means that we randomly discard some packets in the sending end according to the packet loss ratio, that is to say, with a packet loss ratio of 1 %, Alice will discard about most 10 packets if she sends 1000 packets during the transmission. We doesn't evaluate the performance of Girling's and Nair's work in such network because they have a upper application to ensure packet retransmission, otherwise they will fail the communication without the application. The communication is repeated for 100 times. With the packet loss ratio of 0.2 %, 0.5 %, and 1 %, our scheme can all succeed with the covert transmissions.

5.3 Bandwidth and Capacity Analysis

Every steganographic scheme should be analyzed in terms of steganographic bandwidth. One way to calculate the bandwidth is proposed by Mazurczyk in [11] and it can be expressed by means of RBR (Raw Bit Rate), which is defined as a total number of steganogram bits transmitted during one time unit(bit/s). However the bandwidth of our covert channel is mainly decided by the size of the secret message in each sending and the capacity of covert channel would be influenced by the number of shadows for each secret subgroup. We conduct the communication in a 10 M/s network and the bandwidth is averagely 58 KB/s using the formula in [13]. The bandwidth can be acceptable in steganographic communication and such a bandwidth is large enough to transmit secret message in a short duration and small enough not to be detected by the third party.

5.4 Impacts on Overt Communication

Our steganographic scheme barely affect the overt communication. First of all, secret message is encoded in the packet header of the overt packet thus it does not modify the data field of the packet. And such modification of packet header does nothing on the forwarding path of packets that they can reach the destination normally. Secondly all the modification on packets are operated on TCP/IP stack by such as encoding and decoding and these operations does not affect the chat application which is in application layer.

From the results and analysis mentioned above, we can conclude that our scheme performs better than other schemes and is a higher undetectable covert channel.

6   Conclusion

In this paper, we have proposed a secret sharing scheme based covert channel that sends secret data based on the length of the network packets. Different from previous length based covert channel, our channel is based on secret sharing scheme and has a randomly coding technique to imitate the normal traffic better. We have stimulated our scheme on a UDP based network chat application and obtained the length pattern, which follows the normal network flow even after embedding the secret data. Experimental results imply that our scheme is superior than the existing schemes.

Acknowledgment. This work was supported by the National Natural Science Foundation of China (No. 61572456, No. 61379131) and the Natural Science Foundation of Jiangsu Province of China (No. BK20151241, No. BK20151239).

References

1. Shamir, A.: How to share a secret. Commun. ACM **22**(11), 612–613 (1979)
2. Blakley, GR.: Safeguarding cryptographic keys. In: AFIPS 1979 National Computer Conference, vol. 48 (1979)
3. Kothari, S.C.: Generalized linear threshold scheme. In: Blakely, G.R., Chaum, D. (eds.) CRYPTO 1984. LNCS, vol. 196, pp. 231–241. Springer, Heidelberg (1985)
4. Simmons, G.J.: An introduction to shared secret and/or shared control schemes and their application. Contemp. Cryptol.: Sci. Inf. Integrity, pp. 441–497 (1992)
5. Beimel, A.: Secret-sharing schemes: a survey. In: Chee, Y.M., Guo, Z., Ling, S., Shao, F., Tang, Y., Wang, H., Xing, C. (eds.) IWCC 2011. LNCS, vol. 6639, pp. 11–46. Springer, Heidelberg (2011)
6. Sur, A., Nair, A.S., Kumar, A., et al.: Steganalysis of network packet length based data hiding. Circ. Syst. Sig. Process. **32**, 1–18 (2013)
7. Lampson, B.W.: A note on the confinement problem. Commun. ACM **16**(10), 613–615 (1973)
8. Wolf, M.: Covert channels in LAN protocols. In: Local Area Network Security, Springer, Heidelberg, pp. 89–101 (1989)
9. Tsai, C.R., Gligor, V.D., Chandersekaran, C.S.: A formal method for the identification of covert storage channels in source code. In: IEEE Symposium on Security and Privacy, p. 74 (1987)
10. Cabuk, S., Brodley, C.E., Shields, C.: IP covert timing channels: design and detection. In: Proceedings of the 11th ACM Conference on Computer and Communications Security, pp. 178–187. ACM (2004)

11. Mazurczyk, W., Szczypiorski, K.: Steganography in handling oversized IP packets. In: Proceedings of First International Workshop on Network Steganography (IWNS 2009), Wuhan, Hubei, China, 18–20 November 2009 - Co-located with 2009 International Conference on Multimedia Information Networking and Security (MINES 2009), vol. I, pp. 569–572

12. Padlipsky, M.A., Snow, D.W., Karger, P.A.: Limitations of end-to-end encryption in secure computer networks. MITRE CORP BEDFORD MA (1978)

13. Girling, C.G.: Covert channels in LAN's. IEEE Trans. Softw. Eng. **2**, 292–296 (1987)

14. Nair A S, Kumar A, Sur A, et al.: Length based network steganography using UDP protocol. In: 2011 IEEE 3rd International Conference on Communication Software and Networks (ICCSN), pp. 726–730. IEEE (2011)

15. Ji L, Jiang W, Dai B, et al.: A novel covert channel based on length of messages. In: International Symposium on Information Engineering and Electronic Commerce, IEEC 2009, pp. 551–554. IEEE (2009)

16. Ji, L., Liang, H., Song, Y., et al.: A normal-traffic network covert channel. In: 2009 International Conference on Computational Intelligence and Security CIS 2009, vol. 1, pp. 499–503. IEEE (2009)

17. Nair A S, Sur A, Nandi S. Detection of packet length based network steganography. In: 2010 International Conference on Multimedia Information Networking and Security (MINES), pp. 574–578. IEEE (2010)

18. Quan-zhu, Y., Peng, Z.: Coverting channel based on packet length. Comput. Eng. **34**(3), 183–185 (2008)

Discovering Approximate Functional Dependencies from Distributed Big Data

Weibang Li$^{(\boxtimes)}$, Zhanhuai Li, Qun Chen, Tao Jiang, and Zhilei Yin

Northwestern Polytechnical University, Xi'an 710072, China
{liweibang,jiangtao,yinzl2007}@mail.nwpu.edu.cn,
{lizhh,chenbenben}@nwpu.edu.cn

Abstract. Approximate Functional Dependencies (AFDs) discovered from database relations have proven to be useful for various tasks, such as knowledge discovery, query optimization. Previous research has proposed different algorithms to discover AFDs from a centralized relational database. However, none of the proposed algorithms is designed to discover AFDs from distributed data. In this paper, we devise a scalable and efficient approach to discover AFDs from distributed big data and not tied to main memory requirements. To improve the efficiency of AFDs discovery, statistics of local data in each site are collected to filter and prune the candidate AFDs set at first. The AFDs are discovered in parallel after data redistribution. We balance the load as much as possible before the redistribution of data and prune the candidate AFDs set quickly after the redistribution of data. We evaluate our approach using real and synthetic big datasets and the results show that our approach is more efficient and scalable on large relations and the number of nodes.

Keywords: Distributed data · Big data · Knowledge discovery

1 Introduction

Approximate Functional Dependencies (AFDs) [1–3] are rules that denote approximate determinations at attribute level of a relation. An AFD requires the normal Functional Dependency (FD) to be satisfied by most tuples of relation r. There has been a wide range of applications of AFDs discovered from relational databases, for example, query optimization (CORDS) [4] by maintaining correct selectivity estimates, predicting missing values of attributes in relational tables (QPIAD) [5], etc. It is useful to discover AFDs from data. For instance, an AFD in a table of chemical compounds relating various structural attributes to carcinogenicity could be invaluable for biochemists [1].

Work supported by National Basic Research Program 973 of China (No. 2012CB316203), Natural Science Foundation of China (Nos. 61502390, 61472321, 61332006, 61272121), National High Technology Research and Development Program 863 of China (No. 2015AA015307).

© Springer International Publishing Switzerland 2016
F. Li et al. (Eds.): APWeb 2016, Part II, LNCS 9932, pp. 289–301, 2016.
DOI: 10.1007/978-3-319-45817-5_23

As an extension to Functional Dependencies (FDs), AFDs were prevalent in database literature. The discovery of AFDs from data has been well studied [1,2,4]. All these approaches for discovering AFDs are working on centralized databases. When database D is centralized, the AFDs discovering problem is not very hard. However, a relation is often fragmented and distributed across different sites in practice [7]. With the rapid development of cloud computing and the increasing interests in big data, it is quite common to find data fragmented horizontally or vertically and distributed across different sites. In these settings the discovery of AFDs is far more challenging.

Example 1. Consider a relation specified by the schema: CAR (tid, Make, BodyStyle, Country, FuelSystem). Each CAR tuple specifies a car's make, body style, country and fuel system. Here tid is a key of CAR. An instance D of the CAR schema is shown in Fig. 1(a).

tid	Make	BodyStyle	Country	FuelSystem
1	Audi	sedan	Germany	2bbl
2	Audi	wagon	Germany	mpfi
3	Honda	sedan	Japan	mpfi
4	Toyota	hatchback	Japan	2bbl
5	Volvo	SUV	Sweden	mpfi
6	Audi	sedan	Germany	1bbl
7	Toyota	sedan	Japan	mpfi
8	Volvo	turbo	Sweden	mpfi
9	BMW	sedan	Germany	2bbl
10	Honda	hatchback	Japan	1bbl
11	Volvo	turbo	Sweden	1bbl
12	Honda	hatchback	Japan	1bbl

(a) An instance D of relation CAR

D_1 :

tid	Make	BodyStyle	Country	FuelSystem
6	Audi	sedan	Germany	1bbl
10	Honda	hatchback	Japan	1bbl
11	Volvo	turbo	Sweden	1bbl
12	Honda	hatchback	Japan	1bbl

D_2 :

tid	Make	BodyStyle	Country	FuelSystem
1	Audi	sedan	Germany	2bbl
4	Toyota	hatchback	Japan	2bbl
9	BMW	sedan	Germany	2bbl

D_3 :

tid	Make	BodyStyle	Country	FuelSystem
2	Audi	wagon	Germany	mpfi
3	Honda	sedan	Japan	mpfi
5	Volvo	SUV	Sweden	mpfi
7	Toyota	sedan	Japan	mpfi
8	Volvo	turbo	Sweden	mpfi

(b) A horizontal partition of D

Fig. 1. A CAR relation and its horizontal partitions

We want to find all the AFDs in D. When D is a centralized relation, we can employ existing methods to discover AFDs from D. When D is partitioned horizontally or vertically and distributed in different sites, however, it is usually necessary to ship data from one site to the other to discover AFDs in D.

Now suppose that D is horizontally fragmented into three fragments as is shown in Fig. 1(b), and each fragment D_i resides at site S_i. To discover AFDs from D, data shipments between different sites are generally necessary. For example, it is insufficient to verify candidate AFD BodyStyle$\rightsquigarrow$Country only by the fragmented data of D without data shipment. At each site S_i, there are no tuples that violate AFD BodyStyle$\rightsquigarrow$Country. However, it is obvious that tuples t_1, t_6, t_9 and t_3, t_7 violate this AFD. To verify candidate AFD BodyStyle$\rightsquigarrow$Country, one either has to (i) ship tuple t_6 from S_1 to S_3, and tuple t_1 and t_9 from S_2 to S_3, or (ii) ship tuple t_6 from S_1 to S_2, and tuple t_3 and t_7 from S_3 to S_2, etc.

This example tells us that the AFDs discovery algorithms on centralized data no longer work on distributed data. What's more, the number of candidate AFDs is exponential in terms of the number of attributes in the dataset. Thus the total data shipment is considerable if data is shipped when verifying each candidate AFD. We presented a method to discover AFDs from distributed big data in parallel.

Contributions. We make a first effort to investigate the problem of discovering AFDs from horizontally partitioned and distributed data. The main contributions of the paper are listed as follows:

(1) The search strategy and pruning strategies of AFDs discovery from distributed big data are proposed in this paper.
(2) We present an efficient algorithm for discovering AFDs from parallel in horizontally fragmented and distributed data. To improve the efficiency of AFDs discovery, we prune candidate AFDs set at first by making use of fragmented data statistics, then balance the load before data redistribution, and prune the candidate AFDs set again by rapid judgment.
(3) We experimentally validate our algorithm on real and synthetic datasets and compare it with naïve method based on Hadoop framework. The results show that our algorithm is more efficient than the naïve method based on Hadoop framework and is scalable on big data.

Organization. Section 2 presents related work. Section 3 reviews some basic notations. Section 4 presents searching policy and pruning policies for AFDs discovery from distributed data. Section 5 presents the AFDs discovery algorithms. Experimental results are presented in Sect. 6, followed by conclusions in Sect. 7.

2   Related Work

AFDs can be defined as FDs that approximately hold [6]. To measure the approximation of AFDs precisely, the error satisfaction of AFDs is usually calculated. Method to measure information content based on Shannon entropy function was proposed in [7], where the value of conditional entropy is defined between 0 and 1. If functional dependency holds, then the conditional entropy is 0. Three measures were defined for the error of a dependency in [8]. For example, g_3 is the portion of tuples one has to delete to obtain a relation that satisfies the dependency.

There has been work on discovering FDs and AFDs based on agree sets [9]. Two approaches for discovering agree sets were proposed in [9]. A method for mining functional dependency and approximate functional dependency based on partition was presented in [10]. A tool for automatic discovery of soft functional dependencies between columns was proposed in [4]. The definition of soft functional dependencies is similar to that of AFD.

All above works are related to AFDs or other constraints discovering from centralized small data. However, no previous work has studied how to discover AFDs in distributed data, an issue far more challenging than its centralized counterpart. In this paper we discover AFDs from distributed big data that is horizontally fragmented.

3   Preliminaries

This section is devoted to setting the groundwork of AFDs discovery. We give some definitions with regard to AFDs discovery, which are necessary for understanding the proposed method.

Definition 1 (Functional Dependency, FD). Let X, $Y \subseteq \mathrm{attr}(R)$ be two sets of attributes. The functional dependency between X and Y, denoted by $X \rightarrow Y$, holds in R if and only if: $\forall t_1$, $t_2 \in R$, if $t_1(X) = t_2(X)$, then $t_1(Y) = t_2(Y)$.

Definition 2 (Error of FD). Let $X \rightarrow Y$ be a functional dependency defined on relation R, the error of $X \rightarrow Y$ denotes fraction of tuples causing FD violation, denoted by $e(X \rightarrow Y) = \min\{|S| \; |S \subseteq R, \; R \backslash S \models X \rightarrow Y\}/|R|$.

Definition 3 (Approximate Functional Dependency, AFD). Let X, $Y \subseteq \mathrm{attr}(R)$ be two sets of attributes. Given an error threshold ε, $0 \leq \varepsilon \leq 1$, we say that $X \rightsquigarrow Y$ is an approximate functional dependency if and only if $e(X \rightsquigarrow Y) \leq \varepsilon$.

Given a relation R, an $\mathrm{AFD}(\varphi : X \rightsquigarrow Y)$ is about the approximate satisfaction of normal FD. That's to say, AFD φ holding on R still allows a very small portion of tuples of R to violate φ. In $\mathrm{AFD}(\varphi : X \rightsquigarrow Y)$, X is the determining set and Y is called the dependent set.

Definition 4 (Equivalence classes). Let $X \subseteq \mathrm{attr}(R)$ be an attribute set, denote the equivalence classes of a tuple $t \in R$ with respect to the given set X, i.e. $[t]_X(R) = \{p \in R | p[A] = t[A]$ for all $A \subseteq X\}$, where $t[A]$ is the attribute value of tuple t under attribute A.

Example 2. Consider the relation D in Fig. 1(a). Attribute Make has value Audi in tuple 1, 2 and 6, thus they form the equivalence class $\{1,2,6\}$, here we use tids (tuple identifiers) to denote tuples.

Definition 5 (Partition). Let $X \subseteq \mathrm{attr}(R)$ be an attribute set, denote $\Pi_X(R) = \{[t]_X(R) | t \in R\}$ as a partition of R under X.

Definition 6 (Stripped Partition). Let $X \subseteq \mathrm{attr}(R)$ be an attribute set, denote $\Pi'_X(R) = \{[t]_X(R) | t \in R, \ |[t]_X(R)| > 1\}$ as a stripped partition of R under X.

Example 3. In Fig. 1(a), the partition of D under Make is $\Pi_{\mathrm{Make}} = \{\{1,2,6\}, \{3,10,12\},\{4,7\},\{5,8,11\},\{9\}\}$. And according to Definition 5, stripped partition of D under Make is $\Pi'_{\mathrm{Make}} = \{\{1,2,6\},\{3,10,12\},\{4,7\},\{5,8,11\}\}$.

Definition 7 (Error of Key). Let $X \subseteq \mathrm{attr}(R)$ be an attribute set on relation R, error of key means the minimum fraction of tuples to remove for X to be a key, denoted by $e(X) = 1 - |\Pi_X|/|R|$, where $|\Pi_X|$ is the cardinality of Π_X.

Given FD $X \to Y$, any equivalence class $c \in \Pi_X$ is the union of one or more equivalence classes $c'_1, c'_2, \cdots \in \Pi_{X \cup Y}$. The minimum number of tuples to remove equals the size of c minus size of largest c'_i. Thus the error measure of $X \to Y$ can be denoted as follows: $e(X \to Y) = 1 - \Sigma_{c \in \Pi_X} \max\{|c'| \,|\, c' \in \Pi_{X \cup Y} \wedge c' \subseteq c\}/|R|$.

Example 4. Recall the relation D in Fig. 1(a). $\Pi_{\mathrm{Make}} = \{\{1,2,6\},\{3,10,12\}, \{4,7\}, \{5,8,11\},\{9\}\}$, and $\Pi_{\mathrm{Make} \cup \mathrm{BodyStyle}} = \{\{1,6\},\{2\}, \{3\},\{10,12\},\{4\},\{7\},\{5\}, \{8,11\},\{9\}\}$. Thus $e(\mathrm{Make} \leadsto \mathrm{BodyStyle}) = 1 - (2 + 2 + 1 + 2 + 1)/12 = 4/12$.

4   Search and Pruning of Candidate AFDs

4.1   Search Strategy

Let $X = \mathrm{attr}(R)$ be an attribute set of relation R, we only consider the non-trivial candidate AFDs with the form $X \backslash \{A\} \leadsto A$, where A is an attribute of R. Supposed that there are n attributes in R, to discover all non-trivial AFDs on R, we start with candidate AFDs generation. Firstly we generate and verify candidate AFDs with n-1 attribute at the left hand side (LHS), then works its way to less LHS attribute sets through the attribute lattice level by level from top to down. Figure 2 shows a lattice of $\mathrm{attr}(R) = \{A,B,C,D\}$, and the lattice is composed of all candidate AFDs. When searching, the candidate AFDs with 3 attributes at LHS are searched at first. The edge of the lattice between the first node from left at Level-1 and the first node from left at Level-2 represents the candidate AFD $ABC \leadsto D$. To identify all existing AFDs, we can traverse this lattice level by level from top to down. To improve the efficiency of discovering, we prune candidate AFDs as early as possible.

4.2   Pruning

The number of candidate AFDs is exponential to the number of attributes when discovering AFDs. In the case of large-scale data, the verification of each candidate AFD is quite time-consuming. To improve the efficiency of AFDs discovery, it is necessary to prune the candidate AFDs sets timely.

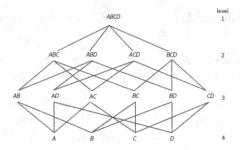

Fig. 2. Attribute lattice with attribute set $= \{A, B, C, D\}$

Theorem 1. For the candidate AFD $X \rightsquigarrow A$, assume that error threshold is ε, if $e(X) \leq \varepsilon$, then $X \rightsquigarrow A$ holds.

Proof: By Definition 7, $e(X)$ is the minimum fraction of tuples to remove for X to be a key. Here $e(X) \leq \varepsilon$, it means that X will be a key after removing no more than ε fraction of tuples. Thus functional dependency $X \rightarrow A$ holds after removing violated tuples. By Definition 3, $e(X \rightarrow A) \leq \varepsilon$, thus $X \rightsquigarrow A$ holds.   □

Theorem 2. For the candidate AFD $X \rightsquigarrow A$, assume that error threshold is ε, if $e(X) - e(XA) > \varepsilon$, then $X \rightsquigarrow A$ does not hold.

Proof: Assumed that Π_X, Π_{XA} are partitions of R under X and XA. For each $c \in \Pi_X$, if c has two or more subsets, the number of equivalence classes need to be removed from Π_{XA} equals $|c| - 1$. To make $X \rightarrow A$ hold, the total number of equivalence classes need to be removed from Π_{XA} equals $|\Pi_{XA}| - |\Pi_X|$. The cardinality of each equivalence class in $|\Pi_{XA}|$ is no less than one. According to Definition 2, $e(X \rightarrow A) \geq (|\Pi_{XA}| - |\Pi_X|)/|R| = 1 - |\Pi_X|/|R| - (1 - |\Pi_{XA}|/|R|) = e(X) - e(XA)$. If $e(X) - e(XA) > \varepsilon$, then $e(X \rightarrow A) > \varepsilon$. By Definition 2, $X \rightsquigarrow A$ does not hold.   □

By Definition 7, we can verify the candidate AFDs quickly by calculating the key error of the LHS and RHS of the candidate AFDs.

Theorem 3. Let X, Y, A be attribute sets on relation R, $Y \subset X$, D is an instance of R, ε is the error threshold. If candidate AFD $X \rightsquigarrow A$ does not hold, then candidate AFD $Y \rightsquigarrow A$ does not hold.

Proof: As is known, $Y \subset X$, supposed that $X = YZ$, $Z \neq \{\}$. Assume that to make functional dependency $X \rightarrow A$ hold, the minimum of tuples need to be removed from D is D_R, and to make functional dependency $Y \rightarrow A$ hold, the minimum of tuples need to be removed from D is D'_R.

Assume that AFD $Y \rightsquigarrow A$ holds and $X \rightsquigarrow A$ does not hold. By Definition 3, $e(Y \rightarrow A) = |D'_R|/|D| \leq \varepsilon$, $e(X \rightarrow A) = |D_R|/|D| > \varepsilon$, and FD $Y \rightarrow A$ holds on $D \backslash D'_R$. By axiom of augmentation in Armstrong axiom system, $YZ \rightarrow AZ$ holds on $D \backslash D'_R$. According to previous assumption, $X = YZ$, thus $X \rightarrow AZ$ holds on $D \backslash D'_R$. Obviously, $X \rightarrow A$ holds on $D \backslash D'_R$. By Definition 3, $e(X \rightarrow A) =$

$|D_R|/|D| < \varepsilon$, which contradicts to previous $e(X{\rightarrow}A) = |D_R|/|D| > \varepsilon$. So the assumption in the beginning of this paragraph does not hold, thus if candidate AFD $X{\rightsquigarrow}A$ does not hold, then candidate AFD $Y{\rightsquigarrow}A$ does not hold, too.   □

Example 5. Take an example of pruning candidate AFDs from the candidate AFDs set with four attributes A, B, C, D. Assume that candidate AFD $ABD{\rightsquigarrow}C$ is verified not hold, $\forall$ AFD φ, if LHS(φ) $\subset ABD$ and RHS(φ) equals C, then by Theorem 3, φ does not hold, too. Obviously, the following candidate AFDs can be pruned by Theorem 3: $AB{\rightsquigarrow}C$, $AD{\rightsquigarrow}C$, $BD{\rightsquigarrow}C$, $A{\rightsquigarrow}C$, $B{\rightsquigarrow}C$, $D{\rightsquigarrow}C$.

5   Algorithm

Algorithm AFDDCet. The first algorithm, AFDDCet, is a naïve approach: it reduces the AFDs discovering problem for horizontally fragmented distributed data to its counterpart for centralized databases. AFDDCet first selects the site with the largest number of tuples as the executive site, then transfers the data distributed at other sites to the executive site, and at which the candidate AFDs are verified.

The number of candidate AFDs is exponential to the number of attributes, and all the data needs to be scanned once when verifying each candidate AFD. In AFDDCet, all the tuples are shipped to execution site, thus the load is imbalance seriously, and there is a bottleneck at the execution site.

Algorithm AFDDPar. To improve the efficiency of AFDs discovery from distributed big data, we propose AFDDPar, an algorithm to discover AFDs from distributed big data in parallel. AFDDPar firstly generates the candidate AFDs set. The number of candidate AFDs is exponential to the number of attributes, and for relation R with n attributes, the number of candidate AFDs equals $n \cdot 2^{n-1} - n$ [11]. After generating the candidate AFDs set, AFDDPar verifies the candidate AFDs preliminarily by the statistics of local data at each site, then removes unsatisfied AFDs and prunes the candidate AFDs set by the result of verification according to Theorem 3.

Theorem 4. Let X, A be attribute sets on relation R, D is an instance of R, D is divided horizontally into n fragments, $D = \{D_1, D_2, \cdots, D_n\}$, ε is the error threshold. Assume that $e_i(X{\rightarrow}A)$ is the error of $X{\rightarrow}A$ on D_i, $e(X{\rightarrow}A)$ is the error of $X{\rightarrow}A$ on D, if $\sum_{i=1}^{n} e_i(X \rightarrow A) \cdot |D_i| > \varepsilon \cdot |D|$, then $X{\rightsquigarrow}A$ does not hold.

Example 6. As is shown in Fig. 1(b), assume that the error threshold $\varepsilon = 0.1$. To verify whether BodyStyle${\rightsquigarrow}$Make holds on D, we can calculate $e(\text{BodyStyle}{\rightarrow}\text{Make})$ and compare the result with ε. If $e(\text{BodyStyle}{\rightarrow}\text{Make}) < \varepsilon$, BodyStyle${\rightsquigarrow}$Make holds on D. Though this method is accurate, it is usually necessary to migrate data between different sites, since data is distributed at different sites. By Theorem 4, we can verify BodyStyle${\rightsquigarrow}$Make quickly by calculating $e_i(\text{BodyStyle}{\rightarrow}\text{Make}) \cdot |D_i|$ at each site in parallel. Here

$\sum_{i=1}^{3} e_i(BodyStyle \leadsto Make) \cdot |D_i| = 0 + 1 + 1 = 2$, $\varepsilon \cdot |D| = 0.1 \cdot 12 = 1.2$.
According to Theorem 4, BodyStyle$\leadsto$Make does not hold on D. By Definition 2, $e(BodyStyle \rightarrow Make) = 5/12 > \varepsilon = 0.1$, BodyStyle$\leadsto$Make does not hold on D obviously. As can be seen from this example, we can verify candidate AFDs quickly by the local data distributed at different site without data shipment according to Theorem 4.

We can verify the candidate AFDs quickly by Theorem 4 and remove these candidate AFDs not hold. In this paper we search the candidate AFDs from the attribute lattice level by level from top to down. The candidate AFDs with largest number of attributes at LHS are verified at first. The result of candidate AFDs verification can be used to prune the candidate AFDs set by Theorem 3.

After the verification of candidate AFDs and the pruning of candidate AFDs set, AFDDPar groups the remaining candidate AFDs. The principle of grouping is the candidate AFDs in each group sharing public attribute and the difference in quantity of candidate AFDs between each group is as large as possible. The grouping method is as shown in Algorithm 1.

Algorithm 1. CandidateAFDsGroup

Input: $D = D_1, D_2, \cdots, D_n$, attribute set X
Output: Grouped candidate AFDs set Θ
1 $\Theta \leftarrow \{\}$, $\theta \leftarrow \{\}$, $\varphi \leftarrow$ null;
2 **for** $X \in attr(R)$ **do**
3   **if** $\varphi \in \Sigma$ **and** $X \in LHS(\varphi)$ **then**
4        $\theta \leftarrow \theta \cup \varphi$;
5     $\Theta \leftarrow \Theta \cup \theta$;

6 **return** Θ.

Assume that the attribute set of R is $attr(R) = \{A, B, C, D\}$. We group the candidate AFDs by the grouping policy of Function 2. The candidate AFDs sharing public attribute A in the LHS is assigned to the same group. As is shown in Fig. 3, the thick solid lines represent the group of candidate AFDs sharing public attribute A in the LHS. The thin solid lines represent the group of candidate AFDs sharing public attribute B in the LHS, long dotted lines represent the group of candidate AFDs sharing public attribute C in the LHS, etc.

This grouping strategy is also conducive to the pruning of candidate AFDs within the group. By Theorem 3, if candidate AFD does not hold, then the candidate AFDs such as $AB \leadsto C$, $AD \leadsto C$, $BD \leadsto C$, $A \leadsto C$, $B \leadsto C$, $D \leadsto C$ can be pruned from the candidate AFDs set.

To verify remaining candidate AFDs precisely, AFDDPar redistributes data by hash function after grouping of the remaining candidate AFDs. Within the same group, the input of the hash function is the value of shared public attribute. The feature of hash function ensures that tuples with potential conflict can be hashed into the same data block.

When redistributing data by hash function, the size of data blocks may differ greatly. To balance the load of each site after data redistribution, we need to

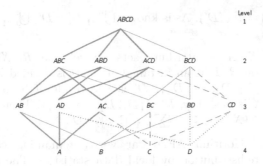

Fig. 3. Groups of candidate AFDs with attribute set $= \{A, B, C, D\}$

allocate the data blocks to sites as uniform as possible. The data blocks redistribution problem is a scheduling on parallel machines problem, which is a NP-hard problem. It is difficult to obtain an exact solution for this problem. We propose an approximate optimal data blocks allocation method. The detail of the method is as shown in Algorithm 2.

Algorithm 2. DataFragmentAllocate

Input: Sites set $S=\{S_1, S_2, \cdots, S_n\}$, $D=\{D_1, D_2, \cdots, D_m\}$
Output: $L(1)$, $L(2)$, $\cdots$, $L(n)$

1   $D' \leftarrow Desc(D)$;//Descending sort elements of D by cardinality
2   **for** $D'_i \in D'$ **do**
3   **for** $j \in [1,n]$ **do**
4   **if** S_j *is with minimal task allocation currently* **then**
5         $L(j) \leftarrow L(j) \cup D'_i$;
6         $D' \leftarrow D' \setminus D'_i$;

7   **return** $L(1)$, $L(2)$, $\cdots$, $L(n)$.

After data redistribution, Algorithm AFDDPar discovers AFDs at each site in parallel. The discovery of AFDs from big data is quite time-consuming. To improve the efficiency of AFDs discovery, we verify the remaining candidate AFDs quickly by local data statistics after data redistribution.

Theorem 5. Let X, A be attribute sets on relation R, D is an instance of R, D is divided horizontally into n fragments, $D = \{D_1, D_2, \cdots, D_n\}$, D' is the result of hash function where attribute value of X is hash key, $D' = \{D'_1, D'_2, \cdots, D'_n\}$. For $\forall D'_i \in D'$, $i \in [1,n]$, $|\Pi_X| = \sum_{i=1}^{n} |\Pi_{iX}|$, where Π_{iX} is the partition of D'_i on attribute set X.

Proof: For $\forall D'_i \in D'$, $i \in [1,n]$, the tuples in D'_i is with the same X attribute value. By the feature of hash function, tuples with same X attribute value share the same hash value. By Definition 4, the tuples with same X attribute value are in the same equivalence class. Thus tuples with the same X attribute value are distributed in the same data block in D'. $\forall [t]_X(D'_i) \in |\Pi_{iX}|$, $\exists [t]_X(D') \in$

$|\Pi_X|$, and $[t]_X(D'_i) = [t]_X(D')$. As is known $\bigcup_{i=1}^{n} D'_i = D'$, $\bigcup_{i=1}^{n} \Pi_{iX} = \Pi_X$, thus $|\Pi_X| = \sum_{i=1}^{n} |\Pi_{iX}|$.   □

Theorem 6. Let X, A be attribute sets on relation R, $X \to A$ is a functional dependency on R. D is an instance of R, D is divided horizontally into n fragments, $D = \{D_1, D_2, \cdots, D_n\}$. D' is the result of hash function where attribute value of X is hash key, $D' = \{D'_1, D'_2, \cdots, D'_n\}$. Then $(\sum_{i=1}^{n} |\Pi_{iXA}| - \sum_{i=1}^{n} |\Pi_{iX}|) \backslash |D| \le e(X \to A) \le 1 - (\sum_{i=1}^{n} |\Pi_{iX}|) \backslash |D|$.

By Theorem 6, algorithm AFDDPar verify remaining candidate AFDs quickly after data redistribution by local data statistics. The results of verification can be used to prune the remaining candidate AFDs set by the pruning policies. The remaining candidate AFDs are checked accurately by Definition 3 at last. When discovering AFDs, if one candidate AFD is verified, then the result will be broadcast to all the sites and the check of this candidate AFD will stop immediately. The early end of candidate AFDs verification can avoid unnecessary check and improve the efficiency of candidate AFDs discovery. The details of AFDDPar are as shown in Algorithm 3.

Algorithm 3. AFDDPar

Input: $S = \{S_1, S_2, \cdots, S_n\}$, $D = \{D_1, D_2, \cdots, D_n\}$, X, error threshold ε
Output: discovered AFDs set Σ'

1   $\Sigma_{tmp} \leftarrow \{\}$, $D' \leftarrow \{\}$;
2   $\Sigma \leftarrow generateAFDs(X)$;
3   **for** *each* $\varphi \in \Sigma$ **do**
4   **if** φ *does not hold* **and** $|LHS(\varphi)| > 1$ **then**
5       $\Sigma \leftarrow \Sigma \backslash \varphi$;
6       $\Sigma \leftarrow \Sigma \backslash pruneAFD(\varphi, \Sigma)$;

7   $\Sigma_{tmp} \leftarrow CandidateAFDsGroup(\Sigma, X)$;
8   $D' \leftarrow HashFragmented(D)$; //data redistribution
9   $DataFragmentAllocate(S, D')$; //banlance load
10 **for** *each* $\xi \in \Sigma_{tmp}$ **do**
11   **for** *each* $\varphi \in \xi$ **do**
12   **if** checkedAFD(φ, ε)==TRUE **then**
13        $\Sigma' \leftarrow \Sigma' \cup \varphi$;
14        $\Sigma \leftarrow \Sigma \backslash \varphi$;
15        $\Sigma \leftarrow \Sigma \backslash pruneAFD(\varphi, \Sigma)$;
16   **else**
17        $\Sigma \leftarrow \Sigma \backslash \varphi$;
18        $\Sigma \leftarrow \Sigma \backslash pruneAFD(\varphi, \Sigma)$;

19 return Σ'.

Function $pruneAFD()$ prunes the candidate AFDs set, the input of function is validated candidate AFD, and the output is the pruned candidate AFDs set. Function $checkedAFD()$ checks candidate AFDs and returns the results. If candidate AFD holds, checkedAFD() returns TRUE, otherwise FALSE.

6   Experimental Results

Experimental Setting. To evaluate algorithms proposed in this paper, we performed several experiments with different data sets. We used a cluster with a dedicated master node and eight workers with 1.7 GHz Intel Xeon 2 processor and 16 GB of RAM, and the operating system of each machine is Ubuntu 10.4. We used Hadoop 1.1.2 and Hama 0.6.4 as the processing framework.

Data. We use two different types of data: (1) Real-life data taken from the United States Department of Transportation [12], referred to as AOTS. We generate instance $aots_6$ of 60 Millions tuples, $aots_{10}$ of one hundred Millions tuples. (2) Synthetic data representing car information, referred to as CAR. We created two instances of CAR containing 60 Millions tuples, one hundred Millions tuples each, referred to as car_6, and car_{10}, respectively.

Experimental results. We conducted two sets of experiments, evaluating the centralized algorithm AFDDCet and distributed algorithm AFDDPar. We varied the size of the distributed data ($|D|$), and the number of sites ($|S|$). All experiments report the average over three runs.

Exp-1: Varying data size. To evaluate the scalability of our algorithms with data size $|D|$, we fixed the number of sites $|S|$ to 6 and increased the size of data

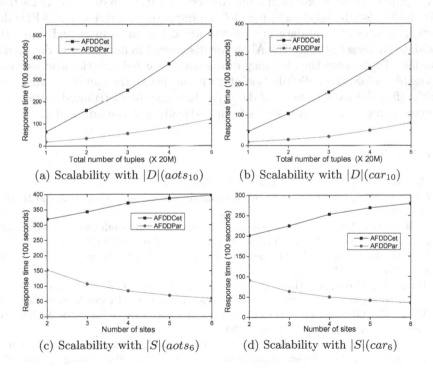

(a) Scalability with $|D|(aots_{10})$ (b) Scalability with $|D|(car_{10})$

(c) Scalability with $|S|(aots_6)$ (d) Scalability with $|S|(car_6)$

Fig. 4. Scalability with $|D|$ and $|S|$

$|D|$ from 20 % to 100 %. We used datasets $aots_{10}$ and car_{10}. Figures 4(a) and (b) show response times for algorithm AFDDCet and algorithm AFDDPar. Obviously, the increasing rate of AFDDCet's response time is much higher than that of algorithm AFDDPar. Figures 4(a) and (b) show that AFDDPar outperforms AFDDCet significantly in response time with data size $|D|$ increasing. With the increase of the number of sites $|S|$, the response time of algorithm AFDDPar less than that of AFDDCet significantly.

Exp-2: Varying the number of sites. To evaluate the scalability of our algorithms with the number of sites, we fixed the total data size and increased the number of sites $|S|$ from 2 to 6. We used datasets $aots_6$ and car_6. Figures 4(c) and (d) show response times for algorithm AFDDCet and AFDDPar. As expected, the response time of AFDDPar decreases as $|S|$ increases significantly. However, the response time of algorithm AFDDCet increases slightly as $|S|$ increases. Since all the verifying tasks of algorithm AFDDCet are executed in one execution site, the increase of $|S|$ has almost no much impact on the response time of algorithm AFDDCet. Figures 4(c) and (d) show that AFDDPar outperforms AFDDCet significantly in response time.

7 Conclusion

In this paper, we have proposed a distributed algorithm to discover AFDs from horizontally distributed data in parallel. To improve the efficiency of AFDs discovery, statistics of local data in each site are collected to filter and prune the candidate AFDs set at first. The AFDs are discovered in parallel after data redistribution. To improve the efficiency of discovery, we balance the load as much as possible before the redistribution of data and prune the candidate AFDs set quickly after the redistribution of data. As data may be distributed vertically, a future research is to discover AFDs from vertically distributed big data.

References

1. Huhtala, Y., Karkkainen, J., Porkka, P., Toivonen, H.: Tane: an efficient algorithm for discovering functional and approximate dependencies. Comput. J. **42**(2), 100–111 (1999)
2. Kalavagattu, A.K.: Mining approximate dependencies as condensed representations of association rules. Master Thesis, Arizona State University (2008)
3. Yao, H., Hamilton, H.J.: Mining functional dependencies from data. Data Min. Knowl. Discov. **16**(2), 197–219 (2008)
4. Ilyas, I.F., Markl, V., Haas, P., Brown, P., Aboulnaga, A.: Automatic discovery of correlations and soft functional fependencies. In: Proceedings of the 2004 ACM SIGMOD International Conference on Management of Data, SIGMOD 2004, New York, NY, USA, pp. 647–658 (2004)
5. Wolf, G., Khatri, H., Chokshi, B., Fan J., Chen, Y., Kambhampati, S.: Query processing over incomplete autonomous databases. In: Proceedings of the 33rd International Conference on Very Large Data Bases, pp. 651–662. VLDB Endowment (2007)

6. Kivinen, J., Mannila, H.: Approximate dependency inference from relations. In: Proceedings of Fourth International Conference on Database Theory (ICDT 1992), pp. 86–98 (1992)
7. Lee, T.: An information-theoretic analysis of relational databases - part I: data dependencies and information metric. IEEE Trans. Softw. Eng. SE **13**(10), 1049–1061 (1987)
8. Kivinen, J., Mannila, H.: Approximate inference of functional dependencies from relations. Theoret. Comput. Sci. **149**, 129–149 (1995)
9. Lopes, S., Petit, J., Lakhal, L.: Functional and approximate dependency mining: database and FCA points of view. J. Exp. Theoret. Artif. Intell. **14**(2), 93–114 (2002)
10. Dalkilic, M.M., Gucht, D.V., Robertson, E.L.: The classifier-estimator framework for data mining. In: Proceedings of the 7th IFIP 2.6 Working Conference on Database Semantics (DS-7), Leysin, Switzerland, October 1997. Chapman and Hall (1997)
11. Weibang, L., Zhanhuai, L., Qun, C., Tao, J., Hailong, L., Wei, P.: Functional dependencies discovering in distributed big data. J. Comput. Res. Dev. **52**(2), 282–294 (2015)
12. United States Department of Transportation. http://apps.bts.gov/xml/ontimesummarystatistics

Research Full Paper: Query Optimization and Scalable Data Processing

Making Cold Data Identification Efficient in Non-volatile Memory Systems

Binbin Wang and Jiwu Shu[✉]

Department of Computer Science and Technology,
Tsinghua University, Beijing, China
wangbb14@mails.tsinghua.edu.cn, shujw@tsinghua.edu.cn

Abstract. Non-volatile memory is emerging as a promising candidate for building efficient data-intensive OLTP systems, due to its advantages in high area density and low energy consumption. Systems now are able to store large datasets in main memory. Because OLTP workloads typically exhibit skew access patterns, the system must maintain an eviction order policy to move the cold data to the economical secondary storage. Existing cold data identification schemes generally employ the linear lists to track the least recently used data. However frequently update cost in these schemes is extremely high which is unsuitable to identify cold data from large scale of memory-resident data. We propose an efficient cold data identification scheme named eLRU. eLRU is a trie-based LRU which is able to fast track billions of tuples. We implemented our eLRU proposal and performed a series of experiments across a range of database sizes, workload skews and read/write mixes. Our results show that eLRU has a 2×–4× performance advantage over the current LRU-based cold data identification schemes.

Keywords: Non-volatile memory · Cold data management · LRU

1 Introduction

For the past few decades, on-line transaction processing (OLTP) applications lead new changes in computer trends. New engines have been optimized both in research systems (H-Store [12], Hyper [13], MonetDB [2], RAMCloud [20]) and commercial systems (VoltDB [26], Memcached [18], TimesTen [25], SAP HANA [10], Microsoft SQL Server [8]). Performance of these memory-oriented systems is affected by the speed to take in information from the storage and make correct decisions.

The challenge to build a large-scale, high performance OLTP system is to overcome the obstacles on energy consumption, capacity, performance and cost. This challenge manifests itself in many ways. Although new engineering technologies have been proposed, DRAM is still expensive than the disk-based storages. For per gigabyte, DRAM price is about 100X higher than HDD, 10X than SSD. Another inherent physical reason that prevents DRAM from building large scale

© Springer International Publishing Switzerland 2016
F. Li et al. (Eds.): APWeb 2016, Part II, LNCS 9932, pp. 305–316, 2016.
DOI: 10.1007/978-3-319-45817-5_24

main memory due to its density limitation [16,19]. However, non-volatile memories (NVM) will overcome these limitations and thereby compete with these two storage media to realize storage class memory in the near future. NVM technologies, including Phase-Change Memory (PCM) [21], Spin-Transfer Torque Random Access Memory (STT-RAM) [9], Resistive Random Access Memory (ReRAM) [1] and Memristor [24], get worldwide attention. These devices support byte-addressable reads and writes with low latency on the same order of magnitude as DRAM, but with persistence writes and high effective areal density as SSDs [3].

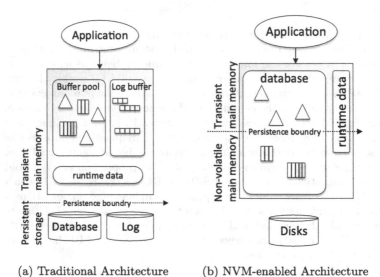

(a) Traditional Architecture   (b) NVM-enabled Architecture

Fig. 1. Database architecture in the presence of NVM.

The next generation storage systems based on NVM technologies make it possible to overcome the limitations on performance and power. Main memory is now the primary storage. The NVM-enabled architecture is shown in Fig. 1. Memory-oriented OLTP systems still face the restriction that all data fit in device. It is unreasonable to warn users not to exceed the scalable of real memory. Once memory is exceeded (or it might be at some point in the future), the inactive data should be gathered and migrated to a larger cluster. "Colder" tuples need to be identified and moved out to free up more space for recently accessed tuples.

It is worth noting that NVMs have much higher area density and consume less power than DRAM [6]. By the next decade, we may have a 1 TB PCM and 100 TB Memristor with comparable price to disk [17]. More than hundreds of billions of tuples can be resident in memory. The current eviction algorithms use one or more linear lists to track the least recently used tuples. The cost of updating a tuple in the common LRU chain increases linearly with the length of the chain list. It is undesirable to use a standard chain list to track cold tuples from large-scale data. Searching a long chain list has extremely high cost which is unacceptable for memory-resident OLTP systems. In this paper, we propose an efficient cold data identification scheme to maintain a single global ordering of tuples to track hot and cold data.

The main contributions are as follows:

- *An Efficient Cold Data Identification:* We introduce a trie-based structure in our design to track the cold data. This prefix tree tracks the recency information fast by directly pointing to the correct location instead of traversing the whole LRU chain list. Unlike the standard LRU based schemes, eLRU supports frequently updates in the long chain. If data access is skewed, the efficient cold identification mechanism has a 2×–4× performance advantage over the traditional designs.
- *Lazy Eviction Strategy:* If a new tuple is inserted, it is added to the hot-end of the chain as the newest tuple, possibly causing other tuples to be evicted. We use a lazy eviction strategy to reduces the potential unnecessary writes. The candidate eviction tuples will just be moved out of the chain but the prefixes still remain.

2 Related Work

For the NVM based main memory, access patterns of data need to be monitored, so that the system can actively promote and demote data to match the suitable hotness level. The cold data definition is supposed to take frequency or recency into account.

The traditional algorithms, such as LRU and CLOCK are by far commonly used for temporal locality workloads. The recently used tuples are simply added to the hot-end of the chain. When the eviction transaction executes, a new block is created by popping tuples from the cold-end. If a tuple already exists in the chain and is read or updated, it will be moved from its original location to the hot-end. It is efficient to add or evict a tuple in LRU chain. But the overhead of moving an existing tuple increases linearly with the length of the chain. In CLOCK, the tuples can be organized as a circle and the eviction algorithm searches through the cycle to find the victim one. Whenever a tuple is accessed, the associate reference bit is set to 1. Research and experiment have shown, CLOCK performs similar to LRU. When identifying the cold data, LRU and CLOCK don't consider the frequency. Moreover, these two algorithms are not scan-resistant. Other algorithms, such as LRU-K, LRFU [14] and FBR [22], consider both recency and frequency.

In NVM-enhanced memory architecture, eviction algorithms should not only consider the hit ratio but also the characteristics of NVM devices. NVM devices exhibit larger write latency. The improved algorithms, such as LRU-WPAM[1] [23], MHR-LRU [4], CLOCK-DWF [15] and D-CLOCK [5], reduce PCM writes

---

[1] LRU-WPAM (LRU-With-Prediction-And-Migration).

by monitoring the access information. LRU-WPAM and MHR-LRU are LRU-based algorithms. LRU-WPAM uses a LRU list and four monitoring queues. The least recently used tuples in DRAM read queue and PCM write queues will be chosen when eviction process are executed. MHR-LRU employs a DRAM write-aware list to monitor the write reference time. It performs data migration between PCM and DRAM to keep write-intensive data in DRAM. CLOCK-DWF and D-CLOCK are CLOCK-based algorithms. They are designed to reduce the write counts on NVM memory too.

All the designs use one or more linear lists to track the cold and hot data in the system. Based on the assumption that memory is much smaller than the secondary storage devices, these designs works well. But in the NVM-enabled systems, more than hundreds of billions of tuples can be resident in memory. It is time to rethink the cold data identification in big memory.

3 An Efficient Cold Data Identification Scheme

In this paper, we introduce an efficient global cold data identification scheme. The main goal of our design is to globally track the cold data to maximize the hit-ratio of skew workloads in memory-oriented systems. And the tracking of the scheme is fast whenever the tuples are accessed, modified, or inserted by a transaction.

Our description of eLRU is broken down into two parts. The first is the architecture of eLRU, i.e. how the components work to support frequency accesses and updates. The second is the process of lazy block eviction, i.e. the steps to use NVM-aware algorithms to evict tuples to disk.

3.1 eLRU Architecture

As shown in Fig. 2, eLRU is composed of a trie [11] and a lightweight LRU order chain. The trie is a prefix tree where the path from root to the leaf represents one key. The shortest unique prefix is able to distinguish a key from others. All the leaf nodes are linked.

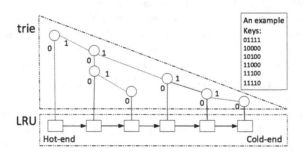

Fig. 2. The eLRU structure.

Similar to the standard LRU, eLRU also has the hot-end and the cold-end. In eLRU, all the keys are stored in the prefix tree, the leaf nodes are linked to work as the LRU chain. The tuples next for eviction are located in the cold-end of the chain. The hot-end of the chain represents the recently frequency accessed tuples. When the eviction transaction executes, the leaf of the coldest tuple is simply removed from the cold-end. In practical systems, the keys can be the block IDs and the value is the data items.

In this design, we use the "left-child right-sibling" representation to build the tree structure of trie. Each node in eLRU contains two 4-byte pointers and one 1-byte element. eLRU has comparable memory overhead with the doubly-linked list. Evicting a tuple is simply removing the front tuple of the chain, which can be done in $O(1)$. Similarly, inserting a tuple to the hot-end of the chain can also be done in constant time. Updating a tuple in traditional LRU chain list involves scanning the entire chain to find the target item. This is an $O(n)$ operation, where n is the number of tuples in the chain. In contrast, eLRU support faster update in $O(k)$, where k is the depth of the prefix of the tuple.

3.2 Insertion, Update and Lookup

To track the frequently data accesses in memory, the cold data identification needs to be quickly adjusted. eLRU supports efficient insertion, lookup and update operations (shown in Fig. 3).

Insertion: If the prefix of the current new key exists in eLRU, the insertion completes after a new node is added into the hot-end of the chain of eLRU as the newest tuple. Otherwise, we need to insert the new prefix into the tree structure of eLRU before the addition in the chain. The pseudo-code of insertion is shown in Algorithm 1.

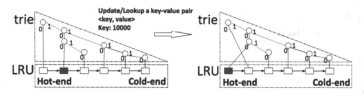

(a) Update or lookup in eLRU.

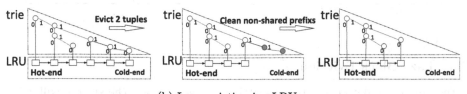

(b) Lazy eviction in eLRU.

Fig. 3. Update and eviction examples in eLRU.

Algorithm 1. *Insertion*

Input: r: root, k: key
Output: SUCCESS/FAILURE
 1: **if** is_full(r) **then**
 2:    evict cold data
 3: **end if**
 4: p=k[0]
 5: **while** *p **do**
 6:    **if** *not_exist(r,p)* **then**
 7:        *new prefix ← CreatePrefix()*
 8:        insert the new prefix to eLRU
 9:    **end if**
10:    ++p
11: **end while**
12: insert a new node to the hot-end of the chain
13: return SUCCESS

Update/Lookup: If a tuple already exists in eLRU and is accessed, it is necessary to "promote" the tuple to the hot-end of the chain, since it is now the most recently used tuple. To do this, we first remove the tuple from the original location. Then, it is simply added to the hot-end of the chain list just as a insertion. The lookup algorithm is similar to the update algorithm. The procedure is shown in Algorithm 2.

Algorithm 2. *Update*

Input: r: root, k: key
Output: SUCCESS/FAILURE
 1: p=k[0]
 2: **while** *p **do**
 3:    **if** *not_exist(r,p)* **then**
 4:        return FALSE
 5:    **end if**
 6:    ++p
 7: **end while**
 8: move the tuple from the original location to the hot-end
 9: return SUCCESS

3.3   Lazy Block Eviction

If the amount of data exceeds the scalable of real memory, the system will evict the cold tuples out of the primary storage. To evict a tuple, we need to update the indexes and eLRU (shown in Algorithm 3). The newly evicted tuples may share the prefixes with other normal non-evicted tuples. Because of this, when evicting a tuple, we can simply move the tuple out of the cold data identifier by

---

Algorithm 3. *Lazy eviction*

Input: r: root
Output: SUCCESS/FAILURE
1: $i = 0$
2: **while** $i <$ MAX_KEY_PER_DB **do**
3:      **if** *is_empty(r)* **then**
4:          break
5:      **end if**
6:      *prefix=get_next_prefix(r)*
7:      **if** *is_isolate(prefix)* **then**
8:          *deleteIsolatePrefix(r, prefix)*
9:      **end if**
        /* *When the execution time limit is reached, the function returns* */
10:     **if** *reach_time_limit()* **then**
11:         return SUCCESS
12:     **end if**
13:     $i+ = 1$
14: **end while**
15: return SUCCESS

---

popping a tuple off the cold-end of eLRU. The original tuple block is moved to disk, the common prefixes of the keys still remains in the cold data identifier.

The prefixes of the evicted keys which are not shared with the non-evicted tuples, will be cleaned up periodically. This cleaning cycle is configured. During a cleaning cycle, a background process will traverse eLRU to find and delete the isolated prefixes as soon as possible. To ensure that the cleaning operation will not take up CPU too much, each cleaning will have a limit time. When the execution time limit is reached, the function returns, not to continue.

4 Experimental Analysis

4.1 Experimental Setup

We conducted the following experiments to evaluate our design. These experiments are conducted on the same hardware configuration.

NVM hardware emulator: The experiments are performed on Intel Lab's NVM hardware emulator. We use the intel xeon integrated memory controller to validate the bandwidth settings. For each experiment, we perform 3 runs and report the average performance across the 3 runs.

Benchmarks: We use the workloads of YCSB, which is a widely-used key-value store benchmark from Yahoo! [7]. For all the experiments, we use a 10 GB YCSB database containing 10 million records. We adopt three different transaction workloads and three types of distributions. Table 1 shows the mixtures of the workloads. We also vary the amount of skews in the experiments. The skew controls how often a tuple is accessed by transactions. And the skew in Zipfian and Latest distributions is configured by the constant s. In our experiment, we use the s between 0.5 and 1.5.

Table 1. Workload characteristics

Traces	Read/Write ratio	Distributions	Skew	Data/Memory
a	100 %reads	Zipfian	0.5	2
b	95 %/5 %	Zipfian	0.5	
c	50 %/50 %	Zipfian	0.5	
d	100 %reads	Latest	0.5	
e	95 %/5 %	Latest	0.5	
f	50 %/50 %	Latest	0.5	
g	100 %reads	Uniform	-	
h	95 %/5 %	Uniform	-	
i	50 %/50 %	Uniform	-	

4.2 Insertion/Update/Eviction Throughput

This subsection compares the throughput of Insertion/Update/Eviction operations in doubly-linked LRU chain, singly-linked LRU chain and eLRU. In this experiment, we first insert 10 to 1000000 tuples, then update each of them and finally evict all of them. The cost of insertion, update or eviction is regardless of the tuple size. Thus we choose fixed-size key-value pairs (e.g. 1024 bytes). The insertion/update/eviction performance with chain length varies from 10 to 1000000 is shown in Fig. 4. The main difference in performance between singly-linked list, doubly-linked list and eLRU is due to the high cost of updating (or searching) a tuple in the chain. As the result shows, update operation overhead of either a singly-linked list or a doubly-linked list increases linearly with the length of the chain list.

4.3 Hit Ratio

We now discuss the results of executing the benchmark in Sect. 4.1 across a range of workloads and data size configurations. The results in Fig. 6 are for running the YCSB benchmark with read-only, read-heavy and write-heavy workloads.

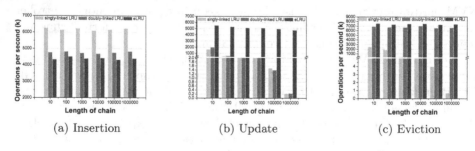

(a) Insertion   (b) Update   (c) Eviction

Fig. 4. Insertion/Update/Eviction throughput comparison

The skew is set to 0.5. For the same skew, we use the three types of workloads when data size is 2× to 16× memory.

In the experiments, we evaluate the performance of eLRU in two LRU-based algorithms, LRU-WPAM [23] and MHR-LRU [4]. Both these two algorithms employ the chain lists to track the access information. We implement these two algorithms with eLRU and standard LRU. As a baseline, we compare against hit ratio with RAND strategy which evicts tuples randomly. Under comparable access performance, eLRU can maintain a longer tracking list than standard LRU. In order to reflect the efficiency of global tracking, the length of standard LRU is 0.01× the length of eLRU.

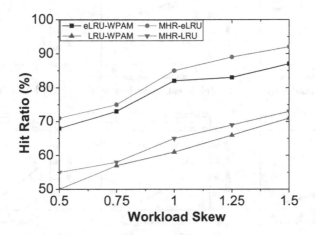

Fig. 5. Hit ratio. Read-only.

These results show that as database size increases relative to the amount of memory, the hit ratio reduces. For larger datasets, more tuples need to be evicted to disk. Thus, with the fixed amount of memory, the system meets more misses. Difference from Uniform and Latest distributions, tuples under Uniform distribution are equally likely to be evicted. In this case, all the cold data identification schemes have similar performance. The hit ratio is only relative to size of database and the amount of memory. We observe that the hit ratio of LRU-WPAM and MHR-LRU algorithms is almost the same with LRU. eLRU outperforms LRU chains by a factor of 2× for Uniform and Latest distributions. To measure the affect of skew, we use the read-only YCSB workload with a database size of 2× memory (shown in Fig. 5). We observe, that for the workload skews between 0.5–0.75, the hit ratio of separate LRU chains is lower than the baseline. The reason is that a extremely hot tuple makes the whole partition hot even if most data in this partition is not actively used. This false hot/cold identification makes the system evict active tuples out of the memory, thus reduce the hit ratio.

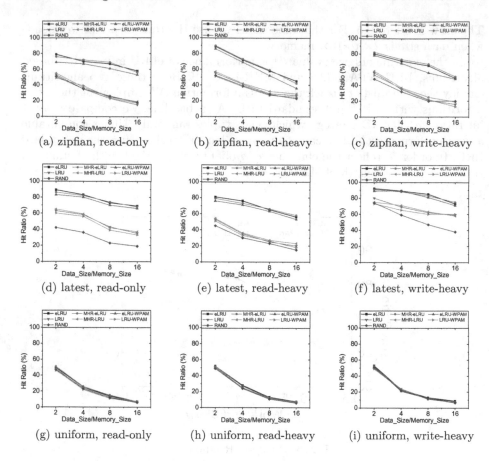

Fig. 6. YCSB experiments. Hit ratio. s=0.5.

4.4 Lazy Eviction Cost

We now compare the relative costs of lazy eviction strategy for fixed size 1 KB tuples. A hash index is used in this experiment. Each run is also repeated 3 times.

When evicting tuples, the engine first copies the coldest tuples into an eviction block, updates the index, writes the block to disk, and then cleans the isolate prefixes in eLRU. This lazy eviction strategy has no need to clean the prefixes of the evicted tuples immediately, but executes the clean operation periodically. The performance improvements are shown in Fig. 7. The results also show that the cost of preparing a eviction block and moving data to disk is linearly relative to the block size. Meaning that I/O operations are the major bottleneck.

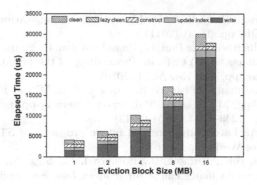

Fig. 7. Eviction cost.

5 Conclusion

Based on the non-volatile memory technologies, the next generation DBMSs seek to overcome the limitation on main memory capacity. Main memory is used as the primary storage device. Cold data needs to be moved to disk as the database grows in size. In this paper, we proposed an efficient cold data identification scheme for non-volatile memory based storage systems. Our scheme is able to maintain a single global ordering of billions of tuples. Thus, for managing datasets that are larger than memory, the system can choose properly identified cold data to evict. We presented an analysis of our design on a popular OLTP benchmark, namely YCSB. For skewed workloads with data 2×–16× the size of memory, eLRU has a 2×–4× performance advantage over the baseline. We conclude that for OLTP workloads, eLRU can outperform traditional designs.

Acknowledgments. This work is supported by the National Natural Science Foundation of China (Grant No. 61232003, 61327902), the Beijing Municipal Science and Technology Commission of China (Grant No. D151100000815003).

References

1. Akinaga, H., Shima, H.: Resistive random access memory (ReRAM) based on metal oxides. Proc. IEEE **98**(12), 2237–2251 (2010)
2. Boncz, P.A., Zukowski, M., Nes, N.J.: MonetDB/X100: hyper-pipelining query execution. CIDR **5**, 225–237 (2005)
3. Burr, G.W., Kurdi, B.N., Scott, J.C., Lam, C.H., Gopalakrishnan, K., Shenoy, R.S.: Overview of candidate device technologies for storage-class memory. IBM J. Res. Dev. **52**(4.5), 449–464 (2008)
4. Chen, K., Jin, P., Yue, L.: A novel page replacement algorithm for the hybrid memory architecture involving PCM and DRAM. In: Hsu, C.-H., Shi, X., Salapura, V. (eds.) NPC 2014. LNCS, vol. 8707, pp. 108–119. Springer, Heidelberg (2014)
5. Chen, K., Jin, P., Yue, L.: Efficient buffer management for PCM-enhanced hybrid memory architecture. In: Cheng, R., Cui, B., Zhang, Z., Cai, R., Xu, J. (eds.) Web Technologies and Applications. LNCS, vol. 9313, pp. 29–40. Springer, Cham (2015)

6. Chen, S., Gibbons, P.B., Nath, S.: Rethinking database algorithms for phase change memory. In: CIDR. pp. 21–31 (2011)
7. Cooper, B.F., Silberstein, A., Tam, E., Ramakrishnan, R., Sears, R.: Benchmarking cloud serving systems with YCSB. In: Proceedings of the 1st ACM Symposium on Cloud Computing, pp. 143–154. ACM (2010)
8. Diaconu, C., Freedman, C., Ismert, E., Larson, P.A., Mittal, P., Stonecipher, R., Verma, N., Zwilling, M.: Hekaton: SQL server's memory-optimized OLTP engine. In: SIGMOD, pp. 1243–1254. ACM (2013)
9. Driskill-Smith, A.: Latest advances and future prospects of STT-RAM. In: Non-Volatile Memories Workshop (2010)
10. Färber, F., Cha, S.K., Primsch, J., Bornhövd, C., Sigg, S., Lehner, W.: SAP HANA database: data management for modern business applications. SIGMOD Rec. **40**(4), 45–51 (2012)
11. Fredkin, E.: Trie memory. Commun. ACM **3**(9), 490–499 (1960)
12. Kallman, R., Kimura, H., Natkins, J., Pavlo, A., Rasin, A., Zdonik, S., Jones, E.P.C., Madden, S., Stonebraker, M., Zhang, Y.: H-store: a high-performance, distributed main memory transaction processing system. VlDB **1**(2), 1496–1499 (2008)
13. Kemper, A., Neumann, T.: Hyper: a hybrid OLTP and OLAP main memory database system based on virtual memory snapshots. In: ICDE, pp. 195–206 (2011)
14. Lee, D., Choi, J., Kim, J.H., Noh, S.H., Min, S.L., Cho, Y., Kim, C.S.: LRFU: a spectrum of policies that subsumes the least recently used and least frequently used policies. TOC **12**, 1352–1361 (2001)
15. Lee, S., Bahn, H., Noh, S.H.: Characterizing memory write references for efficient management of hybrid PCM and DRAM memory. In: MASCOTS, pp. 168–175. IEEE (2011)
16. Mandelman, J.A., Dennard, R.H., Bronner, G.B., DeBrosse, J.K., Divakaruni, R., Li, Y., Radens, C.J.: Challenges and future directions for the scaling of dynamic random-access memory (DRAM). IBM J. Res. Dev. **46**(2.3), 187–212 (2002)
17. Mellor, C.: HP 100 TB memristor drives by 2018 if youre lucky, admits tech titan. The Register (2013)
18. Memcached. http://memcached.org
19. Mueller, W., Aichmayr, G., Bergner, W., Erben, E., Hecht, T., Kapteyn, C., Kersch, A., Kudelka, S., Lau, F., Luetzen, J., et al.: Challenges for the DRAM cell scaling to 40 nm. In: IEDM (2005)
20. Ousterhout, J., Agrawal, P., Erickson, D., Kozyrakis, C., Leverich, J., Mazires, D., Mitra, S., Narayanan, A., Parulkar, G., Rosenblum, M.: The case for RAMClouds: scalable high-performance storage entirely in DRAM. ACM SIGOPS Oper. Syst. Rev. **43**(4), 92–105 (2009)
21. Raoux, S., Burr, G.W., Breitwisch, M.J., Rettner, C.T., Chen, Y.C., Shelby, R.M., Salinga, M., Krebs, D., Chen, S.H., Lung, H.L., et al.: Phase-change random access memory: a scalable technology. IBM J. Res. Dev. **52**(4.5), 465–479 (2008)
22. Robinson, J.T., Devarakonda, M.V.: Data cache management using frequency-based replacement. J. ACM **18**, 134–142 (1990)
23. Seok, H., Park, Y., Park, K.W., Park, K.H.: Efficient page caching algorithm with prediction and migration for a hybrid main memory. ACM SIGAPP Appl. Comput. Rev. **11**(4), 38–48 (2011)
24. Strukov, D.B., Snider, G.S., Stewart, D.R., Williams, R.S.: The missing memristor found. Nature **453**(7191), 80–83 (2008)
25. Oracle TimesTen. http://www.oracle.com
26. VoltDB. http://www.voltdb.com

Preference Join on Heterogeneous Data

Changping Wang, Chaokun Wang[⊠], Hao Wang, Jun Chen, and Xiaojun Ye

School of Software, Tsinghua University, Beijing 100084, China
{wang-cp12,wanghao07,chenjun14}@mails.tsinghua.edu.cn,
{chaokun,yexj}@tsinghua.edu.cn

Abstract. There are different types of join operation dealing with different issues in database research. However, existing join operations cannot meet the increasing demands of the real world. In this paper, we define a new join operation, the *preference join (p-join)*, which introduces the concepts of the personal preference and the satisfaction operator on various data types. We present a general join algorithm (Nested Loop) to deal with the *p*-join, and we also propose an advanced algorithm called MFV for *p*-join. To improve the MFV algorithm, two enhanced mapping methods are employed. A large number of experiments on both real-world and synthetic data sets are conducted. The experimental results demonstrate the effectiveness, efficiency and scalability of our methods, and show the advanced algorithms have advantages over the general algorithms.

1 Introduction

The matchmaking industry generated \$2.2 billion in sales in the U.S. in 2014 [7], and 5.6 million people used online dating sites and apps in the first month of 2015 [10]. The matchmaking problem and some similar ones, such as housing rental and job recruitment, can be summarized as a novel query, called *the preference join (p-join) problem* in this paper. The *p*-join problem focuses upon the join operation on two entity sets with following properties: (1) Entities in the same set share common attributes. (2) Each element of one entity set has individual *preference*, which is composed of a threshold and a list of constraints on the common attributes of the other entity set. (3) The two entities in a matched pair must satisfy each other's preference.

Scenario 1 (Who to date?). *Suppose there are two females {Mary, Lily} and two males {Jim, John} who are willing to date. We have people's attribute values as well as the preference on their dates (see Table 1). The attribute may be of heterogeneous types, e.g., strings, numerical or categorical values. People use preferences to describe their ideal dates. Suppose Mary and Jim are fussy people who set high thresholds of qualifying attributes on their dates. Meanwhile, Lily and John are relatively nice people who set lower thresholds on the qualifying attributes of their dates. The dating problem results in all the matched pairs of dates, i.e. {(John, Lily)}.*

© Springer International Publishing Switzerland 2016
F. Li et al. (Eds.): APWeb 2016, Part II, LNCS 9932, pp. 317–329, 2016.
DOI: 10.1007/978-3-319-45817-5_25

Table 1. The dating information. Each person has several attribute values and a personalized preference which consists of some constraints on their dates and a threshold. **MS** means Martial Status (S—Single).

Name	Age	Height	Education	House	MS	Preference
Mary	23	166	Master	N	S	Age:24~27, House:Y, Height:168~172, 1
Lily	24	167	Bachelor	N	S	Age:25~27, House:Y, Education:Bachelor|Master, 0.8
...	...	...	...	...	...	...

(a) The female set.

Name	Age	Height	Education	House	MS	Preference
Jim	25	174	PhD	Y	S	House:Y, Height:165~167, MS:S, Education:Bachelor|PhD, Age:21~22, 1
John	27	175	Master	Y	S	Age:20~22, House:N, Education:Bachelor, MS:S, 0.75
...	...	...	...	...	...	...

(b) The male set.

The above scenario is a typical case of the *p*-join problem. However, to the best of our knowledge, existing join techniques are not effective enough to deal with this problem due to the following three challenges: (1) As shown in Scenario 1, the attributes of real-world entities are usually of heterogeneous types, such as those listed in Table 1. However, most of the existing join methods are specialized to deal with only one data type, e.g., integer, string or vector, because they use specific optimization methods based on the concrete data type. (2) The thresholds of entities' preferences are usually diversified since people have different personalities. However, most of the existing methods can only employ a unified threshold on the join operation. (3) In the *p*-join problem, we deal with the "satisfaction" relationship, e.g., non-equivalence, inclusion, and Boolean test, between values and constraints on a given attribute of complex entities. However, the existing methods only consider the equivalence or the partial order relations.

In this paper, a novel join operation, namely, *preference join (p-join)*, and the related methods are proposed to address the above challenges. The main contributions of the paper are as follows.

- We present a novel join operation, *preference join (p-join)*, which supports mutual matching from two entity sets of heterogeneous data with personalized preferences.
- We use a simple but effective way (Nested Loop) to deal with *p*-join first, and then propose an advanced method called MFV (i.e. Mapping, Filtering and Verification).

- For the MFV algorithm, this paper brings forward three mapping methods, and each method maps all the heterogeneous data to universal tokens. Among these mapping methods, the heuristic mapping outperforms the injective mapping and the homogeneous mapping considering the time cost.
- Experiments conducted on both the real-world and the synthetic data sets show that our methods are effective and efficient, and can be applied in real-world scenarios. The advanced algorithm has advantages over the general one.

2   Related Work

In this section, the related works on p-join are briefly reviewed. As far as we are concerned, the most related research to our problem is similarity join. There are many existing works of similarity join on specific data types, such as strings [5], vectors [6], graphs [20], sets [2], and time series [18]. Most of these works have their limitations in dealing with our problem, which has been outlined in Sect. 1.

Among these research works, the string similarity join can be seen as one of the most typical representatives, and it has also been among the most active research in all these years. A string similarity join (SSJ) finds out all similar pairs between two sets of strings. The existing SSJ methods can be classified into three categories: The first is the index-based method [8,9,14,15]. These methods employ different index structures on strings, e.g., Trie [14], inverted indices [8], B^+-tree [9], and HS-tree [15], to facilitate the pruning in the search space and generate similar string pairs efficiently. The second is the signature-based method [16,17]. Based on various signature schemes, many effective algorithms, such as FastJoin [16] and VChuckJoin [17], are brought forward to improve the performance of SSJ. The last is the filtering-based method [4,11,19]. Many filters, such as prefix filter [4,11], position filter [19], and suffix filter [19], are designed to avoid computing similarities of all possible pairs of strings. In order to deal with large scale data sets, several SSJ processing methods with MapReduce are also proposed in recent years [1,12,13].

However, the above methods cannot be directly used to deal with the p-join problem. The proposed p-join problem is distinguished by the following aspects: Firstly, most of the studies only concern characters, and few studies consider heterogeneous data. The existing set similarity joins, including exact set similarity joins and approximate set similarity joins, just consider elements in a same category. Secondly, only point-to-point comparison is considered. That is, it does not take into account the interval. Finally, only unified threshold is considered in the above similarity join algorithms.

3   Basic Concepts

In this section, the "satisfaction" operator is firstly defined. Then, the definition of p-join is presented. The symbols used in this paper and their meanings are listed in Table 2.

3.1 The "Satisfaction" Operator

We firstly present the definition of the "satisfaction" operator as below.

Definition 1 (the satisfaction operator). *For an attribute A, the "satisfaction" operator $\propto$ is defined on a value v of A and a constraint c on A. If value v satisfies c, then we say $v \propto c$.*

For different data types, $\propto$ is under different evaluation metrics.

- Let A be a numerical attribute, v a numerical value and c a numerical interval. Then, $v \propto c$ holds *iff* v is within the range of c.
- Let A be a Boolean attribute, and both v and c be Boolean values. Then, $v \propto c$ holds *iff* v equals c.
- Let A be a categorical attribute, and both v and c be sets of categorical values. Then, $v \propto c$ holds *iff* $v \cap c \neq \emptyset$.

In our problem settings, A represents one of the common attributes in one entity set (say R), v represents the value of A of an entity r in R, and c represents a constraint on A of an entity s in the other entity set S.

3.2 The Preference Join

Let each entity in R and S consist of attribute values, constraints and the threshold, i.e.

$$r = \left(v_1^r, v_2^r, \ldots, v_u^r, c_1^r, c_2^r, \ldots, c_p^r, T(r)\right),$$
$$s = \left(v_1^s, v_2^s, \ldots, v_w^s, c_1^s, c_2^s, \ldots, c_q^s, T(s)\right),$$
$$r \in R, s \in S, u \geq q, w \geq p.$$

Here, $(v_1^r, v_2^r, \ldots, v_u^r)$ represents the attribute values, $(c_1^r, c_2^r, \ldots, c_p^r)$ represents the constraints in the preference of r, and $T(r)$ is the threshold of r's preference. The similar interpretations work on $(v_1^s, v_2^s, \ldots, v_w^s)$, $(c_1^s, c_2^s, \ldots, c_q^s)$, and $T(s)$.

Definition 2 (the preference satisfaction). *Let R and S be two entity sets, $r \in R$ and $s \in S$. The preference satisfaction between r and s is defined as $(Sat(r, s), Sat(s, r))$, where*

$$Sat(r, s) = |\{(v_i^s, c_j^r)|v_i^s \propto c_j^r, A(v_i^s) = A(c_j^r), 1 \leq j \leq p\}|/p,$$
$$Sat(s, r) = |\{(v_i^r, c_j^s)|v_i^r \propto c_j^s, A(v_i^r) = A(c_j^s), 1 \leq j \leq q\}|/q.$$

The function $Sat(r, s)$ measures the proportion of the constraints, which are satisfied by s's attribute values, in r's preference. By comparing the values of $Sat(r, s)$ and $T(r)$, we know whether entity s could be matched to entity r. The similar is $Sat(s, r)$. Please notice that $Sat(\cdot, \cdot)$ is asymmetrical, i.e. usually $Sat(r, s) \neq Sat(s, r)$.

Based on the above definition of the preference satisfaction between two entities, the p-join is defined as follows.

Table 2. Symbols and meanings.

Symbol	Meaning
R, S	Entity sets
r, s	Arbitrary entities in entity sets, $r \in R$ and $s \in S$
$\propto$	Satisfaction operator
$\bowtie$	p-join
$Attr(R)$	The set of all attributes of R
$Attr(r)$	The set of all attribute values of r
$Cons(r)$	The set of all constraints in r's preference
$T(r)$	The threshold of r's preference
v	An attribute value in $Attr(r)$
c	An constraint in $Cons(r)$
$A(v)$	The attribute of which r represents the value
$A(c)$	The attribute on which c represents the constraint

Definition 3 (the preference join). *The preference join on entity sets R and S is defined as*

$$R \bowtie S = \{(r, s) \mid Sat(r, s) \geq T(r) \text{ and } Sat(s, r) \geq T(s),$$
$$r \in R, s \in S \}.$$

That is, for each pair $(r, s) \in R \bowtie S$, r and s satisfy the preference of each other. Also, the sets R and S are called dual sets of each other.

For convenience, in the rest of the paper, $Attr(r)$ and $Cons(r)$ represent $(v_1^r, v_2^r, \ldots, v_u^r)$ and $(c_1^r, c_2^r, \ldots, c_p^r)$, respectively. As illustrated in Fig. 1, $Cons(r)$ corresponds to $Attr(s)$, and vice versa.

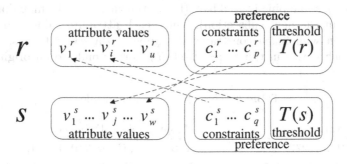

Fig. 1. An illustration of p-join, $r \in R, s \in S$.

4   Algorithms

In this section, we first present how to use the Nested Loop algorithm, the general method for all join problems, to solve the p-join problem. Next, an advanced solution (MFV) specified to the p-join problem is brought forward to address this problem efficiently. At last, we propose two enhanced mapping methods for MFV.

4.1   Nested Loop Algorithms

Due to the challenges of the p-join problem, existing join algorithms for specific join problems, e.g. hash-join algorithms and sort-merge-join algorithms for equivalent join, and the algorithms discussed in Sect. 2 for string similarity join, cannot be used directly to solve our p-join problem.

A simple but effective algorithm for all the join problems is Nested Loop (NL) algorithm, so we discuss how to employ this algorithm at first. The NL algorithm extracts any possible pair of entities from the dual sets, and compares the entities' preference satisfaction and thresholds. If the values of both $Sat(\cdot, \cdot)$ are no smaller than the thresholds, the pair of entities is added to the result set.

4.2   MFV Algorithm

In this subsection, we propose an algorithm called MFV which consists of three steps: **M**apping, **F**iltering and **V**erification.

The mapping step. We integrate the heterogeneous data by employing mappings. All the attribute values and constraints of entities are mapped to universal tokens (Line 3). Table 3 shows an example of the mapping rules (token frequencies are counted considering only attribute values). Tokens are represented by characters as in Table 3, when the size of the token set is small. However, we can easily use integers to represent tokens when the size is large.

The injective mapping is used here (the optimization on the mapping method is discussed in the next subsection). Thus, each attribute value is mapped to a single token while each constraint is mapped to a set of tokens. After the mapping, each entity in the mapped set R_m corresponds to an original entity

Table 3. The injective mapping rule

Attribute value	Mapping token	Frequency
Age $(20, 21, \ldots, 27)$	$A, B, C, \ldots, H$	0, 0, 0, 1, 1, 1, 0, 1
Height $(165, 166, \ldots, 175)$	$I, J, K, \ldots, S$	0, 1, 1, 0, 0, 0, 0, 0, 0, 1, 1
Education (Bachelor, Master, PhD)	T, U, V	1, 2, 1
House (Y, N)	W, X	2, 2
MS (S)	Y	4

Table 4. The injective mapping result

Name	Mapping result	Ordering result
Mary	$DJUXY$ $\{E{\sim}H\}\{L{\sim}P\}\{W\}$	$DJUXY$ $\{L{\sim}P\}\{W\}\{E{\sim}H\}$
Lily	$EKTXY$ $\{F{\sim}H\}\{T{\sim}U\}\{W\}$	$EKTXY$ $\{W\}\{F{\sim}H\}\{T{\sim}U\}$

(a) The female set.

Name	Mapping result	Ordering result
Jim	$FRVWY$ $\{B,C\}\{I{\sim}K\}\{T,V\}\{W\}\{Y\}$	$FRVWY$ $\{B,C\}\{W\}\{T,V\}\{I{\sim}K\}\{Y\}$
John	$HSUWY$ $\{A{\sim}C\}\{T\}\{X\}\{Y\}$	$HSUWY$ $\{A{\sim}C\}\{T\}\{X\}\{Y\}$

(b) The male set.

in R. For instance, the mapped entity corresponding to $John$ is ($HSUWY$, $\{A{\sim}C\}\{T\}\{X\}\{Y\}$) (see Table 4b), where the former part represents $John$'s attribute values and the latter part represents $John$'s constraints. The tokens of each entity's constraints are separately sorted in the order of increasing token frequency, i.e. infrequent tokens on the left while frequent tokens on the right. The token sets corresponding to each entity's constraints are separately sorted in the order of increasing token set frequency, which is the frequency sum of all the tokens in the set.

Then, we build inverted indices on the constraints of these mapped entities (Lines 5–6). According to the filtering principle [3], only the tokens of the first $|Cons(r)| \times (1 - T(r)) + 1$ constraints in the preference of each entity are used to build indices. The sorting step in Algorithm 1 makes the indexed tokens more selective in the filtering, which is essential in the prefix filtering.

The filtering step. In the filtering step, the algorithm performs a two-step filtering strategy to generate candidate pairs (Algorithm 1, Lines 7–21). The first filtering operation finds out all the pairs (r, s) where s is possibly satisfied by r using the indices (Lines 7–13). Then, the second filtering operation filters these pairs again, and only those pairs where r is possibly satisfied by s are preserved (Lines 14–21). Please notice that whether R_m or S_m is filtered first does not affect the generated candidate set.

The verification step. The candidate pairs are verified by comparing the preference satisfactions and the thresholds to obtain the result set in the verification step (Lines 22–26).

4.3 Enhanced Mapping Methods

The injective mapping used in Sect. 4.2 is an accurate and strict mapping method. Supposing the range of one numerical attribute is $[a, b]$, this method

Algorithm 1. MFV Algorithm

Require:
    The dual sets R and S.
Ensure:
    PJ /* The result set of p-join $R \bowtie S$. */
 1: $PJ \leftarrow \emptyset$
 2: $CR_1 \leftarrow \emptyset, CR_2 \leftarrow \emptyset$
    /* Mapping step: Map heterogeneous data to tokens. */
 3: $R_m \leftarrow MAP(R), S_m \leftarrow MAP(S)$
 4: $Sort(R_m, S_m)$
 5: $I_R \leftarrow MFVIndexBuilding(R_m)$
 6: $I_S \leftarrow MFVIndexBuilding(S_m)$
    /* Filtering step: Perform the two-step filtering strategy on mapped entities to generate candidate pairs. */
 7: **for all** $r \in R_m$ **do**
 8:     **for all** $o \in Attr(r)$ **do**
 9:         **for all** $s \in I_S^o$ **do**
10:             $CR_1 \leftarrow CR_1 \cup \{(r, s)\}$ /* Possibly $Sat(s, r) \geq T(s)$. */
11:         **end for**
12:     **end for**
13: **end for**
14: **for all** $(r, s) \in CR_1$ **do**
15:     **for all** $o \in Attr(s)$ **do**
16:         **if** $r \in I_R^o$ **then**
17:             $CR_2 \leftarrow CR_2 \cup \{(r, s)\}$ /* Possibly $Sat(r, s) \geq T(r)$. */
18:             $Break$
19:         **end if**
20:     **end for**
21: **end for**
    /* Verification step: Perform verification on the candidate pairs to generate final matched pairs. */
22: **for all** $(r, s) \in CR_2$ **do**
23:     **if** $Sat(r, s) \geq T(r)$ and $Sat(s, r) \geq T(s)$ **then**
24:         $PJ \leftarrow PJ \cup \{(r, s)\}$
25:     **end if**
26: **end for**
27: **return** PJ

---

maps each distinct attribute value in the interval to a unique token. However, it leads to very large indices and consumes a long index building time. To address this problem, we can divide the interval $[a, b]$ into k (which can be specified for different situations) small intervals, and the values in the same interval are mapped to the same token. Then, we introduce two mapping methods based on different partition strategies.

The Homogeneous Mapping Method. The idea of the homogeneous mapping is to divide the numerical interval equally based on the partition number k. It reduces both the size of indices and the time cost of building indices by mapping multiple values to a same token. However, this method decreases the discrimination power of attribute values, which leads to lower efficiency in filtering. We want to find out a way to minimize this effect and thus, we propose the following mapping method.

The Heuristic Mapping Method. This mapping method employs a heuristic partition strategy which can minimize the decrease of filtering efficiency.

Let us discuss how this decrease happens by example. On Age attribute, Mary has a constraint of 24–27. Supposing we partition the Age range into intervals including [22,25) and [25,28), then Mary's constraint is mapped to the two tokens corresponding to these two intervals. Thus, we extend this constraint to 22–28, in fact. This extension leads to more candidates in filtering step, so the efficiency of filtering decreases.

Based on the above discussion, we propose an optimization objective which finds a partition minimizing the sum of all the extensions:

$$\underset{\prod}{\arg\min} \sum_{r \in R} f_{\prod}(c_A^r)) \\ sub.\ to\ |\textstyle\prod| = k ,$$

where $\prod$ is a partition of attribute A, c_A^r is the constraint of entity r on attribute A, and $f_{\prod}$ is a function which calculates the extension of a constraint under the partition $\prod$.

We employ a partition algorithm based on dynamic programming to achieve the optimization objective. Due to the page limitation, it is not presented here.

Regarding the filtering efficiency, the heuristic mapping is much better than the homogeneous mapping, while it is worse than the injective mapping.

5 Experiments

5.1 Data Sets

Two real-world data sets and two synthetic data sets are used in our experiments.

- **DATING** is a real-world dating data set which consists of 20,261 entities (10,430 male entities and 9831 female entities) crawled from the website http://www.match.com. We ignore the attributes which do not appear in the preferences, and thus each entity in both the male set and the female set has the same eight attributes.
- **JOB** is a real-world job recruitment data set which consists of 18,953 entities (9122 job-seeker entities and 9831 job entities), and it is crawled from the website http://www.resume.com. The same as DATING, attributes not in the preferences are ignored, but different from DATING, the dual sets of JOB are asymmetric. Each job-seeker entity has five different attributes while each job entity has two.
- **L-DATING** is a synthetic dating data set which has 1,000,000 entities (500,000 entities for each of the dual sets, i.e. male/female). It is generated according to the distributions of the attribute values and preferences of **DAT-ING**.
- **L-JOB** is a synthetic recruitment data set which has 1,000,000 entities (500,000 entities for each of the dual sets, i.e. job-seekers/jobs). It is generated according to the distributions of the attribute values and preferences of **JOB**.

5.2 Experimental Results

Comparisons of All the p-join Algorithms. The time cost of all the p-join algorithms is shown in Table 5. Figures 2a and b show the comparisons between

Table 5. Time cost comparison (ms).

n	NL	MFV-I				MFV-O				MFV-R			
		M	F	V	Total	M	F	V	Total	M	F	V	Total
DATING 5000	21,548	393	825	**134**	1352	**266**	579	207	1052	299	**482**	138	**919**
6000	32,016	444	1119	**190**	1753	**308**	762	287	1357	381	**691**	200	**1272**
7000	44,740	498	1569	**274**	2341	**329**	1083	422	1834	419	**939**	279	**1637**
8000	58,175	529	1951	**337**	2817	**349**	1432	508	2289	375	**1249**	350	**1974**
9000	74,324	801	2551	**414**	3766	**409**	1857	676	2942	449	**1667**	493	**2609**
JOB 5000	9786	266	3345	**103**	3714	**177**	903	129	1209	217	**884**	106	**1207**
6000	16,328	305	4528	**141**	4974	**206**	1347	174	1727	235	**1245**	146	**1626**
7000	24,286	327	5643	**188**	6158	**225**	1751	233	2209	268	**1716**	222	**2206**
8000	33,985	346	7711	**193**	8250	**243**	2312	337	2892	284	**1957**	252	**2493**
9000	46,188	370	8376	**251**	8997	**260**	2818	415	3493	318	**2601**	342	**3261**

all the p-join algorithms except NL, which is far slower than others. From these figures and Table 5, we can conclude that: (1) $T_m^O < T_m^R < T_m^I$, $T_f^R < T_f^O < T_f^I$, $T_v^I < T_f^R < T_f^O$, and $T_{total}^R < T_{total}^O < T_{total}^I$. (2) Based on the time cost, MFV-R is the best and MFV-I is the worst among the MFV algorithms. (3) All the MFV algorithms are much better than NL, and MFV-R takes only $\frac{1}{10}$ to $\frac{1}{20}$ time compared with NL on both data sets.

Figures 2c and d illustrate the relationship between the number of candidate pairs after filtering, N_{cp}, and the number of original entities. The values of N_{cp} of the three MFV algorithms increase with the expansion of the size of original entities. The smaller the value of N_{cp} is, the better filtering performance the algorithm achieves. However, notice that without filtering, N_{cp} is the square of the number of the original entities, which is much more than N_{cp} generated by MFV algorithms. Furthermore, these figures also show MFV-I generates the smallest number of candidate pairs since it has the most remarkable performance in filtering out the irrelevant pairs. Besides, the effectiveness of MFV-O is the worst in candidate pair generation, because the mapping step of MFV-R is optimal, which leads to a more selectively inverted index than that of MFV-O.

The Threshold Selection. Figure 3 shows the comparison between the sizes of the result sets when using a unified threshold (all entities share the same threshold) and individual thresholds (each entity has its own threshold), respectively. The unified threshold is set to 0.9 which is the average of the individual thresholds in the data sets. Intuitively, the individual thresholds reflect the user's characters, e.g., *Mary*'s and *Jim*'s picky personality, and *Lily*'s and *John*'s relatively nice characters in Scenario 1. From Fig. 3, we can see that the sizes of the result sets generated by using the unified threshold and the individual thresholds are almost equivalent for **DATING** and **JOB**. For the individual thresholds, the result set of **DATING** contains nearly 200,000 pairs (199,428 pairs), while for the unified threshold, the result set of **DATING** also contains nearly 200,000 pairs (197,958 pairs). However, less than one quarter of pairs in both result sets

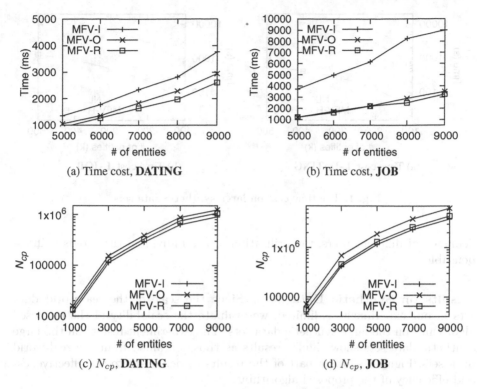

Fig. 2. Comparisons of all the p-join algorithms

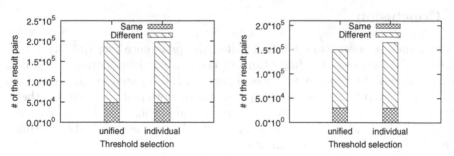

(a) Unified threshold VS individual thresholds, (b) Unified threshold VS individual thresholds,
DATING   **JOB**

Fig. 3. The comparison between unified threshold and individual threshold.

are identical on **DATING**. It indicates that more than 150,000 different result pairs are generated when using the two different kinds of thresholds. For **JOB**, the same situation is observed; the proportion of the identical pairs generated by the two kinds of thresholds is only about $\frac{3}{16}$. This implies that the result set obtained using the unified threshold has a great difference compared with that achieved using individual thresholds. In order to satisfy the personalized

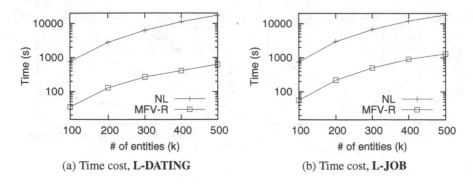

Fig. 4. The time cost on large synthetic data sets.

demands of different users, the utilization of the individual thresholds is indispensable.

Results on Synthetic Data Sets. Since the scale of the real-world data sets in our experiments is limited, we evaluate the scalability of the proposed algorithms on the large synthetic data sets. The experiment results on the large synthetic data sets show similar results as those obtained from the real-world data sets. Figure 4 shows a part of the results to demonstrate the effectiveness and efficiency of the proposed algorithms.

6   Conclusion

In this paper, a novel join operation called the preference join (p-join) is presented, which addresses the limitations of the existing join operations. Also, a general method (NL) and an advanced algorithm (MFV) are brought forward to deal with p-join. A large number of experiments on both real-world and synthetic data sets show that our methods are effective and efficient. In the future, more research will be conducted to improve the scalability of the proposed algorithms, especially in distributed environments.

Acknowledgments. This work was supported in part by the National Natural Science Foundation of China (No. 61373023, No. 61170064).

References

1. Afrati, F.N., Sarma, A.D., Menestrina, D., Parameswaran, A., Ullman, J.D.: Fuzzy joins using MapReduce. In: ICDE, pp. 498–509 (2012)
2. Arasu, A., Ganti, V., Kaushik, R.: Efficient exact set-similarity joins. In: VLDB, pp. 918–929 (2006)
3. Chaudhuri, S., Ganti, V., Kaushik, R.: A primitive operator for similarity joins in data cleaning. In: ICDE, pp. 5–16 (2006)

4. Deng, D., Li, G., Feng, J.: A pivotal prefix based filtering algorithm for string similarity search. In: SIGMOD, pp. 673–684. ACM (2014)
5. Deng, D., Li, G., Feng, J., Li, W.-S.: Top-k string similarity search with edit-distance constraints. In: ICDE (2013)
6. Jacox, E.H., Samet, H.: Metric space similarity joins. TODS **33**(2), 7:1–7:38 (2008)
7. Kelion, L.: Tinder to charge older users more for premium facilities (2015). http://www.bbc.com/news/technology-31700036
8. Li, G., Deng, D., Wang, J., Feng, J.: Pass-join: a partition-based method for similarity joins. PVLDB **5**(3), 253–264 (2012)
9. Lu, W., Du, X., Hadjieleftheriou, M., Ooi, B.: Efficiently supporting edit distance based string similarity search using B+-trees. TKDE **26**(12), 2983–2996 (2014)
10. Molla, R.: The current state of online dating (2015). http://blogs.wsj.com/speakeasy/2015/02/27/the-current-state-of-online-dating/
11. Rong, C., Lu, W., Wang, X., Du, X., Chen, Y., Tung, A.: Efficient and scalable processing of string similarity join. TKDE **25**(10), 2217–2230 (2013)
12. Vernica, R., Carey, M.J., Li, C.: Efficient parallel set-similarity joins using MapReduce. In: SIGMOD, pp. 495–506 (2010)
13. Wang, C., Wang, J., Lin, X., Wang, W., Wang, H., Li, H., Tian, W., Xu, J., Li, R.: MapDupReducer: detecting near duplicates over massive datasets. In: SIGMOD, pp. 1119–1122 (2010)
14. Wang, J., Feng, J., Li, G.: Trie-join: efficient trie-based string similarity joins with edit-distance constraints. PVLDB **3**(1–2), 1219–1230 (2010)
15. Wang, J., Li, G., Deng, D., Zhang, Y., Feng, J.: Two birds with one stone: an efficient hierarchical framework for top-k and threshold-based string similarity search. In: ICDE, pp. 519–530. IEEE (2015)
16. Wang, J., Li, G., Fe, J.: Fast-join: an efficient method for fuzzy token matching based string similarity join. In: ICDE, pp. 458–469 (2011)
17. Wang, W., Qin, J., Chuan, X., Lin, X., Shen, H.: VChunkJoin: an efficient algorithm for edit similarity joins. TKDE **25**(8), 1916–1929 (2013)
18. Xiang, L., Lei, C.: Efficient similarity join over multiple stream time series. TKDE **21**(11), 1544–1558 (2009)
19. Xiao, C., Wang, W., Lin, X., Yu, J.X., Wang, G.: Efficient similarity joins for near-duplicate detection. TODS **36**(3), 15:1–15:41 (2011)
20. Zhao, X., Xiao, C., Lin, X., Wang, W.: Efficient graph similarity joins with edit distance constraints. In: ICDE, pp. 834–845 (2012)

Accelerating Time Series Shapelets Discovery with Key Points

Zhenguo Zhang[1,2], Haiwei Zhang[1(✉)], Yanlong Wen[1], and Xiaojie Yuan[1]

[1] College of Computer and Control Engineering, Nankai University,
38 Tongyan Road, Tianjin 300350, People's Republic of China
{zhangzhenguo,zhanghaiwei,wenyanlong,yuanxiaojie}@dbis.nankai.edu.cn
[2] Department of Computer Science and Technology, Yanbian University,
977 Gongyuan Road, Yanji 133002, People's Republic of China

Abstract. Shapelets are discriminative subsequences in a time series dataset, which provide good interpretability for time series classification results. For this reason, time series shapelets have attracted great interest in time series data mining community. Although time series shapelets have satisfactory performance on many time series datasets, how to fast discover them is still a challenge because any subsequence in a time series may be a shapelet candidate. There are several methods to speed up shapelets discovery in recent years. However, these methods are still time-consuming when dealing with the large datasets or long time series. In this paper, we propose a preprocessing step with time series key points for shapelets discovery which make full use of the prior knowledge of shapelets. Combining with shapelets discovery method based on SAX(Fast-Shapelets), we can find shapelets quickly on all benchmark datasets of UCR archives, while the classification accuracy is almost the same as the current methods.

Keywords: Time series · Shapelets · Classification · Key points

1 Introduction

Time series data mining has attracted significant interest in recent years because the series data are common in a wide range of daily life such as finance, medical treatment, motion, meteorology, etc. As a fundamental research work, classification of time series has been studied most commonly. Many research focus on distance measures for 1-Nearest Neighbor (1-NN) classifiers [1,2]. Although the evidence show that 1-NN classifier with Euclidean Distance (ED) or Dynamic Time Warping (DTW) has good classification accuracy, it requires storing and searching the entire dataset which have high time and space complexity. Moreover, 1-NN classifiers do not give a clear insight to exhibit the most important pattens between two different classes. To solve these problems, a shapelet-based time series classification algorithm is proposed by *Ye et al.* [3].

A shapelet is a time series subsequence with highly discriminative between two classes. The shapelets discovery algorithm searches shapelets on raw data

F. Li et al. (Eds.): APWeb 2016, Part II, LNCS 9932, pp. 330–342, 2016.
DOI: 10.1007/978-3-319-45817-5_26

and calculates the distance between shapelet candidates and time series. By the measure of information gain, the candidates are ordered and then the candidate with highest score is select as a shapelet to split dataset into two parts. For each part, the same process is used to get the corresponding shapelets. Shapelets are interpretable to the problem domain because they can tell us whether a shapelet is a common pattern in one class or not. The advantage has been confirmed by many researchers [4–6]. To some datasets, this algorithm can achieve more accurate classification results than other methods [3,10]. It has been applied to many domains such as medical care [7], gesture recognition [8], electrical power demand [9]. However, shapelets discovery is very time-consuming. All the subsequences with any length in time series dataset are shapelet candidates. Although there are some pruning strategies to speed up shapelets discovery like admissible entropy pruning [3], intermediate result reuse [10] and reducing the distance calculation with SAX method (called Fast-Shapelets) [11], it still needs a lot of time when the dataset is large or time series is very long. For example, when testing all candidates in the dataset *Non-Invasive Fetal ECG Thorax* of UCR time series archives [13], the Fast-Shapelets algorithm still needs over 10 h to get the results.

In this paper, we analyze the nature of time series shapelets and propose a preprocessing stage before searching shapelets to reduce the shapelet candidates. In our algorithm, the shapelet candidates are not all time series subsequences but the subsequences generated from the key points. Combining with the SAX representation of time series to find potential shapelet candidates, we can get the shapelets in less time than other algorithms. The experiments show that our preprocessing method for decreasing shapelet candidates is useful. In general, we make two contributions.

1. We propose a method to find the key points of a time series which is used for searching shapelet candidates. It can fast exclude the obviously impossible candidates from all subsequences.
2. An algorithm of extracting time series subsequences based on key points is presented that allows to generate shapelet candidates without repetition.

The rest of this paper is organized as follows. In Sect. 2, we review the related works about time series shapelets discovery algorithm. We present some notations and definitions of time series in Sect. 3. Section 4 shows our algorithm in details. We demonstrate the performance of our algorithm in Sect. 5 and this work is concluded in Sect. 6.

2   Related Work

The straightforward way for finding shapelets is the brute force algorithm that generates all possible candidates and tests these candidates by information gain. The running time is $O(n^2 m^4)$, where n is the number of time series in dataset D, m is the length of a time series. In the first paper of shapelets discovery, the

author proposes several methods to speed up the searching work, like early aban-
don pruning and entropy pruning technique. However, it is still time-consuming.
An improved algorithm called Logical Shapelets is given by [10]. One technique
is that it precomputes the sufficient statistics to compute the distance between
a shapelet and a candidate of a time series in amortized constant time. It is a
way of trading space for time. The other is to use a novel admissible pruning
technique to skip the costly computation of entropy for the vast majority of
candidates. The worst-case of running time is $O(n^2m^3)$ and it requires a lot of
memory space as large as $O(nm^2)$.

    Rakthanmanon et al. [11] propose a way to find shapelets quickly, where the
raw real-valued and high dimensional data are transformed to a discrete and low
dimensional representation by using Symbolic Aggregate approXimation (SAX)
method [14]. In the discrete representation, they first hash the SAX representation of candidates by the random projecting method and then use the collision
history to give a rough selection for the overall candidate set. By this way, the
candidates that cannot be a shapelet are excluded quickly. The remaining candidates are still confirmed by information gain. The time complexity can reduce
to $O(nm^2)$ according to the paper.

3 Notations and Definitions

3.1 Time Series

A time series T is a sequence of m real-valued variables recorded in temporal
order at fixed intervals of time: $T = (t_1, \ldots, t_i, \ldots, t_m)$, $t_i \in R$. For the problem of time series data mining, a dataset of n time series can be expressed as
$D = \{T_1, T_2, \ldots, T_n\}$. For classification, each time series in the dataset D
has a class label of c. Given a time series T of length m, a subsequence S
is a series of length l $(l < m)$ consisting of contiguous time instants of T:
$S = (t_i, t_i + 1, \ldots, t_{i+l-1})$, where $i \in [1, m - l + 1]$, noted as T_i^l.

3.2 Time Series Key Points

We define the key points of a time series as the *inflection points, local minimum
points, local maximum points* in a time series curve. They are useful for shapelets
discovery and we will discuss them in detail in Sect. 4.

3.3 Distance Measure

In this paper, we take Euclidean distance as the distance measure between two
time series. Suppose S and S' are two time series subsequences with the same
length l, the Euclidean distance is calculated by the following formula:

$$d(S, S') = \sqrt{\frac{1}{l} \sum_{i=1}^{l} (s_i - s_i')^2} \tag{1}$$

The distance between a time series T of length m and a subsequence S of length l ($l < m$) is defined as the minimum distance between the S and all subsequences of T that have the same length with S, i.e.

$$d(S, T) = min \; d(S, T^l) \tag{2}$$

3.4 Time Series Shapelets

As primitives, shapelets can be used to determine the similarity of two time series. Because any subsequence in all time series of a dataset are probable to be a shapelet, finding a shapelet requires us to generate the candidates and to calculate the distances between candidates and a time series. We should also define a measurement to estimate the quality of a shapelet.

Shapelet Candidates. Given a dataset D, a shapelet candidate is the subsequence of length l in a time series. If dataset D contains n time series and the length of a time series is m, we can get $(m - l) + 1$ distinct subsequences in a time series and $n(m - l + 1)$ subsequences in D. We denote the set of all subsequences of length l to be $M_l : \{M_{1,l}, ..., M_{i,l}, ..., M_{n,l}, \}$, where, $M_{i,l}$ is the subsequences of the i-th time series in D. The length l can be changed from 1 to m, so the overall subsequences set is: $M = \{M_1, ..., M_l, ..., M_m\}$.

Generally, if the length of a subsequence is too small or close to m, it may not be a shapelet. So we can give a minimum and maximum length before calculating the distances: $minlen$ and $maxlen$, i.e. $M = \{M_{minlen}, ..., M_l, ..., M_{minlen}\}$. Note that M is very large. The most of shapelet research focus on how to efficiently prune M [3,10,11].

The Quality of a Shapelet. An effective measurement of discriminating the quality of a shapelet is information gain [3]. It involves the concept of *entropy* of a dataset and a *split*. Suppose that the dataset D contains c different classes, the number of time series in class i is n_i, so the entropy of the dataset is $E(D) = - \sum_{i=1}^{c} p_i \log(p_i)$, where $p_i = n_i/n$. A *split* is a tuple $<s, d>$ of a subsequence s and distance threshold d which can separate the dataset into two subsets, D_1 and D_2 with n_1 and n_2 time series, respectively. The information gain of a given split sp in a dataset can be express as

$$IG(sp) = E(D) - \frac{n_1}{n} E(D_1) - \frac{n_2}{n} E(D_2) \tag{3}$$

It is possible that two splits have the same information gain. To solve this problem, the distance between two subsets divided by the given split called a separation *gap* is used as a measurement.

$$gap(sp) = \frac{1}{n_1} \sum_{t_1 \in D_1} d(s, t_1) - \frac{1}{n_2} \sum_{t_2 \in D_2} d(s, t_2) \tag{4}$$

So the quality of a shapelet candidate is measured by the value of information gain and separation gap. The large information gain and separation gap value indicate high discriminatory power of a shapelet candidate.

4   Shapelets Discovery with Key Points

4.1   A Motivating Observation

Note that a shapelet is a subsequence that can discriminate time series from different classes. So if a subsequence of a time series with no variations in its values or gradients is considered as a shapelet, the corresponding subsequence in another time series from a different class must have some variations in values or gradients. Generally, these variations correspond to some actions or movements of observation, which can be regarded as features. Therefore, the subsequence with variations is a more reasonable choice as a shapelet than subsequence with no variations. The results of previous work about shapelets discovery demonstrate that all of extracted shapelets contain some variations. Figure 1 shows some examples from [3].

Fig. 1. (*left*) Coffee dataset and its shapelet; (*right*) Gun point dataset and its shapelet

Form Fig. 1, we find that all shapelets contain one or some of key points in time series. This character can be used to prune unnecessary candidates when searching shaplets.

4.2   Extracting Key Points

The key points of a time series are special points in a time series curve, so we can use mathematical methods to find them. Due to the small changes resulting from the process of data collection, a smooth step is necessary to eliminate these meaningless changes before we extract the key points from a time series. We use a rectangle sliding window along with a time series to do this preprocessing. Figure 2 gives a simple sketch of this method.

The width of rectangle sliding window is preset according to the specific situation. A large width means a rough sketch which only find the obvious changes while a small width means that the tiny changes will be found. For a given width, we scan the points along with time series. If the point is not in the rectangle sliding window, we pause the scan process and fit a straight line using the points

Fig. 2. Time series smoothing. (*left*) Rectangle sliding window; (*right*) The smoothed time series

---

Algorithm 1. Extract_KeyPoints

Input: T: a time series; th: gradient threshold
Output: *keypoints*:the key points in a time series
1: $keypoints = []$
2: $T = TS\_smooth(T)$
3: **for** each *point* in T **do**
4:    $gradient\_1 = Calculate\_Gradient(t)$
5:    **for** each *value* in $gradient\_1$ **do**
6:      **if** $value > th$ **then**
7:        $keypoints \leftarrow$ the *index* of *value*
8:      **end if**
9:    **end for**
10:    $g\_sign\_1 = sign(gradient\_1)$
11:    $gradient\_2 = Calculate\_Gradient(g\_sign\_1)$
12:    $g\_sign\_2 = sign(gradient\_2)$
13:    $index = find(abs(g\_sign\_2)==2)+1$
14:    $keypoints \leftarrow index$
15: **end for**

---

in the sliding window. Then, another sliding window is used to do the same thing until all points of the time series is smoothed. After finishing the smoothing step, we can extract the key points from a time series.

The Algorithm 1 shows the process of extracting key points. Line 2 executes the smooth process that is a key part for finding the key points. Line 4 keeps the variation information of a smoothed time series. The large variation points are stored in lines 6 and 7. The local minimum and maximum points are found from lines 10 to 14. Figure 3 shows the key points of a time series.

Fig. 3. Key points of a time series

Fig. 4. Key points after removing unnecessary points

As we can see, there are too many key points after finishing the previous step and some key points are close enough. It is unnecessary to keep all points because the points that is close to each other will generate overlapped subsequences. So we search each point's neighborhood and remove the other points. By an appropriate neighborhood threshold, we can get the reasonable key points. The result of the above example is expressed in Fig. 4 with a threshold of 10 points.

4.3   Finding the Best Candidates in Key Points Neighborhood

Generating Subsequences in Key Points Neighborhood. A shapelet can be of any length, so all the subsequences that contain a key point should be generated and measured by information gain. In order to avoid subsequences with overlapped parts, we check all key points in the current key point's neighborhood. Suppose that the current key point is p, the length of subsequences to be generated is len, there are two cases when we get shapelets candidates (Fig. 5).

Case 1: If there are no other key points in the area of $[p-len, p]$, the candidates' starting points are from point $p - len$ to p and we can get len candidates.

Case 2: If there are some key points in the area of $[p - len, p]$, the candidates' starting points are from point k to p and we can get $p - k$ candidates. Note that k is the closest point to p.

Using SAX Method to Evaluate the Candidates. If we first calculate the distance of each candidate and all time series in a dataset and then find a split for computing the information gain value, it is very time consuming. *Rakthanmanon et al.* [11] give us a useful way to avoid unnecessary distance computation. Here, we use this method to evaluate the shapelet candidates generated by key points.

We first transform a subsequence to a SAX word. Here, the SAX process will not be discussed in details. Everyone who wants to know the process can refer to [14]. For all SAX words, a random projections method [11] is used to find the $topK$ candidates with the most discriminating ability. The process is shown below.

In line 1, we define a matrix to record the masked SAX word occured in time series. A random projection method is used to mask the SAX words in line 4. We do r times hashing for a SAX word and get the final matrix. In line 10, we

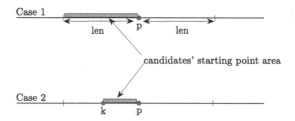

Fig. 5. A visualized representation of candidates' starting point area

---

Algorithm 2. RandomProjections_and_CalculateBestSAX

Input: *SAXlist*: SAX representation of subsequences
    k: the number of symbols when masking
    r: the number of hashing
Output: *topK_candidates*: the *topK* discriminating candidates
1: *discriminate_matrix* = [ ]
2: **for** each *saxword* in *SAXlist* **do**
3:   **for** *i*=1 to *r* **do**
4:     *sub_saxword* ← *RandomProjections*(*saxword*,*k*)
5:     **for** all *saxword* contain *sub_saxword* **do**
6:       *discriminate_matrix*[*saxword*][*ts_label*]++
7:     **end for**
8:   **end for**
9: **end for**
10: *topK_candidates* = *CalculateBestSAX*(*discriminate_matrix*)

---

use the function *CalculateBestSAX* to get the *topK* candidates by counting the occurrence number of a SAX word when we do hashing step in each class. The detailed explanation of *CalculateBestSAX* can be found in [11].

4.4 Shapelets Discovery Algorithm with Key Points

Our shapelet discovery algorithm with key points is presented in Algorithm 3.

The process contains two phase for the subsequence of length *len*. In the first phase (lines 4–15), we first extract all key points of a time series *ts* in line 5 and then keep the necessary points in line 6. In lines 7–9, all the shapelet candidates are generated based on key points. The discrete representation of candidates is gained in lines 11–14. When the first phase finished, the candidates of length *len* from all time series in *D* and their discrete representation are generated. In the second phase (lines 16–21), we used the random projection method to evaluate the quality of a candidates SAX representation. This method is proposed in [11,12], which describe it in detail. We get *topK* candidates by this way in line 16. For the *topK* candidates, we still use the information gain to select the best candidate in lines 18–21. The reason is that the random masking method can effectively reduce the impossible candidates, but it is imprecise to find the best shapelets. After doing all admissible candidate length from *minlen* to *maxlen*, we can get the shapelets for the current dataset *D*.

5 Experimental Evaluation

We will demonstrate two points with our experiments. One is that our shapelets discovery techniques is useful to find shapelets for time series classification. The other is to show that our algorithm is faster than the other algorithms when discovering shapelets. We use the datasets from the UCR Time Series archives [13] to do our experiments. The UCR Time Series archives contains 85 datasets

Algorithm 3. Shapelet_with_Keypoints

Input: D: time series Dataset; $maxlen,minlen$: max and min length of shapelet
Output: sh: shapelet
1: $sh \leftarrow [\ ]$
2: $best\_gain = 0$
3: **for** len from $minlen$ to $maxlen$ **do**
4:     **for** each ts in D **do**
5:         $keypoints = Exact\_KeyPoints(ts)$
6:         $keypoints = Remove\_Closed\_Point(keypoints)$
7:         **for** each p in $keypoints$ **do**
8:             $subseries = Generate\_Subseries(ts,p,len)$
9:         **end for**
10:         $saxlist \leftarrow [\ ]$
11:         **for** each s in $subseries$ **do**
12:             $sax = Generate\_SAX(s,len)$
13:             $saxlist \leftarrow sax$
14:         **end for**
15:     **end for**
16:     $topK\_candidates = RandomProjections\_and\_CalculateBestSAX(saxlist)$
17:     $t\_series = Remap\_to\_ts(topK\_candidates)$
18:     $[gain,shapelet] = Calculate\_Gain(t\_series)$
19:     **if** $gain > best\_gain$ **then**
20:         $best\_gain = gain$ & $sh \leftarrow shapelet$
21:     **end if**
22: **end for**

---

and provides diverse characteristics with various lengths and number of the classes, numbers of time series instances. We also use a decision tree classifier to calculate the classification accuracy by the selected shapelets as the other works do. In the experiments, the width of rectangle sliding window is set to $(max(T) - min(T))/50$, the cardinality of SAX is 4 and the size of key points' neighbourhood is 10.

5.1 Effectiveness of Key Points

Compared with Logical Shapelets. To validate the effectiveness of shapelets generated by key points, we reproduce the paper's [10] algorithm called Logical Shapelets (LS) which is known as the best algorithm for finding time series shapelets and modify it with key points. Because the Logical Shapelets algorithm needs a lot of time to get the results in large datasets (over one day), we compare the classification accuracy between two algorithms using 33 small datasets from the UCR Time Series archives. Figure 6 shows the results.

The points in Fig. 6 are classification accuracy ratios of two algorithms on the same datasets. The horizontal axis is the accuracy by using shapelets from key words and the vertical axis is Logical Shapelets' results. The line in the figure stands for that two algorithms have the same classification accuracy on a

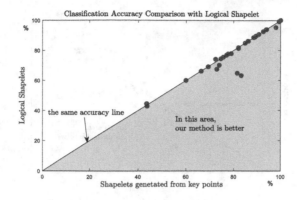

Fig. 6. Classification accuracy of LS and our modified LS by key points

dataset. The lower area under the line states that our algorithm is better. From the results, we can figure out that the shapelets selected by our algorithm is effective to build a classifier. In most datasets, we get the same classification accuracy with Logical Shapelets algorithm. In several datasets, we get better results. Note that this experiment is used to measure the quality of the selected shapelets with key points, it shows that our algorithm does not miss the eligible shapelets although we overlook a lot of candidates when generating the shapelet candidates. So our time series shapelets discovery algorithm with key points is feasible.

Compared with Fast Shapelets. Fast Shapelets (FS) [11] use a symbolic technique to fast prune unnecessary candidates which can speed up the shapelets discovery process and its accuracy is not perceptibly different with Logical Shapelets. To verify the effectiveness of our algorithm in large datasets, we compare our algorithm with Fast Shapelets with the same parameters in all datasets of UCR Time Series archives. The parameters including the number of random projections and the size of the set of potential candidates are referred to [11]. The classification results are shown in Fig. 7. As well as the above experiments, we also take the classification accuracy as axes.

We draw two dot lines on the same accuracy line side. The points between the two lines mean that classification results of two algorithms have little difference. We can find that two algorithms have not quite different difference in most of datasets. In almost half datasets (41 of 85), we get better results than FS algorithm. This experiments demonstrate that our shapelets discovery method with key points is effective even if we use the discrete representation of time series to find shapelets.

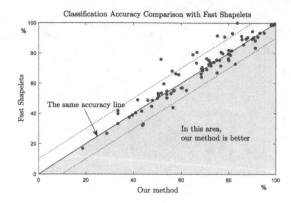

Fig. 7. Classification accuracy of our method and Fast Shapelets on 85 benchmark datasets

5.2   Running-Time Comparison

For all shapelets discovery algorithm, the main factors of finding shapelets are the number of time series in the dataset and the length of time series. To test the time efficiency of our method, we do the experiments on a large time series dataset in UCR time series archives, the $NonInvasiveFetalECGThorax1$ dataset and compare our results with FS algorithm. It contains 42 classes, 1800 training time series of length 750 and 1965 test time series. In this experiments, the parameters are same for two algorithm: the ratio of random projections is set to 0.25 for a SAX word and the size of the set of potential candidates is set to 10, *i.e. topK* in Algorithms 2 and 3 is 10. Due to the long run-time (over 10 h) of FS algorithm when searching all candidates in this dataset, we take 10 points interval when two adjacent candidates are generated as the FS algorithm do. This way don't reduce the performance in finding shapelets [11].

Figure 8 shows the running time changes when the number of time series is varied from 100 to 1800 with a fixed length of 750 for all time series. From the figure, we can find that the running time of FS algorithm increases from 203 s to over 4600 s while our algorithm only need 40 s when the number of time series is 100 and less than 1056 s when the number is 1800. So the speedup factor is 4X-5X in this dataset.

Figure 9 shows that time consumption of our algorithm and the FS algorithm when the length of time series increasing. In our experiments, we fix the number of time series at 1800 and change their length from 50 to 750. The running time of our algorithm is varied form 14 s to 1033 s while the FS algorithm need 26 s at the beginning and over 4600 s when the length is 750.

These results are not surprising because our algorithm decreases the number of shapelet candidates with key points. For all dataset of UCR Time Series archives, the average speedup factor is 3.5 and largest factor is 9.8. The time complexity of our algorithm is the same as the FS algorithm. But given that our

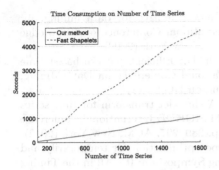

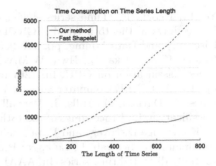

Fig. 8. Time consumption with number of time series

Fig. 9. Time consumption with the time series length

strategy is only to generate necessary shapelet candidates not all candidates, our algorithm is faster for finding shapelets.

6 Conclusions

For most time series, there are many subsequences with no or little changes. When discriminating two time series, most algorithm tend to use the varying subsequences than immutable subsequences. In this paper, we analyze the essential properties of time series shapelets and propose a preprocessing step to speed up the process of shapelets discovery. By using the key points, we decrease the number of shapelet candidates before measuring their quality as a shapelet. We have demonstrated that the subsequences generated by the key points do not miss eligible shapelets in the time series classification experiments compared with the current algorithm. Moreover, our algorithm for finding shapelets is significantly faster in almost all time series dataset of UCR archives than the current algorithm.

Acknowledgments. This work is partially supported by National 863 Program of China under Grant No. 2015AA015401, Tianjin Municipal Science and Technology Commission under Grant No. 14JCQNJC00200, 13ZCZDGX01098, as well as Research Foundation of Ministry of Education and China Mobile Under Grant No. MCM20150507. This work is also partially supported by Jilin NSF Under Grant No. 20130101179JC-18 and YBU development plan 2014–16.

References

1. Ding, H., Trajcevski, G., Scheuermann, P., et al.: Querying and mining of time series data: experimental comparison of representations and distance measures. Proc. VLDB Endowment **1**(2), 1542–1552 (2008)
2. Xi, X., Keogh, E., Shelton, C., et al.: Fast time series classification using numerosity reduction. In: Proceedings of the 23rd International Conference on Machine Learning, pp. 1033–1040. ACM, New York (2006)

3. Ye, L., Keogh, E.: Time series shapelets: a new primitive for data mining. In: Proceedings of the 15th ACM SIGKDD International Conference on Knowledge Discovery and Data Mining, pp. 947–956. ACM, New York (2009)
4. Chang, K.W., Deka, B., Hwu, W.M.W., Roth, D.: Efficient pattern-based time series classification on GPU. In: 12th International Conference on Data Mining, pp. 131–140. IEEE Computer Society, Washington DC (2012)
5. Lines, J., Davis, L.M., Hills, J., Bagnall, A.: A shapelet transform for time series classification. In: Proceedings of the 18th ACM SIGKDD International Conference on Knowledge Discovery and Data Mining, pp. 289–297. ACM, New York (2012)
6. Hartmann, B., Schwab, I., Link, N.: Prototype optimization for temporarily and spatially distorted time series. In: AAAI Spring Symposium: It's All in the Timing (2010)
7. Reiss, A., Weber, M., Stricker, D.: Exploring and extending the boundaries of physical activity recognition. In: 2011 IEEE International Conference on Systems, Man and Cybernetics (SMC), pp. 46–50. IEEE Press, New York (2011)
8. Liu, J., Zhong, L., Wickramasuriya, J., Vasudevan, V.: uWave: accelerometer-based personalized gesture recognition and its applications. Pervasive Mob. Comput. **5**(6), 657–675 (2009)
9. Gordon, D., Hendler, D., Rokach, L.: Fast randomized model generation for shapelet-based time series classification. arXiv preprint arXiv:1209.5038 (2012)
10. Mueen, A., Keogh, E., Young, N.: Logical-shapelets: an expressive primitive for time series classification. In: Proceedings of the 17th ACM SIGKDD International Conference on Knowledge Discovery and Data Mining, pp. 1154–1162. ACM, New York (2011)
11. Rakthanmanon, T., Keogh, E.: Fast shapelets: a scalable algorithm for discovering time series shapelets. In: Proceedings of the Thirteenth SIAM Conference on Data Mining (SDM), pp. 668–676. SIAM (2013)
12. Ulanova, L., Begum, N., Keogh, E.: Scalable clustering of time series with u-shapelets. In: SIAM International Conference on Data Mining (SDM 2015) (2015)
13. Chen, Y.P., Eamonn, E., Hu, B., Begum, N., Bagnall, A., Mueen, A. Batista G.: The UCR time series classification archive (2015). http://www.cs.ucr.edu/~eamonn/time_series_data/
14. Lin, J., Keogh, E., Wei, L., Lonardi, S.: Experiencing SAX: a novel symbolic representation of time series. Data Min. Knowl. Disc. **15**(2), 107–144 (2007)

An Adaptive Partition-Based Caching Approach for Efficient Range Queries on Key-Value Data

Wei Ge, Min Chen, Chunfeng Yuan, and Yihua Huang[✉]

State Key Laboratory for Novel Software Technology,
Nanjing University Collaborative Innovation Center for Novel
Software Technology and Industry of Jiangsu Province,
Nanjing University, Nanjing, Jiangsu, China
gloria.w.ge@qq.com, yhuang@nju.edu.cn

Abstract. Range queries are real demands in big data scenarios, such as analytic and time-traveling queries over web archives. Here we design AdaSI, an adaptive partition-based caching approach for efficient range queries on key-value data. AdaSI partitions data into a number of data slices (consecutive data items). Then the AdaSI Hotscore Algorithm is designed to maximize the cache-hit probability under the limitation of cache space. By measuring *Dutyrate* and *Hotscore* of data slice, the partitioning precision and adjustment sensitivity are pursued by finer partitioning on hot data, whereas the cold data are partitioned with relatively larger granularity to reduce storage overhead and search cost of queries. Our results show that the AdaSI Hotscore Algorithm could obtain a cache hit rate nearly as high as the record-based cache policies, as well as a significant speedup and space reduction, far outperforming record-based policies.

Keywords: Query optimization · Caching policy · Adaptive partitioning

1 Introduction

The increasing speed of big data's volume is greatly accelerated compared to only just a few years back, and expected to increase further. The elastic processing and vast storage requirements has facilitated novel data models to handle massive data of cloud systems. Several NoSQL data models have been devised in today's large-scale, data-intensive applications, such as column-oriented, key-value, document-oriented, and graph-oriented models. Key-value data model is adopted pervasively. It's flexible without strict schema so as to provide high availability and scalability, and support high concurrent throughput that exceeds relational data model in a large degree. Apache's (ex-Facebook) Cassandra, Yahoo's PNUTS and Amazon's Dynamo are all key-value data store that provide foundations for their own web applications successfully and continuously support their expanding business data scales.

© Springer International Publishing Switzerland 2016
F. Li et al. (Eds.): APWeb 2016, Part II, LNCS 9932, pp. 343–354, 2016.
DOI: 10.1007/978-3-319-45817-5_27

Range query scenarios on key-value data stores are typically described as analytic and time-traveling queries over web archives. We focus on supporting efficient range queries by a novel partition-based caching policy. Cache technique is employed popularly in data-intensive workloads because data access is commonly skewed in different fashions. Thus, some research works define data having temperature [1,2]: some data items are "hot" and accessed frequently, while others are "cold" and accessed infrequently. Data accesses are inclined toward certain hotspots and gradually cool down in web applications.

In this paper, we propose an indexing mechanism with adaptive data partition-based caching policy called Adaptive Slice Index (AdaSI). Data set is cut into numerous slices initially. Then *Dutyrate* and *Hotscore* are defined to evaluate the hot degree of data slice, by which data are divided into cold and hot data. Under the limitation of cache space, the most valuable slices for queries are cached according to a fluctuating hot threshold. Furthermore, the adaptive partitioning of AdaSI are introduced to adjust the border and span of slices. To achieve the highest cache-hit probability, the adaptive tuning partitions the hot data finer to pursue the fitting precision and the sensitivity to changing hotspots, while the cold data are partitioned with relatively larger granularity to reduce overhead and search cost of queries.

Our contributions are listed below:

(1) We propose an idea for accelerating range queries on key-value data by expanding the caching granularity as data slices.
(2) We devise a novel and precise measurement to perform our partition-based cache replacement policy.
(3) We propose an adaptive adjustment approach to tune the data partitioning for continuous queries.

The rest of this paper is organized as follows: Sect. 2 provides the related work from three different aspects. Section 3 describes our new indexing mechanism AdaSI, a partition-based skiplist indexing structure. Section 4 presents the AdaSI Hotscore Algorithm and its adaptive partitioning scheme. Section 5 shows the experimental evaluation of AdaSI. In Sect. 6 we conclude the paper and discuss future work.

2   Related Work

A lot of recent research efforts have concentrated on big data indexing, partitioning and caching. We have summarized the works related to our research in three aspects.

2.1   Range Query Optimization on Big Data

Many research groups have concentrated on improving range query performance from various perspectives. Some of the methods drew inspiration from previous relational data model, such as SegmentTree for range query and compressing coding of data warehouse.

[3] proposed a compound interval index comprised of MRSegmentTree (a key-value representation of the Segment Tree which was proposed by [4]) and Endpoints Index (a column family index that stores information for interval end-points) to support the indexing of intervals with logarithmic complexity query-ing. Similarly, [5] supported range queries due to its tree topology. In summary, the search tree is inherently suitable for range queries, but the space overhead of index structures and the time costs on balance adjustment is excessive, especially in big data backgrounds. [6] built the space-frugal index BIDS by coding indexes as bitmaps and compressing the coding using WAH and bit-sliced coding, so that multiple attributes could be indexed to support range and join queries effectively. It is a non-trivial cost for large-scale dynamic data scenario. The update costs of new data arrivals could not be ignored either.

2.2 Data Partitioning on Big Data

In view of the huge volume of big data, the primary challenge of data manage-ment is not the storage, but how to organize and retrieve data efficiently. Several research efforts to organize data as blocks to reduce computation and storage cost. In [7], data were partitioned by a sub-range form of the whole range of the key attribute, and called "full window". Then, "link window" was proposed as a virtual window that had some members belonged to the right window and the other members belonged to the left window for measuring the heat of window as even as possible. In [8], the data column was partitioned into domains and a separate dictionary was defined for each domain. The partitioned domains were called "cells", which was a form of data partitioning. The partitioning mechanism exploited skews to encode more frequent values in fewer bits. By partitioning and encoding, better overall compression was achieved.

2.3 Hot Data Detection and Caching on Big Data

Recently, research work has focused on the data skew characteristics of big data retrieval. A focus on micro-blogging searches could be found in [9,10] as they are time-related skewed query requirements. TI (Tweet Index) [9] wanted to index the data as search results with high probability and delay indexing other data. LSII (Log-Structured Inverted Indices) [10] built a multi-tiered, increasing-sized index. New arrival data were first inserted into the smallest index. They were then moved into the larger index by batches. TI and LSII have considered the time-related, skewed query requirements, such as micro-blogging searches. [2] focused the skewed access patterns obeying the Zipfian distribution. They tried to provide a solution in memory database by identifying hot and cold data using a well-designed parallel backward algorithm. TBF [11] proposed a memory-efficient replacement policy for flash-based cache. The core of TBF was to reduce per-key in-memory index overhead.

3   The AdaSI Design

Skiplist [12] is a data structure that can be used in place of balanced trees by probabilistic balancing rather than strictly enforced balancing. As a result the algorithms for insertion and deletion in a skiplist are much simpler and significantly faster than equivalent algorithms for balanced trees. It is a more natural representation than trees. Skiplist has several remarkable characteristics suitable for big data indexing. First, it is easier to implement. The levels of skiplist nodes were chosen randomly while insert and delete need never change. It is unnecessary to rearrange the node to maintain certain balance conditions that is required in balanced trees. Second, it is space-efficient. Skiplist requires an average of 4/3 pointers per element. The expect search cost is proved to be $\mathcal{O}(\log n)$. Third, the performance of skiplist is stable. Two successive searches for the same element will require the same time cost. Last but most importantly, skiplist has good concurrent properties for insertions or deletions because they require only local modifications. Thus, skiplist is well suited for big data storage systems.

The search of skiplist is high-performance. For example, when we want to locate the key 71 in skiplist, the search path starts from head and moves on the top level. It traverse forward pointers that do not overshoot the aimed key 71. When it has reached the node 21 and finds that the node 85 is overshot, no more progress can be made at the third level of forward pointers. The search moves down to the second level and traverses forward to the node 37. Again, the node 85 is overshot. Further moving down to the bottom level, the search process reaches the node 71 finally. The search is terminated.

On the basis of skiplist, AdaSI organizes data as shown in Fig. 1. Data items with consecutive keys are bundled as data slices. The start key of each data slice are stored in skiplist node, as well as *Itemnum* (the number of data items in data slice), *isCached* (the data slice is cached or not), *Hitcount* (the number of times the data slice was accessed in queries), and *Hotscore* (the evaluation of data slice's access frequency). They are called the **cache metadata** of AdaSI.

The search of AdaSI expands the search of native skiplist. For a range query, for example, when we want to retrieve the key range 41–90, the start key 41 should be located at first. the search goes the path of the node 21, the node 37 and finds the node 71 overshot. Considering range query, the search does not end up. It locates the current data slice $S_5 = \{37, 38, \cdots, 70\}$ as the destination node because 41 is encompassed here. The qualified data items (i.e. the data items those keys between 41 to 70) are loaded into the result set. Then the successive nodes are traversed one by one to cover the range of query. Eventually all data items of the node 71 (i.e. the data items those keys between 71 to 84) and the qualified data items of the node 85 (i.e. the date items those keys between 85 to 90) are loaded into the result set. Then, in AdaSI, the search process moves on to check if the corresponding data slices are cached or not. If the slice isn't cached, it's the reference of data entities that loaded into the result set. Then the disk operation is necessary.

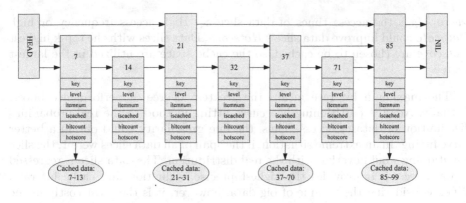

Fig. 1. An example of AdaSI

4   The AdaSI Hotscore Algorithm

Considering the data-skewed access patterns, identifying the hot data and giving them priority in the allocation of computing resources is a common concept. In the AdaSI Hotscore Algorithm, all the data items in one slice share one copy of cache metadata. A visit of data slice probably accesses a portion of data items, so each visit would contribute a 0–1 accumulation to access frequency. Then, it is a ranking problem of access frequency. This is the original idea of the AdaSI Hotscore Algorithm.

4.1   Evaluation of Hotscore

Suppose V is the ordered set of all key-value pairs. We define the hot set as $S^H = \bigcup_{i \in [1,I]} S_i^H$, and the cold set as $S^C = \bigcup_{j \in [1,J]} S_j^C$. Hot set and cold set are a division of V, so there is $\{S_i^H\} \bigcup \{S_j^C\} = V$.

Definition 1. Suppose the probability of the data item v_n in the data slice S_i accessed by a range query is χ_n, and $|S_i|$ is data item number of data slice S_i. We define $Dutyrate(\mathcal{D}r)$ of S_i as the total access probability (also could be seen as the access percentage) of all data items of S_i in a range query. Thus, there is:

$$\mathcal{D}r_i = \sum_{v_n \in S_i} \chi_n / |S_i|$$

$Dutyrate$ of a data slice varies 0–1. When data slices are hit in a range query, we hope they are hit with a large proportion, that is, with high $Dutyrate$.

Definition 2. Define $Hotscore(\mathcal{H})$ of a data slice as accumulated $Dutyrate$ of the data slice in all access occasions. It can be expressed as:

$$\mathcal{H}_i = \sum_{m \in [1,M_i]} \frac{1}{|S_i|} \sum_{v_n \in S_i} \chi_n = \sum_{m \in [1,M_i]} \mathcal{D}r\,(S_i)$$

where M_i is the access times of data slice S_i. High access frequency or high *Dutyrate* could improve data slice's *Hotscore*. Data slices with the top-k highest *Hotscore* are chosen to be cached, so the cache space are utilized in the largest degree.

The maximum hit rate means the greatest degree of performance boost. Actually, AdaSI's partitioning is a curve fitting method to the real probability distribution of dataset access. Thus, the fine granularity could obtain a better curve fitting. In an extreme situation, if the span of all data slices were 1, the slice partitioning would overlap with the real distribution. The data slice is regressed to data record. It provides the highest precise evaluation and cache hit rate. When considering the volume of big data, however, it is the most costly in the space overhead of cache metadata. Thus, the optimization goal of the AdaSI Hotscore Algorithm is expressed as the highest cache-hit probability criterion under limited cache space.

$$\max_{\{iBeg,iEnd\}} \sum_{\text{all } n} \chi_n P\left(v_n \in \bigcup_i S_i^H\right), \ i = 1, 2, \cdots, I$$

$$\text{s.t.} \ \sum_i \text{size}(S_i^H) \leq \text{CACHE}$$

$$\max_j \left\{\sum_m \mathcal{H}\left(S_j^C\right)\right\} \leq \min_i \left\{\sum_m \mathcal{H}\left(S_i^H\right)\right\}, \ j = 1, \cdots, J$$

$$|S_j^C| \leq \text{MaxDataSliceSize}$$

It is hardly possible that the initial partitioning of the data set is coordinated with query distribution. Thus, in our AdaSI Hotscore Algorithm, splitting and merging are adopted to adjust the border and span of slices according to their hot degree and skew degree.

4.2 Splitting and Merging of Data Slice

If some parts of data items are often visited and others not, the data slice is regarded as access-skewed. In another word, data items in a slice seldom caught in the same query. Thus, the data slice should be split. When the average of *Dutyrate* is low enough (less than the configured split threshold) and the size of data slice is larger than *MinSliceSize*, the data slice should be split into several parts by *SplitPartsNum*. The size threshold guarantees the splitting adjustment would not generate a large number of data slice fragments, especially cold fragments, avoiding much unnecessary space and time consumption in cache metadata updating and searching.

For a certain random access distribution, such as Zipfian, the hot spots are sparse in big data scenarios. Therefore, the two adjacent data slices with close access frequencies should be merged because most probably, they are accessed simultaneously. It is expected that two adjacent data slices are merged when they

---

Algorithm 1. DataSliceSkipList Split(DataSlice ds, int $splitnum$)

1: locate ds;
2: **if** $ds.m\_itemnum < MinSliceSize \times SplitPartsNum$ **then**
3:     return;
4: **end if**
5: **if** $ds.m\_hotscore/ds.m\_hitcount < SplitDutyRate$ **then**
6:     remove ds from skiplist;
7:     $splitlen = ds.m_itemnum/SplitPartsNum$;
8:     **for all** $dsPart$ **do**
9:         $key = ds.key + i \times splitlen$;
10:         $len = splitlen$ or remainder length for the last $dsPart$;
11:         $dsPart.m\_hotscore = ds.m\_hotscore$;
12:         $dsPart.m\_hitcount = (ds.m\_hitcount + ds.hotscore)/2$;
13:         insert $dsPart(key, len)$ into skiplist;
14:         insert $dsPart$ into hotscore-ordered queue;
15:     **end for**
16: **end if**

---

are accessed in nearly even frequency, otherwise a splitting would be triggered soon afterwards.

The pseudo of splitting and merging algorithm is shown in Algorithms 1 and 2 respectively.

Algorithm 2. DataSliceSkipList Merge(DataSlice ds)

1: **if** $next\_ds \neq NULL$
    $\&\& \ ds.m\_status == next\_ds.m\_status$
    $\&\& \ ds.key + ds.m\_itemnum == next\_ds.key$
    $\&\& \ ds.m\_itemnum + next\_ds.m\_itemnum < MaxSliceSize$
    $\&\& \ ds.m\_hotscore/ds.m\_hitcount > MergeDutyRate$
    $\&\& \ next\_ds.m\_hotscore/next\_ds.m\_hitcount > MergeDutyRate$
    $\&\& \ abs(ds.m\_hitcount - next\_ds.m\_hitcount) < DiffThreshold$
    then
2:     $ds.m\_itemnum+ = next\_ds.m\_itemnum$;
3:     $avgDr1 = ds.m\_hotscore/ds.m\_hitcount$;
4:     $avgDr2 = next\_ds.m\_hotscore/next\_ds.m\_hitcount$;
5:     $ds.m\_hitcount = \max(ds.m\_hitcount, next\_ds.m\_hitcount)$;
6:     $ds.m\_hotscore = ds.m\_hitcount \times average(avgDr1, avgDr2)$;
7:     $remove(next\_ds)$;
8:     modify order of ds in hotscore-ordered queue;
9: **end if**

---

5 Experiments

In this section, we experimentally evaluate the performance of the AdaSI Hotscore Algorithm (HA) on cache hit rate and execution time respectively.

All experiments were implemented in Java 1.6.0 and run on clusters of 10 nodes (each with 4 Core Intel Xeon E-5620 2.4 GHz and 24G of RAM running Red Hat Enterprise Linux Server 6.0 and HBase 0.94.14). Range query requirements in our experiments consist of 1 million queries on 100 million data records. Each record costs 1 K bytes in average, and data size is 100 GB totally. We generate scrambled Zipfian distribution according to the YCSB Benchmark [13] on 100 million keys as the start points of range queries, where the Zipfian distribution is acknowledged as a common distribution criteria, for example, in [2,11,13]. The spans of range query sequences conform to uniform distribution by configured parameters, such as 100, 200 and 400.

5.1 Hotscore Distribution of HA

As analyzed above, our AdaSI Hotscore Algorithm is a curve fitting method. The data partitioning tries to fit in query distributions. Figure 2 illustrates the close degree of partitioning and distribution. In Fig. 2, 10,000-key distributions are illustrated for clearer demonstration, instead of 100 million key setting in the experiment. The red line represents hot slices and the blue line represents cold slices. The green line represents the query hotspots that are in accordance with random Zipfian distribution. It can be seen that the partitioning of hot slice is finer, and cold slice is coarser. They are both close to the real distribution shown by the green line.

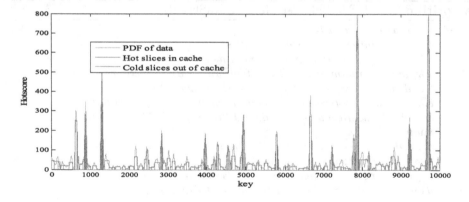

Fig. 2. Partitioning and distribution of 10,000 keys (Color figure online)

5.2 Query Performance Comparisons

We compare the cache hit rate and execution time of HA to LFU (Least Frequency Used), LRU (Least Recently Used)and FIFO (First In First Out) cache replacement policies. HA has three settings: (1) HA with query-cutting initialization, adaptive partitioning (HA-QA), (2) HA with fixed-size initialization,

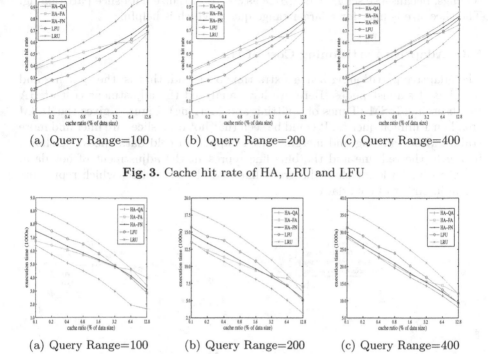

(a) Query Range=100   (b) Query Range=200   (c) Query Range=400

Fig. 3. Cache hit rate of HA, LRU and LFU

(a) Query Range=100   (b) Query Range=200   (c) Query Range=400

Fig. 4. Execution time of HA, LRU and LFU

adaptive partitioning (HA-FA), and (3) HA with fixed-size initialization, non-adaptive partitioning (HA-FN). FIFO is not demonstrated in Figs. 3, 4 because the cache hit rate of LRU and FIFO are pretty close, so the curve would be nearly overlaid. The cache hit rate in Fig. 3 is the average cache hit rate of 1 million queries in a different cache ratio and different query range configurations.

Figures 3, 4 depicts the performance comparisons in three cases of range = 100, 200 and 400. This is evident in our results that the higher cache hit rate contributes to the optimization of query performance since more queries are resolved in memory other than disk access. LRU is the simplest one in our settings and obtains the worst performance. LFU improves on this figure although not by much. HA and LFU are better than LRU because access frequency is considered. HA goes a step further by adjusting the partitioning under the guidance of *Hotscore*, the precise accumulation of access frequencies. HA is advantageous over the current approaches especially in the case of a small cache ratio. It is clear that any algorithm is efficient with abundant cache space. However, considering big data with huge volume, the cache ratio is usually limited by cache space. Thus, HA is better fit for a cloud storing system, since it could obtain a significant improvement with limited cache space. It's depicted that the cache hit rate and execution time of HA-QA and HA-FA are closer in larger range settings, because we set *InitSliceSize* as 300 to initialize data slice partitioning. The slice size is finer for a larger range query, which is helpful.

5.3   Adaptive Partitioning Cost of HA

The adaptive partitioning brings extra time overhead, that is, the splitting and merging of skiplist nodes. Here, we have analyzed the adjustment cost of HA by observing ToSM (Times of Splitting and Merging). Figure 5 gives the ToSM trend on 1 million queries. It could be seen that hot data slices are finer and more stable, while splitting and merging often happen in cold data slices. As shown in Fig. 5, the red line and the blue line represent the adjustment of hot data. They are much lower than the green line and the black line, which represents the adjustment of cold data.

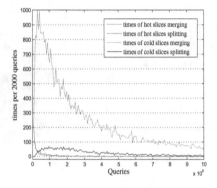

Fig. 5. Splitting and merging time during 1,000,000 queries (Color figure online)

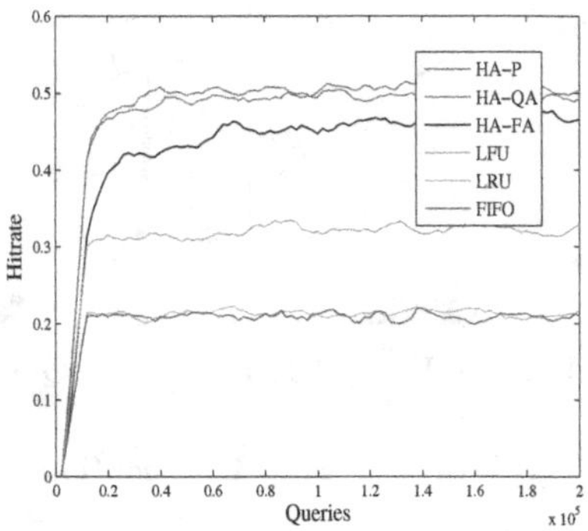

Fig. 6. The ascending cache hit rate of different cache replacement policies (Color figure online)

In extreme cases, if slice size is set to 1, the data slice will regress as data record. Then each data record needs a metadata structure, which we call the record-based cache of the AdaSI Hotscore Algorithm (HA-P). It is used to simulate the performance without data slice partitioning, just like other traditional cache replacement policies. In the case of range = 200, cache ratio = 0.4%, a instance of HA-QA's cold slice number is 369,802, hot slice number is 11,098, so total slice number is 380,900. Compared to record-based cache policy, that has 100 million data items, the space cost is reduced by $100,000,000/380,900 \approx 262$ times. The cache hit rate, however, is just 0.60% lower (the cache hit rate of HA-QA is 0.50293, the cache hit rate of HA-P is 0.505971).

Figure 6 shows the ascending cache hit rate in the case of range = 200, cache radio = 0.4%. It is shown that the cache hit rate of HA-QA (the red line) is nearly approached to HA-P (the pink line). The ascending grade represents the convergence speed of the cache hit rate during continuous queries. The grade of

Table 1. Comparison of query execution time in range = 200 and cache ratio = 0.4 %

Algorithms	200,000 Queries (seconds)			1,000,000 Queries (seconds)		
	DiskTime	MemoryTime	TotalTime	DiskTime	MemoryTime	TotalTime
HA-QA	2,101	42	2,143	10,370	215	10,585
HA-FA	2,244	23	2,267	10,687	86	10,773
HA-FN	2,411	18	2,429	11,897	52	11,949
LFU	2,751	7	2,758	15,919	34	15,953
LRU	3,174	7	3,181	13,343	34	13,377
FIFO	3,191	6	3,197	15,848	35	15,883
HA-P	939	263,849	264,788	/	/	/

HA-QA is close to HA-P and is much better than FIFO (the blue line), LRU (the yellow line) and LFU (the green line).

In the case of range = 200, cache radio = 0.4 %, we also observed the query execution time and distinguished memory and disk access time respectively, as demonstrated in Table 1. Although the ascending grade of cache hit rate on HA-QA and HA-P are very close, the difference between their execution time is very widely. We depicted the comparison in conditions of 200,000 queries and 1,000,000 queries because the time cost of HA-P is too high to execute HA-P in 1,000,000 queries. HA reduce query execution time by two orders of magnitude, and memory operation time by nearly four orders of magnitude. It is more precise of record-based cache policies, but it consumes tremendous search time and space cost. AdaSI meets its goal of providing high cache hit rate with low metadata overhead in memory by expanding the caching granularity as data slices, and the precision improvement by accumulating the access percentage with *Dutyrate* of each query, other than the access counts just like LFU. Furthermore, our adaptive partitioning strategy distinguished on hot and cold data slices is also beneficial for performance improvement.

6   Conclusion and Future Work

This paper takes a first step of range query optimization by adaptive data partitioning in key-value data storing system. Now, the AdaSI Hotscore Algorithm provides a practical adaptive partitioning method by a series of configured parameters. In the future, we will devotes to the study on a more intelligent, feedback-based adaptive partitioning method on AdaSI.

Acknowledgments. This work is funded by China NSF Grants (61223003, 61572250,61362006), Jiangsu Province Science & Technology Research Grant (BE2014131), Guangxi NSF (2014GXNSFBA118288) and Guangxi IBAYT Program (KY2016YB065).

References

1. Canim, M., Mihaila, G.A., Bhattacharjee, et al.: SSD bufferpool extensions for database systems. In: 36th International Conference on Very Large Data Bases, pp. 1435–1446. VLDB Endowment, Singapore (2010)
2. Levandoski, J.J., Larson, P., Stoica, R.: Identifying hot and cold data in main-memory databases. In: 29th IEEE International Conference on Data Engineering (ICDE), pp. 26C–37. IEEE Press, Brisbane (2013)
3. Sfakianakis, G., Patlakas, I., Ntarmos, N., Triantafillou, P.: Interval indexing and querying on key-value cloud stores. In: 29th IEEE International Conference on Data Engineering (ICDE), p. 805–816. IEEE Press, Brisbane (2013)
4. Bentley, J.L.: Solutions to Klee's rectangle problem, Technical report. Carnegie-Mellon University, Pittsburgh (1977)
5. Wu, S., Jiang, D., Ooi, B.C., Wu, K.L.: Efficient b-tree based indexing for cloud data processing. In: 36th International Conference on Very Large Data Bases, pp. 1207–1218. VLDB Endowment, Singapore (2010)
6. Lu, P., Wu, S., Shou, L., Tan, K.L.: An efficient and compact indexing scheme for large-scale data store. In: the 29th IEEE International Conference on Data Engineering, pp. 326–337. IEEE Press, Brisbane, Australia (2013)
7. Feelifl, H., Kitsuregawa, M.: The simulation evaluation of heat balancing strategies for btree index over parallel shared nothing machines. IEIC Technical report (Institute of Electronics, Information and Communication Engineers), vol. 99, pp. 7–12 (1999)
8. Lee, J.G., Attaluri, G.K., et al.: Joins on encoded and partitioned data. In: 40th International Conference on Very Large Data Bases, pp. 1355–1366. VLDB Endowment, Hangzhou, China (2014)
9. Chen, C., Li, F., Ooi, B.C., Wu, S.: TI: an efficient indexing mechanism for real-time search on tweets. In: Proceedings of the ACM SIGMOD International Conference on Management of Data, pp. 649–660. ACM, Athens, Greece (2011)
10. Wu, L., Lin, W., Xiao, X., Xu, Y.: LSII: an indexing structure for exact real-time search on microblogs. In: 29th IEEE International Conference on Data Engineering, pp. 482–493. IEEE Press, Brisbane, Australia (2013)
11. Ungureanu, C., Debnath, B., Rago, S., Aranya, A.: TBF: a memory-efficient replacement policy for flash-based caches. In: 29th IEEE International Conference on Data Engineering Brisbane (ICDE), pp. 1117–1128. IEEE Press, Brisbane (2013)
12. Pugh, W.: Skip lists: a probabilistic alternative to balanced trees. Commun. ACM **33**(6), 668–676 (1990)
13. Cooper, B.F., Silberstein, A., Tam, E., Ramakrishnan, R., Sears, R.: Benchmarking cloud serving systems with YCSB. In: 1st ACM Symposium on Cloud Computing, pp. 143–154, Santa Clara, CA (2010)

Handling Estimation Inaccuracy in Query Optimization

Chiraz Moumen$^{(\boxtimes)}$, Franck Morvan, and Abdelkader Hameurlain

IRIT Laboratory, Paul Sabatier University,
118 Route de Narbonne, 31062 Toulouse Cedex 9, France
{moumen,morvan,hameurlain}@irit.fr

Abstract. Cost-based Optimizers choose query execution plans using a cost model. The latter relies on the accuracy of estimated statistics. Unfortunately, compile-time estimates often differ significantly from run-time values, leading to a suboptimal plan choices. In this paper, we propose a compile-time strategy, wherein the optimization process is fully aware of the estimation inaccuracy. This is ensured by the use of intervals of estimates rather than single-point estimates of error-prone parameters. These intervals serve to identify plans that provide stable performance in several run-time conditions, so called robust. Our strategy relies on a probabilistic approach to decide which plan to choose to start the execution. Our experiments show that our proposal allows a considerable improvement of the ability of a query optimizer to produce a robust execution plan in case of large estimation errors.

Keywords: Query optimization · Robust plans · Estimation errors

1 Introduction

In database query processing, query optimization constitutes a critical stage due to its impact on query processing performance. During this stage, a query optimizer[1] generates several execution plans for a query, estimates the cost of each plan using a cost model and chooses the plan with the lowest estimated cost to execute the query [24]. This plan is called *best plan* [17,19,24]. Best plan selection requires accurate estimates of the costs of alternative plans. Unfortunately, these estimates are often significantly erroneous with respect to values encountered at run-time. Such errors, which may be in orders of magnitude [16], arise due to a variety of reasons [8]. One of the main reasons is the imprecision of statistics about data (e.g., sizes of temporary relations) as well as the use of outdated statistics. Indeed, values of parameters which are important for costing possible query plans may vary unpredictably between compile-time and run-time (e.g., available memory amount).

An adverse effect of estimation errors is to lead the optimizer to a sub-optimal plan choice, resulting in inflated response time. A considerable body of literature

---

[1] We only focus on the Cost-based Query Optimizers.

© Springer International Publishing Switzerland 2016
F. Li et al. (Eds.): APWeb 2016, Part II, LNCS 9932, pp. 355–367, 2016.
DOI: 10.1007/978-3-319-45817-5_28

was dedicated to find solutions to this problem. These solutions include mainly: (i) techniques for better quality of the statistical metadata [7,11,13,22,27,28], (ii) run-time techniques [5,17–20] to monitor a query execution and trigger re-optimization of the plan when a sub-optimality is detected, and (iii) compile-time strategies [1–3,9,12] that permit the optimizer to generate an execution plan, being aware of the imprecision of used estimates.

Accurate cost estimates remain, though, very difficult as this requires a detailed and a prior knowledge about data (e.g., sizes of temporary relations) and run-time characteristics such as resource availability (e.g., system load). Run-time techniques are able to adapt an execution plan to changes in run-time conditions with a risk of an important cost-increase resulting from many possible adaptations of the plan. Compile-time strategies proposed to date suppose that it is often possible to find a single execution plan whose performance remains stable regardless of changes in the run-time conditions compared to the compile-time expectations of the run-time conditions. This assumption is not always valid. In some situations, especially in case of large uncertainty about run-time characteristics, find such a plan becomes hard to achieve. In this paper, we focus on this issue. We propose a compile-time strategy for identifying plans, each of which provides stable performance over a range of possible run-time values of error-prone parameters. Throughout this paper, these plans are said *robust*.

Our optimization method uses an interval of estimates for each uncertain parameter, rather than specifying estimates for single points. Such an interval fully quantifies estimation inaccuracy and is then used to generate robust plans. Note that a plan providing stable performance over the whole interval is a single robust plan. Otherwise, the interval is divided into sub-intervals so as to find execution plans, each of which is associated with the sub-interval within which the plan is robust. Our method relies on a probabilistic approach to decide which plan to choose to start the execution. The major contribution of our optimization method is to maximize the ability of a query optimizer to produce a robust execution plan in the presence of large estimation errors.

The rest of the paper is organized as follows: Sect. 2 presents the problem formulation. Section 3 details our contribution. The experimental evaluation results are highlighted in Sect. 4. Related work is overviewed in Sect. 5. We conclude and outline our future work in Sect. 6.

2   Problem Formulation

In this section, we start with the preliminaries. Then, we provide our problem definition.

2.1   Preliminaries

Motivated by the difficulty of providing precise estimates needed for costing query plans as well as the complexity of cost models, researchers focused their

efforts on introducing new optimization algorithms. Their aim is to avoid significant performance regression caused by the use of erroneous estimates. Researches
on this purpose fall into two main approaches [25]. A first approach, called *Single
Point-based Optimization* [5,17,18,20] consists in monitoring a plan execution
so as to detect estimation errors and a resulting sub-optimalty. This latter is cor-
rected by interrupting the current execution and re-optimizing the remainder of
the plan using up-to-date statistics. At each invocation, the optimizer considers
the used estimates as though they were completely precise and accurate. It uses
specific estimate for each parameter. The generated plan is thus the *best plan*
for specific run-time conditions. The ability of methods relying on this approach
to accurately collect statistics is limited [3]. Consequently, when re-optimizing,
the optimizer may use new erroneous estimates. This may result in several plan
re-optimization and so performance regression.

While *Single Point-based Optimization* uses exclusively single points for estimates ignoring possible estimation errors, a conceptually different approach
called *Multi Point-based Optimization* was introduced. Besides to providing a
solution to estimation errors, this approach avoids performance regression due
to several re-optimizations of a plan. Contrary to the first approach which reacts
after an estimation error is detected, *Multi Point-based Optimization* aims to
predict plan sub-optimalities and anticipate the reaction to this. The methods
proposed as part of this approach consider the possibility of estimation errors
at the optimization phase. They produce plans that are likely to perform reasonably well in many run-time conditions. These methods [1–3,9,14] aim at
preventing rather than *correcting*. The key concept of these methods is the use
of intervals of estimates, which models the estimation inaccuracy. In the literature, there are different techniques for computing such intervals. For instance,
[6] uses strict upper and lower bounds, [3,17] model estimation uncertainty by
means of discrete buckets that depend on the way the estimate was derived, etc.
Once computed, the interval of estimates is used by the optimizer to generate an
execution plan that is likely to provide stable performance within this interval.

2.2   Problem Definition

Existing methods proposed as part of the *Multi Point-based Optimization* approach suppose that it is always feasible to find a single robust plan within an
interval of estimates. We believe that this assumption is not always valid. When
the interval of estimates is large, it becomes difficult to find a plan generating
stable performance over the whole interval. Execution plans may be robust at
only some points of the interval. In the remainder of this paper, we adopt the
following definition for a robust plan: *let V_e be an estimate of an error-prone
parameter, let I be an interval of estimates around V_e exhibiting the uncertainty
about the estimate of this parameter. Let P_{best} be the best plan for a specific
value V_i in I. Finally, let λ be a (user-defined) cost-increase threshold (expressed
in percentage). A plan P_{alt} is robust with respect to estimation errors if:*

$$\forall\, V_i \in I,\ \frac{cost(P_{alt})}{cost(P_{best})} \leq 1 + \frac{\lambda}{100} \qquad (1)$$

For instance, if users tolerate a minor cost increase (λ) of at most 20 %, the cost of P_{alt} is at most 1, 2 times the cost of the best plan.

3   Generation of a Robust Query Execution Plan

In this section, we detail our method called **Identification of Robust Plans (IRP)**, which is part of the *Multi Point-based Optimization* approach. The key concept in our work is the uncertainty of estimates used by a query optimizer to choose an appropriate execution plan along with the difficulty to find a single robust execution plan when the estimation uncertainty is large. We present below an example that underlines our motivations before detailing our method.

3.1   A Motivating Example

We first study the case wherein the size of only one input-relation for a join operation is error-prone. then, we extend it to the general case.

Example. Consider the query Q = *select * from customer as C, order as O where C.customerId = O.customerId and C.country = 'France' and C.city = 'Paris'.* Best plan selection for this query requires accurate estimate of selectivity of the predicates *C.country = 'France'* and *C.city = 'Paris'*. However, even with the existence of histograms on attributes *C.country* and *C.city*, the optimizer can not maintain multi-dimensional histograms for all possible combinations of attributes [23]. It may so assume independence to compute the joint selectivity of the two predicates. This assumption may lead to large estimation errors while costing plans due to an eventual correlation between the predicates. To avoid sub-optimal plan choice, resulting from these errors, we consider $\sigma(C)$, which is the result of *C.country = 'France' and C.city = 'Paris'*, as error-prone. We define an interval of estimates around $\sigma(C)$, denoted I. This interval is then used to select an appropriate execution plan. In this process, one of the following cases may occur: (i) there is a single plan that is robust at all points of I, (ii) there are several plans, each of which is robust at only some points of I.

Existing multi point-based optimization methods assume that the first case is often feasible, that is to say a plan providing stable performance within an interval of estimates can be founded. This assumption is not always valid. Consider the scenario of **Example**, suppose that there is an index on an attribute of the relation O. Through repeated invocations of a query optimizer, plans labelled "Plan1", "Plan2" and "Plan3" (Cf. Fig. 1) are enumerated as possible execution plans for Q.

Based on a cost model, a query optimizer estimates the costs of these plans with respect of variation of $\sigma(C)$ so as to determine the execution plan that offers robust performance within the interval of estimates. This plan must respect robustness definition presented in Sect. 1. Assume that the tolerated cost-increase (λ) is about 20 %. This value is chosen based on the works of [1,3]. Figure 2 plots the plans' costs with respect to variations of $\sigma(C)$ in MB. In this Figure, we observe that for $\lambda = 20$ %, none of plans P1, P2 and P3 is robust

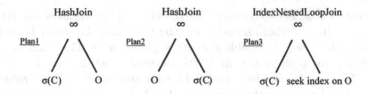

Fig. 1. Possible execution plans for the query $Q = \sigma(C) \bowtie O$

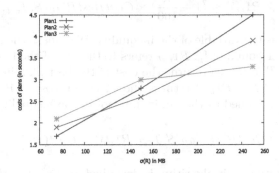

Fig. 2. Costs of plans with respect to variations of $\sigma(C)$

within I. This example highlights the necessity to divide the interval of estimates into sub-intervals and to associate a robust plan with each sub-interval.
An important question concerns the choice of only one plan to start the execution. A solution could be to determine the sub-interval covering the most possible
run-time values and choose the plan associated with this sub-interval to execute
the query.

3.2 Robust Optimization Method

The above mentioned discussion motivates the necessity of incorporating estimation uncertainty in the query optimization process. In order to design and
develop such an optimization method, it is suitable to offer first a method for
one-join operation. This method constitutes the building block for a robust execution of a multi-join query. In this regard, we present a method that consists
of two main modules: (1) Identification of Robust Plans, and (2) Selection of an
Execution Plan. We will examine them in more detail in the subsections below.

Identification of Robust Plans: IRP is a compile-time strategy for identifying robust plans. It relies on intervals of estimates to model the estimation
uncertainty. Such intervals are used to select the plan that minimizes a potential
sub-optimality resulting from possible estimation errors. To conduct our experiments we use the method in [3] to compute these intervals.

Let $Q = \text{Join}(T, R1)$ be a query to join T and R1, where T is a temporary
relation and R1 is a base relation. The size of T is denoted $|T|$. The point estimate

of |T| is denoted T_{est}. Let us assume that the interval of estimates around T_{est} is defined, i.e., $I = [L, U]$, where L and U are the lower and the upper bound of the interval. Let S be a set of possible execution plans for Q. This module consists in determining -for each plan P in S- the interval I_p within which P is robust (Cf. [26]). Calculate I_p may be converted into a numerical solving problem. We vary T_{est} in I and compute the lower bound of I_p by solving the inequality:

$$Cost(P, T_{est}, R1_{val}) \leq (1 + \frac{\lambda}{100}) \times CostBestPlan(S, T_{est}, R1_{val}) \qquad (2)$$

T_{est} is considered as the variable of the inequality. We choose then the minimum root to be the lower bound of I_p. $R1_{val}$ refers to the size of the base relation R1. $CostBestPlan(S, T_{est}, R1_{val})$ returns the cost of the best plan in S for T_{est} and $R1_{val}$ while $Cost(P, T_{est}, R1_{val})$ returns the execution cost of P. Similarly, the upper bound is determined as the minimum root of the inequality below:

$$(1 + \frac{\lambda}{100}) \times CostBestPlan(S, T_{est}, R1_{val}) < Cost(P, T_{est}, R1_{val}) \qquad (3)$$

Due to space limitation, this algorithm is described in greater detail in [26]. The robustness of a plan is checked at different points in the interval. To avoid a large computational complexity, these points are computed based on the secant method, which is very effective and can be extended to n-parameters space [21].

Selection of an Execution Plan: After identifying robust plans, only one plan should be selected to start the query execution. This plan is chosen based on a probabilistic approach. We propose to compute, for each plan, its probability $prob(P)$ to handle run-time variation of estimates. The greater is the probability the higher is the likelihood of the plan to minimize sub-optimality. We process estimates as random variables and compute $prob(P)$ as:

$$prob(P) = \frac{length\ of\ the\ robustness\ range\ associated\ with\ P}{length\ of\ the\ interval\ of\ estimates\ I} \qquad (4)$$

The plan with the higher probability is chosen to start the execution. When two plans have equal probabilities, the plan that maximizes the worst-case performance is selected. The study of this issue is available in [26].

Extension to Multiple Parameters. So far, we have assumed that the size of only one join input-relation is error-prone. This approach can be generalized to the multi parameter case. In this case, the size of each relation is modelled by an interval of estimates. The intersection of these intervals forms a bounding box. We compute -for each plan- a robustness box rather than a robustness range. In addition, the plan associated with the robustness box covering the most space of uncertainty is chosen to start the execution. Notwithstanding these changes, the basic mechanism of the IRP algorithm is similar to the first case. This algorithm can be found in detail in [26].

4   Experiments

In this section we describe the experimental evaluation of our optimization method, termed IRP. We compare IRP with a method that relies on the single point-based optimization approach [24] (termed TRAD) and with RIO [3]. RIO is a method using the multi point-based optimization approach. We study the behaviours of these methods with respect to error in estimates. We consider that an estimation uncertainty may be high, medium or low. We conducted our experiments using a simulation model that we describe below.

4.1   Simulation Model

To perform our experiments, we used a query builder and a simulator. The simulator includes the simulated optimization method and a meta-base. The simulated method consists of two main modules: (1) Query Optimizer, and (2) Query Executor. The meta-base includes information describing the runtime environment (e.g., CPU performance) as well as information about data (e.g., sizes of base relations). The dataset used in the experiments is shown in Table 1.

Table 1. Summary of dataset used in the experiments

Parameter	Value	Parameter	Value
Buffer size	400 MB	CPU performance	100 000 MIPS
Disk bandwidth	100 MB/s	Average disk latency	2 ms
Size of a page on disk	4 KB	Size of a record	1–3 KB
Size of a relation	100–1000 MB	Size of an attribute	10–500 Bytes

In the next subsections, we present the results of our experiments. First, we study the case wherein only one input-relation for a join is prone to an estimation error. Second, we extend our experiments to the case wherein both inputs-relations are prone to estimation errors. During our experiments, we consider that the cost-increase threshold for robustness condition is 20 %.

4.2   Experimental Results

For this experiment, we use a set of 100 queries which we call SQ. Queries in SQ include only one join operation and verify all, the followings conditions: (i) One of the inputs relations is a base relation, denoted R1, and (ii) One of the inputs relations is a temporary relation, denoted T and is resulting from a selection operation on a base relation named R2. The following query Q′ is a sample of queries in SQ:

$$Q' = select * from R1, R2 \, Where \, R1.a = R2.b \, and \, R2.c < V1 \, and \, R2.d > V2$$

Q' can be represented as $Join(R1, T)$, where T is the result of *"select * from R2 Where R2.c < V1 and R2.d > V2"*. Estimates of the temporary relations sizes while costing execution plans for queries in SQ, may be erroneous. This is due to eventual correlations between the selection predicates. For each query, the uncertainty about |T| is modelled by an interval of estimates. To calculate such an interval, we use the method proposed in [3]. We consider three uncertainty levels: high, medium, and low. Note that a high uncertainty in estimates involves a large interval of estimates. This interval is then used by RIO and IRP to generate a robust query execution plan. As for TRAD, it relies on a single-point estimate of |T|, i.e., T_{est}, to select an execution plan expected to be the best.

Impact of Estimation Errors on Execution Times: The first experiment consists in assessing the impact of estimation errors on the execution time of each method. Figure 3 shows the variation of the execution time of plans produced by TRAD, RIO and IRP for queries in SQ. We vary the error between the estimate of the temporary relation size and its actual value. This is repeated for each query in SQ. Figure 3 also shows the execution time of the best plan.

The execution time for a method is computed as the median of the execution times of all queries in SQ by this method. Error in estimates plotted on the x-axis is calculated as $(\frac{|T_{actual}|}{|T_{estimate}|} - 1)$ [3]. A positive error indicates an underestimate while a negative error indicates an overestimate of the actual run-time value compared to the compile-time estimated value. Figure 3 shows that the performance of TRAD remains acceptable when the error on |T|, is very low. However, when the error becomes large, i.e., greater than 2, we observe a significant increase in the execution times of plans produced by TRAD. Indeed, TRAD selects an execution plan whose performance is heavily dependent on the accuracy of compile-time estimates. This plan is optimal for only the single-point estimate of |T|. Plans chosen by TRAD are very sensitive to estimation errors.

Figure 3 shows that RIO generates better performance compared with TRAD. However, the performance of RIO may deteriorate. Indeed, when RIO does not find a single robust plan within the interval of estimates of |T|, it behaves like TRAD. It produces a plan expected to be optimal for the compile-time estimate of |T|. Unlike RIO, IRP remains based on multi-point estimates. It generated an execution plan that provides stable performances within a sub-interval of the interval of estimates. Using a sub-interval of estimates rather than a single-point estimate for |T| allows to IRP to produce plans whose execution times are usually more stable compared with RIO and TRAD. We also notice in Fig. 3 that when the uncertainty is low, the performance of TRAD and IRP becomes close. This is because when the interval of estimates is narrow, it becomes more feasible for RIO to find a plan that is robust within an interval.

Impact of Estimation Errors on the Consistency of Methods: The consistency of a method refers to its ability to cope with estimation errors and/or changes in run-time conditions compared to compile-time expectation of run-time conditions. A method is said highly consistent if its performance does not

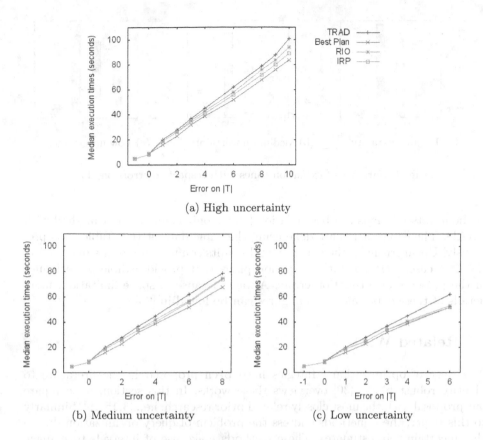

(a) High uncertainty

(b) Medium uncertainty   (c) Low uncertainty

Fig. 3. Variation of execution times with respect to errors on |T|

degrade significantly in the presence of estimation errors [29]. To measure the consistency of TRAD, RIO and IRP, we compute the variance of performance of each method and compare it to the variance in case of best performance. Note that the consistency is inversely proportional to the variance [29]. To achieve this experiment, we use the same set of queries previously used, i.e., SQ. The results of this evaluation are shown in Fig. 4.

As we can see in Fig. 4, the variances of RIO and IRP are close when the estimation uncertainty is low. However, when the uncertainty is large (high or medium), the gap between the variances of methods increase. A high variance means a significant dispersion of execution times. Figure 4 demonstrates that plans generated by IRP are more stable than those generated by RIO. This figure also proves that estimation errors have a pronounced impact on the consistency of TRAD. This figure confirms the evaluation results previously obtained.

When Both Inputs-Relations are Prone to Estimation Errors. We performed more experiments under the assumption that both join-inputs relations, denoted T1 and T2, are prone to estimation errors. As RIO assumes that at least one

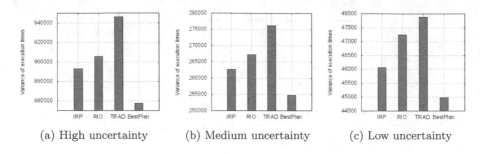

(a) High uncertainty   (b) Medium uncertainty   (c) Low uncertainty

Fig. 4. Variance of execution times with respect to errors on |T|

of the inputs-relations is a base relation [3], we only compared our method with TRAD. The experiments are done using the same dataset (Cf. Table 1) as previously. Unsurprisingly, the experimental results confirm the results obtained in the first case. IRP generates execution plans that provide stable performance in the presence of estimation errors. Always because of space limitation, more details of these experiments and figures can be found in [26].

5   Related Work

A variety of optimization strategies have been proposed in the literature to identify robust plans. [30] overviews these works. In this section, we compare our proposal with the most closely related prior research, i.e., [3, 15, 19]. Similarly to this paper, these methods address the problem of query optimization due to the uncertainty in estimates. These methods make use of intervals to manage estimation uncertainty. These intervals are then used by the optimizer to pick an appropriate execution plan. However, our work defers from these works on several aspects.

In the initial optimization phase of Rio [3], if the plan chosen by the optimizer at the lower and the upper bounds of the interval is the same as that chosen for the single-point estimate, then the plan is assumed *robust* within the whole interval. In our method, we adopt a different way to check the robustness of a plan. We verify the robustness of a plan at different points in the interval. These points are determined using our modified secant method, which provides reduced computation complexity and precise results. In addition, Rio assumes that at least one of the join inputs-relations is a base relation. We extend this by considering the case wherein both relations are prone to estimation errors.

Among previous related research, the work by Markl et al. [19]. In [19], and later in [4], validity ranges are computed for each sub-plan of a query execution plan. The best plan is first chosen based on single-point estimates. The upper and lower bounds wherein the plan remains the best plan are then computed. These methods aim for optimal performance while IRP aims for robustness. Robust plans choices are especially important to safely process queries when the uncertainty about estimates is large.

The use of intervals may seem similar to the principle of parametric optimization [10, 15] where different plans are generated for different intervals of the optimization parameters. The main limitation of this approach is the large number of plans to generate, store and compare at the optimization phase. We reduce this complexity by computing robustness ranges. We accept a minor cost-increase that reduces the number of plans generated. In addition, parametric optimization is particularly attractive in the case of queries that are compiled once and executed several times, possibly with minor modifications of parameters. Unlike our work, its objective being to avoid compiling multiple instances of the same query, but not to ensure robustness against estimation errors.

6 Conclusion and Future Work

This paper proposes an optimization strategy to handle uncertainty in compile-time estimates. Uncertainty is modelled by means of intervals of estimates. These intervals are then used to identify robust plans. We characterize a plan as robust if its cost remains stable and acceptable in case of estimation errors. Our method relies on a probabilistic approach to select the plan by which to start the execution. The experiments show that the proposed strategy may improves the ability of a query optimizer to generated a robust execution plan, especially in the presence of large estimation errors.

The proposed method performs in a uniprocessor environment. As future work, we plan to extend it to a multiprocessor environment. It would be interesting to study the importance of the parallelism degree choices over the identification of a robust query execution plan.

References

1. Abhirama, M., Bhaumik, S., Dey, A., Shrimal, H., Haritsa, J.R.: On the stability of plan costs and the costs of plan stability. Proc. VLDB Endow. **3**, 1137–1148 (2010)
2. Babcock, B., Chaudhuri, S.: Towards a robust query optimizer: a principled and practical approach. In: Proceedings of ACM SIGMOD International Conference on Management of Data, pp. 119–130 (2005)
3. Babu, S., Bizarro, P., DeWitt, D.: Proactive re-optimization. In: Proceedings of ACM SIGMOD International Conference on Management of Data, pp. 107–118 (2005)
4. Bizarro, P., Bruno, N., DeWitt, D.J.: Progressive parametric query optimization. IEEE Trans. Knowl. Data Eng. **21**, 582–594 (2009)
5. Bruno, N., Jain, S., Zhou, J.: Continuous cloud-scale query optimization and processing. PVLDB **6**, 961–972 (2013)
6. Chaudhuri, S., Narasayya, V., Ramamurthy, R.: Estimating progress of long running SQL queries. In: Proceedings of ACM SIGMOD International Conference on Management of Data, pp. 803–814 (2004)
7. Chen, C.M., Roussopoulos, N.: Adaptive selectivity estimation using query feedback. In: Proceedings of ACM SIGMOD International Conference on Management of Data, pp. 161–172 (1994)

8. Christodoulakis, S.: Implications of certain assumptions in database performance evaluation. ACM Trans. Database Syst. **9**, 163–186 (1984)
9. Chu, F.C., Halpern, J.Y., Seshadri, P.: Least expected cost query optimization: an exercise in utility. In: Proceedings of the Eighteenth ACM SIGACT-SIGMOD-SIGART Symposium on Principles of Database Systems, Philadelphia, pp. 138–147 (1999)
10. Cole, R.L., Graefe, G.: Optimization of dynamic query evaluation plans. In: Proceedings of ACM SIGMOD International Conference on Management of Data, pp. 150–160 (1994)
11. Deshpande, A., Garofalakis, M.N., Rastogi, R.: Independence is good: dependency-based histogram synopses for high-dimensional data. In: ACM SIGMOD Conference, pp. 199–210 (2001)
12. Dutt, A., Neelam, S., Haritsa, J.R.: Quest: an exploratory approach to robust query processing. Proc. VLDB Endow. **7**, 1585–1588 (2014)
13. Getoor, L., Taskar, B., Koller, D.: Selectivity estimation using probabilistic models. In: Proceedings of ACM SIGMOD International Conference on Management of Data, pp. 461–472 (2001)
14. Harish, D., Pooja, N.D., Jayant, R.H.: Identifying robust plans through plan diagram reduction. Proc. VLDB Endow. **1**, 1124–1140 (2008)
15. Hulgeri, A., Sudarshan, S.: Parametric query optimization for linear and piecewise linear cost functions. In: Proceedings of the 28th International Conference on Very Large Data Bases, pp. 167–178. VLDB Endowment (2002)
16. Ioannidis, Y.E., Christodoulakis, S.: On the propagation of errors in the size of join results. In: Proceedings of SIGMOD International Conference on Management of Data, pp. 268–277 (1991)
17. Kabra, N., DeWitt, D.J.: Efficient mid-query re-optimization of sub-optimal query execution plans. In: Proceedings of ACM SIGMOD International Conference on Management of Data, pp. 106–117 (1998)
18. Karanasos, K., Balmin, A., Kutsch, M., Ozcan, F., Ercegovac, V., Xia, C., Jackson, J.: Dynamically optimizing queries over large scale data platforms. In: Proceedings of ACM SIGMOD International Conference on Management of Data, pp. 943–954 (2014)
19. Markl, V., Raman, V., Simmen, D., Lohman, G., Pirahesh, H., Cilimdzic, M.: Robust query processing through progressive optimization. In: Proceedings of ACM SIGMOD International Conference on Management of Data, pp. 659–670 (2004)
20. Neumann, T., Galindo-Legaria, C.A.: Taking the edge off cardinality estimation errors using incremental execution. In: DBIS, Germany, pp. 73–92 (2013)
21. Papakonstantinou, J.M., Tapia, R.A.: Origin and evolution of the secant method in one dimension. Am. Math. Mon. **120**(6), 500–518 (2013)
22. Poosala, V., Haas, P.J., Ioannidis, Y.E., Shekita, E.J.: Improved histograms for selectivity estimation of range predicates. In: Proceedings of ACM SIGMOD International Conference on Management of Data, pp. 294–305 (1996)
23. Poosala, V., Ioannidis, Y.E.: Selectivity estimation without the attribute value independence assumption. In: Proceedings of 23rd International Conference on Very Large Data Bases, pp. 486–495 (1997)
24. Selinger, P.G., Astrahan, M.M., Chamberlin, D.D., Lorie, R.A., Price, T.G.: Access path selection in a relational database management system. In: Proceedings of ACM SIGMOD International Conference on Management of Data, pp. 23–34 (1979)
25. Moumen, C., Morvan, F., Hameurlain, A.: Estimation error-aware query optimization: an overview. Int. J. Comput. Syst. Sci. Eng. (2016, in press)

26. Moumen, C., Morvan, F., Hameurlain, A.: Handling estimation inaccuracy in query optimization. Research report (2016). www.irit.fr/~Riad.Mokadem/report%20Chiraz%20Moumen.pdf
27. Tzoumas, K., Deshpande, A., Jensen, C.S.: Lightweight graphical models for selectivity estimation without independence assumptions. In: PVLDB (2011)
28. Tzoumas, K., Deshpande, A., Jensen, C.S.: Efficiently adapting graphical models for selectivity estimation. VLDB J. **22**, 3–27 (2013)
29. Wiener, J.L., Kuno, H., Graefe, G.: Benchmarking query executionrobustness. In: TPC Technology Conference on Performance Evaluation and Benchmarking, pp. 153–166 (2009)
30. Yin, S., Hameurlain, A., Morvan, F.: Robust query optimization methods with respect to estimation errors: a survey. SIGMOD Rec. **44**, 25–36 (2015)

Practical Study of Subclasses of Regular Expressions in DTD and XML Schema

Yeting Li[2], Xiaolan Zhang[1,2], Feifei Peng[1,2], and Haiming Chen[1(✉)]

[1] State Key Laboratory of Computer Science, Institute of Software,
Chinese Academy of Sciences, Beijing 100190, China
{zhangxl,pengff,chm}@ios.ac.cn
[2] University of Chinese Academy of Sciences, Beijing, China
liyeting@snnu.edu.cn

Abstract. DTD and XSD are two popular schema languages widely used in XML documents. Most content models used in DTD and XSD essentially consist of restricted subclasses of regular expressions. However, existing subclasses of content models are all defined on standard regular expressions without considering counting and interleaving. Through the investigation on the real world data, this paper introduces a new subclass of regular expressions with counting and interleaving. Then we give a practical study on this new subclass and five already known subclasses of content models. One distinguishing feature of this paper is that the data set is sufficiently large compared with previous relevant work. Therefore our results are more accurate. In addition, based on this large data set, we analyze the different features of regular expressions used in practice. Meanwhile, we are the first to simultaneously inspect the usage of the five subclasses and analyze different reasons dissatisfying the corresponding definitions. Furthermore, since W3C standard requires the content models to be deterministic, the determinism of content models is also tested by our validation tools.

Keywords: XML · DTD · XML schema · Interleaving · Counting

1 Introduction

As a main file format for data exchange, the eXtensible Markup Language (XML) has been widely used on the web [1]. The presence of a schema provides a lot of conveniences and advantages for various applications such as data processing, automatic data integration, static analysis of transformations and so on [3,12,20–23,27,31]. DTD (Document Type Definitions) and XSD (XML Schema Definitions) are two popular schema languages recommended by W3C (World Wide Web Consortium) [30]. Most content models used in DTD and XSD essentially consist of restricted subclasses of regular expressions. Therefore for practical purpose many researches focus on the study of subclasses practically used.

Work supported by the National Natural Science Foundation of China under Grant Nos. 61472405, 61070038.

© Springer International Publishing Switzerland 2016
F. Li et al. (Eds.): APWeb 2016, Part II, LNCS 9932, pp. 368–382, 2016.
DOI: 10.1007/978-3-319-45817-5_29

In 2004, Bex et al. proposed a subclass called the simple regular expressions after analyzing 109 DTDs and 93 XSDs from Cover Pages [6], which became the basis of later work. Martens et al. discussed the complexity of decision problems for eCHARE, an extension of simple regular expression, and still call it simple regular expression [24]. Later eCHARE was called CHARE in [25]. In 2005, Bex et al. discussed the expressiveness of XSDs based on 819 XSDs harvested from the web in [5]. In this corpus only 225 XSDs remained no errors and 85 % were in fact structurally equivalent to a DTD. In 2006, after an analysis of 819 DTDs and XSDs gathered from Cover Pages as well as from the web, Bex et al. proposed two new subclasses: single occurrence regular expression (SORE) and chain regular expression (CHARE) [7]. In [7], CHARE is defined as a subclass of SORE and it has a weaker expressiveness than the CHARE defined in [25]. Using 966 DTDs and XSDs, Feng et al. extended CHARE [7] to Echare to cover more content models [15]. Echare allows two kinds of base symbols a and a^+ where $a \in \Sigma$.

The above discussion reveals some shortcomings of existing work. First, it is clear that the names of different subclasses of regular expressions above are quite confusing in the literature. Therefor we rename these subclasses and use new names in this paper. The relations between the new and the old names are shown in Table 1. Second, the scale of data sets was far from enough for analysis. One distinguishing feature of this paper is that the data set is sufficiently large compared with previous relevant work. So it will be helpful to get a more accurate result. Using techniques such as proxies, disguised as a browser, multi-threading, we gather a large sample of 2427 DTD and 4859 XSD files from Google after removing duplicate schemas by MD5. The data set covers different fields such as education, agriculture, science, economics, engineering, sports and so on. The whole list for our data set and tools used in this paper can be found in http://lcs. ios.ac.cn/~zhangxl/project.html. Besides, in existing work the above subclasses were separately discussed in different papers using relatively small data sets, while in this paper we simultaneously analyze these subclasses using a new larger data set.

Table 1. Relations between new and old restricted subclass names

New names	Old names
CHARE	Simple regular expression [6]
eCHARE	CHARE [25]
SORE	SORE [7]
Simplified CHARE	CHARE [7]
eSimplified CHARE	Echare [15]

In addition, these subclasses are all defined on standard regular expressions. However, counting and interleaving have already been used in XSDs. Björklund et al. made an incremental evaluation of regular expressions with counters based on a data set from three libraries: RegExLib, Snort and XML Schema on the web in [10]. The results show that almost half of the regular expressions use non-trivial counters which exclude the forms of $a^{[0,0]}$, $a^{[0,1]}$ and $a^{[1,1]}$. They also found that the vast majority is the simple form of CHAINs: $e_1 \cdot e_2 \cdots e_m$ where $e_i = (a_1 + a_2 + \cdots + a_n)^{[k,l]}$. In [17], Ghelli et al. proposed a restricted subclass defined by $T:: = \varepsilon|a^{[m,n]}|T+T|T \cdot T|T\&T$ where $m \in N\backslash\{0\}$ and $n \in N\backslash\{0\}\cup\{^*\}$. In $L(T)$, each alphabet symbol can appear at most once. Counters can only occur as a constraint for terminal symbols. For example, $(a? \cdot (b + c + d)^{[1,100]})$ is not allowed. In 2008 they introduced a linear-time membership algorithm [18] for this subclass. Furthermore, determinism of regular expressions with counting and interleaving has been discussed by many researchers [13,16,19,28]. In this paper, we extend CHARE with counting and interleaving operators. We conduct a series of experiments with result of a more popular use of this new subclass (94 %) in real world. We also analyze the determinism of it together with other five commonly used subclasses based on our data set. The main contributions of this paper are listed as follows.

- Considering numerical occurrence constraints and interleaving, we extend CHARE to a new restricted class called *extended CHARE with counting and interleaving* (eCICHARE). To the best of our knowledge, we are the first to analyze the usage of regular expressions with counting and interleaving together through real world data.
- Based on the large data set, we inspect the properties of different subclasses used in practice including eCICHARE by different measures. Particularly, the proportions of all subclasses used in practice are analyzed. In addition, we are the first to analyze the usage of eCHARE using real world data.

The rest of paper is organized as follows. Section 2 gives the definitions used in this paper. Then we introduce the data set and experiments in Sect. 3. The related work is discussed in Sect. 4 and Sect. 5 gives the conclusion.

2   Definitions

2.1   Regular Expression with Counting and Interleaving

Let Σ be a finite alphabet of terminal symbols. Each string consists of a finite sequence of symbols in Σ. Σ^* means the set of all strings over Σ. A *regular expression with counting and interleaving* over Σ is $\emptyset, \varepsilon$, or $a \in \Sigma$, or is the union $r_1 + r_2$, the concatenation $r_1 \cdot r_2$, the interleaving $r_1 \& r_2$, the Kleene-star r_1^*, the choice $r_1?$, the counting $r_1^{[m,n]}$ with $m \leq n$ and $n > 0$, or the plus r_1^+ where r_1 and r_2 are both regular expressions. $r_1 \cdot r_2$ is also written as $r_1 r_2$. Let s be a string in Σ^* and $|s|$ denotes its size. We use $s_1 \& s_2$ to denote the set of strings obtained by s_1 and s_2 in every possible way. For $s \in \Sigma^*$, $s \& \varepsilon = \varepsilon \& s = s$ and $a \cdot s_1 \& b \cdot s_2 = (a \cdot (s_1 \& b \cdot s_2)) \cup (b \cdot (a \cdot s_1 \& s_2))$. For example, strings accepted by $a \& b \& c$ are $\{abc, acb, bac, bca, cab, cba\}$. Counting is the numerical occurrence constraint which defines the minimal and maximal number of times for a certain symbol in a regular expression. For $E = (a^{[1,3]} + b)^{[2,3]}b$, it means that two-to-three repeats of a choice between a sequence of one-to-three a elements or a single b, followed by a single b. By $RE(\&, \#)$ we represent regular expressions extended with interleaving and counting. We use $RE(\&)$ and $RE(\#)$ to denote the regular expressions with interleaving and counting respectively. $\#$ and $\&$ operators help us to express the content models more succinctly and concisely.

2.2 eCICHARE

We introduce a new subclass which extends CHARE with interleaving and counting.

Definition 1 (extended CHARE with counting and interleaving (eCICHARE)). *Base symbols are a, $a?$, a^*, or a^+ where $a \in \Sigma$. A factor is of the form e, $e?$, e^*, $e^{[m,n]}$, or e^+ where e is a disjunction of base symbols of the same kind. An eCICHARE is $\emptyset$, ε, a concatenation of factors, or an unordered sequence of factors with interleaving among them.*

In the definition, each eCICHARE cannot contain both concatenation and interleaving operators at the same time. Numerical occurrence can only be the constraint to factors. This restriction is severe, but the experiment result shows that it is actually met by most of regular expressions in real world data. In addition, two forms of eCICHARE: $a^{[0,1]}$ and $a^{[1,1]}$ are always substituted by $a?$ and a in practice.

For instance, $E_1 = (a^* + b^*)?(a + b)^*(c^+ + d^+)e?$, $E_2 = a^{[1,3]}b?$ and $E_3 = (a^* + b^*)?\&(a + b)^*\&(c^+ + d^+)\&e?$ are both eCICHAREs while $E_4 = (a^* + b^*)?(a + b^{[0,3]})^*(c^+ + d^+)\&e?$ is not.

2.3 Determinism

Determinism is required by W3C specification for content models of DTDs and XSDs. It is also called the unique particle attribution (UPA) property by W3C Recommendation. It has the same meaning with one-ambiguity [11], and weak determinism [16,29]. Suppose that we match a string s against a given regular expression E from left to right. If we always know definitely the next symbol we will match without looking ahead in the string, E is deterministic. The formal definition of determinism is as follows.

Definition 2 (Determinism [4]). *An expression E is deterministic if and only if for all words $uxv, uyv \in L(E')$ where $|x| = |y| = 1$, if $x \neq y$ then $x' \neq y'$.*

In this definition, E', x' and y' are the marked forms of E, x and y. For example. $E = (a + b)^*a?b^*$, then $E' = (a_1 + b_2)^*a_3?b_4^*$ and $(E')' = E$. $E = aa^*$ is a deterministic regular expression while $E = a^*a$ is not.

2.4   Definitions for Other Five Subclasses

Definition 3 *(CHARE* [6]). *Base symbols are a, $a?$, a^*, a^+ where $a \in \Sigma$. A factor is of the form e, $e?$, e^*, e^+ where e is a disjunction of base symbols of the same kind. A simple regular expression is $\emptyset$, ε, or a concatenation of factors.*

For instance, $(a^* + b^*)(a + b)?b^*(a + b)^*$ is a CHARE while $(abc + c?)(a^* + b^*)d?$ is not.

Definition 4 *(eCHARE* [25]). *Base symbols are s, $s?$, s^*, s^+ where s is a non-empty string. A factor is of the form e, $e?$, e^*, e^+ where e is a disjunction of base symbols of the same kind. That is of the form $(s_1 + \cdots + s_n)$, $(s_1^* + \cdots + s_n^*)$, $(s_1? + \cdots + s_n?)$, $(s_1^+ + \cdots + s_n^+)$, where $n \geq 1$ and s_i is non-empty string. An eCHARE is $\emptyset$, ε or a concatenation of factors.*

For example, $((abc)^* + b)(a + b)?(ab)^+(ac + b)^*$ is an eCHARE.

Definition 5 *(SORE* [7]). *Let Σ be a finite alphabet. A single-occurrence regular expression (SORE) is a regular expression over Σ in which every terminal symbol occurs at most once.*

For instance, $(((a + b), c)?d^+)^*e$ is a SORE while $(a^* + b^*)a$ is not.

Definition 6 *(Simplified CHARE* [7]). *A Simplified CHARE is a SORE over Σ of the form $f_1 \cdots f_n$ where $n \geq 1$. Every factor f_i is an expression of the form $(a_1 + \cdots + a_n)$, $(a_1 + \cdots + a_n)?$, $(a_1 + \cdots + a_n)^*$, $(a_1 + \cdots + a_n)^+$ where $n \geq 1$ and every a_i is a terminal symbol.*

For example, $(a + b)?c^*$ is a Simplified CHARE while $(a + b)?a$ is not.

Definition 7 *(eSimplified CHARE* [15]). *An eSimplified CHARE is a SORE over Σ of the form $f_1 \cdots f_n$ where $n \geq 1$. Every factor f_i is an expression of the form $(a_1 + \cdots + a_n)$, $(a_1 + \cdots + a_n)?$, $(a_1 + \cdots + a_n)^*$, $(a_1 + \cdots + a_n)^+$ where $n \geq 1$ and every a_i is a terminal symbol or the form of a_i^+.*

For example, $(a + b^+)c?$ is an eSimplified CHARE.

CHARE and eCHARE require the base symbols in each factor must be the same kind while eSimplified CHARE can be different.

3   Experiments

3.1   Data Set

Data Preprocess. There are two steps in our data preprocess. First, get the DTD and XSD files. Repositories of data set such as GSML, DMTF, DSML, DWML, FACETMAP, GITHUB, GRAPHML, HAPMAP, IOP, KAIST, NCBI, BIOXSD, CORBA, CSML and so on are well-formed. We can harvest DTDs and XSDs from their official websites directly. But others need to be crawled through Google by queries: filetype:dtd or filetype:xsd. Using these two queries, we get many URLs for DTDs (or XSDs). However, not all these URLs point to a DTD (or XSD) directly but to some HTML files. Such HTML file may have a link or path to a DTD (or XSD), or just some information on the website. For links, we make a recursive resolving to download these related DTDs and XSDs. For information on the websites, we use a specific script tool: $html\_parser$ which is written in Python to analyze the information and obtain the schema. Second, do data cleaning. We remove duplicate files with same URLs. More precisely, we use the technique MD5 to analyze whether two files with different URLs have the same content. If so, redundant files are also removed. At last, we obtain 2427 DTD files and 4859 XSD files and extract 64249 and 67255 regular expressions from DTDs and XSDs files respectively.

For the convenience of discussion, we introduce a uniform syntax to denote subclasses of restricted regular expression by specifying the allowed factors used in [25]. The base symbols are denoted by a, $a?$, a^*, a^+. s means a string in Σ^*. The disjuncts are denoted by $(a_1 + a_2 + \cdots + a_n)$, $(a_1? + a_2? + \cdots + a_n?)$, $(a_1^* + a_2^* + \cdots + a_n^*)$, $(a_1^+ + a_2^+ + \cdots + a_n^+)$ which can be also extended with choice, Kleene-star, and plus respectively. These factors can be abbreviated by the form of $(+\cdots)$. We use $RE((+a?), a^*)$ to illustrate the subclass whose factors can be in the form of $(a_1? + a_2? + \cdots + a_n?)$ where $a_i \in \Sigma$ and $n \geq 1$ or the form of a^* for some $a \in \Sigma$. We list some possible factors in Table 2.

Table 2. Possible factors in subclasses of regular expressions and their abbreviations [25]

Factor	a	$a?$	a^*	a^+	$s?$	s^*	s^+
Abbr.	a	$a?$	a^*	a^+	$s?$	s^*	s^+

Factor	Abbr.	Factor	Abbr.
$(a_1 + \cdots + a_n)$	$(+a)$	$(s_1 + \cdots + s_n)$	$(+s)$
$(a_1 + \cdots + a_n)?$	$(+a)?$	$(s_1 + \cdots + s_n)?$	$(+s)?$
$(a_1 + \cdots + a_n)^*$	$(+a)^*$	$(s_1 + \cdots + s_n)^*$	$(+s)^*$
$(a_1 + \cdots + a_n)^+$	$(+a)^+$	$(s_1 + \cdots + s_n)^+$	$(+s)^+$
$(a_1^* + \cdots + a_n^*)$	$(+a^*)$	$(s_1^* + \cdots + s_n^*)$	$(+s^*)$
$(a_1^+ + \cdots + a_n^+)$	$(+a^+)$	$(s_1^+ + \cdots + s_n^+)$	$(+s^+)$

Based on 64249 and 67255 regular expressions from DTDs and XSDs, we analyze the occurrence types of regular expressions in practice. We treat the form of a^+ operator as aa^*. The result is shown in Table 3. From Table 3, we can find that the form of $RE(a, (+a)^*)$ accounts for the most proportion (34.6 %) for DTDs while in XSDs, forms of $RE(a, a?)$ and $RE(a, a^*)$ are more popular. In addition, the vast majority of regular expressions belongs to the subclass of eCICHARE with proportion of 90.3 % for DTDs and 94.1 % for XSDs respectively.

Table 3. Occurrence of types

	% of DTDs	% of XSDs
RE(a)	18.71	19.28
RE(a,a?)	6.18	22.56
RE(a,a*)	17.02	23.34
RE(a,a?,a*)	3.16	9.49
RE(a,(+a))	1.74	3.07
RE(a,(+a)?)	0.36	1.09
RE(a,(+a)*)	34.63	5.33
RE(a,(+a)?,(+a)*)	3.72	1.60
RE(a,(+a*)*)	3.59	0.16
RE(#)	0	4.06
RE(&)	0	2.07
RE(#,&)	0	≈ 0
Others	0.97	1.96
Total eCICHARE	90.08	94.00

3.2 Definitions for Measures

In this section, we first introduce some definitions of measures used in the experiment such as star height, nesting depth, density and so on. Then, using these measures, we analyze the properties and complexity of regular expressions from different aspects.

Definition 8 (Star Height [2]**).** *The star height of a regular expression E over the alphabet Σ, denoted by $h(E)$, is a nonnegative integer defined recursively as follows:*

1. *$h(E) = 0$, if $E = \emptyset$ or a for $a \in \Sigma$,*
2. *$h(E) = max\{h(E_1), h(E_2)\}$, if $E = (E_1 + E_2)$ or $E = (E_1 \cdot E_2)$,*
   where E_1 and E_2 are regular expressions over Σ,
3. *$h(E) = h(E_1) + 1$ if $E = (E_1)^*$ and E_1 is a regular expression over Σ.*

The star height [2] reflects the maximum nesting depth of Kleene-star occurring in a regular expression. It is an illustration for the complexity of DTDs and XSDs. We give the star height of DTDs and XSDs respectively. We use some substitutions in our experiment of computing the star height. For example, $E_1 = a^+$ and $E_2 = a^{[m,\infty]}$ can be rewritten as the form of $E_1' = aa^*$ and $E_2' = a^*$. We treat interleaving similarly as the operator of concatenation. In fact, other forms of counting and interleaving do not influence the results of star height. From Table 4, we observe that the result of distributions for DTDs and XSDs have no significant differences. Content models with star height larger than 2 are very

Table 4. Star height observed in DTDs and XSDs

Star height	0	1	2	3	4
% of DTDs	27.71	70.66	1.58	0.05	0
% of XSDs	53.62	44.94	1.15	0.29	≈ 0

rare. For XSDs, the proportion with the star height equal to 1 is higher in our result than that in [6] which is 38 % and 17 % respectively.

Besides, we also consider the counter height for regular expressions of XSDs. Counter height is defined just like the star height. The result indicates that the counter height with 1 of regular expressions with counting accounts for almost 89.3 %. Star height and counter height are both illustration of iteration depth of regular expressions. In this paper, we introduce Nesting depth to measure the complexity of a regular expression. For example, the star height for regular expressions $E_1 = (a + b?)?c^{[2,4]}d$ and $E_2 = a$ are both zero. But the complexity for E_1 and E_2 should be different. Nesting depth considers all operators possible in regular expressions including counting and interleaving. For example, let $E = 1^+ + (2? \cdot 3^+)^* + 4^{[1,3]}$. Its corresponding nesting depth is 2. We show proportions of DTDs and XSDs by nesting depth in Table 5. From Table 5, we can find that both DTDs and XSDs with nesting depth lower than 2 are more than 95 %. That means schemas with complex structures are very rare.

Definition 9 (Nesting Depth). *The nesting depth of a regular expression E over Σ, denoted by $ND(E)$, is a nonnegative integer defined recursively as follows:*

1. $ND(E) = 0$, if $E = \emptyset$, ε or a for $a \in \Sigma$,
2. $ND(E) = max\{ND(E_1), ND(E_2)\}$, if $E = (E_1 + E_2)$, $E = (E_1 \& E_2)$ or $E = (E_1 \cdot E_2)$, where E_1 and E_2 are regular expressions over Σ,
3. $ND(E) = ND(E_1) + 1$, if $E = (E_1)^*$, $E = (E_1)^+$, $E = (E_1)?$ or $E = (E_1)^{[m,n]}$ for E_1 is a regular expression over Σ.

Then we consider the density distribution of DTDs and XSDs respectively based on the real world data. The density [6] can be another measure to illustrate the complexity degree of rules in content models.

Definition 10 (Density [6]). *The density of a schema is defined as the number of elements occurring in the right hand side of its rules divided by the number of elements. The formula is $d = \frac{1}{N} \sum_{i=1}^{N} |A_i|$ where N is the total number of*

Table 5. Nesting depth observed in DTDs and XSDs

Nesting depth	0	1	2	3	4
% of DTDs	20.59	76.38	2.49	0.50	0.04
% of XSDs	22.86	72.82	3.91	0.37	0.04

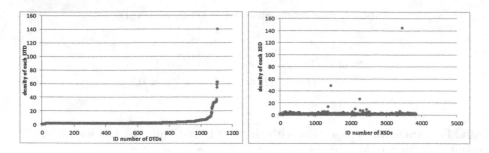

Fig. 1. ID number of DTDs (left) and XSDs (right) versus their density

element definitions occurring in this schema, A_i is the string in the right hand of a rule, $|A_i|$ denotes the size of A_i.

The density [6] can be a measure to illustrate the complexity degree of regular expressions. From Definition 10, we can easily find that the larger the value is, the more sophisticated rules a schema will have. In our experiment, we treat counting as unary operator and the interleaving operator as concatenation. Theses processings do not influence the results of density. Based on the large data set, we give the distributions of density in Fig. 1 for DTDs and XSDs respectively. From Fig. 1, it is easy to find that the density less than 10 for DTDs and XSDs is 95.8 % and 94.4 %.

3.3 XML Schema Features Used in Practical

Although DTD is simple, it develops with some shortcomings such as no modularity, limited expressiveness for new domains, limited basic types and so on. The content model of an element in DTD depends only on the element name. In contrast, XML Schema is based on type definitions and allows the content model to depend on the context in which the element is used. With stronger expressiveness, the usage of XML Schema grows gradually though it is complicated to some extent. Table 6 shows the features of XSDs used in practice.

3.4 Determinism

We consider the determinism of the regular expressions on our large data set. We use our own tools to check the determinism of regular expressions for DTD and XSD respectively. The result is shown in Table 7. From Table 7, we observe that these subclasses almost all satisfy the deterministic requirement. Take the first number 58536/64249 for example. 64249 means the total number of regular expressions for DTDs in the whole data. 58536 means the number of deterministic regular expressions in the whole data set.

Table 6. XML schema features used in the corpus

Features	% of XSDs
SimpleType extension	5.62
SimpleType restriction	13.85
ComplexType extension	10.27
ComplexType restriction	1.69
Abstract attribute	9.57
Final attribute	0.86
Block attribute	0.29
Fixed attribute	2.94
SubstitutionGroup	8.87
Redefine	1.54
xs:all	3.56
Occurrence	4.48
Namespace	96.29
Import	19.16
Key/keyref	0.95
Unique	1.61

Table 7. Determinism of regular expressions

Google	Whole data	CHARE	eCHARE	eCICHARE
DTDs	58536/64249	58343/58404	58513/58579	58343/58404
XSDs	67218/67255	59061/59078	59222/59247	60492/60509

3.5   Usage of Subclasses of Regular Expressions in Practice

In this section, we mainly investigate the usage and proportions of six subclasses in practice based on our large data set. In particular, k-ORE [4] is discussed with different values of k at the same time. k-ORE means each alphabet symbol in regular expression occurs at most k times. The reasons and proportions dissatisfying the corresponding definitions are also discussed. In our experiment, we call regular expressions with counting and interleaving as non-standard regular expressions. They are analyzed separately because counting and interleaving are specific features for XSDs. This is a little different from that in [6,7]. In [6,7], they rewritten regular expressions with counting as a new form using choice operator ?. For example, $E = a^{[1,3]}$ is transformed to $E' = aa?a?$. This transformation is not reasonable enough which influences the results of CHARE and eCHARE. The result in Table 8 indicates that the existing five subclasses still have been used in a large scale (more that 80 %). However, the usage of regular expressions with counting and interleaving increases gradually. They are called

Table 8. Proportions of subclasses of regular expressions

Subclasses	% of DTDs	% of XSDs
CHARE	90.08	87.88
Non-CHARE: non-terminal symbols	8.84	3.52
Non-CHARE: different unary operators	1.08	1.93
Non-CHARE: non-standard expressions	0	6.67
eCHARE	90.37	88.13
Non-eCHARE: not a string	8.31	3.26
Non-eCHARE: different unary operators	1.32	1.94
Non-eCHARE: non-standard expressions	0	6.67
SORE	95.01	92.45
Non-standard expressions	0	6.67
2-ORE	3.62	0.65
3-ORE	0.37	0.10
4-ORE	0.31	0.07
5-ORE	0.52	0.01
6-ORE	0.10	0.05
$k \geq 7$	0.07	≈ 0
Simplified CHARE	88.10	86.10
Non-Simplified CHARE: not a SORE	4.99	0.88
Non-Simplified CHARE: non-terminal symbols	5.00	2.82
Non-Simplified CHARE: different unary operators	1.91	3.52
Non-Simplified CHARE: non-standard expressions	0	6.67
eSimplified CHARE	89.12	86.73
Non-eSimplified CHARE: not a SORE	4.99	0.88
Non-eSimplified CHARE: non-terminal symbols	5.00	2.82
Non-eSimplified CHARE: the unary operator * or ?	0.89	2.90
Non-eSimplified CHARE: non-standard expressions	0	6.67
eCICHARE	90.08	94.00
Non-eCICHARE: non-terminal symbols	8.84	3.68
Non-eCICHARE: different unary operators	1.08	1.94
Non-eCICHARE: regular expressions with counting	0	0.37

non-standard expressions in our experiment and accounts for 6.67 %. After we extend CHARE to eCICHARE, the proportion rises to 94.00 % from 87.88 %.

Bex et al. concluded that 92 % of DTDs and 97 % of XSDs were CHARE based on 109 DTDs and 93 XSDs [6]. In [7], Bex et al. found that 99 % regular expressions were Simplified CHARE, which were also SORE based on 819 DTDs and XSDs. While in our experiment, the proportions for CHARE, SORE,

Simplified CHARE, eSimplified CHARE are lower. The main reason is the difference of data sets. Our data set is larger and more comprehensive, which leads to more accurate results and is closer to the real situation. Another reason is the different process of counting and interleaving operators. The transformation of counting in [6, 7] such as the substitution $aa?a?$ of $a^{[1,3]}$, increases the proportions of CHARE and eCHARE. Besides, we are the first to analyze the usage of eCHARE through real world data. The proportions of eCHARE are 90.37 % for DTDs and 88.13 % for XSDs. Reasons dissatisfying the corresponding definitions are shown in the table clearly. For example, non-terminal symbols mean using expressions such as the forms of ab and $a?b^*$ where $a, b \in \Sigma$ as base symbols in a disjunction. They are not allowed in four subclasses: CHARE, Simplified CHARE, eSimplified CHARE and eCICHARE. For CHARE, they account for 8.84 % and 3.52 % for DTDs and XSDs respectively. The proportions for other three subclasses can be found in Table 8. However, strings like the form of ab as the base symbols is valid for eCHARE. The form of $(a?b^* + c)$ is not allowed in eCHARE. We call it as *extended strings*. The proportion of *extended strings* is not small and they will be considered in our future work. The proportion of SORE (95.01 % for DTDs and 92.45 % for XSDs) means that symbols in the vast majority of regular expressions only occur once most of the time. Simplified CHARE and eSimplified CHARE are both subclasses of SOREs. In [15], eSimplified CHARE accounts for 84.8 % based on 2009 regular expressions while in our experiment the proportion is 89.12 % for DTDs (64249 regular expressions) and 86.73 % for XSDs (67255 regular expressions). The unary operators $*$ and $?$ cannot be constraints for base symbols in eSimplified CHARE, which account for 0.89 % and 2.9 % for DTDs and XSDs respectively. For all six subclasses, proportions of other reasons are not large. For example, different unary operators depict the use of the form $(a? + b^*)c$ as factors which is not valid in the six subclasses. The unary operator $*$ or $?$ is the form of $(a^* + b^*)$ which is not allowed in eSimplified CHARE. In the definition of eCICHARE, numerical occurrence constraints cannot be nested. The reason of regular expressions with counting is the form of $(a^{[1,3]} + b)^{[1,2]}$ which is not valid for eCICHARE.

4 Related Work

Early in 2002, Choi [14] made an experiment about DTDs. 60 DTDs were extracted from the XML.org DTD repository [32]. He analyzed the features of DTDs and proposed measures to make a deep study of their structural properties such as local properties including syntactic complexity, determinism, ambiguity and global properties including reachability, recursion, simple path and simple cycle, chain of stars, hubs. He found that the majority of DTDs used in real world is the form of chain. This result has provided important suggestion for later study in this field.

Based on 109 DTDs and 93 XSDs crawled from Cover Pages, Bex et al. [6] proposed simple regular expression which was named CHARE in this paper. They found that 92 % and 97 % of all element definitions in DTDs and XSDs are CHAREs. Martens et al. extended CHARE to eCHARE in [25]. Using an improved data set crawled from Cover Pages and the web, about 819 DTDs and XSDs, Bex et al. introduced two new subclasses of regular expressions: SORE and Simplified CHARE in [7]. The results revealed that more than 99 % of the regular expressions occurring in practical schemas are Simplified CHARE (therefore also SORE). In 2006, Bex et al. in [8] proposed the concept of k-occurrence. A regular expression is k-occurrence if every alphabet symbol occurs at most k times in it [8]. According to the same data set in [7], they concluded that regular expressions occurring in practical schemas are such that every alphabet symbol occurs at most k times and actually, in 98 % of the cases $k = 1$. Martens et al. in [26] concluded that in more that 98 % of the XSDs occurring in practice the content model of an element depends only on the label of the element itself, the label of its parent, and (sometimes) the label of its grand-parent after an examination of 225 XSDs gathered from the Cover Pages. Later in 2007, Bex et al. [9] introduced the concept of k-local. An XSD is k-local if its content models depend only on labels up to the k-th ancestor [9]. For most cases in real world, the value of k is 1. This result is conformed with that in [8]. In 2014, inspired by Simplified CHARE, Feng et al. proposed eSimplified CHARE whose base symbol allows the forms of a and a^+ where $a \in \Sigma$. The data set used in [15] consists of 966 valid DTDs and XSDs which were rewritten as 2009 regular expressions. Based on this data set, the cover ratio of eSimplified CHARE reached 84.8 % from 79.5 % for Simplified CHARE. In addition, two inference algorithms for eSimplified CHARE were given.

In addition, regular expressions with counting and interleaving have been studied by many researchers. Björklund et al. were the first to make an incremental evaluation of regular expressions with counters. They gathered about 3024, 458 and 8830 regular expressions respectively from three libraries: RegExLib, Snort, and XML Schema on the web in [10]. The results show that there are 1705 out of 3024 (about 56.3 %), 270 out of 458 (about 58.9 %) regular expressions use non-trivial counters in RegExLib and Snort. Regular expressions with non-trivial counters are the forms excluding three specific ones: $a^{[0,0]}$, $a^{[0,1]}$ and $a^{[1,1]}$. In addition, the proportions of CHAINs are 73.3 %, 85.1 % and 86 % for the data of three libraries above. In [17], Ghelli et al. proposed a restricted subclass defined by $T::= \varepsilon|a^{[m,n]}|T{+}T|T{\cdot}T|T\&T$ where $m \in N\backslash\{0\}$ and $n \in N\backslash\{0\}\cup\{^*\}$. In $L(T)$, each alphabet symbol can appear at most once. Counter can only occur as a constraint for terminal symbols. For example, $(a? \cdot (b+c+d+e+f)^{[1,100]})$ is not allowed. Based on this subclass, they first proposed a linear-time translation algorithm of the translation of each regular expression into a set of constraints in [18]. Then they introduced a linear-time membership algorithm to check whether a word satisfies the resulting constraints.

5   Conclusion and Future Work

In this paper, we introduce a new restricted subclass of regular expression with counting and interleaving. The experiment results show that this subclass can cover more content models in real world. Then different features of five sub-classes of content models, i.e., CHARE, eCHARE, SORE, Simplified CHARE, eSimplified CHARE together with eCICHARE are also analyzed using different measures. We inspect their usages and give the corresponding proportions based on the real world data. Our data set is much larger than previous work which leads to more accurate results. We believe that our work will be helpful to the applications and further study of DTDs and XSDs. One future work is the study of inference algorithms and complexity problems related to eCICHARE. The strong and weak determinism of eCICHARE will also be considered. Besides, based on our experiment results, it is possible to propose other useful subclasses of content models.

References

1. Abiteboul, S., Buneman, P., Suciu, D.: Data on the Web: From Relations to Semi-structured Data and XML. Morgan Kaufmann, Burlington (2000)
2. Bala, S.: Intersection of regular languages and star hierarchy. In: Widmayer, P., Triguero, F., Morales, R., Hennessy, M., Eidenbenz, S., Conejo, R. (eds.) ICALP 2002. LNCS, vol. 2380, pp. 159–169. Springer, Heidelberg (2002)
3. Benedikt, M., Fan, W., Geerts, F.: XPath satisfiability in the presence of DTDs. J. ACM (JACM) 55(2), 8 (2008)
4. Bex, G.J., Gelade, W., Neven, F., Vansummeren, S.: Learning deterministic regular expressions for the inference of schemas from XML data. ACM Trans. Web (TWEB) 4(4), 14 (2010)
5. Bex, G.J., Martens, W., Neven, F., Schwentick, T.: Expressiveness of XSDs: from practice to theory, there and back again. In: Proceedings of the 14th International Conference on World Wide Web, pp. 712–721. ACM (2005)
6. Bex, G.J., Neven, F., Van den Bussche, J.: DTDs versus XML schema: a practical study. In: Proceedings of the 7th International Workshop on the Web and Databases: Colocated with ACM SIGMOD/PODS 2004, pp. 79–84. ACM (2004)
7. Bex, G.J., Neven, F., Schwentick, T., Tuyls, K.: Inference of concise DTDs from XML data. In: Proceedings of the 32nd International Conference on Very Large Data Bases, pp. 115–126. VLDB Endowment (2006)
8. Bex, G.J., Neven, F., Schwentick, T., Vansummeren, S.: Inference of concise regular expressions and DTDs. ACM Trans. Database Syst. (TODS) 35(2), 11 (2010)
9. Bex, G.J., Neven, F., Vansummeren, S.: Inferring XML schema definitions from XML data. In: Proceedings of the 33rd International Conference on Very Large Data Bases, pp. 998–1009. VLDB Endowment (2007)
10. Björklund, H., Martens, W., Timm, T.: Efficient incremental evaluation of succinct regular expressions. In: Proceedings of the 24th ACM International on Conference on Information and Knowledge Management, pp. 1541–1550. ACM (2015)
11. Brüggemann-Klein, A., Wood, D.: One-unambiguous regular languages. Inf. Comput. 140(2), 229–253 (1998)
12. Che, D., Aberer, K., Özsu, M.T.: Query optimization in XML structured-document databases. VLDB J. 15(3), 263–289 (2006)
13. Chen, H., Lu, P.: Checking determinism of regular expressions with counting. Inf. Comput. 241, 302–320 (2015)
14. Choi, B.: What are real DTDs like. Technical reports (CIS), p. 17 (2002)

15. Feng, X.Q., Zheng, L.X., Chen, H.M.: Inference algorithm for a restricted class of regular expressions. Comput. Sci. **41**(4), 178–183 (2014)
16. Gelade, W., Gyssens, M., Martens, W.: Regular expressions with counting: weak versus strong determinism. In: Královič, R., Niwiński, D. (eds.) MFCS 2009. LNCS, vol. 5734, pp. 369–381. Springer, Heidelberg (2009)
17. Ghelli, G., Colazzo, D., Sartiani, C.: Efficient inclusion for a class of XML types with interleaving and counting. In: Arenas, M. (ed.) DBPL 2007. LNCS, vol. 4797, pp. 231–245. Springer, Heidelberg (2007)
18. Ghelli, G., Colazzo, D., Sartiani, C.: Linear time membership in a class of regular expressions with interleaving and counting. In: Proceedings of the 17th ACM Conference on Information and Knowledge Management, pp. 389–398. ACM (2008)
19. Kilpeläinen, P.: Checking determinism of XML schema content models in optimal time. Inf. Syst. **36**(3), 596–617 (2011)
20. Koch, C., Scherzinger, S., Schweikardt, N., Stegmaier, B.: Schema-based scheduling of event processors and buffer minimization for queries on structured data streams. In: Proceedings of the 30th International Conference on Very Large Data Bases, vol. 30, pp. 228–239. VLDB Endowment (2004)
21. Manolescu, I., Florescu, D., Kossmann, D.: Answering XML queries on heterogeneous data sources. VLDB **1**, 241–250 (2001)
22. Martens, W., Neven, F.: Typechecking top-down uniform unranked tree transducers. In: Calvanese, D., Lenzerini, M., Motwani, R. (eds.) ICDT 2003. LNCS, vol. 2572, pp. 64–78. Springer, Heidelberg (2002)
23. Martens, W., Neven, F.: Frontiers of tractability for typechecking simple XML transformations. In: Proceedings of the Twenty-Third ACM SIGMOD-SIGACT-SIGART Symposium on Principles of Database Systems, pp. 23–34. ACM (2004)
24. Martens, W., Neven, F., Schwentick, T.: Complexity of decision problems for simple regular expressions. In: Fiala, J., Koubek, V., Kratochvíl, J. (eds.) MFCS 2004. LNCS, vol. 3153, pp. 889–900. Springer, Heidelberg (2004)
25. Martens, W., Neven, F., Schwentick, T.: Complexity of decision problems for XML schemas and chain regular expressions. SIAM J. Comput. **39**(4), 1486–1530 (2009)
26. Martens, W., Neven, F., Schwentick, T., Bex, G.J.: Expressiveness and complexity of XML schema. ACM Trans. Database Syst. (TODS) **31**(3), 770–813 (2006)
27. Papakonstantinou, Y., Vianu, V.: DTD inference for views of XML data. In: Proceedings of the 19th ACM SIGMOD-SIGACT-SIGART Symposium on Principles of Database Systems, pp. 35–46. ACM (2000)
28. Peng, F., Chen, H., Mou, X.: Deterministic regular expressions with interleaving. In: Rueda, C., et al. (eds.) ICTAC 2015. LNCS, vol. 9399, pp. 203–220. Springer, Heidelberg (2015). doi:10.1007/978-3-319-25150-9_13
29. Sperberg-McQueen, C.: Applications of Brzozowski derivatives to XML schema processing. In: Extreme Markup Languages®, Citeseer (2005)
30. Thompson, H.S., Beech, D., Maloney, M., Mendelsohn, N.: XML schema part 1: structures. 2nd edn. W3C Recommendation (2004)
31. Wang, G., Liu, M., Yu, G., Sun, B., Yu, G., Lv, J., Lu, H.: Effective schema-based XML query optimization techniques. In: 2003 Proceedings of Seventh International Database Engineering and Applications Symposium, pp. 230–235. IEEE (2003)
32. XML: XML.org Registry (2002). http://www.xml.org/xml/registry.jsp/

Research Short Paper

Chinese Microblog Sentiment Analysis Based on Sentiment Features

Weiwei Li[✉], Yuqiang Li, and Yan Wang

School of Computer Science and Technology,
Wuhan University of Technology, Wuhan 430063, China
13517248639@163.com, 17954183@qq.com

Abstract. As the microblog has increasingly become an information platform for netizens to share their ideas, the study on the sentiment analysis of microblog has got scholars' wide attention both at home and abroad. The primary goal of this research is to improve the accuracy of microblog sentiment polarity classification. With a view to the characteristics of microblog, a new method of semantically related feature extraction is proposed. Firstly, the Chinese word features are selected by text presentation in VSM and computing the weight by TF*IDF. Secondly, the proposed eight microblog semantic features are extracted, including sentence sentiment judgment based on emotional dictionary. Finally, three kinds of machine learning methods are used to classify the Chinese microblog under the feature vector combining the two methods. The experimental results indicate that the proposed feature extraction method outperforms the state-of-the-art approaches, and for this feature extraction algorithm, the classification performance is best when using the Naïve Bayes algorithm.

Keywords: Semantic feature · Sentiment classification · Machine learning · Microblog

1 Introduction

With the rapid development of social media on the Web, microblog is highly favored by users, and increasingly becomes the public network platform for people to share opinions because of its convenience, simplicity and real-time. Analyzing the subjective text aims to understand the people's attitude, viewpoint and reviews of entities, such as products, event and topics.

In this paper, we propose a sentiment classification model combining semantic features and lexical features of microblog. After studying on component elements of microblog, like links, emoticons, emotional words, sentences, etc., we select eight Chinese microblog features as the semantic features. Then the traditional feature vector is chosen by text presentation in vector space model and TF-IDF is adopted to calculate feature weight [1]. Finally the feature set obtained by two kinds of algorithm is combined for microblog sentiment classification. In order to examine the effect of the proposed feature extraction method, classification experiments are done using Support Vector Machine (SVM), Naïve Bayes and Decision Tree, and then their performances are

© Springer International Publishing Switzerland 2016
F. Li et al. (Eds.): APWeb 2016, Part II, LNCS 9932, pp. 385–388, 2016.
DOI: 10.1007/978-3-319-45817-5_30

compared. Experimental results show that our proposed feature selection method outperforms state-of-the-art semantic analysis methods.

2   Chinese Microblog Feature Research

The Chinese microblog introduces many unique attributes, such as emoticon, picture, hyperlink, hashtag etc., so it has rich semantic features. Making full use of the microblog specific features is helpful to improve the accuracy of classification. Our feature extraction is done in two steps: (1) Microblog-specific features are extracted, (2) Document representation based on Vector Space Model.

2.1   Microblog-Specific Features Extraction

Apart from the characteristics of short content, rich of unknown words and colloquial expressions, the microblog text has rich semantic features compared with normal text. So we selected eight Chinese microblog semantic features as shown in Table 1.

Table 1. The Chinese microblog features

Feature name	Feature description
The number of positive emoticons	Positive emoticons
The number of negative emoticons	Negative emoticons
Punctuation	"?" and "!"
Sentence number	Sentence number in a post
Number of positive sentence	Positive sentence in a post
Number of negative sentence	Negative sentence in a post
Emotion of the first sentence	The first sentence polarity: positive, negative and neutral
Emotion of the last sentence	The last sentence polarity: positive, negative and neutral

In the process of feature extraction on microblog sentence composition rules, we need to compute the number of positive and negative emotional sentences and judge the polarity of the first sentence and the tail sentence in the post. In the face of vast amount of microblog information, relying on the manually annotation will consume huge manpower and financial resources. We proposed a method referring emotional tendency of clause based on sentiment dictionary. The feature extraction that computes the emotional polarity of clause based on sentiment dictionary is done in three steps: (1) construction of sentiment dictionary, (2) construction of degree dictionary, (3) calculation of sentiment value in clause.

Extension and Improvement of HowNet-Based Sentiment Dictionary. The main source of emotional words in this paper is HowNet Sentiment Dictionary. The emotional words with stable polarity are selected to form an emotional dictionary. We add another category dictionary collected manually from internet. The final sentiment dictionary contains 8554 positive words and 9660 negative words.

Construction of Degree Dictionary. In Chinese text, emotion words are often accompanied by a lot of adverbs. The original sentiment of text is determined by the emotion words and subsequently modified by adverbs. In order to improve the accuracy of emotional computing, we establish the adverb of degree dictionary provided by LinHuang [2], and the negative adverb dictionary provided by Wang et al. [3].

Feature Extraction of Microblog Sentence Composition Rules. The sentiment analysis algorithm of clause is a lexicon-based method that depends on the extended emotion dictionary mentioned above.

2.2 Text Representation in Vector Space Model

After taking into account the semantic features of microblog, they are removed from microblog, microblog can be then considered as a normal text. The feature extracted by representing the text in vector space model can help to identify microblog's emotional polarity.

3 Experimental Results and Analysis

In our experiments, the corpus is taken from Chinese microblog emotion tagging dataset provided by 3rd CCF Conference on Natural Language Processing & Chinese Computing (NLP&CC2014)[1]. In order to verify the validity of the proposed feature extraction method, we use lexicon-based method, machine learning based on text representation using VSM and machine learning based proposed feature extraction method, respectively, to conduct the experiment. This experiment uses SVM as the classification algorithm. Figure 1 show the classification results.

Experimental results show that the F-Score of the proposed method are higher than those of the other two methods. It indicates that the added semantic related features of

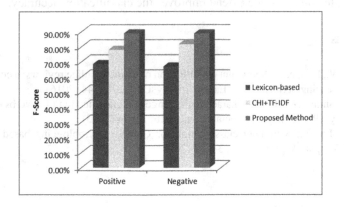

Fig. 1. The comparison of F-Score

---

[1] http://tcci.ccf.org.cn/conference/2014/pages/page04_tdata.html.

microblog aid in better sentiment analysis, in addition to the basic words features, semantic features also has a better performance. The classification performance of sentiment dictionary is low. This is because the content of microblog is colloquial, ambiguous and filled with acronym, internet vocabulary, the sentence structure is also irregular.

In order to study the classification performance of three classifiers, SVM, Naïve Bayes and decision tree, under two kinds of feature selection method. We conduct a comparison experiment using two kinds of feature extraction method. Table 2 shows that decision tree gets the worst classification performance compared with SVM and Naïve Bayes algorithm. When adopting the word feature selection method, the classification performance of SVM algorithm is optimal. When using proposed method, the Naïve Bayes attains the superior result, which can reach more than 89 %.

Table 2. Performance comparison of different classifiers in sentiment analysis

Classifier	Feature selection method			
	CHI + TF – IDF		Composite feature	
	Positive	Negative	Positive	Negative
SVM	78.0 %	81.7 %	88.9 %	88.8 %
Naïve Bayes	74.5 %	77.7 %	90.3 %	89.4 %
Decision Tree	54.2 %	70.5 %	66.9 %	68.5 %

4   Conclusion

In this paper, we study the Chinese microblog characteristics and provide eight semantic features. Considering the large amount of workload in judging the polarity of the sentence of microblog, this paper presents a clause emotion judgment approach based on emotion dictionary, which improves the speed of feature extraction. After that, the semantic features and basic features extracted by text representation using vector space model are integrated into one composite feature vector. Experimental results show that the semantic features to some extend improves the classification accuracy.

References

1. Hongwei, W., Lijuan, Z.: Sentiemnt classification of Chinese online reviews: a comparison of factors influencing performances. Enterp. Inf. Syst. **10**, 228–244 (2016)
2. Huang, L., Shuhui, G.: On the characteristic, range and classification of adverbs of degree. J. Shanxi Univ. (Philosophy and Social Science) **26**, 71–74 (2003)
3. Yong, W., Lu, X.: Sentiment classification for Chinese microblogging based on polarity lexicons. Comput. Appl. Softw. **1**, 34–37 (2014)

FVBM: A Filter-Verification-Based Method for Finding Top-k Closeness Centrality on Dynamic Social Networks

Yiyong Lin[1,2], Jinbo Zhang[1,2], Yuanxiang Ying[1,2],
Shenda Hong[1,2], and Hongyan Li[1,2(✉)]

[1] Key Laboratory of Machine Perception, Peking University, Ministry of Education,
Beijing 100871, China
{linyiyong,hongshenda,lihy}@cis.pku.edu.cn
[2] School of Electronics Engineering and Computer Science, Peking University,
Beijing 100871, China
cszhangjinbo@gmail.com, yingyuanxiang34@sina.com

Abstract. Closeness centrality is often used to identify the top-k most prominent nodes in a network. Real networks, however, are rapidly evolving all the time, which results in the previous methods hard to adapt. A more scalable method that can immediately react to the dynamic network is demanding. In this paper, we endeavour to propose a filter and verification framework to handle such new trends in the large-scale network. We adopt several pruning methods to generate a much smaller candidate set so that bring down the number of necessary time-consuming calculations. Then we do verification on the subset; which is a much time efficient manner. To further speed up the filter procedure, we incremental update the influenced part of the data structure. Extensive experiments using real networks demonstrate its high scalability and efficiency.

Keywords: Closeness centrality · Filter-Verification · Dynamic social network

1 Introduction

Finding Top-k Closeness Centrality in a network is within the most fundamental tasks in network analysis. Lots of previous work has devoted to decreasing the unique number of shortest path calculation [2,3]. Moreover, to make the calculation scalable, lots of previous work makes attempts in sacrificing precision to speed up the procedure. However, some new trends turn these previous work meaningless: Networks in real application are dynamic evolving in nature. It is unnecessary, unwise,impossible and unrealistic to recalculate from scratch every time the network changes as it will consume significant unnecessary calculation resource, which is precious in most applications.

In conclusion, we need a more efficient method to calculate the *top-k* closeness centrality on dynamic network, where it can quickly react to the modification of

© Springer International Publishing Switzerland 2016
F. Li et al. (Eds.): APWeb 2016, Part II, LNCS 9932, pp. 389–392, 2016.
DOI: 10.1007/978-3-319-45817-5_31

the network and return the results in a real-time manner. Lastly, the proposed method should be scalable to large scale network with millions or billion of nodes, which is often the case in nowadays applications. In this work, we propose FVBM (Filter-Verification-Based Method) to conduct real-time top-k closeness centrality search on dynamic evolving social networks.

2   Related Work

Δ−PFS [2] makes an attempt to share the intermedia computation result and also reduce redundant computation. This method, though, makes better advantage of the computation history, the time complexity remains the same of magnitude. The boost of big data analytic platform also shed light on this area, such as, P-$APSP$ firstly absorbed the distributed manner into computing the shortest path, which largely improve the complexity into $\mathcal{O}(n^3/p+logn)$. For the dynamic setting, $STREAMER$ [3] makes the first attempt, where it uses an internal data stream computing framework (DataCutter) to update the closeness centrality of influenced nodes in dynamic networks. $CENDY$ [4] also updates the closeness centrality of influenced nodes in dynamic networks. In most cases, the number of influenced nodes is sadly too large, sometimes even can be up to 75 %, which is almost the whole network.

3   Filter-Verification-Based Method (FVBM)

At time T_1 (initial time), we use the *Flajolet Martin Algorithm* [1] to generate the initial FM-Sketch Coding String, based on which we will generate the returned *Top-k* results. Whenever the network changes, we will first calculate the initially FM-Sketch affected vertices set and filter out a much smaller candidate subset to update the previous FM-Sketch. After updating FM-Sketch, we estimate closeness centrality of all nodes. We use the approximate closeness centrality to filter out a candidate set. Then, the distributed verification algorithm will be invoked to regenerate the results. There are some mature methods which can get the top-k closeness centrality in a small candidate set. So the distributed verification algorithm uses the mature method. This paper focuses on how to filter a small candidate set efficiently and scalably.

Incremental Optimization: When the network sharply evolves, the naive framework needs to update the *FM-Sketch* of all the vertices, which will turn into the bottleneck of the whole framework. However, when the network changes from $G_{\tau-1}$ to G_τ, only part of the *FM-Sketch* changes correspondingly. So we only updates part of the *FM-Sketch* that the changed edges influences.Based on this idea, we propose the incremental filter and verification version (FVBM-IO).

Distributed Optimization: For the distributed version (FVBM-IO-DO), we first partition the big network into smaller networks and then distribute the partitions into different machines. Each machine has its only copy. Then we modify the Algorithm FVBM-IO to distinguish between the local copy and the global version, where communication is conducted when necessary.

Algorithm 1. Filter-Verification-Based Method (FVBM)

Input: k, Network snapshort $G_{\tau-1}$ and Network snapshort G_τ
Output: Vertices link list $L(k, G_t)$ of Top-k Closeness Centrailty of G_t
for $i \leftarrow 1$ *to* D **do**
   for *each* $v \in V_t$ **do**
      for $u \in From(v, 1)$ **do**
         $\lfloor$ $FM_{\leq i}(v) \leftarrow FM_{\leq i}(v)$ OR $FM_{\leq i-1}(u)$
      $num \leftarrow Count(FM_{\leq i}(v)) - Count(FM_{\leq i-1}(v));$
      $S(v) \leftarrow S(v) + i \times num;$
      $R(v) \leftarrow R(v) + num;$

;   `/* Estimate Closeness Centrality */`
for *each* $v \in V$ **do**
   $\lfloor$ $\overline{C(v)} \leftarrow \frac{R(v)}{S(v)};$
$V_{k'} \leftarrow AfftedVertices(k, G_{\tau-1}, G_\tau);$
$L_c(k', G_t) \leftarrow Top(V_{k'}, \overline{C(*)});$
$L(k, G_t) \leftarrow Verification(L_c(k', G_t), Gt);$
return $L(k, G_t)$

---

4 Performance Studies

Dynamic Setting: To manually construct the dynamic network, we randomly select 600 edges from ego-Twitter and divide into 6 batches. Each batch has 100 edges (80 edges for insertion and 20 edges as deletion).

Precision: As the result in Fig. 1(a) shows, as the k' increases, the precision will increase. This is mainly because the larger k' is, the more vertex will be selected into the candidate set and the results is, of course, more accurate. Moreover, when k'/k is greater than 50, the precision is up to 100 %.

Overall Performance: For the details, as Table 1 shows, FVBM-IO-DO can search top-k closeness centrality efficiently on different networks. On most

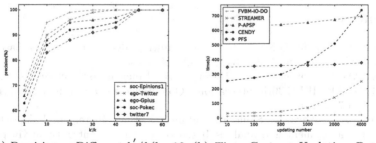

(a) Precision vs Different k'/k(k=10,   (b) Time Cost vs Updating Batch
updating batch size=100)   Size(k=10,network=ego-Twitter)

Fig. 1. Experiment result

Table 1. Time cost per updating batch on different networks (k = 10, updating-number = 100, in seconds)

	soc-Epinions1	ego-Twitter	ego-Gplus	soc-Pokec	twitter7
PFS	341	368	476	5301	> 24 h
CENDY	266	283	349	4980	> 24 h
P-APSP	624	638	1410	> 24 h	> 24 h
STREAMER	22	37	76	102	265
FVBM	186	195	213	289	335
FVBM-IO	93	101	180	245	270
FVBM-IO-DO	**5**	**13**	**26**	**36**	**44**

datasets, our *FVBM-IO-DO* is superior to other methods in magnitude. As Fig. 1(b) shows, when the updating batch size increases, the time cost of CENDY and STREAMER increases quickly. This is mainly because the number of vertices which CENDY and STREAMER need to deal will also increase quickly. P-APSP and $\Delta-$PFS remain "stable" even though the ratio increase. This is mainly because these two methods recompute closeness centrality of every node, moreover, the time cost to deal with network updating is relatively small. So the total time cost of P-APSP and $\Delta-$PFS only has a little increase.

5   Conclusion

In this work, we propose the *FVBM-IO-DO* to handle the problem of top-k closeness centrality search on a large-scale dynamic network. Several pruning methods are adopted in the traditional filter and verification framework. Comprehensive experiments on real network prove the applicability and scalability of the proposed method.

Acknowledgments. This work was supported by Natural Science Foundation of China (No. 61170003).

References

1. Flajolet, P., Martin, G.N.: Probabilistic counting algorithms for data base applications. J. Comput. Syst. Sci. **31**(2), 182–209 (1985)
2. Olsen, P.W., Labouseur, A.G., Hwang, J.H.: Efficient top-k closeness centrality search. In: 2014 IEEE 30th International Conference on Data Engineering (ICDE), pp. 196–207 (2014)
3. Sariyuce, A.E., Kaya, K., Saule, E., Catalyurek, U.V.: Incremental algorithms for network management and analysis based on closeness centrality, arXiv preprint (2013). arXiv:1303.0422
4. Yen, C.C., Yeh, M.Y., Chen, M.S.: An efficient approach to updating closeness centrality and average path length in dynamic networks. In: 2013 IEEE 13th International Conference on Data Mining (ICDM), pp. 867–876. IEEE (2013)

Online Hot Topic Detection from Web News Based on Bursty Term Identification

Chao Wang, Xue Zhao, Ying Zhang$^{(\boxtimes)}$, and Xiaojie Yuan

Nankai University, Tianjin 300353, People's Republic of China
{wangchao,zhaoxue,zhangying,yuanxiaojie}@dbis.nankai.edu.cn

Abstract. With the increment in the volume of information, it's almost impossible for people to assimilate all the news in time. A method to automatically detect hot topics from web news is strongly desired. Existing solutions take different perspectives ranging from identifying frequencies of terms to terms' distribution or part-of-speech characteristics. However, most of them are either too simplistic or unfitting to the properties of hot topics. Therefore, this paper presents a hot topic detection approach based on bursty term identification. We propose a new bursty term identification approach which considers both frequency and topicality properties to detect the bursty terms and hot topics. A series of experiments have demonstrated that our proposed approach has good performance compared with baseline methods.

Keywords: Bursty term · Weighting scheme · Hot topic detection

1 Introduction

Due to the development of internet, how to gain information is no more a problem. But the explosion of information has brought us another problem, information overload. The increment in the volume of digitized information has been far more beyond the capacity of humans to assimilate news.

The Topic Detection and Tracking (TDT) community has been trying to come up with a practical solution for many years. Most of TDT solutions use a term weighting scheme to capture representative terms. [1] proposed a scheme named TF*PDF, which assigns higher weights to terms that occur frequently in many documents and lower weights to those rarely mentioned. But it doesn't consider variation in the popularity over time. He *et al.* [2] detected the periodicity of terms with periodogram estimator, but only considered the period characteristic which is not enough to tell a term bursty. Lee *et al.* [3] selected several features and combined them together to develop a simple score function. But it is still too simplistic to represent characteristics of bursty terms.

This paper develops a novel bursty term identification approach. We present a bursty term weighting scheme assessing weight to terms from both perspectives of frequency and topicality. A series of experimentations have demonstrated that our proposed approach has good performance compared with baseline methods.

© Springer International Publishing Switzerland 2016
F. Li et al. (Eds.): APWeb 2016, Part II, LNCS 9932, pp. 393–397, 2016.
DOI: 10.1007/978-3-319-45817-5_32

2   Hot Topic Detection Approach

Terms are the basic components of news story, and topics are composed of many related news stories. Thus, changes in the topics will lead to variation of the usage of key terms necessarily, namely, bursty terms. In this paper, **frequency** and **topicality** are considered to identify them accurately:

2.1   Frequency Property

In order to describe the frequencies of terms, TF*PDF [1] is introduced:

Definition 1 (Term Weight). *Let $tf_{t,d}$ represent the frequency of term t at day d, $n_{t,d}$ represent the number of documents at day d where term t occurs, N_d represent the total number of documents at day d, D and T represent the total sets of date and term respectively. The term weight W_t of a given term t is*

$$W_t = \sum_{d \in D} |tf_{t,d}| exp(\frac{n_{t,d}}{N_d}), |tf_{t,d}| = tf_{t,d} / \sqrt{\sum_{t \in T} tf_{t,d}^2}; \tag{1}$$

2.2   Topical Property

In the following section, we introduce three features to indicate bursty terms.

Skewness: Bursty terms appear along with the related events. Taking the news propagation mode into account, news always appear in large numbers when the corresponding event occurs, and then taper and fade out slowly as time goes by. Accordingly, we assume the shape of the bursty term distribution will have a big head on the left and a long tail on the right. To model this property, we use the skewness. Skewness is a measure to describe the asymmetry of the probability distribution. Positive skew indicates that the tail on the right side is longer and thinner than the left side, the mass of the distribution is concentrated on the left, and vice versa. The definition of skewness score is shown as follow:

Definition 2 (Skewness Score). *The skewness score of a given term t is*

$$skew(t) = E(X(t) - \mu(X(t)))^3 / \sigma(X(t))^3, \tag{2}$$

where $X(t) =< tf_{t,1}, tf_{t,2}, \cdots, tf_{t,|D|}, >$, μ is the mean and σ is the deviation.

Timeliness: The occurrences of a bursty term is in accordance with its corresponding event, which leads to the term frequency changes greatly over time. Meanwhile some other terms appear frequently and the distributions don't change much. We call these terms absolute noise [4]. To reduce the impact of absolute noise, we introduce the timeliness feature of the term distribution. The more greatly the term frequency varies over time, the higher timeliness value it gets. The timeliness score is defined by χ^2 test as follow [5]:

Definition 3 (Timeliness Score). *The timeliness score of a given term t is*

$$timl(t) = \max_{d \in D} timl(t, d) = \max_{d \in D} \frac{(mq - np)^2}{(m+n)(m+p)(n+p)(n+q)}, \tag{3}$$

where m and n are the numbers of news stories containing and not containing term t at day d, p and q are the numbers of news stories containing and not and not term t at days but d.

Periodicity: Periodicity is important to determine whether a term is bursty, periodic terms are less likely to be bursty terms [2]. Hence, we decompose $X(t) = < tf_{t,1}, tf_{t,2}, \cdots, tf_{t,|D|} >$ into the sequence of $|D|$ complex numbers $< \hat{x}_1(t), \hat{x}_2(t), \cdots, \hat{x}_{|D|}(t) >$ via the Discrete Fourier Transform (DFT):

$$\hat{x}_k(t) = \sum_{d=1}^{|D|} X(t) \exp(-\frac{2\pi i}{|D|} k(d-1)), k = 1, 2, \cdots, |D|, \tag{4}$$

where $\hat{x}_k(t)$ denotes the amplitude of the sinusoid with frequency $k/|D|$.

The period of the original term frequency can be indicated by the dominate frequency, which can be determined by periodogram estimator [6]. The periodogram is defined as $< \|\hat{x}_1(t)\|^2, \|\hat{x}_2(t)\|^2, \cdots, \|\hat{x}_{|D|}(t)\|^2 >$, in which $\|\hat{x}_k(t)\|^2$ represents the signal power at frequency $k/|D|$. Thus, the dominant period can be chosen as the inverse of the frequency with the highest signal power. Accordingly, the periodicity score is defined as follow:

Definition 4 (Periodicity Score). *The periodicity score of a given term t is*

$$peri(t) = \frac{1}{\arg \max_k \|\hat{x}_k(t)\|^2}. \tag{5}$$

2.3   Term Weighting Scheme Based on Logistic Regression

In this section, logistic regression model is applied to combine the scores of different features together to build the final term weighting scheme as follow:

Definition 5 (Bursty Score). *The bursty score of a given term t is*

$$burst(t) = h_\Theta(X(t)) = \frac{1}{1 + exp(-\Theta^T X(t))}, \tag{6}$$

where $\Theta = \{\theta_1, \theta_2, \theta_3, \theta_4, b\}$, $X(t) = \{\log skew(t), \log timl(t), \log peri(t), \log W_t, 1\}$.

The values of Θ can be determined by maximizing the log-likelihood function (i.e., Quasi-Newton optimization method in this paper).

Table 1. Top-10 bursty terms

	TF*PDF	He	Lee	Proposed
1	peopl	presid	rodney	angele
2	writ	govern	davo	claim
3	articl	spac	govern	presid
4	wrong	angele	car	uk
5	who	offic	angele	riot
6	will	uiuc	evidenc	year
7	god	kill	reason	u.s.
8	govern	fir	bill	rodney
9	read	christ	key	fir
10	lot	u.s	kill	law

Table 2. Performances of top-k bursty term

Top-k	Method	Precision	Recall	F_1
Top-5	TF*PDF	0.1400	0.0425	0.0652
	He	0.2800	0.0928	0.1394
	Lee	0.3000	0.0905	0.1390
	Proposed	0.3400	0.1022	0.1572
Top-10	TF*PDF	0.1200	0.0722	0.0902
	He	0.2900	0.1781	0.2207
	Lee	0.3900	0.2238	0.2844
	Proposed	0.4200	0.2658	0.3256
Top-20	TF*PDF	0.1050	0.1366	0.1187
	He	0.3900	0.5223	0.4465
	Lee	0.3800	0.5601	0.4528
	Proposed	0.4250	0.5611	0.4837

3   Experiment

We use the 20 Newsgroups data set for experiments. The data set is a collection of 19,997 news stories collected from UseNet. For the purpose of evaluation, we labeled hot topics manually. We compare our proposed approach with three existing methods presented in [1–3].

Table 1 shows the extracted bursty terms by four considered methods. TF*PDF focuses on terms with high frequency rather than terms that really describe a topic. As a result, there are few named entities and noun phrases extracted. He's method pays more attentions to the terms with long period and high frequency, but it still tends to choose the time-related terms like 'christ'. Lee's method has a slightly better result since it uses several features to model the basic characteristics of bursty terms. But it's still too simplistic and unpersuasive, such as its periodicity feature is estimated manually. In contract, our proposed approach is able to filter out those non-bursty terms and find out the real bursty ones effectively. Terms 'angele', 'riot' and 'rodney' are related to a hot topic about *1992 Los Angeles riots*, the term 'presid' is a stemmed form of 'president' which is related to a hot topic about *1992 US Presidential Election*.

The performances are shown in Table 2. He's method is better than TF*PDF since it takes not only abruptness but also periodicity into consideration. Lee's approach tends to identify terms which have dramatic variation in frequency. It also takes account of the periodicity even if in a very simple way. Thus Lee's is the second best. Our proposed method uses the TF*PDF weights to represent the pervasive property of a term and three different features, skewness, timeliness and periodicity, to describe the term's topicality. The combination of TF*PDF and topicality improves the quality of bursty term identification.

Experiment results indicate that our proposed method outperform the others.

4 Conclusions

In this paper, we propose a novel methodology for hot topics detection. We develope a bursty term weighting scheme assessing the terms burstiness from the perspectives of pervasiveness and topicality. The experiments show that our approach yields more accurate results than traditional approaches.

Acknowledgement. This work is supported by Tianjin Municipal Science and Technology Commission under Grant No. 13ZCZDGX01098, 15JCTPJC62100 and 16JCQNJC00500 and the Fundamental Research Funds for the Central Universities of China.

References

1. Bun, K.K., Ishizuka, M.: Topic extraction from news archive using TF*PDF algorithm. In: International Conference on Web Information Systems Engineering, pp. 73–73. IEEE Computer Society (2002)
2. He, Q., Chang, K., Lim, E.P.: Analyzing feature trajectories for event detection. In: Proceedings of the 30th Annual International ACM SIGIR Conference on Research and Development in Information Retrieval, pp. 207–214. ACM (2007)
3. Lee, S., Lee, S., Kim, K., Park, J.: Bursty event detection from text streams for disaster management. In: Proceedings of the 21st International Conference Companion on World Wide Web, pp. 679–682. ACM (2012)
4. Li, H., Zhang, H., Qin, P., Yu, M., Liu, J.: Keywords based hot topic detection on internet [c]. In: Proceedings of the 5th CCIR, pp. 134–143 (2009)
5. Swan, R., Allan, J.: Automatic generation of overview timelines. In: Proceedings of the 23rd Annual International ACM SIGIR Conference on Research and Development in Information Retrieval, pp. 49–56. ACM (2000)
6. Welch, P.: The use of fast fourier transform for the estimation of power spectra: a method based on time averaging over short, modified periodograms. IEEE Trans. Audio Electroacoust. **15**(2), 70–73 (1967)

Grouped Team Formation in Social Networks

Ze Lv[1,2], Jianbin Huang[1(✉)], Yu Zhou[1,2], Heli Sun[3], and Xiaolin Jia[3]

[1] School of Software, Xidian University, Xi'an, China
jbhuang@xidian.edu.cn
[2] School of Computer Science and Technology, Xidian University, Xi'an, China
[3] Department of Computer Science and Technology, Xi'an Jiaotong University,
Xi'an, China

Abstract. Given an expert collaboration social network and a task, the team formation problem in social networks aims at forming a team which satisfies the skill requirements of the task with efficient collaboration. Different communication cost functions have been proposed in the existing work, but the grouped organization structure inside the team is not considered. With a novel communication cost function as objective function, we define the Grouped Team Formation problem. We propose an exact algorithm for solving the problem, and evaluate it by experiments.

1 Introduction

Lappas et al. [4] first bring the social network to the team formation problem. In the existing work, all the communication cost functions consider the cost among all the experts in the team, ignoring the internal grouped organization structure. In this paper, we propose the Grouped Team Formation problem with a novel communication cost function which considers the intra-costs among the group members in the same group and the inter-costs among the leaders. A group member only needs to communicate with his/her group leader and the other group members in the same group. A group leader only needs to communicate with his/her group members and the other leaders. To solve the problem, an exact algorithm is proposed. This algorithm is based on a method of solving a subproblem which can be constructed into an assignment problem, and a pruning method is proposed to filter the subproblems.

2 Problem Definition

Let $A = \{a_1...a_m\}$ be a set of m skills. Let $G = (V, E)$ to be an expert collaboration social network. Each expert node $v \in V$ owns a set of skills $\chi_v \subseteq A$; each edge $(u, v) \in E$ represents the previous collaboration, and the weight on the edge represents the communication cost between the two experts. We assume $P = \{g_i = (s_i, k_i) | 1 \leq i \leq q\}$ to be a project. g_i represents a group; $s_i \subseteq A$ represents the skillset required by the group; k_i is an integer, representing the number of experts that g_i requires; q is the number of groups. We assume l_P

© Springer International Publishing Switzerland 2016
F. Li et al. (Eds.): APWeb 2016, Part II, LNCS 9932, pp. 398–401, 2016.
DOI: 10.1007/978-3-319-45817-5_33

to be a project leader of project P, also the team leader. A grouped team is a set of q expert subsets, denoted by $T = \{V(g_1)...V(g_q)\}$. Each expert subset $V(g_i) \subseteq V$ represents a set of experts including a group leader $l_{g_i} \in V(g_i)$ in group g_i. For two expert nodes u and v, we define the communication cost between them as the shortest path distance of them, denoted by $d(u,v)$. We compute the shortest path distances between each pair of experts on the offline. The Grouped Leader Distance function is formulated as

$$C(T) = \sum_{g_i \in P} \sum_{v \in V(g_i)} d(v, l_{g_i}) + \sum_{g_i \in P} d(l_{g_i}, l_P) . \tag{1}$$

The total communication cost consists of two parts. To measure the communication costs among the group members in same group and among the leaders, we make reference to the Leader Distance proposed in [3].

The Grouped Team Formation problem can be formulated as

$$\min_{T} C(T) \tag{2}$$

$$\text{s.t.} : \forall v \in V, |\{g_i | g_i \in P \ and \ v \in V(g_i)\}| \leq 1$$

$$\forall 1 \leq i \leq q, |\{v | v \in V(g_i) \ and \ s_i \subseteq \chi_v\}| \geq k_i .$$

The first constraint is the expert assignment constraint. It ensures that each expert can be assigned to at most one group. The second constraint is the group cardinality constraint with the skill cover constraint which requires that each group has a specific number of experts with corresponding skills.

3   Assignment and Pruning Algorithm (AP)

In this section, we propose an exact algorithm called AP with its framework shown in Fig. 1.

Consider the subproblem: when given the group leaders for each group, how to find its optimal team with the minimum communication cost. In fact, this subproblem can be constructed into an assignment problem from the candidate group members to the groups. We form a matrix where rows represent candidate group members and columns represent groups as shown in Fig. 1(c). For each group, we add at most $|T| - 1$ candidate group members closest to the group leader, while the others will never be chosen in the optimal assignment. For each group g_i, we accordingly add $k_i - 1$ columns representing the groups. With the matrix as input, we can find the optimal assignment for the given group leaders by calling the auction algorithm [1].

We can find the optimal solution if we traverse every possible combination of the group leaders and get the optimal assignment. Next we construct the pruning lists $L = \{L_1...L_q\}$ to filter the combinations of group leaders as shown in Fig. 1(a). A list L_i corresponds to a group g_i. Given a group g_i and a candidate expert $v \in S(g_i)$, the group leader expectation cost $EC(v, g_i)$ is the minimum communication cost caused by v when v is the group leader l_{g_i}, i.e., $EC(v, g_i) = d(v, l_P) + \min_{V(g_i)} \sum_{u \in V(g_i)} d(v, u)$. Each tuple $t = (v, EC)$ consists of an expert

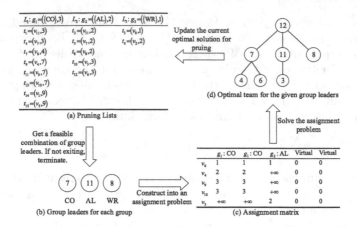

Fig. 1. Illustration of the procedure of the AP algorithm.

node v and the value of the group leader expectation cost $EC\,(v, g_i)$. The tuples in the same list are sorted in ascending order of E. We can guess that the optimal combination of group leaders should be in the upper layers of L. We adopt a pruning strategy which is similar to the upper bound algorithm in [2] to traverse the pruning lists layer by layer and filter the combinations of the group leaders. A set of p tuples $\{t_1...t_p\}$ is called a feasible combination, denoted by c, if each tuple is from a distinct list in L and has a distinct expert. A feasible combination c is called a full combination if $p = q$. The d-th layer is called depth d of L. Let $x_1^d...x_q^d$ be the values in the tuples each from $L_1...L_q$ at depth d of L. If there is no tuple in L_i at depth d, then $x_i^d = +\infty$. Suppose that here is a feasible combination c, and $L_{i_1}...L_{i_j}$ are the lists from which we have not selected any tuple for c. Then the lower bound communication cost of c at depth d is $lower(c)^d = \sum_{t=(v,EC)\in c} EC + x_{i_1}{}^d + ... + x_{i_j}{}^d$. At depth d, we can prune c if $lower(c)^d \geq C_{min}$, where C_{min} is the communication cost of the current optimal solution.

Initially, there is an empty combination. We traverse the tuples in L layer by layer and check if the tuple can be combined with any existing combination to become a new feasible combination. On each layer, for a new formed feasible combination, we can compute a lower bound for it, and check if it can be pruned. And for a full combination, we can compute its optimal assignment and update the current optimal solution. When no combinations are left, the algorithm terminates.

4   Experimrntal Evaluations

There are 100 nodes and 1,000 random linked edges in the dataset. Each node owns 5 skills that randomly selected from a total of 50 skills. We evaluate the efficiency of the AP algorithm by comparing with the exhaustive method (Eh)

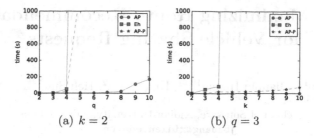

(a) $k = 2$ (b) $q = 3$

Fig. 2. The running time of the AP algorithm (AP), the exhaustive method (Eh) and the AP algorithm without pruning (AP-P).

and the AP algorithm without pruning (AP-P). Though the AP algorithm is far more efficient than the exhaustive method, it becomes time consuming when the input scale rises. So the comparison with the exhaustive method is on the small scale dataset. We can see that the exhaustive method and the AP algorithm without pruning work badly even on a small scale dataset. And the effect of the pruning method in the AP algorithm is very clear (Fig. 2).

5   Conclusions

Considering the grouped organization inside team, we define the Grouped Team Formation problem. The basic idea of our exact algorithm is as follows. When given the group leaders, we can transform it into a classical assignment problem. A pruning strategy is proposed to improve the efficiency of selecting the group leaders. Experiments verify the efficiency of our algorithm.

Acknowledgements. This work was supported by Natural Science Foundation of China (61474299, 61540008).

References

1. Bertsekas, D.P.: The auction algorithm: a distributed relaxation method for the assignment problem. Ann. Oper. Res. **14**(1), 105–123 (1988)
2. Feng, K., Cong, G., Bhowmick, S.S., Ma, S.: In search of influential event organizers in online social networks. In: Proceedings of the 2014 ACM SIGMOD International Conference on Management of Data, pp. 63–74. ACM (2014)
3. Kargar, M., An, A.: Discovering top-k teams of experts with/without a leader in social networks. In: Proceedings of the 20th ACM International Conference on Information and Knowledge Management, pp. 985–994. ACM (2011)
4. Lappas, T., Liu, K., Terzi, E.: Finding a team of experts in social networks. In: Proceedings of the 15th ACM SIGKDD International Conference on Knowledge Discovery and Data Mining. pp. 467–476. ACM (2009)

Profit Maximizing Route Recommendation for Vehicle Sharing Requests

Zhiqiang Zhao[1,2], Jianbin Huang[1(✉)], Hua Gao[1,2], Heli Sun[3], and Xiaolin Jia[3]

[1] School of Software, Xidian University, Xi'an, China
jbhuang@xidian.edu.cn
[2] School of Computer Science and Technology, Xidian University, Xi'an, China
[3] Department of Computer Science and Technology, Xi'an Jiaotong University,
Xi'an, China

Abstract. Vehicle sharing is a popular and important research in the knowledge discovery community and data mining. In this paper, we proposed a problem that recommends a group of requests to the driver to acquire the maximum profit. Simultaneously, these requests must satisfy some constraints, e.g. the request compatibility and the vehicle capacity. The request compatibility means all the requested routes can be merged into one common route without interruption. The solution to this problem which has three phases including Combination and Pruning, Compatibility Pruning and Recommendation can lead to the optimal result. Extensive experimental results show the effectiveness of problem and the value to the environment protection and economic profits.

1 Introduction

The dynamic ride sharing systems attract the interests from daily life [1,3]. Consequently, a number of dynamic ride sharing systems are available currently. In traditional ride sharing service, the passengers will make a request of their destination and some constraints such as waiting time and cost. The common objective of these systems is to maximize the driver earning and minimize the passenger cost. In this paper, we proposed a vehicle sharing route recommendation problem that considers route assignment for users' requests. The primary objective is to increase driver income and to decrease user cost. The problem aims to recommend an optimal group to the driver. In the group, all the requests should be compatible with each other which means their route sequence from source to destination could be combined into one sequence. In addition, our method can recommend a complete route sequence to the driver, rather than rough driving direction.

A good price function is meaningful for both taxi drivers and passengers. The price model should have following properties: (1) taxi fare for drivers is higher than that fare with single request; (2) taxi fare for each request is lesser than that single request pays. Based on the above two properties, we adopt the following price model [2]. For each request, the profit of him, denoted by $RFare$

© Springer International Publishing Switzerland 2016
F. Li et al. (Eds.): APWeb 2016, Part II, LNCS 9932, pp. 402–405, 2016.
DOI: 10.1007/978-3-319-45817-5_34

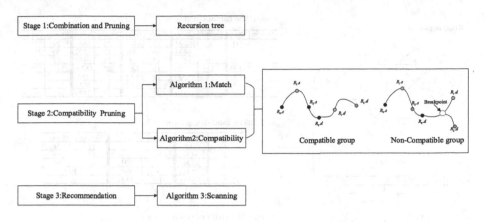

Fig. 1. Framework of the proposed approach.

which is defined in formula 1 where d_m is distance shared by m passengers and c is the number of sharing request. Let α be a float parameter which set by control center and p be the price per mile.

$$RFare = p\left(d_1 + \sum_{m=2}^{c}(1+\alpha)*d_m/m\right) \quad \alpha \in (0,1) \tag{1}$$

For a driver, the profit of him, denoted by $TFare$ which is defined in formula 2, is calculated as the following function which add the $RFare$ of each request. D_n is total distance and D_r is sharing distance.

$$TFare = D_n + (1+\alpha)\,D_r \quad \alpha \in (0,1) \tag{2}$$

In this study, we have considered a dynamic ridesharing route recommendation problem. Given a set of taxi T traveling on the road, where each $t \in T$ has a current origin location, a destination, and a set of requests R. We aim to find an optimal combination of requests R^* which is defined in formula 3, where $ST(R)$ represents the compatible group.

$$R^* = \arg\max\left(TFare\left(ST\left(R\right)\right)\right) \tag{3}$$

2   The Proposed Approach

The framework of our proposed approach is shown in the Fig. 1. It contains three stages: (1) Combination and Pruning; (2) Compatibility Pruning; (3) Recommendation. In the first stage, The inputs are the set of requests and a vehicle that provides sharing service. The output is the combination table which is the possible groups of requests. When passengers request to share the vehicle, they will request their source, destination and the number of passengers. Then the

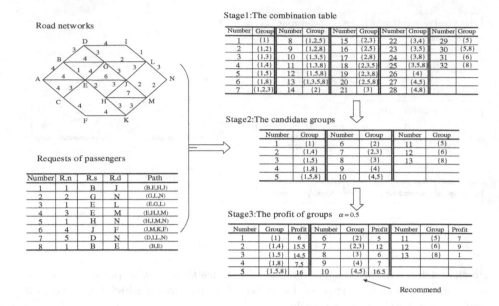

Fig. 2. The example of the proposed approach.

sequence of route will be produced by Dijkstra algorithm. We all know that the capacity of one vehicle is finite. Consequently, the sum of passengers in one group should be less than the capacity of a vehicle. In the second stage, the input is the combination table and the output is the candidate group table. Through the first stage, a set of groups through combining the requests have been produced. However, not all the groups can be served by a vehicle because some of them are not compatible. Consequently, the unsatisfied groups would be removed in this section. In order to figure out whether a group is a Compatibility Group, a Match algorithm and Compatibility algorithm can be introduced. In the last stage, the input is the candidate group and the output is the optimal group. Figure 2 is a complete example of the proposed approach. The road networks have 14 vertexes and 25 edges. The weight above the edge is the distance of two vertexes. The results of each stage have been shown in the Fig. 2.

3   Performance Evaluation

We evaluate the effectiveness of the problem based on the real world network and synthetic data. The real part comes from the road network of San Francisco, CA, USA, containing 223,606 edges and 175,343 nodes. In experiment, we chose a small area of the real network data which contains 1,000 nodes and 1,169 edges. The synthetic data contain 1,000 nodes and 1,200 edges.

This section studies the distance of sharing (SD) and non-sharing (NSD) when the requests are served by a vehicle. SD represents total distance among passengers who share a vehicle and NSD represents the sum of each single distance of passengers who share a vehicle. In order to show the effectiveness, the experiment vary the capacity from 4 to 13, while fixing the number of requests to 150. The requests are generated with uniformly distribution. We calculate the ratio which indicates the extent of reduction in the distance around the vehicle sharing by $\frac{NSD-SD}{NSD}$. The results have been showed in the Table 1.

Table 1. The ratio of the reducing distance for vehicle sharing in two dataset.

Capacity	4	5	6	7	8	9	10	11	12	13
Real	0.315	0.304	0.287	0.322	0.32	0.339	0.313	0.332	0.307	0.312
Synthetic	0.359	0.326	0.346	0.354	0.347	0.358	0.378	0.342	0.334	0.338

In Table 1, the minimum ratio is 0.287 and 0.326 and the average ratio is 0.315 and 0.348 in real networks and synthetic data respectively. Consequently, the distance of vehicle driving has been reduced, which obviously shows the value of the sharing problem for urban environment protection and energy saving.

4   Conclusion

This paper proposed a vehicle sharing route recommendation problem, which can accept multiple requests. Drivers who are willing to share the vehicle can register their current locations and the capacity of vehicles. Meanwhile, the riders request the sharing system and their current location, destination and the number of passengers. Then given the price function the system will recommend an optimal group to the driver according to the principle of the Combination and Pruning, Compatibility Pruning and Recommendation. Extensive experiment evaluation shows the effectiveness of this problem.

Acknowledgement. This paper is supported by the Natural Science Foundation of China (No. 61474299 and No. 61540008).

References

1. Huang, J., Huangfu, X., Sun, H., Li, H.: Backward path growth for efficient mobile sequential recommendation. IEEE Trans. Knowl. Data Eng. **27**(1), 46–60 (2013)
2. Ma, S., Zheng, Y., Wolfson, O.: T-share: a large-scale dynamic taxi ridesharing service. In: IEEE 29th International Conference on Data Engineering (ICDE), pp. 410–421. IEEE (2013)
3. Qian, S., Cao, J., Moul, F.L., Le, R., Sahel, I., Li, M.: SCRAM: a sharing considered route assignment mechanism for fair taxi route recommendations. In: ACM SIGKDD Conference on Knowledge Discovery and Data Mining, pp. 955–964 (2015)

Ontology-Based Interactive Post-mining of Interesting Co-location Patterns

Xuguang Bao, Lizhen Wang[✉], and Hongmei Chen

Department of Computer Science and Engineering,
School of Information Science and Engineering, Yunnan University,
Kunming, 650091, China
bbaaooxx@163.com, lzhwang@ynu.edu.cn

Abstract. Spatial co-location patterns represent the subsets of spatial features whose instances are frequently located together in geographic space. Common frameworks for mining co-location patterns generate numerous redundant co-location patterns. Thus, several methods were proposed to overcome this drawback. However, most of these methods do not guarantee that the extracted co-location patterns are interesting for the user because they are generally based on statistical information. Thus, it is crucial to help the decision-maker choose interesting co-location patterns with an efficient interactive procedure. This paper proposed an interactive approach to prune and filter discovered co-location patterns. First, ontologies were used to improve the integration of user knowledge. Second, an interactive process was designed to collaborate with the user to find the interesting co-location patterns efficiently. The experimental results on a real data set demonstrated the effectiveness of our approach.

Keywords: Spatial data mining · Co-location pattern mining · Interesting patterns · Interactive feedback

1 Introduction

Discovering interesting co-location patterns is an important task in spatial data mining [1]. However, a common problem in most co-location mining algorithms is that there are too many co-location patterns in the output while only a few of them is really interesting to a user. To overcome this drawback, several methods were proposed in the literature such as co-location pattern concise representations [2, 3], redundancy reduction [4], and post processing [5, 6]. However, most of these methods don't guarantee that the extracted co-location patterns are interesting for the user, since pattern interestingness strongly depends on user knowledge and goals. Thus, the pattern-finding methods should be imperatively based on a strong interactivity with the user.

The representation of user knowledge is an important issue. In the Semantic Web field, ontology [5] is considered as the most appropriate representation to express the complexity of the user knowledge, and several specification languages were proposed.

This paper proposed a new interactive approach, OICPP (Ontology-based Interesting Co-location Patterns Post-mining), to find interesting co-location patterns.

© Springer International Publishing Switzerland 2016
F. Li et al. (Eds.): APWeb 2016, Part II, LNCS 9932, pp. 406–409, 2016.
DOI: 10.1007/978-3-319-45817-5_35

Instead of requiring the user to explicitly express his real interesting co-location patterns, we alleviate the user's burden by only asking him to choose a small set of sample co-location patterns according to his interest.

As shown in Fig. 1, OICPP takes a set of candidates as input. In each round, a small collection (e.g., 10) of sample co-location patterns is selected from the candidates and is asked for the user's preference. The user chooses his interested co-location patterns, and the feedback information will be used by the ontology to update the candidates, and then OICPP decides which co-location patterns to be selected for the next feedback. The interaction continues for several rounds. Finally, the interesting co-location patterns will be output.

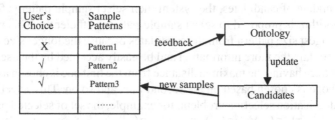

Fig. 1. Interactive process description

2 Problem Statement

Suppose F is a set of spatial features, S is a set of their instances and R is a spatial neighbor relationship over S. If the Euclidean metric is used as the neighbor relationship R, two spatial instances are neighbors if the ordinary distance between them is no more than a given threshold d. A **co-location pattern** c ($c \subseteq F$) is a subset of spatial features whose instances form cliques under the neighbor relationship frequently. **Participation index (PI)** is often used as a measure of the prevalence of a co-location pattern. Given a minimum prevalence threshold min_prev, a co-location c is a **prevalent co-location** if $PI(c) \geq min\_prev$ holds.

Formally, an ontology is a quintuple $O = \{C, R, I, H, A\}$, C is a set of *concepts* and R is a set of relations defined over *concepts*. I is a set of instances of *concepts* and H is a Directed Acyclic Graph defined by the inclusion relation (*is-a* relation, $\leq$) between *concepts*. A is a set of axioms bringing additional constraints on the ontology.

This paper focuses on finding interesting co-location patterns interactively with the ontologies.

3 Strategies for Updating Candidates and Selecting Samples

Definition 1. Given two co-location patterns c_1 and c_2, the semantic distance between c_1 and c_2 is defined as $SD(c_1, c_2) = 1 - \dfrac{|C(c_1) \cap C(c_2)|}{|C(c_1) \cup C(c_2)|}$, where $C(c)$ is the generalization of c and demonstrates the set of *concepts* which contain c directly.

Once the user feedbacks his preference on the provided samples, the semantic distance measure then is used to find the "similar" co-location patterns from the candidates for each co-location pattern from the user's feedback.

Definition 2. Given a co-location c and the semantic distance threshold sdt $(0 \leq sdt \leq 1)$, if there exists a co-location c' that $SD(c, c') \leq sdt$, we call c' is a *similarity* of c, particularly, if $SD(c, c') = 0$, c' is called a *hard similarity* of c.

To update candidates, the system first discovers all the *similarities* of each co-location pattern by the user's feedback, and then moves the interesting co-location patterns and their *similarities* to the result set, discards uninteresting patterns and their *similarities* from the candidates.

After the update of candidates, the system then select sample patterns for the user. A greedy algorithm is proposed to select sample patterns efficiently. This method first chooses the longest size pattern from the candidates because the longer size co-location patterns may contain the more information and be easily accepted by the user. Then, the co-location pattern having the maximal distance from the first one is chosen as the second one; the third one is chosen based on the second one, and so on. This selection method may cause a duplication selection problem, for example, the set of selected patterns may be $s = \{c_1, c_2, c_3, c_4\}$, $C(c_1) = C(c_3)$, $C(c_2) = C(c_4)$, the selection doesn't satisfy our selection criteria that the selected patterns should contain more ontology *concepts*, thus, the current chosen co-location pattern c should satisfy that it has the maximal distance from the previous chosen co-location pattern and $C(c)$ is not equal to the generalization of any of the already-chosen co-location patterns.

4   Experiments

4.1   Experiment Setting

We used points of interest (POI) in Beijing for this experiment. The number of features is 16, the number of instances is 90458, the range is 18 km * 18 km, the prevalence threshold is set as low as 0.1 because we want some rare co-location patterns, and the neighbor distance is 100 m.

In this experiment, OICPP takes all the closed co-location patterns as the input. 92 closed co-location patterns are generated from the real dataset. The default k value is 5 and the default sdt value is 0.1. The user selects 23 interesting co-location patterns.

4.2   Experiment Results

We mainly measure the accuracy and efficiency of selecting strategy in this section. The accuracy measure is defined as $Accuracy = \dfrac{P \cap Q}{P \cup Q}$, where P is the interesting co-location patterns selected by the user, Q is the interesting co-location patterns discovered by OICPP interactively with the user. In order to test the efficiency of selecting method, we use a random method which selects k candidate randomly in each round for comparison.

Figure 2(a) and (b) show the accuracy of OICPP, Fig. 2(c) and (d) show the efficiency of selecting method. From Fig. 2 we can see that our approach OICCP has a high accuracy and efficiency of selecting strategy.

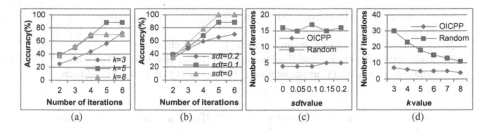

Fig. 2. Experiment results

5 Conclusion

In this paper, we proposed an interactive approach to find interesting co-location patterns based on the preference of the user. We measured the semantic distance between two co-location patterns and gave an efficient method to reduce the number of candidates. Furthermore, a selection strategy is proposed to select the best sample co-location patterns for the feedback. The experimental results show the high accuracy and efficiency of our algorithm.

Acknowledgements. This work was supported in part by grants (No. 61472346, No. 61262069) from the National Natural Science Foundation of China and in part by a grant (No. 2016FA026, No. 2015FB149, and No. 2015FB114) from the Science Foundation of Yunnan Province.

References

1. Huang, Y., Shekhar, S., Xiong, H.: Discovering co-location patterns from spatial data sets: a general approach. IEEE Trans. Knowl. Data Eng. (TKDE) **16**(12), 1472–1485 (2004)
2. Yoo, J.S., Bow, M.: Mining top-k closed co-location patterns. In: IEEE International Conference on Spatial Data Mining and Geographical Knowledge Services, pp. 100–105 (2011)
3. Wang, L., Zhou, L., Lu, J., et al.: An order-clique-based approach for mining maximal co-locations. Inf. Sci. **179**(2009), 3370–3382 (2009)
4. Xin, D., Shen, X., Mei, Q., et al.: Discovering interesting patterns through user's interactive feedback. In: 12th ACM SIGKDD International Conference on Knowledge Discovery and Data Mining, pp. 773–778 (2006)
5. Marinica, C., Guillet, F.: Knowledge-based interactive postmining of association rules using ontologies. IEEE Trans. Knowl. Data Eng. (TKDE) **22**(6), 784–797 (2010)
6. Bao, X., Wang, L., Fang, Y.: OSCRM: a framework of ontology-based spatial co-location rule mining. J. Comput. Res. Dev. **52**(Suppl), 74–80 (2015)

AALRSMF: An Adaptive Learning Rate Schedule for Matrix Factorization

Feng Wei, Hao Guo, Shaoyin Cheng$^{(\boxtimes)}$, and Fan Jiang

School of Computer Science and Technology,
University of Science and Technology of China, Hefei 230027, China
{enable,ghao}@mail.ustc.edu.cn, {sycheng,fjiang}@ustc.edu.cn

Abstract. Stochastic gradient descent (SGD) is an effective algorithm to solve matrix factorization problem. However, the performance of SGD depends critically on how learning rates are tuned over time. In this paper, we propose a novel per-dimension learning rate schedule called AALRSMF. This schedule relies on local gradients, requires no manual tunning of a global learning rate, and shows to be robust to the selection of hyper-parameters. The extensive experiments demonstrate that the proposed schedule shows promising results compared to existing ones on matrix factorization.

Keywords: Adaptive learning rate · SGD · Matrix factorization

1 Introduction

Let $R \in \mathbb{R}^{m \times n}$ be a target matrix, matrix factorization finds two latent factor matrices $P \in \mathbb{R}^{k \times m}$ and $Q \in \mathbb{R}^{k \times n}$ which accurately approximate the original matrix R. The problem can be formulated as the following objective function:

$$min \sum_{(u,v) \in \Omega} (r_{u,v} - p_u^T q_v)^2 + \lambda(\|p_u\|^2 + \|q_v\|^2). \tag{1}$$

SGD is a basic solution to matrix factorization, it randomly picks a (u, v)-th entry and iteratively updates the learning parameters by the product of the gradient and a pre-specified learning rate η, the gradient is listed as follows:

$$g_u = -e_{u,v} q_v + \lambda p_u, \ h_v = -e_{u,v} p_u + \lambda q_v. \tag{2}$$

As far as we know, the performance of SGD relies heavily on its learning rate schedule. There exist several learning rate schedules for SGD-based matrix factorization including fixed schedule (FS) [3], monotonically decreasing schedule (MDS) [4], bold driver schedule (BDS) [2], and reduced per-coordinate schedule (RPCS) [1]. The major practical drawback of all these schedules is that they are all required to tune a global learning rate manually.

In this paper, based on ADADELTA [5], we propose a novel per-dimension learning rate schedule called AALRSMF for matrix factorization. The proposed schedule does not require to set a global learning rate manually and appears robust to the selection of hyper-parameters.

F. Li et al. (Eds.): APWeb 2016, Part II, LNCS 9932, pp. 410–413, 2016.
DOI: 10.1007/978-3-319-45817-5_36

2 Our Schedule

2.1 AALRSMF

In order to dramatically reduce the memory usage and computational complexity compared with ADADELTA, we decide to apply the same learning rate to all elements in p_u and q_v. Specifically, if $r_{u,v}$ is sampled, our schedule adjusts the decaying average $E[g_u^2]$ and $E[h_v^2]$ at each iteration by the following rules:

$$E[g_u^2] = (1 - \rho)E[g_u^2] + \rho \frac{g_u^T g_u}{k}, \ E[h_v^2] = (1 - \rho)E[h_v^2] + \rho \frac{h_v^T h_v}{k}, \qquad (3)$$

where $E[g_u^2]$ is the u-th element of $E[g^2]$ and $E[h_v^2]$ is the v-th element of $E[h^2]$. For $E[\Delta p_u^2]$ and $E[\Delta q_v^2]$, we use the same approach to update them.

The learning rates of p_u and q_v are shown below. Because $E[\Delta p_u^2]$ and $E[\Delta q_v^2]$ for the current iteration are not known, we decide to use $E[\Delta p_u^2]$ and $E[\Delta q_v^2]$ of the previous iteration. Note that $E[\Delta p_u^2]$ and $E[\Delta q_v^2]$ are initialized to zero.

$$\eta_u = \frac{RMS[\Delta p_u]}{RMS[g_u]}, \ \eta_v = \frac{RMS[\Delta q_v]}{RMS[h_v]}. \qquad (4)$$

2.2 An Annealing Scheme

We conduct some experiments on three datasets to validate the proposed schedule. The result shows that the loss function Eq. (1) exhibits high oscillatory and difficult convergence. We observe that the learning rate at the beginning is too high for each dataset, and even worse, increases over time.

To alleviate this problem, we add the squared learning rate to the denominator of Eq. (4) at each iteration, the detail is listed as follows:

$$\eta_u = \frac{RMS[\Delta p_u]}{RMS[g_u] + S_u}, \ \eta_v = \frac{RMS[\Delta q_v]}{RMS[h_v] + W_v}, \qquad (5)$$

where S_u and W_v are the accumulation of the squared learning rates at each iteration. The detail of our schedule is shown in Algorithm 1.

3 Experiments

3.1 Experimental Setting

Throughout our experiments, we evaluate the proposed schedule on three datasets: Netflix, Yahoo! Music, and MovieLens latest. We compare all schedules in terms of root mean square error (RMSE) on the test set.

3.2 Comparison on Convergence Speed Among Schedules

We evaluate the proposed schedule by comparing it with existing ones. Figure 1 shows the RMSE change of each schedule on three datasets. These results show that the proposed schedule achieves better convergence quality at faster speed than other schedules.

Algorithm 1. One iteration of SGD algorithm when AALRSMF is applied.

1   $e_{u,v} = r_{u,v} - p_u^T q_v$
2   $\overline{E} = 0,\ \overline{F} = 0,\ \overline{G} = 0,\ \overline{H} = 0,\ g_u,\ h_v$
3   **for** $d = 1, \cdots, k$ **do**
4     $(g_u)_d = -e_{u,v}(q_v)_d + \lambda(p_u)_d$
5     $(h_v)_d = -e_{u,v}(p_u)_d + \lambda(q_v)_d$
6     $\overline{E} = \overline{E} + (g_u)_d^2,\ \overline{F} = \overline{F} + (h_v)_d^2$
7   **end**
8   $E[g_u^2] = (1 - \rho)E[g_u^2] + \rho\overline{E}/k,\ E[h_v^2] = (1 - \rho)E[h_v^2] + \rho\overline{F}/k$
9   calculate η_u and η_v by Equation (5)
10   **for** $d = 1, \cdots, k$ **do**
11     $(p_u)_d = (p_u)_d - \eta_u(g_u)_d$
12     $(q_v)_d = (q_v)_d - \eta_v(h_v)_d$
13   **end**
14   $\overline{G} = \eta_u^2\overline{E},\ \overline{H} = \eta_v^2\overline{F}$
15   $E[\Delta p_u^2] = (1 - \rho)E[\Delta p_u^2] + \rho\overline{G}/k,\ E[\Delta q_v^2] = (1 - \rho)E[\Delta q_v^2] + \rho\overline{H}/k$
16   $S_u = S_u + \eta_u^2,\ W_v = W_v + \eta_v^2$

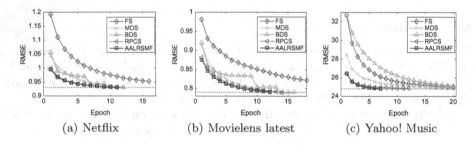

(a) Netflix   (b) Movielens latest   (c) Yahoo! Music

Fig. 1. Comparison among schedules for SGD-based matrix factorization.

3.3 Comparison on Hyper-Parameters Selection

In Table 1, we vary the initial learning rate for each schedule and show RMSE after 20 epochs of training on Netflix dataset. For FS, MDS, BDS, and PRCS, the initial learning rate needs to be set to the correct order of magnitude, above

Table 1. RMSE on Netflix after 20 epochs of training for various hyper-parameter settings using FS, MDS, BDS, and RPCS.

	FS	MDS	BDS	RPCS
$\eta_0 = 1e - 1$	1.1296	**0.9298**	1.1296	**0.9244**
$\eta_0 = 1e - 2$	**0.9340**	0.9573	**0.9268**	0.9904
$\eta_0 = 1e - 3$	1.0382	1.1557	0.9997	1.2542
$\eta_0 = 1e - 4$	1.3311	1.5700	1.2468	2.7222

Table 2. RMSE on Netflix after 20 epochs for various hyper-parameter settings using AALRSMF.

	$\rho = 0.5$	$\rho = 0.1$	$\rho = 0.05$
$\epsilon = 1e - 2$	0.9302	0.9298	0.9299
$\epsilon = 1e - 3$	0.9253	0.9251	0.9247
$\epsilon = 1e - 4$	0.9249	0.9247	**0.9245**
$\epsilon = 1e - 5$	0.9308	0.9278	0.9275

which the results obviously diverge and below which the training process proceeds slowly. We can see these results are highly various for each schedule, especially compared to AALRSMF in Table 2.

4 Conclusion

In this paper, we propose a novel per-dimension learning rate schedule called AALRSMF for matrix factorization, which requires no manual tunning of learning rates and shows to be robust to the selection of hyper-parameters. From our extensive evaluations, we conclude that AALRSMF outperforms existing ones.

Acknowledgement. The numerical calculations in this paper have been done on the supercomputing system in the Supercomputing Center of University of Science and Technology of China.

References

1. Chin, W.-S., Zhuang, Y., Juan, Y.-C., Lin, C.-J.: A learning-rate schedule for stochastic gradient methods to matrix factorization. In: Cao, T., Lim, E.-P., Zhou, Z.-H., Ho, T.-B., Cheung, D., Motoda, H. (eds.) PAKDD 2015. LNCS, vol. 9077, pp. 442–455. Springer, Heidelberg (2015)
2. Gemulla, R., Nijkamp, E., Haas, P.J., Sismanis, Y.: Large-scale matrix factorization with distributed stochastic gradient descent. In: Proceedings of the 17th ACM SIGKDD International Conference on Knowledge Discovery and Data Mining, pp. 69–77. ACM (2011)
3. Koren, Y., Bell, R., Volinsky, C., et al.: Matrix factorization techniques for recommender systems. Computer **42**(8), 30–37 (2009)
4. Yun, H., Yu, H.F., Hsieh, C.J., Vishwanathan, S., Dhillon, I.: Nomad: non-locking, stochastic multi-machine algorithm for asynchronous and decentralized matrix completion. Proc. VLDB Endowment **7**(11), 975–986 (2014)
5. Zeiler, M.D.: Adadelta: an adaptive learning rate method. arXiv preprint arXiv:1212.5701 (2012)

A Graph Clustering Algorithm for Citation Networks

Bo Zhang, Tiezheng Nie[✉], Derong Shen, Yue Kou, Ge Yu, and Ziwei Zhou

Northeastern University, Shenyang, China
zhangbo@stumail.neu.edu.cn,
{nietiezheng,shendr,kouyue,yuge}@mail.neu.edu.cn,
zhouziweivivian@gmail.com

Abstract. In this paper, we propose a novel network clustering algorithm, called CPSCAN for detecting the communities of citation network based on the temporal feature. Firstly, with combining temporal interval and citation path, a structural similarity model and a clustering algorithm is proposed. Then we propose a new modularity for measuring the quality of clustering on citation network. In empirical evaluation, we compare our method with existing methods on real world datasets. The experimental results demonstrates our algorithm has better performance on citation network than others methods.

Keywords: Network clustering · Temporal feature · Citation network · Modularity

1 Introduction

Many existing approaches have been proposed for addressing the problem of community detection, such as *betweenness* [3], *scan* [4], *modularity* [1], scan ++ [2], linkscan* [5] and so on. However, all of existing approaches cannot be used for detecting clusters of citation networks since they have has a great deal of differences with social networks. In this paper, we propose a new approach for clustering citation networks called CPSCAN (Citation Path Structural Clustering Algorithm for Networks). Our approach focuses on discovering clusters in a large time interval. We consider both the temporal feature of vertices and citation paths between two connected vertices as similarity criteria when clustering vertices. Therefore, vertices that have more connecting paths in a long time span will be grouped into the same cluster. It makes sense because techniques expressed by papers or patents always have a long life cycle. We will also propose a new modularity model for measuring the clusters of citation network based on the temporal feature.

2 The Notion of Clustering on Citation Network

We first model a citation network as a directed graph $G = \{V, E, T\}$, where V is a set of vertices mapping to nodes, E is the set of edges which connect distinct vertices, and T is set of time span for each edge $e \in E$. In our method, we consider both connectivity and temporal interval between vertices. When a set of vertices are in the same cluster,

F. Li et al. (Eds.): APWeb 2016, Part II, LNCS 9932, pp. 414–418, 2016.
DOI: 10.1007/978-3-319-45817-5_37

they should have higher structure similarity than that out of the cluster. Therefore, we define the *Structural Similarity* of vertices v and u as $Sim(v, u)$ which is calculated with Formula (1), where $SP(v, u)$ denotes the *Path Similarity* from *disjoin indirect citing path*, and $SN(v, u)$ denotes the *Neighbors Similarity*. The *Path Similarity* is computed based on citation neighbors of two vertices. Here, we only consider *disjoin indirect citing path* which has no overlap node with each other and is the shortest path among overlapping paths. The *Path Similarity* is calculated with Formula (2), where $\Gamma(v)$ denotes the neighborhood of v and $P(v, u)$ is the set of *disjoin indirect citing paths*. For $SN(v, u)$, we consider the effect of temporal feature. So we define a function $d(x)$ to calculate the temporal influence of a neighbor w in Formula (3), where λ is the parameter for affecting ratio, and $t(w)$ is the temporal coefficient from w and citation pair(v, u). Then $SN(v, u)$ is defined as in Formula (4).

$$Sim(v, u) = SP(v, u) + SN(v, u) \tag{1}$$

$$SP(v, u) = \frac{|P(v, u)|}{\sqrt{|\Gamma(v)||\Gamma(u)|}} \tag{2}$$

$$d(w) = \begin{cases} e^{\lambda t(w)} & x < 0 \\ 2 - e^{-\lambda t(w)} & x \geq 0 \end{cases} \tag{3}$$

$$SN(v, u) = \frac{\sum_{w \in \Gamma(v) \cap \Gamma(u)} d(w)}{\sqrt{|\Gamma(v)||\Gamma(u)|}} \tag{4}$$

In this paper, we also propose a new model of modularity as the quality measure of citation network clustering, which is defined as follows.

Definition 1 (Modularity). Let $G = \{V, E\}$ be a graph of citation network, A_{ij} denote the edge from v_i to v_j, and m is the number of edges; the modularity is defined as:

$$CQ = \sum_{ij} \left[\frac{A_{ij}}{m} - \frac{k_i^{out} k_j^{in}}{(m)^2} T\left(x(t_{ij})\right) \right] \delta_{c_i, c_j} \tag{5}$$

where k_i^{out} is the out-degree of v_i, k_j^{in} is the in-degree of v_j, $T(x(t_{ij}))$ is the punish function for time span. The punish function is defined as:

$$T(x(t_{ij})) = \begin{cases} e^{x(t_{ij})} & x(t_{ij}) < 0 \\ 2 - e^{-x(t_{ij})} & x(t_{ij}) \geq 0 \end{cases} \tag{6}$$

where t_{ij} is the time span, and $x(t_{ij}) = t_{ij} - T_{avg}$. Therefore, a larger value of t_{ij} will lead to larger modularity.

3   Algorithms for Clustering

In this section, we proposed an algorithm CPSCAN (Algorithm 1) for citation network which considered the model of calculating similarity between adjacent vertices in citation graph. In the algorithm, for each v in citation graph, v is a core vertex with respect to ε and μ if $|N_\varepsilon(v)| \geq \mu$. $N_\varepsilon(v)$ is the set of all nodes that have at least ε similarity with v, is defined as $N_\varepsilon(v) = \{w \in V | Sim(v, w) \geq \varepsilon \ or \ Sim(w, v) \geq \varepsilon\}$.

Algorithm 1 CPSCAN

```
Input: G = (V,E,T), parameter ε μ
Output: cluster
1   for each e_ij ∈ E do
2       calculate Sim(v_i,v_j)
3   for each unclassified core vertex v∈V do
4       Generate a new clusterID and Add N_ε(v) to a queue Q;
5       While Q ≠ 0 do
6           y = first element in Q;
7           R = {x∈V|x is directly reachable from y};
8           for each x∈R do
9               if x is unclassified or neutral then
10                  Assign current clusterID to x;
11              If x is unclassified then
12                  Insert x into queue Q;
13          Remove y from Q;
```

Then for the computation of similarity between vertices, we use the algorithm *find-DisjointPath(v_i, v_j)* to find disjoint paths of adjacent vertices v_i and v_j. Here, we set up a parameter δ to limit the maximum length of disjoint path, otherwise it might cause high cost on similarity computation. We omit the detail of the algorithm for the length of the paper.

4   Experimental Results and Conclusion

In this section, we evaluate our approach on two real world datasets of citation network, DBLP and Physical paper. We extract the timestamp for each node of the dataset to add the temporal feature.

We first compare the modularity CQ on clustering results of our approach with SCAN in different size of citation network. In this experiment, we set μ to be 2, δ to be 3. The results are shown in Fig. 1. We can see CPSCAN is better when the size of data becomes larger. This is because the temporal feature has a weakly effect on the similarity of nodes in a small network.

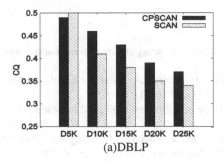
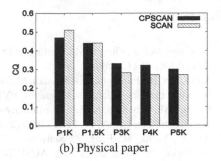

(a)DBLP	(b) Physical paper

Fig. 1. CQ result on different size of citation network

The threshold of similarity ε is important for the clustering result. So we take experiments to evaluate the effection of ε on our CPSCAN. Figure 2 shows the modularity CQ values on varying the value of ε in DBLP and physical paper respectively. We can see that CQ values didn't changed monotonously when varying the value of ε.

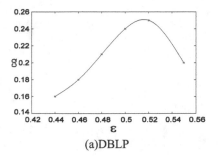
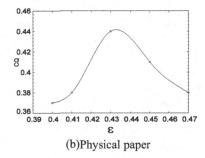

(a)DBLP	(b)Physical paper

Fig. 2. CQ values on varying ε

We also evaluate the quality of clustering on citation networks with different time span. We extract data from DBLP dataset and generate five citation networks with time span from 10 years (2001–2010) to 30 years (1981–2010). With time span increasing, the CQ of both algorithms is decrease, and CPSCAN has more superior over SCAN.

In this paper, we propose an approach called CPSCAN to cluster citation networks as a graph. CPSCAN uses not only the structural of neighbors, but also the temporal feature to assign two vertices into a cluster. The empirical evaluation demonstrates the superior performance of our algorithms on the modularity CQ proposed for clustering citation network.

Acknowledgment. The National Natural Science Foundation of China under Grant No. 61402213, 61472070, the Fundamental Research Funds for the Central Universities No. N150408001-3, N150404013

References

1. Newman, M.E.J., Girvan, M.: Finding and evaluating community structure in networks. Phys. Rev. E **69**, 026113 (2004)
2. Shiokawa, H., Fujiwara, Y., Onizuka, M.: SCAN ++: efficient algorithm for finding clusters, hubs and outliers on large-scale graphs. PVLDB **8**(11), 1178–1189 (2015)
3. Girvan, M., Newman, M.E.J.: Community structure in social and biological networks. Proc. Natl. Acad. Sci. U.S.A. **99**, 7821–7826 (2002)
4. Xu, X., Yuruk, N., Feng, Z., Schweiger, T.A.J.: Scan: a structural clustering algorithm for networks. In: Proceedings 13th ACM SIGKDD International Conference on Knowledge Discovery and Data Mining, pp. 824–833 (2007)
5. Lim, S., Ryu, S., Kwon, S., Jung, K., Lee, J.G.: LinkSCAN*: Overlapping Community Detection Using the Link-Space Transformation. In: ICDE (2014)

A Distributed Frequent Itemsets Mining Algorithm Using Sparse Boolean Matrix on Spark

Yonghong Luo[1], Zhifan Yang[1], Huike Shi[1], and Ying Zhang[1,2](✉)

[1] College of Computer and Control Engineering, Nankai University, Tianjin, China
{luoyonghong,yangzhifan,shihuike,zhangying}@dbis.nankai.edu.cn
[2] College of Software, Nankai University, Tianjin, China

Abstract. Frequent itemsets mining is one of the most important aspects in data mining for finding interesting knowledge in a huge mass of data. However, traditional frequent itemsets mining algorithms are usually data-intensive and computing-intensive. Take Apriori algorithm, a well-known algorithm in finding frequent itemsets for example, it needs to scan the dataset for many times and with the coming of big data era, it will also cost a lot of time over GB-level data. In order to solve those problems, researchers have made great efforts to improve Apriori algorithm based on distributed computing framework Hadoop or Spark. However, the existing parallel Apriori algorithms based on Hadoop or Spark are not efficient enough over GB-level data. In this paper, we proposed a distributed frequent itemsets mining algorithm by sparse boolean matrix on Spark (FISM). And experiments show FISM has better performance than all others existing parallel frequent itemsets mining algorithms and can also deal with GB-level data.

Keywords: Frequent itemsets mining · Apriori algorithm · Spark · Sparse matrix · FISM

1 Introduction

In order to solving the problem of association rules mining in the era of big data, many researchers have proposed lot of algorithms that using parallel computing technologies to deal with association rules mining. Yang X Y [4] proposed a distributed Apriori algorithm using Hadoop's MapReduce framework, however, this algorithm would cause a large number of I/O operations. Qiu H [3] proposed a parallel frequent itemsets mining algorithm with Spark called YAFIM (Yet Another Frequent Itemset Mining). And experiments show that, compared with the algorithms implemented with Hadoop, YAFIM achieved 18× speedup in average for various benchmarks [3]. But, YAFIM will also cost many hours when processing GB-level data. In the year of 2015, Zhang F proposed DPBM, a distributed matrix-based pruning algorithm based on Spark [1]. Experiments

© Springer International Publishing Switzerland 2016
F. Li et al. (Eds.): APWeb 2016, Part II, LNCS 9932, pp. 419–423, 2016.
DOI: 10.1007/978-3-319-45817-5_38

show that it is faster than YAFIM. But this algorithm may have a critical problem: if the total transactions are large enough, the computing speed will be very slow and not efficient.

In order to solve these problems, this paper proposed FISM, a new distributed frequent itemsets mining algorithm by using sparse boolean matrix on Spark. First, we use sparse matrix to replace the boolean-matrix of DPBM. Next, we also improved the generating process of candidate frequent itemsets. The experiments show that FISM outperforms DPBM, YAFIM, Parallel FP-Growth and other algorithms in both speed and scalability performance.

2   Distributed Frequent Itemsets Mining Algorithm by Using Sparse Boolean Matrix on Spark (FISM)

Assume that there are n transactions in the dataset D, and we call it $T = T_1$, $T_2, \cdots, T_n$. Next, assume that there are m items in the dataset called $I = I_1, I_2, \cdots, I_m$. The following are main steps of FISM:

First scan the dataset and for every item I_i ($i = 1, 2, 3, \cdots$, m), get the vector V_i ($b_{i1}, b_{i2}, b_{i3}, \cdots, b_{in}$) where

$$b_{ij} = \begin{cases} 1 & I_i \in T_j \\ 0 & I_i \notin T_j \end{cases}$$

($j = 1, 2, 3, \cdots$, n; $i = 1, 2, 3, \cdots$, m), and if $b_{ij} = 0$ then transaction T_j does not contain item I_i. There are total m vectors which build up the boolean matrix B. The number of columns of boolean matrix B is n and the number of rows of boolean matrix B is m. Sometimes, the number of transactions are very large, so the boolean matrix B will be implemented by sparse matrix. Figure 1 shows the actual data structure of B. For example, the i-th member of vectors is a list $(4, 7, 68, 6457, \cdots)$, so it means that transaction $4, 7, 68$ and 6457 all contain Item I_i and other transactions do not contain Item I_i. According to boolean matrix B, we can easily get the support number of Item I_i .

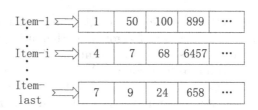

Fig. 1. Sparse Boolean matrix B

Definition 1. The "AND" result of one itemset E is defined as follows:

Suppose that E consists of h items, they are I_1, I_2, I_3, ..., I_h, and suppose that these items' corresponding rows in sparse boolean matrix B are V_1, V_2, V_3, ..., V_h, and suppose that $V_{result} = V_1$ & V_2 & V_3, ..., & V_h. Define that V_{result} is the "AND" result of E. The symbol "&" means "AND" operation between two vectors.

It is easy to generate (k + 1)-candidate frequent itemsets by k-frequent itemsets. After we get all k-frequent itemsets, for every k-frequent itemset, we stored its "AND" result (see Definition 1) in a list.Obviously, every (k + 1)-candidate frequent itemset consists of a new item I_{new} and a k-frequent itemset called $itemset_{old}$, and the "AND" result of $itemset_{old}$ is called V_{old}. Then V_{old} will be reused in the "AND" operation of V_{old} and I_{new}'s corresponding row of sparse boolean matrix B. In this way, we needn't to do k+1 times "AND" operations for the confirming of every one of (k + 2)-frequent itemset. We just do one time of "AND" operation of V_{old} and I_{new}'s boolean vector. Especially, we implemented the "AND" operation with multithreading technology.

In general,traditional frequent itemsets finding algorithms will scan the dataset for many times and this cause large amount of I/O operations while FISM only needs to scan dataset for once and use sparse matrix to accelerate the procedure of frequent itemsets finding. In theoretically, FISM outperforms traditional Apriori algorithm by up to one order of magnitude. And the experiments will confirm it.

3   Experiments Results

In this section, we compared FISM with MRApriori [4], YAFIM [3], DPBM [1] and PFP-growth (parallel FP-Growth [2] algorithm implemented by Spark team) to evaluate its performance. We implemented all the algorithms in Spark1.5.0. All the datasets are stored in HDFS and the cluster consists of 4 nodes and each node has 4 Intel Xeon cores with 2.60 GHz, 22.5 GB memory and 1 TB disk. The running system is CentOS6.5 and the version JDK is 1.7. Last but not least, **the correctness of all the algorithms mentioned above are exactly same.** Table 1 are characteristics of the datasets. Table 2 are results of experiments.

We can see that in every repetition and dataset, FISM outperforms all the others algorithms. For all the datasets, FISM is 1.8× faster than PFP-growth, 20× faster than YAFIM and 10× faster than DPBM.

Table 1. Detail properties of datasets

Dataset	Number of items	Number of transactions	Size
T10I4D100K	870	100000	3.9 MB
T40I10D100K	1000	100000	14.8 MB
Webdocs	1000	1692300	1.37 GB

Table 2. Experiments results

Dataset	YAFIM	MRApriori	DPBM	FP_Growth	FISM
T10I4D100K sup = 0.01	340 s	1920 s	75 s	32 s	15 s
T40I10D100K sup = 0.01	167 min	>12 h	43 min	6.6 min	2.9 min
Webdocs sup = 0.2	Out of memory	Out of memory	Out of memory	7.0 min	5.1 min

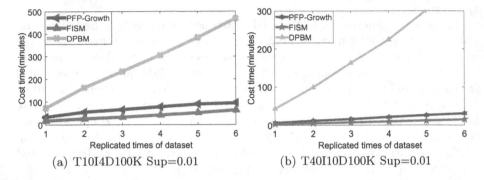

(a) T10I4D100K Sup=0.01   (b) T40I10D100K Sup=0.01

Fig. 2. The sizeup performance of each algorithms

Figure 2 shows the sizeup performance of all algorithms. The x-axis shows the number of replicated times of dataset. FISM is always faster than any other algorithms. MRApriori and YAFIM are not included in Fig. 2 because they spend more than 24 h. At the same time, we can see that with the increasement of dataset, the FISM 'cost time are increasing in a nearly linear way.

4   Conclusions

In order to accelerate frequent itemsets mining in big data era, this paper proposed FISM, a new distributed sparse boolean-matrix based frequent itemsets mining algorithm with Spark. It generate a sparse boolean matrix which shows whether one item is included in one transaction or not, then use the matrix to get all frequent itemsets. In this way, we only need to scan the dataset once. The experiments show that FISM is 1.8× faster than PFP-growth, 20× faster than YAFIM and 10× faster than DPBM. In addition, the FISM also has a better performance in scalability.

Acknowledgements. This work is partially supported by National 863 Program of China under Grant No. 2015AA015401, as well as the Research Foundation of Ministry and China Mobile under Grant No. MCM20150507. This work is also partially supported by Tianjin Municipal Science and Technology Commission under Grant No. 13ZCZDGX01098 and No. 16JCQNJC00500.

References

1. Gui, F., Ma, Y., Zhang, F., Liu, M., Li, F., Shen, W., Bai, H.: A distributed frequent itemset mining algorithm based on Spark. In: 2015 IEEE 19th International Conference on Computer Supported Cooperative Work in Design (CSCWD), pp. 271–275. IEEE (2015)
2. Han, J., Pei, J., Yin, Y.: Mining frequent patterns without candidate generation. In: ACM SIGMOD Record, vol. 29, pp. 1–12. ACM (2000)
3. Qiu, H., Gu, R., Yuan, C., Huang, Y.: YAFIM: a parallel frequent itemset mining algorithm with Spark. In: 2014 IEEE International Parallel and Distributed Processing Symposium Workshops (IPDPSW), pp. 1664–1671. IEEE (2014)
4. Yang, X.Y., Liu, Z., Fu, Y.: Mapreduce as a programming model for association rules algorithm on Hadoop. In: 2010 3rd International Conference on Information Sciences and Interaction Sciences (ICIS), pp. 99–102. IEEE (2010)

A Simple Stochastic Gradient Variational Bayes for the Correlated Topic Model

Tomonari Masada[1(✉)] and Atsuhiro Takasu[2]

[1] Nagasaki University, 1-14 Bunkyo-machi, Nagasaki, Japan
masada@nagasaki-u.ac.jp
[2] National Institute of Informatics, 2-1-2 Hitotsubashi, Chiyoda-ku, Tokyo, Japan
takasu@nii.ac.jp

Abstract. This paper proposes a new inference for the correlated topic model (CTM) [3]. CTM is an extension of LDA [4] for modeling correlations among latent topics. The proposed inference is an instance of the stochastic gradient variational Bayes (SGVB) [7,8]. By constructing the inference network with the diagonal logistic normal distribution, we achieve a simple inference. Especially, there is no need to invert the covariance matrix explicitly. We performed a comparison with LDA in terms of predictive perplexity. The two inferences for LDA are considered: the collapsed Gibbs sampling (CGS) [5] and the collapsed variational Bayes with a zero-order Taylor expansion approximation (CVB0) [1]. While CVB0 for LDA gave the best result, the proposed inference achieved the perplexities comparable with those of CGS for LDA.

1 Introduction

Topic modeling is one of the outstanding text mining techniques that are based on unsupervised machine learning and have a wide variety of applications. After the proposal of LDA [4], many extensions are provided by considering more realistic situations. Especially, LDA cannot model correlations among latent topics. Therefore, the correlated topic model (CTM) has been proposed [3]. However, the inference for CTM is a bit complicated, because the logistic normal prior is not conjugate to the multinomial distribution. This paper proposes a new variational Bayesian inference for CTM. The main contribution is that we make the inference simple with the stochastic gradient variational Bayes (SGVB) [7,8]. While the proposed inference adopts a gradient-based optimization similar to the original one [3], no explicit inversion of the covariance matrix is required.

We briefly describe the variational inference for topic models. Let $\boldsymbol{x}_d = \{x_{d1}, \ldots, x_{dN_d}\}$ be the multiset of the words in the document d. z_{dn} denotes the topic to which the word token x_{dn} is assigned. Then the log evidence of the document d is lower bounded as $\log p(\boldsymbol{x}_d) \geq \mathbb{E}_{q(\boldsymbol{z}_d,\boldsymbol{\theta}_d)}[\log p(\boldsymbol{x}_d|\boldsymbol{z}_d, \boldsymbol{\Phi})p(\boldsymbol{z}_d|\boldsymbol{\theta}_d)p(\boldsymbol{\theta}_d)] - \mathbb{E}_{q(\boldsymbol{z}_d,\boldsymbol{\theta}_d)}[\log q(\boldsymbol{z}_d, \boldsymbol{\theta}_d)]$, where $\boldsymbol{\theta}_d$ is the topic probability distribution of the document d. $\boldsymbol{\Phi} = \{\boldsymbol{\phi}_1, \ldots, \boldsymbol{\phi}_K\}$ is the set of the per-topic word probability distributions, where K is the number of topics, and is MAP-estimated in this paper. Even when the posterior $q(\boldsymbol{z}_d, \boldsymbol{\theta}_d)$ is assumed to factorize as $q(\boldsymbol{z}_d)q(\boldsymbol{\theta}_d)$, closed

© Springer International Publishing Switzerland 2016
F. Li et al. (Eds.): APWeb 2016, Part II, LNCS 9932, pp. 424–428, 2016.
DOI: 10.1007/978-3-319-45817-5_39

form updates cannot be obtained for several parameters in CTM. Therefore, a gradient-based optimization is required. Further, since $\boldsymbol{\theta}_d$ is drawn from the logistic normal prior in CTM, the covariance matrix makes the inference complicated. The implementation by [3] explicitly inverts the covariance matrix. This paper proposes a simpler inference as an instance of SGVB [7,8]. The proposed method does not invert the covariance matrix explicitly thanks to SGVB.

2 Our Proposal: SGVB for CTM

In CTM, the per-document topic probabilities $\boldsymbol{\theta}_d$ are drawn from the logistic normal distribution, which is parameterized by the mean parameter $\boldsymbol{m}$ and the covariance matrix $\boldsymbol{\Sigma}$. We assume that the variational posterior $q(\boldsymbol{\theta}_d; \boldsymbol{\mu}_d, \boldsymbol{\sigma}_d)$ is the diagonal logistic normal, where the variational mean and standard deviation parameters for each pair (d, k) are referred to by μ_{dk} and σ_{dk}, respectively. The main contribution of this paper is that the inference is made simple by applying SGVB. A sample from $q(\boldsymbol{\theta}_d)$ is computed as $\theta_{dk} \propto \exp(\epsilon_{dk}\sigma_{dk} + \mu_{dk})$ with the reparameterization technique [7], where ϵ_{dk} is a noise distribution sample. The ELBO, i.e., the variational lower bound of the log evidence, is obtained as $\mathbb{E}_{q(\boldsymbol{z}_d; \boldsymbol{\gamma}_d)}\big[\log p(\boldsymbol{x}_d | \boldsymbol{z}_d; \boldsymbol{\Phi})\big] + \mathbb{E}_{q(\boldsymbol{z}_d; \boldsymbol{\gamma}_d)q(\boldsymbol{\theta}_d)}\big[\log p(\boldsymbol{z}_d | \boldsymbol{\theta}_d)\big] + \mathbb{E}_{q(\boldsymbol{\theta}_d)}\big[\log p(\boldsymbol{\theta}_d | \boldsymbol{m}, \boldsymbol{\Sigma})\big] - \mathbb{E}_{q(\boldsymbol{z}_d; \boldsymbol{\gamma}_d)}\big[\log q(\boldsymbol{z}_d; \boldsymbol{\gamma}_d)\big] - \mathbb{E}_{q(\boldsymbol{\theta}_d)}\big[\log q(\boldsymbol{\theta}_d)\big]$ for the document d, where $q(\boldsymbol{z}_d; \boldsymbol{\gamma}_d)$ is assumed to be the discrete distribution. As no significant improvement was achieved by increasing the number of samples, the number of noise distribution samples was set to one in our experiment.

To estimate parameters, we take the partial derivatives of the ELBO L. The partial derivatives with respect to μ_{dk} and τ_{dk}, defined by $\tau_{dk} \equiv \log(\sigma^2)$, are $\frac{\partial L}{\partial \mu_{dk}} = \sum_{n=1}^{N_d} \gamma_{dnk} - N_d \exp(\epsilon_{dk}\sigma_{dk} + \mu_{dk})/\zeta_d - 2\sum_{k'=1}^{K-1} \Lambda_{kk'}(\epsilon_{dk}\sigma_{dk} + \mu_{dk} - m_{k'})$ and $\frac{\partial L}{\partial \tau_{dk}} = \frac{1}{2} + \frac{1}{2}\epsilon_{dk}\exp(\frac{\tau_{dk}}{2})\big[\sum_{n=1}^{N_d} \gamma_{dnk} - N_d \exp\{\epsilon_{dk}\exp(\frac{\tau_{dk}}{2}) + \mu_{dk}\}/\zeta_d - 2\sum_{k'=1}^{K-1} \Lambda_{kk'}\{\epsilon_{dk'}\exp(\frac{\tau_{dk'}}{2}) + \mu_{dk'} - m_{k'}\}\big]$, where $\zeta_d = 1 + \sum_{k=1}^{K-1}\exp(\epsilon_{dk}\sigma_{dk} + \mu_{dk})$. We skip the derivation due to the space limitation. Adam [6] is used for updating μ_{dk} and τ_{dk} repeatedly. Since the precision matrix $\boldsymbol{\Lambda}$, i.e., the inverse of the covariance matrix, appears only in the multiplication with a vector, the Cholesky decomposition can make the explicit inversion unnecessary. However, the analytic shrinkage [2] is used to make the computation stable. The $\boldsymbol{\phi}_k$s are MAP-estimated. The other parameters are updated by $\gamma_{dnk} \propto \theta_{dk}\phi_{kx_{dn}}$, $\boldsymbol{m} = \sum_{d=1}^{D}(\boldsymbol{\mu}_d + \boldsymbol{\epsilon}_d \circ \boldsymbol{\sigma}_d)/D$, and $\boldsymbol{\Sigma} = \sum_{d=1}^{D}(\boldsymbol{\epsilon}_d \circ \boldsymbol{\sigma}_d + \boldsymbol{\mu}_d - \boldsymbol{m})(\boldsymbol{\epsilon}_d \circ \boldsymbol{\sigma}_d + \boldsymbol{\mu}_d - \boldsymbol{m})^{\top}/2D$, where $\circ$ is the element-wise product.

3 Evaluation Experiment

The evaluation experiment was conducted over the four English document sets in Table 1. NYT is the first half of the New York Times news articles in "Bag of Words Data Set" of the UCI Machine Learning Repository.[1] MOVIE is the

---

[1] https://archive.ics.uci.edu/ml/datasets.html.

Table 1. Specifications of the four document sets used in the experiment

	# documents	# vocabulary words	# word tokens	Average doc. length
NYT	149,890	46,650	50,528,379	337.1
MOVIE	27,859	62,408	12,788,477	459.0
NSF	128,818	21,471	14,681,181	114.0
MEDLINE	125,490	42,830	17,610,749	140.3

set of movie reviews known as "Movie Review Data."[2] NSF is "NSF Research Award Abstracts 1990-2003 Data Set" of the UCI Machine Learning Repository. MEDLINE is a subset of the MEDLINE®/PUBMED®.[3] For all document sets, we applied the Porter stemming and removed high- and low-frequency words.

We compared the proposed SGVB for CTM to the collapsed Gibbs sampling (CGS) for LDA [5] and also to the collapsed variational Bayes with a zero-order Taylor expansion approximation (CVB0) for LDA [1]. The original VB for CTM has already been compared to LDA in [3]. Therefore, we did not repeat the comparison. However, CVB0 could not be considered in [3]. Therefore, we picked CVB0 up. The evaluation measure was the predictive perplexity. We first ran each compared method on the randomly selected 90 % training documents. We then ran each method on a randomly selected one third of the word tokens of each test document to obtain an estimation of the topic probabilities, where the per-topic word probabilities were never updated. By using the rest two thirds, the perplexity was computed by $\exp\left\{-\frac{1}{N_{\text{test}}}\sum_{d\in\mathcal{D}_{\text{test}}}\sum_{i\in\mathcal{I}_d}\log(\sum_{k=1}^{K}\theta_{dk}\phi_{kx_{di}})\right\}$, where $\mathcal{D}_{\text{test}}$ denotes the test document set, $\mathcal{I}_d$ the set of the indices of the test word tokens in the dth document, and N_{test} the total number of the test tokens. For each data set, the methods were compared for $K = 50, 100$, and 150.

Figure 1 depicts the mean and standard deviation of the perplexities obtained from ten different training/test random splits. The horizontal axis gives K, and the vertical axis the perplexity. CVB0 was better than the other methods. A similar result has been given in [1], though only the inferences for LDA is considered. With respect to the comparison of SGVB for CTM to CGS for LDA, the former was better than the latter for the following five cases: $K = 50$ and 100 for NYT (resp. $p = 0.00072$ and 0.001), $K = 50$ and 100 for MOVIE (resp. $p = 7.6 \times 10^{-7}$ and 0.00013), and $K = 50$ for MEDLINE ($p = 9.1 \times 10^{-6}$). The latter was better for the following five cases: $K = 150$ for MOVIE, $K = 50, 100$, and 150 for NSF, and $K = 150$ for MEDLINE. The former was comparable with the latter for the other two cases. The p values were obtained by the paired two-tailed t-test. In sum, the proposed SGVB for CTM was as good as CGS for LDA. However, the proposed method was far worse only for NSF, where it is suspected that the gradient-based optimization did not work well. Figure 2 gives topic correlations obtained from NYT for $K = 100$.

---

[2] http://www.cs.cornell.edu/people/pabo/movie-review-data/.
[3] We used the XML files from medline14n0770.xml to medline14n0774.xml.

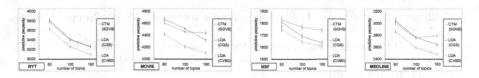

Fig. 1. Evaluation results in terms of predictive perplexity.

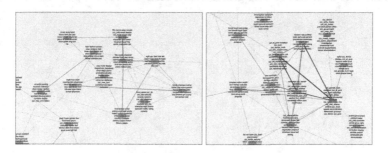

Fig. 2. Topic correlations obtained by our method from the NYT data set. The left and right panels contain the topics seemingly relating to art and politics, respectively.

The graph is drawn by Cytoscape[4]. The edge thickness represents the magnitude of the corresponding entry of the covariance matrix. The left panel contains the topics seemingly relating to art, and the right those seemingly relating to politics.

4   Conclusion

This paper proposed a new inference for CTM. We apply SGVB to CTM and obtain a set of simple updating formulas. The experiment showed that the proposed method was comparable with CGS for LDA in terms of perplexity, though CVB0 for LDA was the best for all settings. While the proposed method was as good as CGS for LDA, it did not work well for some cases. A further elaboration seems required for a more effective gradient-based optimization.

References

1. Asuncion, A., Welling, M., Smyth, P., Teh, Y.W.: On smoothing and inference for topic models. In: UAI, pp. 27–34 (2009)
2. Bartz, D., Müller, K.R.: Generalizing analytic shrinkage for arbitrary covariance structures. In: NIPS 26, pp. 1869–1877 (2013)
3. Blei, D.M., Lafferty, J.D.: Correlated topic models. In: NIPS, pp. 147–154 (2005)
4. Blei, D.M., Ng, A.Y., Jordan, M.I.: Latent dirichlet allocation. JMLR **3**, 993–1022 (2003)

---

[4] http://www.cytoscape.org/.

5. Griffiths, T.L., Steyvers, M.: Finding scientific topics. PNAS **101**(Suppl 1), 5228–5235 (2004)
6. Kingma, D.P., Ba, J.: Adam: a method for stochastic optimization. In: ICLR (2015)
7. Kingma, D.P., Welling, M.: Stochastic gradient VB and the variational auto-encoder. In: ICLR (2014)
8. Rezende, D.J., Mohamed, S., Wierstra, D.: Stochastic backpropagation and approximate inference in deep generative models. In: ICML, pp. 1278–1286 (2014)

Reasoning with Large Scale OWL 2 EL Ontologies Based on MapReduce

Zhangquan Zhou[1](✉), Guilin Qi[1], Chang Liu[2], Raghava Mutharaju[3], and Pascal Hitzler[3]

[1] Southeast University, Nanjing, China
{quanzz,gqi}@seu.edu.cn
[2] Carnegie Mellon University, Pittsburgh, USA
liuchang2005acm@gmail.com
[3] Wright State University, Dayton, USA
{mutharaju.2,pascal.hitzler}@wright.edu

Abstract. OWL 2 EL, which is underpinned by the description logic $\mathcal{EL}$, has been used to build terminological ontologies in real applications, like biomedicine, multimedia and transportation. On the other hand, there have been techniques that allow developers and users acquiring large scale ontologies by automatically extracting data from different sources or integrating different domain ontologies. Thus the issue of handling large scale ontologies has to be tackled. In this short paper, we report our work on classification of OWL 2 EL ontologies using MapReduce, which is a distributed computing model for data processing. We discuss the main problems when we use MapReduce to handle OWL 2 EL classification and how we address these problems. We implement the algorithm using Hadoop, and evaluate it on a cluster of machines. The experimental results show that our prototype system achieves a linear scalability on large scale ontologies.

Keywords: OWL 2 EL · Description logics · Classification · MapReduce

1 Introduction

Among different profiles of OWL 2[1], OWL 2 EL (EL for short), which is based on description logics $\mathcal{EL}$ family, stands out for its positive complexity results and sufficient expressivity. EL is mainly used in biomedicine. One of the most popular biomedicine ontology expressed in EL, SNOMED CT [5], is supposed to be a large scale ontology, i.e., nearly six hundred thousand axioms (or statements) are involved. Thus, to make EL play a better role in real applications, it is necessary to give efficient solutions for handling large scale ontologies. One of the most important reasoning services in EL, called *classification* [2], is the task of computing a subsumption hierarchy between concept descriptions.

---

[1] www.w3.org/TR/owl2-overview/.

© Springer International Publishing Switzerland 2016
F. Li et al. (Eds.): APWeb 2016, Part II, LNCS 9932, pp. 429–433, 2016.
DOI: 10.1007/978-3-319-45817-5_40

In order to handle large ontologies efficiently, several works employ parallel or distributed computing techniques. ELK [3] is the first reasoner that exploits multi-core techniques to enhance the efficiency of classification in EL. Although the experimental results show that these in-memory reasoners can be scalable to some extent, they are restricted to the main memories of the utilized machines. On the other hand, a distributed approach based on Redis is also proposed in [4] for EL classification. This work also verifies that distributed techniques can handle classification on large scale EL ontologies.

In this paper, we report our work on classification of EL ontologies using MapReduce, which is conducted parallelly with the work [4]. We briefly discuss the main problems when we use MapReduce to handle OWL 2 EL classification and how we address these problems. We implement a prototype system and evaluate it on a Hadoop cluster. The experimental results show that our system has a linear scalability on large real ontologies. The details of this work can be found in our technical report which is available at this address[2].

2 The Problems of Performing EL Classification on MapReduce and Our Solutions

Due to the page limit, we refer the readers to our technical report for the formalism of EL and the work mechanism of MapReduce. In this part, we briefly discuss the main problems when we use MapReduce to handle OWL 2 EL classification and how we address these problems.

• **Translating ontologies and rules to MapReduce languages.** Since MapReduce programs can only handle key/value pairs. We should first translate EL axioms to key/value pairs. However this is not a trivial task when considering the issue of performance. We consider representing EL axioms using relational tables. For example, for axioms of the form $A \sqsubseteq \exists r.B$, we introduce a relational table of the form $R(A, r, B)$ where R is the name of this table, (A, r, B) is the table schema. In this way, applying classification rules can be transformed to joins of relational tables. Finally, it is relatively easy to map operations of relational tables to MapReduce programs.

• **Reducing the number of jobs.** If we use MapReduce programs to perform EL classification, several jobs are needed to apply different rules until a fix-point is reached. This delays computation due to the inherent overheads of platforms. Thus reducing the number of jobs can significantly improve running performance. We apply some optimizations to reach this goal. For example, we carefully decide rule application order. We also combine some MapReduce jobs into one.

• **Handling multi-way joins.** It is easy for MapReduce to handle a two-way join using one job. For a multi-way join, we can partition it into several two-way joins to adapt to MapReduce programs. In the case of EL classification, there are three rules which have multiple joints in their preconditions. However, it is

---

[2] https://drive.google.com/drive/folders/0ByjKIQyCPHldSDNrWnlZN3pRdTg.

challenging to partition it into several two-way joins and, the performance of the whole computation can also be guaranteed. We analyze and compare different partition methods. The details can be found in our technical report.

3   Evaluation

We implement a prototype system based on Hadoop[3], which is an open-source Java implementation of MaReduce. We conduct all the experiments in a cluster that consists of 14 nodes, and each node has a 16 GB RAM and two quad-core processors. We use two famous medical ontologies: Galen and SNOMED CT. In order to validate the scalability of our system, we evaluate it on different copies of Galen and SNOMED CT. The Galen copies are renamed by n-Galen, where n is the number of copies in n-Galen. SNOMED CT is processed similarly.

Table 1. The results of scalability tests

Datasets	#input axioms(k)	#derived axioms(k)	Time(min)	♯jobs
1-Galen	91	6,941	143.4	93
2-Galen	182	13,808	179.8	98
4-Galen	364	27,626	242.7	94
8-Galen	728	55,209	319.5	94
1-SCT[a]	1,151	18,980	479.5	94
2-SCT	2,302	38,795	976.8	93
3-SCT	3,453	60,715	1589.5	94

^{a} Abbr. of SNOMED CT.

Scalability tests. We assign 52 units (a unit represents a logic core) of the cluster to perform classification on different ontology copies. The experimental results are collected in Table 1. We use Fig. 1 to give a graphical representation of the relevant contents with respect to Galen in Table 1, where the curve in the left part (resp. right part) shows the relation between the classification time and the number of copies (resp. the inverse of the number of units we set to perform classification on 1-Galen). From Fig. 1, we can see that the classification time is approximately linear in the number of copies, and inversely linear in the number of units. The experiments on SNOMED CT have the similar results.

From Table 1, we can estimate the maximum speedup of the evaluated ontology based on the view of Amdahl's Law[4] that indicates the speedup of a computation task only depends on the fraction of sequential computation part with *the assumption of infinite processors being allocated* (IPA). According to the

---

[3] http://hadoop.apache.org/.

[4] Amdahl argues in [1] that the speedup s of a computation task depends on the fraction of sequential computation part, i.e., $s \leq 1/(f + \frac{1-f}{p})$, where s is the speedup, f is the fraction of sequential computation part and p is the number of processors.

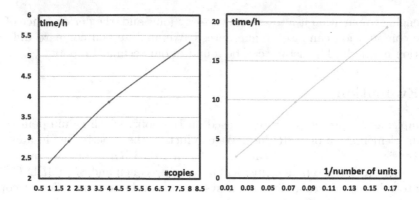

Fig. 1. Classification time on different copies and different units.

Amdahl's Law, with IPA, the run-time of each job is $O(1)$ in theory, while the whole run-time depends on the number of jobs[5]. This helps estimate the maximum speedup to be $O(T_1/n)$, where T_1 is the run-time units of classification on one processor and n is the number of jobs (or the run-time units) on infinite processors. For example, the number of jobs for classifying 1-SNOMED ($d(\mathcal{O}_{1-SNOMED})$) is 94 and more than 18 million axioms are derived (see Table 1). Thus the maximum speedup of classification on 1-SNOMED can be estimated to be $max - speedup(\mathcal{O}_{1-SNOMED}) \geq \frac{|\mathcal{C}_O|}{d(\mathcal{O}_{1-SNOMED})} (= 191,489)$. Similarly, the maximal speedup for classifying 1-Galen is greater than 70,000. It indicates that classifying these two ontologies has a high degree of parallelism. That is to say if we set more computing units to our system, it should have a higher performance.

4   Conclusions

In this paper, we reported our work on classification of EL using MapReduce. We discussed the main problems and gave our solutions. We implemented a prototype system using Hadoop. The experimental results showed that our system has a linear scalability on real medical ontologies and their copies.

References

1. Amdahl, G.M.: Validity of the single processor approach to achieving large scale computing capabilities. In: Proceedings of AFIPS, pp. 483–485 (1967)
2. Baader, F., Brandt, S., Lutz, C. Pushing the $\mathcal{EL}$ envelope. In: Proceedings of IJCAI, pp. 364–369 (2005)

---

[5] This conclusion also holds with a polynomial number of processors, since the classification on EL ontologies is in P.

3. Kazakov, Y., Krötzsch, M., Simancik, F.: Concurrent classification of EL ontologies. In: Proceedings of ISWC, pp. 305–320 (2011)
4. Mutharaju, R., Hitzler, P., Mateti, P., Lécué, F.: Distributed and scalable OWL EL reasoning. In: Gandon, F., Sabou, M., Sack, H., d'Amato, C., Cudré-Mauroux, P., Zimmermann, A. (eds.) ESWC 2015. LNCS, vol. 9088, pp. 88–103. Springer, Heidelberg (2015)
5. Stearns, M.Q., Price, C., Spackman, K.A., Wang, A.Y.: SNOMED clinicalterms: overview of the development process and project status. In: Proceedings of AMIA Symposium, pp. 662–666 (2001)

Purchase and Redemption Prediction Based on Multi-task Gaussian Process and Dimensionality Reduction

Chao Wang, Xiangrui Cai, Zhenguo Zhang, and Yanlong Wen[(✉)]

College of Computer and Control Engineering, Nankai University,
38 Tongyan Road, Tianjin 300350, People's Republic of China
{wangc,caixr,zhangzg,wenyl}@dbis.nankai.edu.cn

Abstract. Predicting future cash flows based on historical data is an essential and hard problem of financial business. Most of the previous works have attempted to convert cash flows into time series to make prediction. However, real-valued datasets are mostly multidimensional sources with complex curves, large amplitude and high frequency. The handful of research efforts that consider those truths have met with limited success. This paper proposes an algorithm based on Multi-task Gaussian Process model to predict cash flows in funds. Purchase refers to cash inflow, while redemption refers to cash outflow. MTGP can learn the correlation within multiple time series and make regression on each time series simultaneously. Furthermore, motif discovery is used for dimensionality reduction in data before MTGP to improve the accuracy. Experimental results on real-world data demonstrate the advantages of our proposed algorithm.

Keywords: Time series · Gaussian process · Motif discovery · Prediction

1 Introduction

Internet finance (ITFIN) has a rapid development in last five years and plays an important role in modern economy. With the development of e-commerce and mobile internet, there is a lot of cash floating around banks, funds and stock market. It is easy to understand that prediction of cash flow's trend has been a topic of great interest in finance. Most financial institutions take advantage of users' historical purchase and redemption logs to predict future cash flows.

Many previous models and algorithms try to convert complex finance data into time series and formulate the prediction as regression problem. However, they fail when applied to irregularly sampled data unless strong assumptions are made in the underlying data source. Park [1] can find rules in time series and predict the following data. However, it is only evaluated for speed and they only use random walk data as dataset. Wu [2] also use a piecewise linear representation to predict time series, but their algorithm has the same performance

© Springer International Publishing Switzerland 2016
F. Li et al. (Eds.): APWeb 2016, Part II, LNCS 9932, pp. 434–438, 2016.
DOI: 10.1007/978-3-319-45817-5_41

on dataset and purn random walk data. Their work [3] suggests their original results did not outperform random guessing. These methods fail because there are inflow, outflow and rate between each financial institution in business, these data are actually multivariate time series, and they work on a single time series, while fail to take advantage of other related time series data. This paper proposes an algorithm D-MTGP to predict the purchase and redemption in fund based on multi-task Gaussian process model. MTGP uses the correlation within multiple time series to estimate parameters instead of considering each time series separately. To avoid over optimizing hyperparameters and analyze the original datasets. We first reduce the time series' dimensionality by motif discovery, discover the repeated fragments (*motif*) which appear n times ($n \geq T$), and replace these motifs in the original time series by new label before MTGP.

2   Methods

D-MTGP can be divided into two parts: MTGP [4] models to learn the correlation within time series and dimensionality reduction to simplify time series.

MTGP is the evolution of STGP. The general STGP framework may be extended to model m time series simultaneously. MTGP can models multiple time series simultaneously, which makes use of the covariance in related time series (tasks) to reduce uncertainty in the inferred time series.

Given a set $T = \{X_n, Y_n\}$, let $X_n = \{x_{n_j}^j | j = 1, 2, \ldots, m\}$ be our train data and $Y_n = \{y_{n_j}^j | j = 1, 2, \ldots, m\}$ be the observations for the m tasks, where task j has n_j number of training data. We consider the regression model $\boldsymbol{y_n} = f(\boldsymbol{x_n}) + \varepsilon$. In MTGP, there is a labels vector $L_n = \{l_{n_j}^j | j = 1, 2, \ldots, m\}$ with $l_{n_j}^j = j$, the vector is used as the additional input to MTGP model to specify that training data $x_{n_j}^j$ and observation and observation $y_{n_j}^j$ belong to task j, training data may be observed at different times for different tasks.

To model the correlation between tasks and the temporal behavior of the tasks, two independent covariance functions can be assumed, hyperparameters would learn the implicit relations within m tasks. The complete covariance matrix for the m tasks K_{MTGP} is

$$K_{MTGP}\left(\tilde{x}_n, \tilde{l}_n, \theta^c, \theta^t\right) = K^c\left(\tilde{l}_n, \theta^c\right) \otimes K^t\left(\tilde{x}_n, \theta^t\right) \tag{1}$$

where $\otimes$ is the Kronecker product, K^c and K^t represent the correlation and temporal covariance functions, respectively. Note that K^t depends only on (x, x') and the size is $\sum_{j=1}^m n_j \times \sum_{j=1}^m n_j$, and K^c depends only on (l, l'), the size of K^c is $m \times m$. θ^c and θ^t are vectors containing hyperparameters for K^c and K^t, respectively.

Computational cost of the MTGP is $O\left(m^3, n^3\right)$, the cost compared with $m \times O\left(n^3\right)$ for STGP is a limitation that can not be ignored. This limitation is remarkable when dealing with densely-sampled time series data, and these data are very common in financial field. To reduce the computational work, we propose a method to discretize time series, and guarantee time series utility before and after dimensionality reduction.

A time series TS could be consider as a single point in multi-dimensional space. For example, $TS = t_1, t_2, \ldots, t_m$ is a single point in m-dimensional space, the label in each t_i is the value in i-dimension. Time series in Finance, Biology and mechanical control have an important characteristic: some subsequences will periodic or aperiodic repeat in time series, we call these subsequences motifs. If we set a suitable threshold T, the time series can be divided into limited subsequences, each subsequence belongs to one class of motif. Then we replace each class of motif with a specified label and predict on the new time series. By prediction on the new-label time series after motif replacement, we get an efficient and useful dimensionality reduction.

3   Experiment

We evaluate our algorithm on real-word data in fund field [5]. The datasets from AFSG contain about 2.8 million records including the purchase and redemption behaviors during July 1, 2013 to August 31, 2014. Each record in this sheet has users Id, report date, total purchase and total redemption. We first group all records by their report date and sum all records which have the same date. Then, we normalize the results. All experiments are carried out on a PC with Intel core i5 CPU 3.10 GHz, 4 GB memory and Microsoft Windows 10. To evaluate the goodness of fit of the model we also report the root mean squared error (RMSE) and the absolute mean error (MAE). The RMSE is defined as

$$RMSE = \sqrt{\frac{1}{N} \sum_{i}^{N} (y_i - \hat{y}_i)^2}, MAE = \frac{1}{N} \sum_{i}^{N} |y_i - \hat{y}_i| \tag{2}$$

where y_i is the target value, $\hat{y}_i$ predicted value, and N the number of test examples.

We predict purchase and redemption by D-MTGP. After grouping and summation, there are 427 data points in the time series. To take advantage of the implicit correlation between purchase and rate of Yields and Shibor, we take purchase, *mfd_daily_yield* and *Interest_O_N* into MTGP and model the three time series simultaneously. We do the same for redemption. We discover 34 classes of motifs in purchase and redemption. Then we simplify these time series and replace motifs with new labels and predict the new label time series by D-MTGP. Figure 1 illustrates the effectiveness of our algorithm.

The experimental results show that D-MTGP has a great performance. Because there are many motifs in purchase and redemption, dimensionality reduction can greatly reduce the complexity of time series. The confidence interval has great credibility before the point 390 and has a terrible result after 390 due to the length of train time series and limitation of mean function of MTGP.

We compare our algorithm (D-MTGP) with original MTGP, STGP and random walk. Tables 1 and 2 shows the performance of our method, we note that

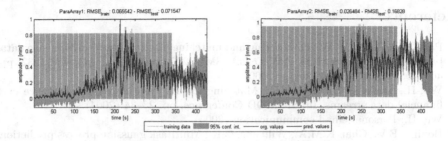

Fig. 1. Prediction using MTGP and dimensionality reduction. ParaArray1 is the purchase and ParaArray2 is the redemption

Table 1. Prediction in purchase

	D-MTGP	MTGP	STGP	Random walk
RMSE	0.071547	0.085495	0.12655	0.32942
MAE	0.048131	0.057514	0.085133	0.22161

Table 2. Prediction in redemption

	D-MTGP	MTGP	STGP	Random walk
RMSE	0.16828	0.20967	0.58712	0.70854
MAE	0.11321	0.14183	0.39497	0.47665

our method can obtain a good accuracy both in purchase and redemption. Compared with STGP, we consider the correlation within the rate, purchase and redemption, and learn hyperparameters from these time series simultaneously.

4   Conclusion

Prediction is the essential but pretty hard problem in both finance and data mining. We propose a method to predict purchase and redemption by using MTGP, MTGP uses correlation between multiple time series to estimate parameters instead of considering each time series separately. Furthermore, we achieve dimensionality reduction on time series by motif discovery to improve accuracy. One interesting future work is reduce computational cost of characterize time series in MTGP.

Acknowledgments. This work is supported by Tianjin Municipal Science and Technology Commission under Grant No. 14JCQNJC00200, and the Research Foundation of Ministry of Education and China Mobile under Grant No. MCM20150507.

References

1. Park, S.-H., Chu, W.W.: Discovering and matching elastic rules from sequence databases. In: Ohsuga, S., Raś, Z.W. (eds.) ISMIS 2000. LNCS (LNAI), vol. 1932, pp. 400–408. Springer, Heidelberg (2000)
2. Wu, H., Salzberg, B., Zhang, D.: Matching, online event-driven subsequence over financial data streams. In: SIGMOD Conference, pp. 23–34 (2004)
3. Wu, H.: Personal email communication (2005)
4. Bonilla, E.V., Chai, K.M.A., Williams, C.K.: Multitask gaussian process prediction. In: Advances in Neural Information Processing Systems, 153–160 (2007)
5. Purchase_Redemption_Forecasts. https://tianchi.aliyun.com/datalab/index.htm

Similarity Recoverable, Format-Preserving String Encryption

Yijin Li and Wendy Hui Wang[✉]

Department of Computer Science,
Stevens Institute of Technology, Hoboken, NJ, USA
{yli134,hwang4}@stevens.edu

Abstract. Format-preserving encryption (FPE) encrypts a plaintext of some specified format into a ciphertext of identical format. In this paper, we consider FPE on string data, and design new FPE algorithms that preserve the string similarity for a large number of string metrics. The experiments show that our encryption algorithm is efficient and robust against the frequency analysis attack.

1 Introduction

During the recent years, format-preserving encryption (FPE) has emerged as a useful tool in applied cryptography. Intuitively, FPE deterministically encrypts a plaintext X into a ciphertext Y that has the same format as X. For example, a 9-decimal-digit plaintext, say a U.S. social-security number, is encrypted into a ciphertext that is again a 9-decimal-digit number. FPE has been used in a wide range of applications. In this paper, we consider format-preserving encryption on *string* data. Format-preserving encryption of string data is in great need in many applications, for example, in the Platform-as-a-Service (PaaS) and Software-as-a-Service (SaaS) cloud computing paradigms that have strict requirement of users' input format, and in the crowdsourcing computing paradigm that human workers only can deal with textual data for some specific tasks (e.g., entity resolution [9]). In these applications, one type of computations is to measure string similarity. When the similarity measurement involves sensitive data, it is important to protect the sensitive data while enabling the string similarity measurement. However, most of the existing privacy-preserving techniques (e.g., [6,8]) transform the input strings into non-string formats.

A straightforward solution is to apply an one-to-one mapping on the strings to map each character to another character in the same alphabet. However, it is weak against the attacks based on the statistics analysis. For example, the attacker can easily break the encryption by mapping plaintext and ciphertext characters based on their frequency.

To address this issue, we design $SOAR$, a novel Similarity recoverable, format-preserving string encryption framework. $SOAR$ extends the existing FPEs by ensuring the ciphertext is still of string format, while the ciphertext strings preserve the similarity of the plaintext strings. It supports a number of string

© Springer International Publishing Switzerland 2016
F. Li et al. (Eds.): APWeb 2016, Part II, LNCS 9932, pp. 439–443, 2016.
DOI: 10.1007/978-3-319-45817-5_42

similarity metrics, including edit distance, q-grams, to name a few. The experiments show that $SOAR$ is robust against the frequency analysis attack.

The rest of the paper is organized as follows. Section 2 presents our algorithms. Section 3 reports the experimental results. Section 4 discusses the related work. Section 5 summarizes the paper.

2   Approach

For a given set of plaintext strings $PT = \{pt_1, \ldots, pt_n\}$, PT is transformed to a set of string pairs $PP = \{(pt_i, pt_j) | \forall pt_i, pt_j \in PT, i \neq j\}$. $SOAR$ encrypts each pair in PP. We must note that $SOAR$ is not an one-to-one mapping, since the same string in multiple string pairs can be encrypted to different ciphertext strings. Next, we explain the encryption algorithm of $SOAR$.

Given the alphabet Σ, the **Enc** algorithm takes as input pp, sk, and a plaintext string pair (pt_i, pt_j), where both pt_i, $pt_j \in 2^\Sigma$. It outputs a ciphertext string pair (ct_i, ct_j), where both ct_i, $ct_j \in 2^\Sigma$. We require that $|pt_i| = |ct_i|$, and $|pt_j| = |ct_j|$ (i.e., the encryption preserves the string length). Before we explain the algorithm details, we first define some notions. Given two strings pt_i and pt_j, we use $pt_i \cap pt_j$ to denote the set of unique characters in both pt_i and pt_j, and $pt_i - pt_j$ be the substring of pt_i that only contains the characters that appear in pt_i but not in pt_j. Our algorithm consists of the following three steps.

Step 1. The algorithm computes $p_1 = pt_i - pt_j$ and $p_2 = pt_j - pt_i$. The algorithm applies a FPE algorithm to encrypt p_1 (p_2, resp) to string c_1 (c_2, resp) s.t. $|p_1| = |c_1|$ ($|p_2| = |c_2|$, resp.). Let U, V, $\hat{U}$, and $\hat{V}$ be the set of unique characters of p_1, p_2, c_1, and c_2 respectively. We denote the number of unique characters of p_1, p_2, c_1, c_2 as $|U|, |V|, |\hat{U}|, |\hat{V}|$ respectively. We use the BPS cipher that implements FPE [5]. We require that $BPS(pp, sk, p_1)$ and $BPS(pp, sk, p_2)$ iterate until $\hat{U} \cap \hat{V} = \emptyset$ and $|U| + |V| = |\hat{U}| + |\hat{V}|$.
Step 2. The algorithm generates a bijective mapping $F : \Sigma - U - V \rightarrow \Sigma - \hat{U} - \hat{V}$, where Σ is the given alphabet.
Step 3. The algorithm generates ct_i (i.e., the ciphertext of pt_i) by: (1) replacing p_1 with c_1, and (2) convert the remaining characters (i.e., the characters shared by pt_i and pt_j) in pt_i by the bijective mapping F in Step 2. Similarly, the algorithm generates ct_j (i.e., the ciphertext of pt_j) by concatenating c_2 with the converted remaining characters in pt_j.

We have the next theorem to show that $SOAR$ encryption can preserve similarity for a number of widely-used similarity metrics.

Theorem 1. *For any two strings pt_i and pt_j, let ct_i and ct_j be their encrypted strings by SOAR. Then: (1) For any character-based distance metric $Dist()$ that is based on Levenshtein distance (only insertion and deletion are allowed), $Dist(pt_i, pt_j) = Dist(ct_i, ct_j)$; and (2) For any q-gram based distance metric $Dist()$ that only uses intersection of tokens, $Dist(pt_i, pt_j) = Dist(ct_i, ct_j)$.*

The proof of Theorem 1 is omitted due to limited space.

3   Experiments

3.1   Setting

We implemented *SOAR* in C++. All experiments were executed on a PC with a
1.7 GHz Intel Core i5 CPU and 8 GB memory. We use the BPS implementation[1]
for FPE algorithm. We use two real-world datasets: (1) *LastName* dataset[2] from
US Census Bureau. *LastName* dataset contains 88799 last names which are chosen from the character domain A-Z. The longest length is 13, while the shortest
length is 2; (2) *Adult* dataset[3] from UCI Machine Learning Repository. *Adult*
dataset contains 48842 records.

3.2   Performance

Encryption Time. Figure 1 reports the encryption time of *SOAR* on *LastName*
and *Adult* datasets respectively with various string lengths. We observe that the
encryption is very fast; for string length 4–10, the average encryption time of a
single string is no more than 11 ms. Even for long strings of length 30–50, the
encryption takes no more than 60 ms.

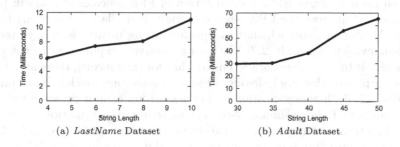

(a) *LastName* Dataset   (b) *Adult* Dataset

Fig. 1. Encryption time

Similarity. We measure the similarity of both plaintext and ciphertext strings
according to a few similarity metrics[4] It turns out *SOAR* can preserve Levenshtein distance, normalized Levenshtein distance, weighted Levenshtein distance, Damerau distance, and Jaro-Winkler distance. It also preserves longest
common subsequence (LCS) and metric LCS. But it does not preserve Cosine,
Jaccard, and Sorensen Dice distance. However, for Cosine, Jaccard, and Sorensen
Dice metrics, the similarity difference is negligible. In particular, for *LastName*
dataset, the difference of Cosine, Jaccard, and Sorensen Dice similarity between
plaintext and ciphertext is at most 0.062, 0.036, and 0 (for 2-grams). While

---

[1] https://github.com/miracl/MIRACL.
[2] http://www2.census.gov/topics/genealogy/1990surnames/dist.all.last.
[3] http://archive.ics.uci.edu/ml/datasets/Adult.
[4] Similarity metrics: https://github.com/tdebatty/java-string-similarity.

for *Adult* dataset, the difference of Cosine, Jaccard, and Sorensen Dice similarity between plaintext and ciphertext is at most 0.033, 0.011, and 0.013 (for 3-grams). We omit the details due to the limited space.

3.3 Robustness Against Frequency Analysis Attack

We measure the *successful ratio* of the frequency analysis attack. In particular, given n strings PT, let PC_P be the set of all pairs of PT (i.e., $\binom{n}{2}$ pairs in total), and PC_C be the ciphertext of PC_P by $SOAR$. Let K be the total number of characters of PC_P, and k be the total number of characters in PC_C that are correctly mapped by the frequency analysis attack. The *successful ratio* P of the attack is measured as $P = \frac{k}{K}$. We observe that the successful rate of the frequency analysis attack is very small. It never exceeds 3.9 % for *LastName* dataset and 5 % for the *Adult* dataset. This is encouraging as it shows that the attacker has little chance to break the encryption by $SOAR$.

4   Related Work

Format-preserving encryption [3] is used in a broad range of real-world applications. [4] defines integer FPE. Formalization of FPE schemes appears in [3]. In this paper, we improve the FPE scheme to make it similarity preserving for the strings. A different approach that can protect string privacy is anonymization. The k-anonymity approach [2,7] has been extensively explored in recent years. Its key idea is to generalize data to ensure that for each record, there are at least $k-1$ other records that are indistinguishable. Another approach is based on condensation [1]. Both anonymization and condensation cannot preserve the exact string similarity. Related similarity-preserving string transformation mainly convert strings to non-string formats. The strings can be transformed as hash values by using substitution encoding (e.g., [6] and data points in Euclidean space by using embedding [8]. However, the non-string formats violate the format-preserving requirements.

5   Conclusion

In this paper, we designed $SOAR$, a format-preserving encryption method for string data that preserves the similarity for a number of widely-used string similarity metrics. Our experiments show that $SOAR$ is fast and robust against the frequency analysis attack.

Acknowledgment. This material is based upon work supported by the U.S. National Science Foundation under Grant No. 1464800.

References

1. Aggarwal, C.C., Philip, S.Y.: On anonymization of string data. In: SDM, pp. 419–424 (2007)
2. Bayardo, R.J., Agrawal, R.: Data privacy through optimal k-anonymization. In: Proceedings of the 21st International Conference on Data Engineering, pp. 217–228 (2005)
3. Bellare, M., Ristenpart, T., Rogaway, P., Stegers, T.: Format-preserving encryption. In: Jacobson Jr., M.J., Rijmen, V., Safavi-Naini, R. (eds.) SAC 2009. LNCS, vol. 5867, pp. 295–312. Springer, Heidelberg (2009)
4. Black, J.A., Rogaway, P.: Ciphers with arbitrary finite domains. In: Preneel, B. (ed.) CT-RSA 2002. LNCS, vol. 2271, pp. 114–130. Springer, Heidelberg (2002)
5. Brier, E., Peyrin, T., Stern, J.: BPS: a format-preserving encryption proposal (2010). http://csrc.nist.gov/groups/ST/toolkit/BCM/documents/proposedmodes/bps/bps-spec.pdf
6. Churches, T., Christen, P.: Some methods for blindfolded record linkage. BMC Med. Inf. Decis. Mak. **4**(1), 1 (2004)
7. Samarati, P.: Protecting respondents identities in microdata release. IEEE Trans. Knowl. Data Eng. **13**(6), 1010–1027 (2001)
8. Scannapieco, M., Figotin, I., Bertino, E., Elmagarmid, A.K.: Privacy preserving schema and data matching. In: Proceedings of the ACM SIGMOD International Conference on Management of Data, pp. 653–664 (2007)
9. Wang, J., Kraska, T., Franklin, M.J., Feng, J.: Crowder: crowdsourcing entity resolution. Proc. VLDB Endowment **5**(11), 1483–1494 (2012)

RORS: Enhanced Rule-Based OWL Reasoning on Spark

Zhihui Liu[1,2], Zhiyong Feng[1,2], Xiaowang Zhang[1,2(✉)], Xin Wang[1,2], and Guozheng Rao[1,2]

[1] School of Computer Science and Technology, Tianjin University, Tianjin, China
xiaowangzhang@tju.edu.cn
[2] Tianjin Key Laboratory of Cognitive Computing and Application, Tianjin, China

Abstract. In this paper, we present an approach to enhancing the performance of the rule-based OWL reasoning on Spark based on a locally optimal executable strategy. Firstly, we divide all rules (27 in total) into four main classes, namely, *SPO rules* (5 rules), *type rules* (7 rules), *sameAs rules* (7 rules), and *schema rules* (8 rules) since, as we investigated, those triples corresponding to the first three classes of rules are overwhelming (e.g., over 99% in the LUBM dataset) in our practical world. Secondly, based on the interdependence among those entailment rules in each class, we pick out an optimal rule executable order of each class and then combine them into a new rule execution order of all rules. Finally, we implement the new rule execution order on Spark in a prototype called RORS. The experimental results show that the running time of RORS is improved by about 30% as compared to Kim & Park's algorithm (2015) using the LUBM200 (27.6 million triples).

1 Introduction

Owing to the explosion of the semantic data, the number of RDF triples in large public knowledge bases, e.g., DBpedia has increased to billions [2]. Therefore, to improve the performance of OWL reasoning becomes a core problem. This paper focuses on the rule-based OWL reasoning using OWL-Horst rules [1].

Recently, Urbani et al. proposed a MapReduce-based parallel reasoning system with OWL-Horst rules called WebPIE [5]. It can deal with large scale ontologies on a distributed computing cluster. However, WebPIE exhibits poor reasoning time. Cichlid [2] has greatly improved the performance of OWL reasoning on Spark as compared with the state-of-the-art distributed reasoning systems, but it only considers parts of OWL rules and does not analyze the interdependence of rules. Although Kim & Park [4] has also implemented parallel reasoning algorithms with an executable rule order, but lacked the evidence to prove that the strategy is optimal.

In this paper, we present an approach to enhancing the performance of the rule-based OWL reasoning based on a locally optimal executable strategy and implement the new rule execution strategy on Spark in a prototype called RORS. The major contributions and novelties of our work are summarized as follows:

© Springer International Publishing Switzerland 2016
F. Li et al. (Eds.): APWeb 2016, Part II, LNCS 9932, pp. 444–448, 2016.
DOI: 10.1007/978-3-319-45817-5_43

- According to the characteristic of dataset which is divided into SPO triples, sameAs triples and type triples, the OWL-Horst rules are classified into four classes.
- We respectively analyze the rule interdependence of each class, and find the optimal executable strategies.
- Based on the locally optimal strategies, we pick out an optimal rule execution order of each class and then combine them into a new rule execution strategy of all rules and implement the new rule execution strategy on Spark.

The rest of this paper is organized as follows. Section 2 presents our locally optimal strategies. Section 3 implements our proposed strategy on Spark and Sect. 4 evaluates our method on the LUBM dataset. In Sect. 5, we conclude this paper.

2   Locally Optimal Strategy

Based on the statistics of LUBM dataset, we divide the OWL-Horst rules [1] into four classes as follow:

- Rules whose conditions or consequences have triples whose predicates are rdf:type, including R4, R5, R6, O13, O14, O15, O16. The dependency graph of type rules is shown in Fig. 2.
- Rules whose conditions or consequences have triples whose predicates are owl:sameAs, including O1, O2, O5, O6, O8, O9, O10. The dependency graph of sameAs rules is shown in Fig. 3.
- Rules whose conditions or consequences have triples of SPO class, including R3, O3, O4, O7a, O7b. The dependency graph of SPO rules is shown in Fig. 1, which the rule O7a and O7b are classified as O7.
- The remainder is classified as a class, including R1, R2, O11a, O11b, O11c, O12a, O12b, O12c. The dependency graph of schema rules is shown in Fig. 4.

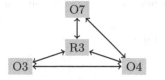

Fig. 1. SPO rules dependency graph

Fig. 2. Type rules dependency graph

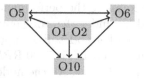

Fig. 3. sameAs rules dependency graph

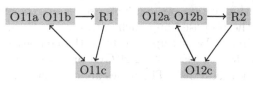

Fig. 4. Schema rules dependency graph

Table 1. The optimal strategies of each class

SPO rules	type rules
	R4 → R6 → O14 → O13 → O15 → O16
O3 → R3 → O7 → O4	R4 → R6 → O14 → O13 → O16 → O15
O7 → R3 → O3 → O4	R5 → R6 → O14 → O13 → O15 → O16
	R5 → R6 → O14 → O13 → O16 → O15
schema rules	sameAs rules
O11a, O11b → R1 → O11c	O1 → O10 → O2 → O6 → O5
O12a, O12b → R2 → O12c	O2 → O10 → O1 → O6 → O5

For each class, based on the dependency of rules, we use the *depth-first-search* (DFS) algorithm to find all possible executable orders among the rules and acquire the optimal executable orders. Through the experiments, there is a large number of orders for each class, but we should choose the longest orders that contain rules as many as possible. The optimal executable orders for each class are listed in Table 1.

3   Distributed Rule-Based OWL Reasoning on Spark

After proposing the locally optimal orders for each class, we design the overall strategy of the OWL reasoning shown in Fig. 5. The reasoning of instance triples uses the output of schema triples, so the schema reasoning should be executed firstly. There are several orders for each class of instance triples. We can choose any order and then combine a new executable strategy. In this paper, we select the first order for each class to perform reasoning.

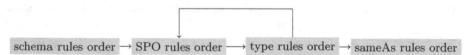

schema rules order → SPO rules order ⟶ type rules order → sameAs rules order

Fig. 5. The reasoning strategy

4   Evaluation

In this section, we conduct a series of experiments to compare the performance of the proposed approach with the KP [4] and Cichlid [2] under the same environment. We set up a cluster with one master and four worker nodes. Each node has 48 Xeon E5-4607 2.20 GHz processors, 64 GB memory and 10TB 7200 RPM SATA hard disk. The nodes are connected with Gigabit Ethernet. All the nodes run on a 64-bit Ubuntu 12.04 LTS operating system. The version of the Spark is 1.0.2. And the corresponding Hadoop v2.2 with Java 1.7 is installed on this cluster. LUBM [3] is a widely used standard benchmark for semantic programs. Due to the limitation of hardware, we use the data generator UBA to generate 5 sets of data with different universities: LUBM_10, LUBM_50, LUBM_100, LUBM_150 and LUBM_200 in our experiments.

We evaluate the performance of our method with KP and Cichlid, where KP adopts the executable strategy in [4]. All experiments run three times and the average value is listed as follows.

Figure 6 displays the reasoning time of our approach and KP with different scale of data sets. The reasoning time includes the time of dividing input data and eliminating duplicate triples. The result shows that our approach is better than the KP under the same environment. The performance of reasoning is improved by 30 % approximately. Figure 7 shows the number of inferred triples per second. Our method can infer more implicit triples than Cichlid and the performance is improved by 26 % approximately.

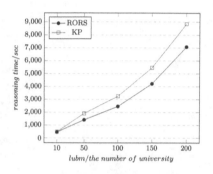

Fig. 6. The runtime of RORS and KP   **Fig. 7.** The number of triples per sec

5   Conclusions and Future Work

In this paper, we present an approach to enhancing the performance of the rule-based OWL reasoning on Spark based on a locally optimal executable strategy. Our method performs better than KP in the reasoning strategy. In the future work, we plan to implement the above reasoning strategy on more diverse datasets and design some algorithms to find the optimal strategy for global rules.

Acknowledgments. This work is supported by the program of the National Key Research and Development Program of China (2016YFB1000603) and the National Natural Science Foundation of China (NSFC) (61502336, 61373035, 61572353). Xiaowang Zhang is supported by Tianjin Thousand Young Talents Program.

References

1. Liu, C., Qi, G., Wang, H., Yu, Y.: Large scale fuzzy pD* reasoning using mapreduce. In: Aroyo, L., Welty, C., Alani, H., Taylor, J., Bernstein, A., Kagal, L., Noy, N., Blomqvist, E. (eds.) ISWC 2011, Part I. LNCS, vol. 7031, pp. 405–420. Springer, Heidelberg (2011)
2. Gu, R., Wang, S., Wang, F., Yuan, C., Huang, Y.: Efficient large scale RDFS/OWL reasoning with Spark. In: Proceedings of IPDPS 2015, pp. 700–709. IEEE (2015)
3. Guo, Y., Pan, Z., Heflin, J.: LUBM: a benchmark for OWL knowledge base systems. J. Web Seman. **3**(2–3), 158–182 (2005)
4. Kim, J., Park, Y.: Scalable OWL-horst ontology reasoning using Spark. In: Proceedings of BigComp 2015, pp. 79–86 (2015)
5. Urbani, J., Kotoulas, S., Maassen, J., Van Harmelen, F., Bal, H.: OWL reasoning with WebPIE: calculating the closure of 100 billion triples. In: Aroyo, L., Antoniou, G., Hyvönen, E., ten Teije, A., Stuckenschmidt, H., Cabral, L., Tudorache, T. (eds.) The Semantic Web: Research and Applications. LNCS, vol. 6088, pp. 213–227. Springer, Heidelberg (2010)

A Hadoop-Based Database Querying Approach for Non-expert Users

Yale Chai, Chao Wang, Yanlong Wen$^{(\boxtimes)}$, and Xiaojie Yuan

College of Computer and Control Engineering, Nankai University,
38 Tongyan Road, Tianjin 300350, People's Republic of China
{chaiyl,wangc,wenyl,yuanxj}@dbis.nankai.edu.cn

Abstract. Increasingly complex relational databases are challenging for non-expert users to query. To facilitate querying, keyword search and form-based interface methods have been proposed, but they both have limitations. Besides, existing methods rely on single machine, leading to insufficient memory and long time consuming for large-scale data. This paper proposes a Hadoop-based database querying method combining the advantages of form-based interface and keyword search. We first generate a set of query forms according to the database schema, and build distributed indexes offline. At online query stage, aiming at the keywords that users input, we return top-k related forms. After users choose and fill out the form, we can automatically create a structured query and return results to them. Experiments show that our method can not only provide an easy way to query databases for non-expert users, but also achieve better performance in time efficiency.

Keywords: Keyword search · Query form · Hadoop · Relational database

1 Introduction

With the development of Internet, more and more information from various fields is stored in databases. Many ordinary users are eager to access these databases directly. In order to get intended information over relational databases, users need to be familiar with structured query language (e.g., SQL) and understand complex database schema.

Research on querying databases can mainly be divided into keyword search and form-based interface. Keyword search method allows users to query databases the same way as they search on the Internet. However, it may lead nonideal results for complex query tasks such as join operations [1]. Form-based interface provides users with forms. The forms will be equal to structured queries after completed. By filling in and submitting the form, users can access databases easily. But [2] is unable to provide users with particular forms according to their queries. [3] proposes to combine form-based interface and keyword search. The basic idea is to find out users' potential query intentions before displaying forms.

© Springer International Publishing Switzerland 2016
F. Li et al. (Eds.): APWeb 2016, Part II, LNCS 9932, pp. 449–453, 2016.
DOI: 10.1007/978-3-319-45817-5_44

Unfortunately, there is neither an anticipation of related entities selection, nor an effective strategy to retrieve forms. What's more, the existing methods rarely mention about how to deal with large-scale databases, they may result in out of memory, huge I/O overhead, long time consuming and so on.

This paper firstly proposes a method for querying relational databases on Hadoop called HRMO(Hadoop-based database Retrieve Method for Ordinary users). HRMO combines the advantages of form-based interface and keyword search. Compared to [3], HRMO considers the range of form generation as well as the correlation between query keywords and forms, so as to return more precise forms. Besides, HRMO introduces parallel processing into database querying.

2 Our Approach

The process of HRMO can be divided into two parts: offline indexing stage and online querying stage. Throughout the process, users only need to input keywords, fill the form and submit it.

Indexing Strategy: Forms are consist of schema terms such as tables or attributes. However, sometimes the keywords that users input match data values in the database rather than schema terms, may leading no result when retrieving forms. Thus, we build indexes on both data collection and forms collection called DataIndex and FormIndex, so as to modify the query with relevant schema terms using DataIndex before retrieving forms. For each data value in the database, DataIndex records the tables/attributes it belongs to.

As for FormIndex, creating forms for all entities will be difficult for users to choose. Therefore we only build FormIndex for entity combinations with high query intentions. The procedure contains three steps:

(1) Given a target database, we construct a weighted schema graph in which each node represents an entity and each edge represents a foreign key between two entities. Single entity's query intention depends on its data cardinality and connectedness.

$$W_V = \frac{AC(e) \times EC(e)}{\sum_e AC(e) \times EC(e)} \qquad (1)$$

where $AC(e)$ is the absolute cardinality of the entity. Absolute cardinality is the number of times each element in the schema graph (node or edge) is instantiated in the data. $EC(e)$ is the number of attributes that entity is connected via referential links. The query intention of multiple related entities depends on their own query intentions as well as the edges between them:

$$W_{(V_i,V_j...V_m)} = W_{V_i} \times W_{V_j} \times ... \times W_{V_m} \times \frac{\sum W_{E_{(V_i,V_j)}} + ... + \sum W_{E_{(V_i,V_m)}}}{m!} \qquad (2)$$

where $\sum W_E$ are the absolute cardinality of the edges between two entities.

(2) In order to find all the combination of entities, each node is used as a source vertex to get all accessible points through breadth first search. We sort the entity combinations by their query intentions.

(3) We determine the threshold of forms number k_f and threshold of attributes number on per form k_a. We sort attributes by their query intentions [2]. Then we build FormIndex for top-k_f entities, each create various forms based on different query operations. And we select top-k_a attributes to display on each form.

Querying Strategy: The procedure contains four steps:

(1) HRMO provides users with a search interface where users can input keywords, then modifies keywords with relevant schema terms using DataIndex.
(2) We calculate the correlation between keywords and related schema terms to improve the ranking of forms that related to user's query. The IR score calculation method is introduced in [4].
(3) To find related forms, we probe FormIndex by calculating the score between query and forms. We determine the final score based on both original query and IR score, as bellow.

$$score(Q, F) = scorefiled(F) \times queryNorm(F) \times boost \times score^*(Q, F) \quad (3)$$

where $scorefield$ is the IR score, $queryNorm(F)$ is to normalize the query, $boost$ is a motivating factor (usually 1), and $score^*(Q, F)$ is the basic Lucene score.
(4) We display related forms in groups: the first-level is distinguish by forms' entities, the second-level are forms with different query operations ordered by their scores. Users can scan the first-level title to find the entities, and then unfold it for the exact form. HRMO can automatically create structured query according to the form that user fills out, and finally return results.

3   Experiments

We set up a four-node Hadoop cluster, one of them as master and the other three as slaves. Each one is a virtual machine on Ubuntu 14.04 with 2.33 GHz CPU and 4 GB memory. We implement HRMO in java with Eclipse, use MySQL as the database, and an Apache Tomcat Web Server to host the service.

We perform experiments over three datasets: two benchmark synthetic datasets TPCC, TPCH and a real-world dataset ZI-Database. ZI-Database has part of electronic product information of ZhongGuanCun. TPCC has 9 tables, 92 attributes. TPCH has 8 tables, 25 attributes. ZI-Database has 14 tables, 74 attributes with an average of 469 rows of tuples per table.

Measuring a system's quality is to see whether it can meet the needs of users easily and quickly. Therefore we pose several typical query needs and interact with system by simulating users' behavior. Assuming that users have the ability to submit keywords and select the correct form. We will observe the location of correct forms and whether users can get correct results eventually. Experiments evaluate the system from both the efficiency and effectiveness.

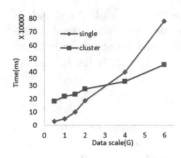

Fig. 1. Comparison of time

To test the time performance of HRMO, we generate 500M, 1G, 2G, 4G, 6G size of the data based on TPCC, compare indexing time of our Hadoop cluster with single-node machine. As shown in Fig. 1, when the scale of data is small, the advantage of Hadoop cluster is not obvious due to the transmission time and other reasons. When the data set becomes larger than 3G, Hadoop cluster performs better. Experiments show that our method can scale up from a single node to multiple machines.

To test the query effect of HRMO, we compare it with [3] which uses ordinary Lucene(OL) to retrieve forms. We adopt two evaluating indicators: Mean Reciprocal Rank(MRR) and Mean Precision@N(MPN). MRR only concerns about the position of the best answer. MRR(f) is to evaluate the ranking of the right form, and the MRR(g) is to evaluate the ranking of the group contains the right form. MPN concerns about the precision in first N positions. P@N is 1 if correct form appears in first N groups, else P@N is 0. In this paper, we set $N = 2$.

$$MRR = \frac{\sum_{i=1}^{n} \frac{1}{r_{best}}}{n}, \quad MPN = \frac{\sum_{i=1}^{n} P@N}{n} \tag{4}$$

where r_{best} is the ranking of the right form or group.

The comparison between OL and HRMO can be seen in Figs. 2, 3 and 4. Obviously, MRR and MPN are larger in HRMO than that in OL. It proves that in HRMO right forms ranking better and easier for users to find. In other words, our method provides a better retrieve strategy than before.

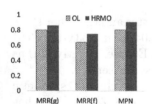

Fig. 2. Result in TPCC

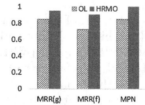

Fig. 3. Result in TPCH

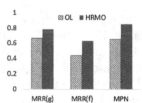

Fig. 4. Result in ZIDatabase

4   Conclusion

In this paper, we present an approach for ordinary users with little knowledge of structured query language or schema to query relational databases easily. HRMO combines the advantages of form-based interfaces and keyword search, and can be parallelized on Hadoop. Experiments show that our approach performs well on both efficiency and effectiveness.

Acknowledgments. This work is supported by Tianjin Municipal Science and Technology Commission under Grant No. 14JCQNJC00200, the Research Foundation of Ministry of Education and China Mobile under Grant No. MCM20150507.

References

1. Baid, A., Rae, I., Li, J., et al.: Toward scalable keyword search over relational data. Proc. VLDB Endowment **3**(1–2), 140–149 (2010)
2. Jayapandian, M., Jagadish, H.V.: Automated creation of a forms-based database query interface. Proc. VLDB Endowment **1**(1), 695–709 (2008)
3. Chu, E., Baid, A., Chai, X., et al.: Combining keyword search and forms for ad hoc querying of databases. In: Proceedings of the 2009 ACM SIGMOD International Conference on Management of data. ACM, pp. 349–360 (2009)
4. Hristidis, V., Gravano, L., Papakonstantinou, Y.: Efficient IR-style keyword search over relational databases. In: Proceedings of VLDB, pp. 850–861 (2003)

A Collaborative Join Scheme on a MIC-Based Heterogeneous Platform

Kailai Zhou[1,2], Hui Sun[1(✉)], Hong Chen[1], Tianzhen Wu[1], and Cuiping Li[1]

[1] Key Lab of Data Engineering and Knowledge Engineering of MOE,
and School of Information, Renmin University of China, Beijing, China
sun_h@ruc.edu.cn
[2] School of Computer and Information, Southwest Forestry University,
Kunming, China

Abstract. Join is one of the most important operations in data analytics systems. Prior works focus mainly on join optimization using GPUs, but little is known about performance impact on the MICs. In order to investigate potential benefits of the use of MIC accelerators in improving performance of join operation, in this paper we design a join scheme with a CPU and MICs working collaboratively. This scheme includes task partitioning, a data transfer mode, join algorithm design. Experimental results show that our collective join scheme is effective for a heterogeneous platform with two Xeon Phi cards, and can improve performance by up to 30 % over the CPU-only platform.

Keywords: Join · Heterogeneous platform · CPU-MIC · Xeon Phi · Coprocessor

1 Introduction

Due to excellent computing abilities and lower power consumption, Many Integrated Cores (MICs) are widely used in many compute-intensive applications [1]. Join is one of the most important operations in database systems. Accordingly, researchers have made great efforts to improve its performance using accelerators, such as GPUs. Although the field of traditional GPU-oriented join optimization has made great achievements, and the potential of using GPUs to accelerate join operation is shown to be significant, it is hard to transplant these optimization methods directly to MICs due to obvious hardware architectural differences between them [2]. Therefore, a redesign of the join scheme is required in order to fully exploit MIC-based hybrid platforms. To this end, in this work we carry out an in-depth study of join optimization issues in a heterogeneous system consisting of multi-core CPUs and MICs. In order to take full advantage of each component's hardware features, we propose a collaborative join scheme to accommodate each device's underlying hardware and keep them all busy during the whole join process.

F. Li et al. (Eds.): APWeb 2016, Part II, LNCS 9932, pp. 454–458, 2016.
DOI: 10.1007/978-3-319-45817-5_45

2   Collaborative Join Scheme for Hybrid MIC/CPU Platform

2.1   Workload Partitioning on the Host

As the MIC and CPU have separate memory address spaces, we need to partition input tables into smaller disjoint tables (referred to as sub-tables), such that each sub-table can reside in its own device memory. There is a large body of data partitioning techniques [3], which can partition workload efficiently. We partition the two input tables to obtain n partitions, where n is the number of processing units. Due to the asymmetry in parallel processing capabilities of the CPU and MIC, the key issue is how to decide the size of each partition. We exploit the following computational formula to estimate the size of each partition for the MIC and CPU.

$$
\begin{cases}
P_{CPU} = \dfrac{1}{(n-1)\beta + 1} \cdot |DS| \\[2mm]
P_{MIC} = \dfrac{1}{(n-1)\beta + 1} \cdot |DS|
\end{cases}, \quad
\beta = \frac{Cost_{cpu}}{Cost_{MIC}} \tag{1}
$$

where $|DS|$ is the size of the input dataset. The cost for the CPU or MIC can be estimated using the cost function proposed by Karnagel et al. [4].

2.2   Data Transfer

In order to reduce the data volume to be transferred, we only transfer the join attributes from the host to the MIC, whereas other attributes are hidden in the form of row indices. Similarly, the MIC only returns the join index to the host, and lets the host construct join results. There are several communication channels between the host and the MIC. Among these channels, the SCIF has the highest communication bandwidth according to literature [5]. In this paper we thus use the SCIF for data transfer.

2.3   Join Algorithm Design

After partitions are transferred to different processing units, join operations can run on processors simultaneously. Thus, the next key issue to be addressed is how to implement join operation efficiently. For modern CPUs, there is a large body of efficient join algorithms. Since the *parallel radix join* is reported to have the highest performance [6], we select this algorithm to implement join operation on the CPU. While the MIC has a different architecture compared to the CPU, according to previous studies [2], the join algorithm should carefully tailor to the MIC's underlying hardware. Thus, we propose a MIC-oriented hardware-conscious join method, named the MIC-aware hash join.

The pseudocode of our hash join algorithm is described in Algorithm 1. Line 1 shows the initialization operations such as memory allocation. As for MIC, the cross-core access cache leads to longer latency. Therefore, in order to avoid this problem, the next step (lines 2–3), we partition the two input tables R_{PA}, S_{PA} as evenly as possible into n non-intersecting sub-tables (assuming there are n cores), denoted by

$R_1, ..., R_n$ and $S_1, ..., S_n$, respectively, where each pair R_i and S_i is mapped to one core i. Lines 4–14 show the whole hash join process on each core. The code for building the hash table is shown in line 5. We select sub-table S_i as the inner table (assuming S_i is the smaller), and use the same optimization methods as the previous studies [2] to build the hash table HT_i, such as SIMD vectorizing and data prefetching. Lines 6–13 show the probe phase. In this phase, we start m threads on each core and divide sub-table R_i evenly among these threads into m segments, i.e. $R_{i1}, R_{i2}, ..., R_{im}$. Then, for each segment R_{ik}, we assign a thread T_k to scan its tuples. For each tuple in R_{ik}, we probe it in the hash table HT_i to find matches tuple (line 10). For each matched tuple, we merge the tuple to the output array $JoinRs$ (line 11). Finally, when a thread finishes its task, this thread can transfer its join result $JoinRs$ to the host via PCIe (line 15).

Algorithm 1. MIC-oriented hash join algorithm on Xeon Phi

input: $R_{PA}[], S_{PA}[]$ /* The input relations accepted from the host */
output: $JoinRs$ /* join results */
01. *initialize ()*   /* memory allocate and initialization */
02. $\{R_1[], R_2[], \cdots, R_n[]\} \leftarrow PartitionByCores (R_{PA}[], n)$
03. $\{S_1[], S_2[], \cdots, S_n[]\} \leftarrow PartitionByCores(S_{PA}[], n)$
  /* arrange each partition pair $<R_i, S_i>$ to one core C_i */
04. for each core C_i where $1 \leqslant i \leqslant n$ parallel-do
05. $HT_i \leftarrow build\_hashtable(S_i)$ /*build hash table, S_i is the inner table */
06. $\{R_{i1}[], R_{i2}[], \cdots, R_{im}[]\} \leftarrow$ split R_i into m sub-tables
07. $\{T_1, T_2, \cdots, T_m\} \leftarrow fork\_threads(m)$ //fork m threads per core
08.   for each thread T_k where $1 \leqslant k \leqslant$ m parallel-do
09.   for each tuple $\in R_{ik}$
10. $Rs \leftarrow probe\_hashtable (HT_i, tuple)$
11   if Rs is not null then $JoinRs \leftarrow \{ Rs \} \times tuple$
12.   end for
13.   end for
14. end for
15. *TransferbackToHost(JoinRs)*
16. end for

---

3   Experimental Evaluation

3.1   Experimental Setup

We conduct our experiments on a server equipped with two MIC coprocessors (Intel Xeon Phi 5110P). Xeon Phi runs Linux 2.6 as an embedded OS, and the Haswell machine has Linux CentOS (kernel version 2.6.32). The code is compiled with optimization level 3 using the Intel compiler ICC 15.

3.2 Performance Evaluation

To better observe parallel scalability of the hash join variants running on the MIC and CPU, we carry out experiments to test the impact of the number of threads on *non-partitioned hash join*, *parallel radix hash join* and our algorithm. In order to avoid interference of data communication, the following implementations all run on Xeon Phi or Haswell as native programs. We perform the equal-join on two relations R and S, both with 128 million tuples. All tuples are randomly generated in memory, and each tuple consists of 4-byte integer *key* and 4-byte *payload*.

We can conclude from Fig. 1(b) that both the PRO and NPO have good thread scalabilities and the optimal performance is achieved with 32 threads. The performance of the PRO is much better than that of the NPO and the speedup is 1.5–2.5X. By comparing Fig. 1(a) with (b), we can observe the fact that, for the same workload, the optimal performance (0.5 s) using our method on Xeon Phi is up to 37 % faster than the PRO (0.8 s) on Haswell.

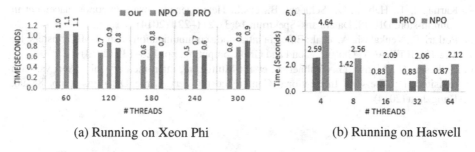

(a) Running on Xeon Phi   (b) Running on Haswell

Fig. 1. Thread scalability of multiple hash join variants running on Xeon Phi/Haswell

In order to learn profits we get from using MIC accelerators, we carry out a CPU-only join experiment with the same workload (456 M ⋈ 456 M tuples). In this case, the execution time is 5.4 s. Compared to our result 3.8 s, the co-processing join method achieves up to 30 % performance improvement over the CPU-only system (see Fig. 2). Since we have two Xeon Phi cards, each card provides 15 % performance gain.

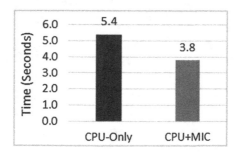

Fig. 2. Performance comparison between CPU-only and CPU+MIC

Acknowledgment. This work is supported by National Basic Research Program of China (973) (No. 2014CB340403, No. 2012CB316205), National High Technology Research and Development Program of China (863) (No. 2014AA015204) and NSFC under the grant No. 61272137, 61033010, 61202114 and NSSFC (No. 12&ZD220), and the Fundamental Research Funds for the Central Universities, and the Research Funds of Renmin University of China (15XNH113, 15XNLQ06).

References

1. Stuart, O., Brian, R., Ziliang, Z.: SQLPhi: a SQL-based database engine for Intel Xeon Phi coprocessors. In: Proceedings of the 2014 International Conference on Big Data Science and Computing, pp. 1–6. ACM Press, New York (2014)
2. Jha, S., He, B., Lu, M., et al.: Improving main memory hash joins on Intel Xeon Phi processors: an experimental approach. VLDB Endowment **8**(6), 642–653 (2015)
3. Kim, C., Sedlar, E., Chhugani, J., et al.: Sort vs. hash revisited: fast join implementation on modern multicore CPUs. VLDB Endowment **2**(2), 1378–1389 (2009)
4. Karnagel, T., Habich, D., Schlegel, B., et al.: Heterogeneity-aware operator placement in column-store DBMS. Datenbank-Spektrum **14**(3), 211–221 (2014)
5. Potluri, S., Venkatesh, A., et al.: Efficient intra-node communication on intel-MIC clusters. In: The 13th IEEE/ACM Cluster, Cloud and Grid Computing, pp. 128–135 (2013)
6. Balkesen, C., Alonso, G., Teubner, J., et al.: Multi-core, main-memory joins: sort vs. hash revisited. In: The 40th International Conference on Very Large Data Bases, pp. 85–96, Hangzhou (2014)

Pairwise Expansion: A New Topdown Search for mCK Queries Problem over Spatial Web

Yuan Qiu[✉], Tadashi Ohmori, Takahiko Shintani, and Hideyuki Fujita

Graduate School of Information Systems,
The University of Electro-Communications,
1-5-1 Chofugaoka, Chofu, Tokyo 182-8585, Japan
kyuu_genn@hol.is.uec.ac.jp, {omori,shintani,fujita}@is.uec.ac.jp

Abstract. This paper focuses on the problem of m-Closest Keywords (mCK) queries over spatial web objects. An mCK query is to find the optimal set of objects (object-set) in the sense that they are the spatially-closest records and satisfy m user-given keywords. We propose a new approach called *Pairwise Expansion* to find an exact solution of mCK queries based on topdown search of an on-the-fly quad-tree. This approach first enumerates object-pairs in a topdown way, then picks up each 'closer' object-pair and expands it into candidate object-sets. Experimental results show that this approach is more efficient than existing topdown search strategies and applicable for real spatial web data.

Keywords: Spatial web data · Spatial keyword query

1 Introduction

Today, various queries to find the objects that satisfy the requirements of both contents and locations in spatial web data such as Twitter and Flickr have been studied [1,2,4]. As one of them, according to [1,2], an *m-closest keywords* (or, mCK) query is to find the spatially closest set of objects (*object-set*) such that these objects must satisfy m keywords given as a query. Formally, given an object-set $O = \{o_1, o_2, ..., o_l\}$ and a distance $dist(o_x, o_y)$ for two objects in O, let $diam(O) = \max_{o_x, o_y \in O} dist(o_x, o_y)$ $(x \neq y)$ be termed a *diameter* of O [1]; then the mCK query of m given keywords Q is defined as a query to find the object-set O_{opt} where O_{opt} has the smallest diameter δ^*_{opt} among all possible object-sets O that contain all keywords of Q.

Because the number of possible object-sets is in an exponential order of data size, some existing methods [1,4] use topdown exploration approaches taking advantage of hierarchical indices, which firstly enumerate sets of internal-/leaf- nodes (termed *node-sets*) of a hierarchical index, and prune unnecessary node-sets to reduce generation of object-sets. However, the number of necessary node-sets may be too large on some data distributions, which leads to a high exploration cost. To overcome this problem, we propose a new topdown approach called *Pairwise Expansion*, which generates no node-sets and still

© Springer International Publishing Switzerland 2016
F. Li et al. (Eds.): APWeb 2016, Part II, LNCS 9932, pp. 459–463, 2016.
DOI: 10.1007/978-3-319-45817-5_46

keeps a strong pruning ability. By definition of mCK queries, the diameter of an object-set is an object-pair (a pair of two objects). And the result of an mCK query is the closest object-pair p among all the object-pairs such that there exists an object-set O satisfying all given keywords Q and the diameter of O is p. Thus we first enumerate object-pairs in ascending order of their distances by topdown exploration of node-pairs. Next for each 'closer' object-pair, we expand it into object-sets in a small shuttle area to test if it is a diameter of any 'correct' object-sets satisfying Q. Finally we return the closest object-pair which passes the test.

2   Proposal

2.1   On-the-Fly Creation of a Quad-Tree

Practical cases of an mCK query may use different datasets (like Twitter and Flickr) in its result. Thus we employ to create an on-the-fly quad-tree structure for mCK queries. When Q, as an mCK query keywords, is given, we only load necessary objects associated with Q from one or multiple datasets and then create a quad-tree for these objects. Next we start the search process from this quad-tree.

2.2   Pairwise Expansion Method

According to the description of *Pairwise Expansion* in Sect. 1, We can accomplish the search process in two stages: *Object-Pair Generation* and *Object-Pair Check*.

Stage1: *Object-Pair Generation.* In this stage, we want to give a higher priority to a closer object-pair. Thus we employ an algorithm OPG, which is a branch-and-bound method by topdown exploration of node-pairs, according to the method to the Closest Pair Query (CPQ) problem [3]. In the algorithm OPG, there are two global variables: δ^* and $Queue$. δ^* is an upper bound of the optimal diameter δ^*_{opt}. $Queue$ is a priority queue to hold node-pairs in ascending order according to $Mindist$ of these node-pairs. ($Mindist$ is the minimum distance between two MBRs of a node-pair). Then we give an outline of OPG as follows:

1. Set $\delta^* := \infty$; Put the root node-pair into $Queue$.
2. Dequeue the node-pair $\langle n_i, n_j \rangle$ with the smallest $Mindist$ from $Queue$. If either n_i or n_j is an internal node, then add all the child-node pairs into $Queue$; if both n_i and n_j are leaves, then for each object-pair $\langle o_a, o_b \rangle$ in $\langle n_i, n_j \rangle$, check if it can be the diameter of a 'correct' object-set. (The 'correct' object-set means that Q is satisfied).
3. If the object-pair $\langle o_a, o_b \rangle$ passes the test and $dist(o_a, o_b) < \delta^*$, then update δ^* with $dist(o_a, o_b)$.
4. Repeat the step 2 and 3 until $Queue = \emptyset$ or the $Mindist$ of the head node-pair of $Queue$ is larger than δ^*. Then δ^* is the solution.

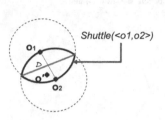

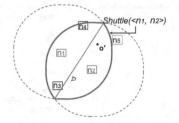

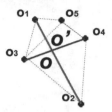

Fig. 1. Shuttle area of object-pair

Fig. 2. Shuttle area of node-pair $\langle n_1, n_2 \rangle$

Fig. 3. Division of object-set

In OPG, we add two rules of pruning. (i) Given an object-pair $\langle o_1, o_2 \rangle$, if it is the diameter of an object-set O, then each object $o' \in O$ must satisfy that $dist(o', o_1) \leq dist(o_1, o_2)$ and $dist(o', o_2) \leq dist(o_1, o_2)$. This constraint forms an area of $Shuttle(\langle o_1, o_2 \rangle)$ (Fig. 1). If $Shuttle(\langle o_1, o_2 \rangle)$ does not satisfy Q, then $\langle o_1, o_2 \rangle$ will never be a diameter of any 'correct' object-set, and we can skip it. (ii) Similarly, given a node-pair $\langle n_1, n_2 \rangle$, for any object-pair p in $\langle n_1, n_2 \rangle$ such that p is the diameter of an object-set O, then each object $o' \in O$ must satisfy that $Mindist(o', n_1) \leq Maxdist(n_1, n_2)$ and $Mindist(o', n_2) \leq Maxdist(n_1, n_2)$. ($Maxdist$ is the maximum distance between two MBRs of a node-pair). This constraint forms an area of $Shuttle(\langle n_1, n_2 \rangle)$ (Fig. 2). Thus if $Shuttle(\langle n_1, n_2 \rangle)$ does not satisfy Q, then no object-pair belonging to $\langle n_1, n_2 \rangle$ can be the diameter of a 'correct' object-set, and we can skip it. Furthermore the maximum distance $\mathcal{D}$ in the shuttle area of either an object-pair or a node-pair can be an upper bound of δ^*_{opt} (Figs. 1 and 2). Thus we use $\mathcal{D}$ to update δ^* in order to reduce $Queue$.

Stage2: *Object-Pair Check.* In this stage, we will expand an object-pair $\langle o_1, o_2 \rangle$ into object-sets by adding some other objects in $Shuttle(\langle o_1, o_2 \rangle)$. We use a recursive algorithm OPC to check if an object-pair $\langle o_1, o_2 \rangle$ can be the diameter of a 'correct' object-set. In Fig. 3, a 'correct' object-set $O = \{\mathbf{o_1}, \mathbf{o_2}, o_3, o_4, o_5\}$ consists of two parts: diameter $\langle o_1, o_2 \rangle$ and a sub object-set $O' = \{\mathbf{o_3}, \mathbf{o_4}, o_5\}$ of O except o_1 and o_2. Note that O' needs to contain the remaining keywords except the keywords of o_1 and o_2, and $\langle o_3, o_4 \rangle$ is the diameter of O'. Thus the check of $\langle o_1, o_2 \rangle$ is to test if such a sub object-set of O' exists in $Shuttle(\langle o_1, o_2 \rangle)$. Similarly, O' also consists of two parts: O''s diameter $\langle o_3, o_4 \rangle$ and a sub object-set $O'' = \{o_5\}$ of O'. Then for $\langle o_3, o_4 \rangle$, we recursively check if such a correct sub object-set O'' exists in $Shuttle(\langle o_3, o_4 \rangle)$. Then we give a general description of $OPC(o_a, o_b)$ as follows:

1. Enumerate all the object-pairs in $Shuttle(\langle o_a, o_b \rangle)$ in ascending order according to their distances.
2. For each object-pair $\langle o_i, o_j \rangle$, do the following: If there are no remaining keywords, then return true. If $Shuttle(\langle o_i, o_j \rangle)$ contains all the remaining

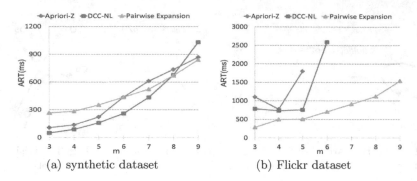

(a) synthetic dataset   (b) Flickr dataset

Fig. 4. Performance comparison

keywords except those of already enumerated object-pairs, then invoke $OPC(o_i, o_j)$ recursively. Otherwise go to the next object-pair. When all object-pairs are checked, return false.

Furthermore, we use a pruning strategy according to the *Minimum Enclosing Circle* of [2] to skip the unnecessary long list of object-pairs in each recursion.

As a preliminary evaluation, we tested our algorithm over a synthetic uniform dataset (with 10^5 points) and Flickr datasets (with 0.4 million Flickr photos in Tokyo area). As a comparison, we use two existing topdown algorithms: Apriori-Z (Zhang's Apriori-based algorithm of [1]) and DCC-NL in [4], both of which enumerate node-sets. All the algorithms are implemented in Java.

Figure 4 is ART(Average Response Time) vs. the number of keywords m as a query. We can see Pairwise Expansion method outperforms Apriori-Z and DCC-NL when m increases in both synthetic dataset and real dataset.

3   Summary

In this work, we proposed *Pairwise Expansion* approach for the issue of mCK query. Compared with the existing topdown search methods, *Pairwise Expansion* can avoid the unstable factor of enumeration of node-sets. The performance of synthetic and real datasets demonstrated that our approach keeps stable and better performance for different distribution of data, including real spatial web datasets.

References

1. Zhang, D.X., Chee, Y.M., Mondal, A., Tung, A.K.H., Kitsuregawa, M.: Keyword search in spatial databases: towards searching by document. In: IEEE ICDE, pp. 688–699 (2009)
2. Guo, T., Cao, X., Cong, G.: Efficient algorithms for answering the m-closest keywords query. In: ACM SIGMOD, pp. 405–418 (2015)

3. Corral, A., Manolopoulos, Y., Theodoridis, Y., Vassilakopoulos, M.: Closest pair queries in spatial databases. In: ACM SIGMOD, pp. 189–200 (2000)
4. Qiu, Y., Ohmori, T., Shintani, T.,Fujita, H.: A new algorithm for m-closest keywords query over spatial web with grid partitioning. In: IEEE/ACIS SNPD, pp. 507–514 (2015)

Mentioning the Optimal Users
in the Appropriate Time on Twitter

Zhaoyun Ding$^{(\boxtimes)}$, Xueqing Zou, Yueyang Li, Su He, Jiajun Cheng,
Fengcai Qiao, and Hui Wang

College of Information System and Management,
National University of Defense Technology,
Changsha 410073, People's Republic of China
{zyding,xqzhou,liyueyang14,hesu,jiajun.cheng,fcqiao,huiwang}@nudt.edu.cn

Abstract. Nowadays, Twitter has become an important platform to
expand the diffusion of information or advertisement. Mention is a new
feature on Twitter. By mentioning users in a tweet, they will receive notifications and their possible retweets may help to initiate large cascade
diffusion of the tweet. In order to maximize the cascade diffusion, two
important factors need to be considered: (1) The mentioned users will
be interested the tweet; (2) The mentioned users should be online. The
second factor was mainly studied in this paper. If we mention users when
they are online, they will receive notifications immediately and their possible retweets may help to maximize the cascade diffusion as quickly as
possible. In this paper, an unbalance assignment problem was proposed
to ensure that we mentioned the optimal users in the appropriate time.
In the assignment problem, constraints were modeled to overcome the
overload problems on Twitter. Further, the unbalance assignment problem was converted to a balance assignment problem, and the Hungarian
algorithm was took to solve the above problem. Experiments were conducted on a real dataset from Twitter containing about 2 thousand users
and 5 million tweets in a target community, and results showed that our
method was consistently better than mentioning users randomly.

Keywords: Mention · Time · Assignment problem · Twitter

1   Introduction

On Twitter, people can carry out social interactions by reading, commenting and
forwarding the latest updates from others. However, due to the large amount
of user-generated content, people become easily distracted during information
seeking. As a result, the effect of social media marketing (SMM) (also known as
marketing practices on social media platforms), such as efforts to gain attentions

This work was supported by National Natural Science Foundation of China
(No.71331008).

F. Li et al. (Eds.): APWeb 2016, Part II, LNCS 9932, pp. 464–468, 2016.
DOI: 10.1007/978-3-319-45817-5_47

and encourage participation by posting attractive contents, can be very limited if the right audiences cannot be properly identified in time.

Fortunately, as a new feature on Twitter, Mention can help ordinary users to improve the visibility of their tweets and go beyond their immediate reach in social interactions. Mention is tagged as @username. All the users mentioned by a tweet will receive a mention notification. By using Mention, one can draw attention from specific users. Properly using mention can quickly help an ordinary user spreading his tweets.

Due to the significance of the mention feature, Mention Recommendation has attracted some studies in previous works. The representative work was proposed by Wang et al. [1]. In their work, features of users' interest and influence were considered to recommend users by mentioning. However, the temporal pattern of mentioned users was neglected in their work. If they mentioned users who were not online, these users would be not notified as quickly as possible. Users usually preferred to retweet the instant information. If users are online in next time and they discover the information is outdated, the probability of retweets would become lower. Moreover, due to the overload problems on Twitter, large number of other mentioned tweets maybe drown earlier mentioned tweets and caused these earlier mentioned tweets to be read with a lower probability.

In order to expand the diffusion of tweets by @ recommendation on Twitter, it is important to consider whether mentioned users are online. For an account on Twitter, if we want to expand the diffusion of tweets by @ recommendation by this account, the naive method is to mention as many users as possible in each hour. However, there is an overload problem on Twitter. If an account posts too many tweets in each hour, it is likely to be treated as a spam and the Twitter Service Provider will seal this account. Moreover, if the account posts tweets automatically by the API of Twitter, it will be restricted by the frequency of Twitter API. Usually, only 60 tweets at most are allowed to publish in an hour by the API of Twitter.

The above problem was modeled as a linear programming problem in this paper. Time windows for an account in a day were divided into 24 windows. Due to the overload problem and limits of Twitter API, the mentioned users in each window were lower than a threshold. Moreover, in order to avoid the account be treated as a spam by the Twitter Service Provider, the threshold in each window should be different. The number of mentioned users in each window for an account was set up in advance. We should find such a number of users on Twitter and put them to each window in order to maximize the probability of online.

2   Methods

In this paper, only the temporal pattern was considered as the behavior of a user. If we want to spread topics or events on Twitter, the target audience would be found according to the interest similarity. A series of keywords $\{w_1, w_2, ..., w_i\}$ were constructed to represent a topic or an event. Then, the target audience

$\{u_1, u_2, ..., u_n\}$ who were interested in the topic or the event could be retrieved by the series of keywords $\{w_1, w_2, ..., w_i\}$. Also, if you want to give your product advertising on Twitter, the target audience $\{u_1, u_2, ..., u_n\}$ could be retrieved by the series of keywords $\{w_1, w_2, ..., w_i\}$ which represented the idea of the advertising.

For an account on Twitter, if we want to expand the diffusion of tweets by @ recommendation by this account, the naive method was to mention as many users as possible in each hour. However, there was an overload problem on Twitter. If an account posts too many tweets in each hour, it would be likely to be treated as a spam and the Twitter Service Provider would seal this account. Moreover, if the account posts tweets automatically by the API of Twitter, it would be restricted by the frequency of Twitter API. Usually, only 60 tweets at most were allowed to publish in an hour by the API of Twitter. So, in order to maximize the benefit in the constraints of the number of tweets in each hour, we should mention users when they would be online. Then, more users could accept the topic or the event as quickly as possible and the topic or the event may be spread as quickly as possible.

So, for an account on Twitter, the optimization model was defined as follows.

$$\max z = \sum_{i=1}^{m} \sum_{j=1}^{n} c_{ij} x_{ij}$$

$$\begin{cases} \sum_{i=1}^{m} x_{ij} \leq 1, j = 1, 2, ..., n \\ \sum_{j=1}^{n} x_{ij} \leq \sigma_i, i = 1, 2, ..., m \\ \sum_{i=1}^{m} \sum_{j=1}^{n} x_{ij} \leq \delta \\ x_{ij} = 0 \ \ or \ \ 1 \end{cases} \tag{1}$$

Here, The x_{ij} indicated whether the j user was mentioned in the i hour window. If the j user was mentioned in the i hour window, the x_{ij} was 1; otherwise, it was 0. c_{ij} indicated the probability of online, and it was the element of the matrix P^T. The m indicated the number of windows, and the maximum number of the m was equal to 24. The n indicated the number of all users. In order to overcome the overload problem, the number of tweets in each hour was limited as σ_i; that was $\sum_{j=1}^{n} x_{ij} \leq \sigma_i$. Moreover, the total number of users mentioned was less a threshold δ, that was $\sum_{i=1}^{m} \sum_{j=1}^{n} x_{ij} \leq \delta$. In this paper, only a user was mentioned in a tweet to overcome the overload problem; that was $\sum_{i=1}^{m} x_{ij} \leq 1$. A tweet mentioning a lot of users was likely to be treated as a spam tweet, which would decrease others interest in retweeting it.

The above optimization model was called as an unbalanced assignment problem. Here, users mentioned by the virtual account were called as the tasks. And windows for each hour were called as the agents. The target function was to maximize the benefit when each task was assigned the most suitable agent. The first constraint indicated each task was accomplished only by an agent. The second constraint indicated each agent could accomplish multiple tasks and the number of tasks was less than a threshold σ_i. The third constraint indicated the total number of tasks was less a threshold δ. Because the total number of users mentioned by a virtual account was usually less than the number of users in a target community, parts of tasks were not accomplished by agents. The fourth constraint indicated a task was accomplished or was not accomplished. The above optimization model was called as the balanced assignment problem. In this paper, the Hungarian algorithm was took to solve the above problem.

3 Experiments

In order to evaluate the effective of our model, we divided the data set into two parts. The time stamps from 2015 January to 2015 October. We compared the mention users list by recommending users randomly. Our model was named as AP, and the randomized method was named as RM. The experimental results were illustrated in Fig. 1.

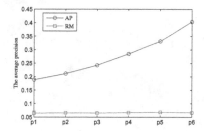   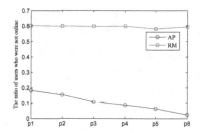

Fig. 1. The average precision of all users

Fig. 2. The ratio of users who were not online

We could find our method was more effective. The average precision of the method by recommending users randomly was lower. Most of users recommended by randomly were not even online in a whole 2015 November.

Moreover, the precision of these users who were not online in whole November was equal to 0. The more these users were mentioned, the worse results to expand the diffusion of tweets in time. Because these users may be not online in whole month. We counted the average ratio of these users who were not online in whole November. The experimental results were illustrated in Fig. 2. According to experimental results, we could find that the ratio of users who were not online for our method was lower than the method by recommending users randomly, and more valuable users were mentioned by our method.

4   Conclusions

In order to find the optimal users to mention in the appropriate time on Twitter, an unbalance assignment problem was proposed in this paper. In the future work, besides for the time-series pattern of users, other factors such as interests of users, behaviors of users, et al. will be considered in order to find the better users to mention on Twitter.

Reference

1. Wang, B., Wang, C., Bu, J., Chen, C., Zhang, W.V., Cai, D., He, X.: Whom to mention: expand the diffusion of tweets by @ recommendation on micro-blogging systems. In: Proceedings of the 22nd International Conference on World Wide Web (WWW 2013), Rio de Janeiro, Brazil, pp. 1331–1340, May 2013

Historical Geo-Social Query Processing

Xiaoying Chen[1,2], Chong Zhang[1,2]([✉]), Yanli Hu[1,2], Bin Ge[1,2],
and Weidong Xiao[1,2]

[1] Science and Technology on Information Systems Engineering Laboratory,
National University of Defense Technology, Changsha 410073, China
chenxiaoying1991@yahoo.com, leocheung8286@yahoo.com,
smilelife1979@163.com, gebin1978@gmail.com, wilsonshaw@vip.sina.com
[2] Collaborative Innovation Center of Geospatial Technology, Wuhan, China

Abstract. Nowadays, more and more people would like to share their
locations and check-in information through social networks – geo-social,
which is also a hotspot in academic community. Previous works focus on
query to geo-social network at current time, however, historical query
is an important tool for analyzing and essential for system-centric purpose. This paper addresses two historical geo-social queries on group
problem. One is for check-in historical query and the other is for life
pattern query. We build sophisticated structures to index the data and
propose algorithms to process queries. Experimental evaluation is carried
out on real-synthetic hybrid dataset, and the results verify our approach
is effective and efficient.

Keywords: Spatio-Temporal · Geo-Social · Group query

1 Introduction

A Geo-Social Network (GeoSN) includes four components: a social network, a
set of points pf interest, the checkins of users to the places and the trajectory
of users. The social network is an undirected graph $G = (U, E)$, where U is
the set of all users and each edge $(u_i, u_j) \in E$ indicates that users u_i, $u_j \in U$
are friends. Set P contains the set of all places of interest (POI) visited by
users, in the form of $< x, y, text >$ GPS points with a set of keywords. Set
$CK = \{< u_i, p_k, t_r > | u_i \in U$ and $p_k \in P\}$ includes all checkins which means that
user u_i visit place p_k at time t_r. For a place p_k and a time interval $[t_s, t_e]$, the set
U_{p_k, t_s, t_e} is defined by $U_{p_k, t_s, t_e} = \{u_i | < u_i, p_k, t_p > \in CK$ and $t_p \in [t_s, t_e]\}$, which
returns a set of users that checkin p_k in $[t_s, t_e]$. Besides these, user would form
a trajectory formed as $TJ = \{< x_1, y_1, t_1 >, < x_2, y_2, t_2 >, ..., < x_n, y_n, t_n >\}$,
where $\exists < x_i, y_i, t_i > \in TJ$, $< x_i, y_i > \notin P$. We can find that this kind of social
network contains two important branches: spatial social network (or geo-social

This work is supported by NSF of China grant 61303062 and 61302144.

F. Li et al. (Eds.): APWeb 2016, Part II, LNCS 9932, pp. 469–473, 2016.
DOI: 10.1007/978-3-319-45817-5_48

network) [1,3] and temporal social network (or dynamic social network) [2,4]. Then, we address two useful historical Geo social network queries as follows:

Given a spatial range $R = (c, r)$, where c is centroid, and r is radius, a time interval $[t_s, t_e]$, and a keyword set W_q, **Spatio-Temporal Group of Interest (STGI)** query aims to find groups of users, where each group satisfies the following conditions, every member checks in POIs spatially bounded by R, temporally bounded by $[t_s, t_e]$, and the checked-in POIs cover set W_q.

Given a spatial range $R = (c, r)$, where c is centroid, and r is radius, a time interval $[t_s, t_e]$, a time slot (or granularity) t_g, an integer k, and a keyword set W_q, **Spatio-Temporal Group Life Pattern (STGLP)** query aims to find groups of users, where each group satisfies the following conditions, for every t_g during $[t_s, t_e]$, every member enters range R at least k times, and the POIs checked in by the group members cover set W_q.

2 Index and Query Processing

HCKI: HCKI is an HR-tree like index, i.e., the tree root is a time axis containing entries, in which each entry is a time interval valid for a spatial area, the following subtrees are actually R-trees. Different from HR-tree, each tree node of R-trees in the HCKI is augmented with keywords. For a leaf node, the entry is formatted in $< x, y, t, W_p >$, pointing to a check-in record representing that user u_i checks in POI p_k associated with location (x, y) and keyword set W_p. While for an internal node, the entry is $< MBR, BF >$, where MBR represents the spatial minimum bounding rectangle of subtree, and BF represents the Bloom Filter built from the keywords of the subtrees. Such a design takes spatial, temporal and textual attributes into account, which is easy for querying these three dimensions at the same time. Figure 1 illustrates an example of HCKI structure, where internal node A and leaf node a are labeled.

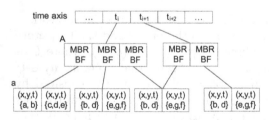

Fig. 1. An illustration of HCKI

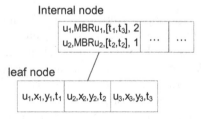

Fig. 2. An illustration of partial UTI

UTI: UTI is also built based on HR-tree structure, however, considering for answering life pattern query, more sophisticated design is needed. The entry in leaf node is in format $< uid, x, y, t, W_p >$, where uid is user identifier, and (x, y) is user's location valid at time t, and W_p is the keyword set of POI with location (x, y) checked in by uid at time t, while, when W_p is null, (x, y, t)

only represents user's historical location record. The entry in internal node is in format $< MBR, t_{ns}, t_{ne}, BF, U >$, where MBR means the spatial minimum bounding rectangle of subtree, and $[t_{ns}, t_{ne}]$ means the valid temporal interval for MBR, and U is a set containing the statistics of users from subtree, and each element in U is in format $< uid, MBR_u, t_{us}, t_{ue}, num >$, where uid is identifier of user, and MBR_u is MBR for user uid valid in temporal interval $[t_{us}, t_{ue}]$, and num is the total number of records from subtree. Figure 2 shows an example of partial UTI structure, where only one leaf node and one internal node are presented. The leaf node contains user u_1 and u_2's historical locations, and the internal node aggregates the user location records, and stores the MBR, time interval and records amount as statistical information.

STGI Query Processing: For a STGI query $(R, [t_s, t_e], W_q)$, where $R = (c, r)$, the basic workflow for processing the query is, first predicate $[t_s, t_e]$ is used against the root of HCKI, and a set of candidate R-tree nodes are obtained, and then the R-tree searching algorithm is executed, and for each node Bloom Filter is inspected whether it contains the querying keyword set, recursively, the satisfied leaf node entries are found, and group forming algorithm is carried out, then the final results are obtained. Each subtree is examined by function $search()$ which is presented in Procedure $search$ and satisfied users are added into temporary result $Uset$.

Procedure $search(node, R, t_s, t_e, W_q)$

if $node$ is internal node then
    for each entry $e \in node$ do
        if $e.MBR \cap R \neq \phi \wedge e.timeintv \cap [t_s, t_e] \neq \phi \wedge e.BF.cover(W_q)$ then
            $search(e.node, R, [t_s, t_e], W_q)$;

else
    for each entry $e \in node$ do
        if $e.(x, y) \in R \wedge e.t \in [t_s, t_e] \wedge e.W_p.cover(W_q)$ then
            return e;

STGLP Query Processing: For a STGLP query The difference from STGI query is, for an entry in internal node, besides the MBR, valid time interval and keyword set is inspected, users' statistics are also examined, in particular, a user, say u_i, its time interval is divided by t_g into segments, and u_i's record amount num is normalized by the number of segments, say $avgc$, if $avgc$ is less than k, then u_i is discarded in the succeeding search, i.e., the coming subtree's MBR is shrunk by minus the u_i's MBR, and same for valid time interval, and other nodes (neighbor nodes) containing u_i. If all users in an entry are unsatisfied with condition above specified by t_g and k, the subtree pointed by the entry is discarded. Another difference is that, when traversing the tree, W_q is not used as a filter, because STGLP query does not require all users should check in the POI with W_q. After descending to the leaf nodes, the users satisfied with life pattern predicate are retrieved, and then connected graph formation operation is carried out, and only the graph which is able to cover W_q is returned. The

---

Procedure search($node$, R, t_s, t_e, t_g, k)

---

if *node is internal node* then
  for *each entry $e \in node$* do
    if $e.MBR \cap R \neq \phi \wedge e.[t_{ns}, t_{ne}] \cap [t_s,$
    $t_e] \neq \phi$ then
      for *each $u \in e.U$* do
        if
        $u.num/((u.t_{ue}-u.t_{us})/t_g)<k$
        then
          $e.node=e.node.minus(u.MBR_u)$;

      $search(e.node, R, [t_s, t_e], t_g, k)$;

else
  for *each entry $e \in node$* do
    if $e.(x, y) \in R \wedge e.t \in [t_s, t_e] \wedge$
    $e.lifePattern(t_g, k)$ then
      return e;

---

distinguished code is from 5 to 7, where user's statistics information is checked and if the user is unsatisfied, function $minus()$ shrinks the corresponding node's MBR. In line 10, function $lifePatter()$ checks whether a user satisfies with life pattern predicate specified by t_g and k.

3 Experimental Evaluation

The two indexes and query processing algorithms are implemented in Java and run on a synthetic-real-hybrid dataset. To make comparison, the non-index approach is used as a baseline, i.e., straight forward and exhausted search is carried out. We vary the query parameters and at each testing point, 30 queries are issued to collect the average results.

STGI Query: First we vary parameters on testing STGI query processing. As spatio-temporal selectivity is increased, we can find that both the response time also increase, which can be explained a larger spatial and temporal range would raise more nodes to be involved, and increase more time to process. And from the results, it is apparent that our HCKI and searching algorithm are effective. Next, when the number of querying keywords is increased, the response time also increases. This can be explained that, more keywords would incur more nodes to be covered. Similar, our acceleration method is also more efficient than non-index approach.

STGLP Query: There are four parameters to test. First, we increase spatio-temporal selectivity and the result is similar to that of STGI query, and similar for the number of keywords. Next we increase $t_g/(t_e - t_s)$, and the results are opposite, a larger t_g would decrease the response time, which can be explained a smaller t_g would cut $[t_s, t_e]$ into more segments, which would increase the number of comparisons and incur more time to be consumed. Then we increase k, and the result is similar to that of t_g, and this can be explained that a larger k would filter more users, and accelerate the process.

4   Conclusion

This paper addresses two historical queries in geo-social networks. One is Spatio-Temporal Group of Interest (STGI) query, the other is Spatio-Temporal Group of Life Pattern (STGLP) query, which are essential for system-centric analysis. We propose indexes and query algorithms to solve the problem and conduct experiments to prove our method is effective and efficient. In the future, we will exploit the methods to answer more historical queries on geo-social network.

References

1. Armenatzoglou, N., Ahuja, R., Papadias, D.: Geo-social ranking: functions and query processing. VLDB J. Int. J. Very Large Data Bases **24**(6), 783–799 (2015)
2. Chen, X., Zhang, C., Ge, B., Xiao, W.: Temporal social network: storage, indexing and query processing. CEUR Workshop Proc. **1558** (2016)
3. Doytsher, Y., Galon, B., Kanza, Y.: Querying geo-social data by bridging spatial networks and social networks. In: Proceedings of International Workshop on Location Based Social Networks, LBSN 2010, San Jose, CA, USA, 2 November 2010, pp. 39–46 (2010)
4. Yang, D.N., Chen, Y.L., Lee, W.C., Chen, M.S.: On social-temporal group query with acquaintance constraint. Proc. VLDB Endowment **4**(6), 397–408 (2011)

WS-Rank: Bringing Sentences into Graph for Keyword Extraction

Fan Yang, Yue-Sheng Zhu$^{(\boxtimes)}$, and Yu-Jia Ma

Communication and Information Security Lab, Shenzhen Graduate School,
Institute of Big Data Technologies, Peking University, Shenzhen, China
{yangfan0705,zhuys}@pku.edu.cn

Abstract. Graph-based method is one of the most efficient unsupervised ways to extract keyword from a single web text. However, rarely did the previous graph-based methods consider the sentence importance. In this paper, we propose a graph-based keyword extractor WS-Rank which brings sentences into graph where sentences are distinctively treated according to their importance. The candidate keywords are extracted through the voting mechanism between words and sentences. To evaluate the experiment, we compare our method with TextRank, a graph-based method which uses the logic distribution relationship only between words. Experiment on 13702 web texts carried out shows that WS-Rank achieves more ideal results with an average F-score of 25.20 %.

1 Introduction

Due to the explosion of web information, it becomes more difficult to search and manage the network resources. Keyword extraction aims to select a set of words from a text as its short summary, which can help people to identify whether they are interested quickly. Since keyword extraction in manual way is very expensive and time-consuming, many studies have been done for keyword extraction.

The keyword extraction method based on graph is simple and robust and has been used in many ways [2,7]. A representative method is TextRank [4] which uses a syntactic graph, where vertices represent words and edges represent word co-occurrence within a fixed window. After that, some varietal methods of TextRank are proposed to extract keyword such as Tag-TextRank [5] and TimedTextRank [6]. Besides, recent years have seen a lot of applications of keyword extraction using graph-based methods, especially in the field of social networks [1,3]. In this paper, sentence importance is brought into graph. We provide a method for keyword extraction using a graph where vertices represent words and sentences. The edges in graph represent the word existence in corresponding sentence. The graph is constructed according to the relationship between words and sentences. Besides, WS-Rank is proposed as a ranking algorithm to extract keyword, which will be introduced in the following section.

© Springer International Publishing Switzerland 2016
F. Li et al. (Eds.): APWeb 2016, Part II, LNCS 9932, pp. 474–477, 2016.
DOI: 10.1007/978-3-319-45817-5_49

2   WS-Rank: Sentence-Based Extractor

WS-Rank is an unsupervised, graph-based and language-independent keyword extractor. In this section, the candidate keyword graph and the words ranking algorithm of WS-Rank will be described.

WS-Rank uses a graph where vertices stand for words and sentences which are connected by undirected edges. We consider a word vertex and a sentence vertex are directly connected if the sentence contains the corresponding word. After the candidate keyword graph of WS-Rank is constructed, the score associated with each word vertex is set to an initial value of 1 and the score of each sentence vertex is set to 0. The ranking algorithm of WS-Rank runs on the graph for several iterations until a convergence is reached. The algorithm contains two steps: the transfer of score from word vertices to sentence vertices according to the corresponding edge weight which is determined by the importance of sentences and the transfer of score from sentence vertices to corresponding word vertices according to the importance of words. The score of the word vertex W_i is defined by the recursive formula (1):

$$WS(W_i) = (1 - d) + d *$$

$$\sum_{S_m \in In(W_i)} \left(\frac{\mu_{mi}}{\sum_{W_k \in Out(S_m)} \mu_{mk}} * \sum_{W_j \in In(S_m)} \frac{\omega_{jm}}{\sum_{S_n \in Out(W_j)} \omega_{jn}} WS(W_j) \right) \tag{1}$$

where $In(W_i)$ and $In(S_m)$ represent the set of vertices that points to W_i and S_m respectively. $Out(W_j)$ and $Out(S_m)$ represent the set of vertices that W_i and S_m points to respectively. μ_{mi} represents the edge weight from S_m to W_i (the importance of W_i is measured by μ_{mi}). ω_{jm} is the edge weight from W_j to S_m (the importance of S_m is measured by ω_{jm}). d is a damping factor that gives the probability of jumping from a vertex to another random vertex in the graph. Usually, the damping factor is set to 0.85 [4]. In the formula, each word vertex gives its score to the adjacent sentence vertices and each sentence vertex gives its score back to the adjacent word vertices according to the corresponding edge weight. To evaluate the effect of sentence importance separately, we consider the words have the same importance, which means formula (1) can be described as:

$$WS(W_i) = (1-d) + d * \sum_{S_m \in In(W_i)} \frac{\sum_{W_j \in In(S_m)} \frac{\omega_{jm}}{\sum_{S_n \in Out(W_j)} \omega_{jn}} WS(W_j)}{|Out(S_m)|} \tag{2}$$

where $|Out(S_m)|$ represents the number of the edges which connect S_m. In this formula, each word vertex gives its score to the adjacent sentence vertices according to the corresponding edge weight and each sentence vertex gives its score back to the adjacent word vertices equally. A sample of WS-Rank in a short Chinese text is shown in Fig. 1 where the edge weight is represented by arrows with different size (Suppose $S1$ is more important than $S2$ and $S3$). The score of each vertex is computed iteratively until a convergence is reached. The top ranked vertices are extracted as the keywords.

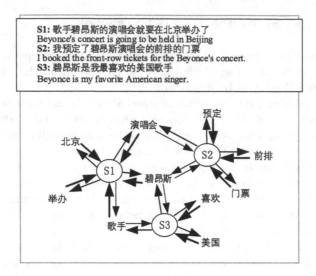

Fig. 1. Sample graph for keyword extraction with WS-Rank

3   Experiment and Discussion

To investigate experimental result when introducing the sentence importance, we give different values to the edge weight from word vertices to sentence vertices.

The experiment is based on a large collection of Chinese texts[1] from 163.com. In this dataset, there are 13702 texts. The average keywords number of the dataset is 2.4 words. The word segmentation method used in our experiment is based on NLPIR[2]. The stop words are removed after word segmentation. To evaluate our method, we use precision($P = |CA \cap CB|/|CB|$), recall($R = |CA \cap CB|/|CA|$) and the macro-average F-score($F = |2 \times P \times R|/(P+R)$) where CA represents the set of keywords extracted in manual way and CB represents the set of keywords extracted by the method used in this paper.

As we know, the first or the last sentence usually plays a summative role of the corresponding paragraph. In the experiment, these sentences are selected as important sentences while the others are treated as normal sentences. The value of ω_{ij} (as is mentioned above, ω_{ij} represents the edge weight from W_i to S_j) is set to k if S_j is important sentence. Otherwise, ω_{ij} is set to 1. The keywords are extracted according to different value of k, the result of which is shown in Table 1. The highest F-score is achieved when k is 1.97, which means when the first and the last sentences are given 1.97 times the importance of other sentences, the experimental result is the optimum. It is also shown in Table 1 that if we neglect or give an overemphasis on the important sentences, the F-score will decrease. We also compare our method with TextRank which merely considers the relationship between words. We can see the performance of WS-Rank is better than that of

---

[1] http://nlp.csai.tsinghua.edu.cn/~lzy/#Data.

[2] http://ictclas.nlpir.org/.

Table 1. Results of WS-Rank and TextRank when keyword number is 3

Method	k	Prescion(%)	Racall(%)	F-score(%)
WS-Rank	1.00	22.26	27.91	24.77
	1.50	22.55	28.27	25.09
	1.95	22.63	28.37	25.18
	1.97	22.65	28.39	25.20
	2.00	22.63	28.37	25.17
	2.50	22.48	28.18	25.01
TextRank	-	20.91	26.22	23.26

TextRank. In particular, when k is set to 1(the important and normal sentences are equally treated like TextRank), WS-Rank is still better than TextRank, which illustrates WS-Rank is meaningful even without the consideration of edge weight. In TextRank, a word with wide distribution means the corresponding word vertex has more adjacent vertices, from which the word can get more score through the rank algorithm. Like TextRank, word distribution is also considered in WS-Rank for that if a word has wide distribution, it can get more votes from the adjacent sentence vertices. That is to say, WS-Rank has the advantage of TextRank and gets a better performance.

In the end, it needs to be stressed that the highlighted sentence of WS-Rank is not limited to the first or the last sentence in a paragraph but also the sentence which can be manually labeled, which is more helpful to improve the experimental result. Through the experiment and comparison, we can draw a conclusion that the introduction of sentence importance is considerable in the graph-based keyword extraction.

References

1. Abilhoa, W.D., de Castro, L.N.: A keyword extraction method from twitter messages represented as graphs. Appl. Math. Comput. **240**, 308–325 (2014)
2. Boudin, F.: A comparison of centrality measures for graph-based keyphrase extraction. In: International Joint Conference on Natural Language Processing (IJCNLP), pp. 834–838 (2013)
3. Li, L., Su, C., Sun, Y., Xiong, S., Xu, G.: Hashtag biased ranking for keyword extraction from microblog posts. In: Zhang, S., Wirsing, M., Zhang, Z. (eds.) KSEM 2015. LNCS, vol. 9403, pp. 348–359. Springer, Heidelberg (2015). doi:10.1007/978-3-319-25159-2_32
4. Mihalcea, R., Tarau, P.: Textrank: Bringing order into texts. In: Association for Computational Linguistics (2004)
5. Peng, L., Bin, W., Zhiwei, S., Yachao, C., Hengxun, L.: Tag-textrank: a webpage keyword extraction method based on tags. J. Comput. Res. Dev. **11**, 014 (2012)
6. Wan, X.: Timedtextrank: adding the temporal dimension to multi-document summarization. In: Proceedings of the 30th Annual International ACM SIGIR Conference on Research and Development in Information Retrieval, pp. 867–868. ACM (2007)
7. Wan, X., Xiao, J.: Single document keyphrase extraction using neighborhood knowledge. AAAI **8**, 855–860 (2008)

Efficient Community Maintenance
for Dynamic Social Networks

Hongchao Qin[1,2(✉)], Ye Yuan[2], Feida Zhu[1], and Guoren Wang[2]

[1] School of Information Systems, Singapore Management University,
Singapore, Singapore
[2] School of Computer Science and Engineering,
Northeastern University, Shenyang, China
qhc.neu@gmail.com

Abstract. Community detection plays an important role in a wide
range of research topics for social networks. The highly dynamic nature
of social platforms, and accordingly the constant updates to the underlying network, all present a serious challenge for efficient maintenance
of the identified communities—How to avoid computing from scratch
the whole community detection result in face of every update, which
constitutes small changes more often than not. To solve this problem,
we propose a novel and efficient algorithm to maintain the communities
in dynamic social networks by identifying and updating only those vertices whose community memberships are affected. The complexity of our
algorithm is independent of the graph size. Experiments across varied
datasets demonstrate the superiority of our proposed algorithm in terms
of time efficiency and accuracy.

Keywords: Community detection · Dynamic · Heuristic · Modularity

1 Introduction

Communities are groups of network nodes, within which the links connecting
nodes are dense but between which they are sparse. In many applications, the
structure of networks evolves over time, so do the communities in them. Most
existing works [1,3] identify communities in a static network. They define an
objective function Q which measures the quality of the communities. As the
number of all possible partitions of nodes [5] is exponential of the network size,

This research is partially funded by the National Research Foundation, Prime
Ministers Office, Singapore under its International Research Centres in Singapore
Funding Initiative and Pinnacle Lab for Analytics at Singapore Management University, National Natural Science Foundation of China (No. 61332006, 61332014,
61328202, 61572119, U1401256) and the Fundamental Research Funds for the Central Universities of China (No. N150402005, N130504006).

© Springer International Publishing Switzerland 2016
F. Li et al. (Eds.): APWeb 2016, Part II, LNCS 9932, pp. 478–482, 2016.
DOI: 10.1007/978-3-319-45817-5_50

it is NP-hard to obtain the communities with the maximum Q. Heuristic algorithms, such as [2], are therefore usually proposed as a solution. Yet those algorithms typically need to recompute the entire result for each network change, even through the change affects only a small number of vertices, which renders the detection computationally unaffordable for graphs that are large and evolve fast.

We propose a novel algorithm to maintain communities in dynamic social network by identifying and updating only those communities affected in graph evolution. The complexity of our algorithm is independent of the graph size. We conduct extensive experiments on various datasets and demonstrate the efficiency and accuracy of our approach.

Consider an undirected graph $G = (V, E)$. Let $n = |V|$ and $m = |E|$ be the number of vertexes and the number of edges in G. The most general method of detecting communities is to maximize the quality function Q [3].

Definition 1 (Modularity). *Given a social network G and the partition $C_G = \{C_G^1, C_G^2...C_G^k\}$. The **modularity** is a quality function of the community structure, it can be denoted by Q and*

$$Q(C_G) = \sum_{i=1}^{k} (\frac{l(C_G^i)}{m} - (\frac{d(C_G^i)}{2m})^2) \tag{1}$$

where $l(C_G^i)$ is the total number of edges joining vertexes inside community C_G^i and $d(C_G^i)$ is the sum of the degrees of all the vertexes $\{V_i \| V_i = C_G^i \bigcap V\}$ in G. And $Q(C_G^k)$ denotes single modularity of C_G^k.

Definition 2 (Dynamic Social Network). *A dynamic network $\mathcal{G}$ is represented by a series of time dependent network snapshots*

$$\mathcal{G} = \{G_0, G_1, G_2, ...G_t, G_{t+1}, ...\}$$

where $G_{(t)} = (V_t, E_t)$ is the snapshot of the network at the time point t. The differences between two consecutive snapshots G_{t+1} and G_t is denoted by ΔG_t, which contain nodes' insertions(deletions) and edges' insertions(deletions).

Problem Definition. The goal of the community maintenance problem is to update the communities of all the vertexes in G when the graph G is changed by inserting or deleting edges (all the nodes' insertions or deletions can transform into edges' insertions or deletions). It means that given the dynamic social network $\mathcal{G} = \{G_0, G_1, ...G_t, G_{t+1}, ...\}$ and the communities at t time C_{G_t}, the community structure at time point $t + 1$ is detected based on the community structure at the time point t and the changes ΔG_t. And ΔG_t is split into a sequence of only one edge's insertion or deletion so it is easy to consider.

2   The Proposed Algorithm

Algorithm for Edge Insertion: There are three kinds of edge insertion.

When one vertex of the edge is a new vertex, we can simply join this node to the community of the other side's. There may have new edges containing the new node, but we can match the later insertion with the following sub-algorithms.

When both sides of the new edge are in one community, if we don't change the community partition, the modularity will increase. It can be proved that we can not divide the community to have higher modularity.

When the two vertexes of the edge are in different communities, after the insertion, the communities may not change, or the two vertexes will be added to a new community. We can mark the most possible common community as $C_{common}(u, v)$, and it has the largest number of same neighbours for vertex u and v, i.e., $C_{common}(u, v) = \{C^i_{G_t} \in \mathcal{C}_{G_t} \| argmax(|N(u) \cap C^i_{G_t}| + |N(v) \cap C^i_{G_t}|)\}$.

After the insertion of edge (u,v), we can join both u and v in $C_{common}(u, v)$ and the modularity will increase if there holds a inequation and the complexity checking it is independent with the graph size. If the modularity increase after joining both u and v in $C_{common}(u, v)$ (marked as C_c), then we have

$$Q(C_c \cup u \cup v) + Q(C_{G_t}(u) \backslash u) + Q(C_{G_t}(v) \backslash v) > Q(C_c) + Q(C_{G_t}(u)) + Q(C_{G_t}(v)) \quad (2)$$

Considering the definition of Q, formula 2 can be transformed to an simple inequation and the complexity checking it is independent with the graph size.

If the inequations in formula 2 do not hold, the modularity will not increase whatever community the vertex u and v join in.

Algorithm for Edge Deletion: There are two kinds of edge deletion.

When the two vertexes of the edge are in different communities, we can simply remain the communities unchanged. The number of the inside edges of $C_{G_t}(u)$ and $C_{G_t}(v)$ will not change after the deletion of edge (u, v), and the sum degree of all the vertexes in $C_{G_t}(u)$ and $C_{G_t}(v)$ will increase, so the modularity will increase if the communities don't change.

When vetex u, v are in the same community C^*, after the deletion of edge (u,v), we split the community into two parts $\{C^*_1, C^*_2\}$ and the modularity will increase if the number of edges between the two partions is not larger than one equation. Suppose that the number of edges between the two partions is $d(C^*_1, C^*_2)$. If we need to split the community, it's obvious that u, v are in different community of $\{C^*_1, C^*_2\}$ and there holds $Q(C^*_1) + Q(C^*_2) > Q(C^*)$. Then we can take

$$d(C^*_1, C^*_2) > \frac{d(C^*_1)d(C^*_2) - d(C^*_1) - d(C^*_2) + 1}{2(m-1)} - 1 \quad (3)$$

The formula 3 is easy to calculate. If it holds, we can split the community C^* into two parts, otherwise we need not to change it.

3   Experiments

In the experiments, we collect four real world datasets named Blogs (6,803 edges), ego-Facebook (88,234 edges), soc-Epinions1 (508,837 edges) and com-DBLP

(1,049,866 edges). We get all the datasets from [4] and conduct the experiments on a server with Intel Core i7 @3.6 GHz CPU and 16 GB RAM.

We can use the FNM (Fast Newman) algorithm [5] to get the communities of G_0 in the static network. And we compare our method, marked as CM (Community Maintenance), with FNM and the QCA (Quick Community Adaptive) [6,7] algorithm in dynamic network. To maintain the communities in the network, we need to emulate changes to the graph. Suppose that from G_t to G_{t+1}, five percent numbers of the nodes change. So we randomly insert or delete nodes to modify the graph as large as 5% between each point.

Accuracy: We investigate the accuracy of the algorithm. As the FNM algorithm compute completely in every step, the modularity of the graph is a little much higher than the other algorithms. But they can only get a approximation value of the modularity and the difference is small in both algorithms. From the result, we can know that our algorithm have good accuracy.

Efficiency: In terms of efficiency, we report the time cost of maintaining communities from G_t to G_{t+1}. Figure 1 shows the running time of the algorithms. We can see that the speed of our method CM is faster than the QCA algorithm. But the cost of FNM is not acceptable if the original graph is large.

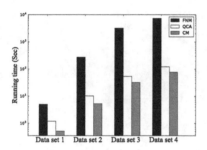
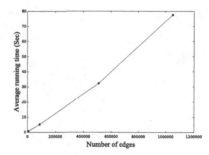

Fig. 1. Running time   **Fig. 2.** Scalability of our algorithm CM

Scalability: To show the scalability of our algorithm, we average the running time of updating the communities from G_t to G_{t+1} with t ranges from 1 to 10 in different datasets. From Fig. 2, we can see that our algorithm CM perform well with the size of the dataset becomes larger.

4   Conclusion

We propose an efficient algorithm for maintaining the communities in dynamic social networks. We split the changes of the graph into edge insertions or deletions and discuss how they influence the communities. Extensive experiments demonstrate the efficiency and accuracy of our algorithm.

References

1. Blondel, V.D., Guillaume, J.L.: Fast unfolding of communities in large networks. J. Stat. Mech. Theor. Exp. **2008**(10), P10008 (2008)
2. Brandes, U., Delling, D., Gaertler, M.: On modularity clustering. IEEE Trans. Knowl. Data Eng. **20**(2), 172–188 (2008)
3. Fortunato, S., Castellano, C.: Community structure in graphs. In: Meyers, R.A. (ed.) Computational Complexity, pp. 490–512. Springer, New York (2012)
4. Leskovec, J., Krevl, A.: SNAP datasets: Stanford large network dataset collection, June 2015. http://snap.stanford.edu/data
5. Newman, M.E.: Fast algorithm for detecting community structure in networks. Phys. Rev. E **69**(6), 066133 (2004)
6. Nguyen, N.P., Dinh, T.N., Shen, Y., Thai, M.T.: Dynamic social community detection and its applications. PloS one **9**(4), e91431 (2014)
7. Nguyen, N.P., Dinh, T.N., Xuan, Y., Thai, M.T.: Adaptive algorithms for detecting community structure in dynamic social networks. In: 2011 Proceedings of IEEE INFOCOM, pp. 2282–2290. IEEE (2011)

Open Sesame! Web Authentication Cracking via Mobile App Analysis

Hui Liu[✉], Yuanyuan Zhang, Juanru Li, Hui Wang, and Dawu Gu

Computer Science and Engineering Department,
Shanghai Jiao Tong University, Shanghai, China
hui.liu803@gmail.com

Abstract. Web authentication security can be undermined by flawed mobile web implementations. Mobile web implementations may use less secure transport channel and enforce less strict brute-force-proof measures, making web authentication services vulnerable to typical attacks such as password cracking. This paper presents an in-depth penetration testing based on a comprehensive dynamic app analysis focusing on vulnerable authentication implementations of Android apps. An analysis of Top 200 apps from China Android Market and Top 100 apps from Google Play Market is conducted. The result shows that 71.3 % apps we analyze fails to protect users' password appropriately. And an experiment carried out among 20 volunteers indicates that 84.4 % passwords can be cracked with the knowledge of password transformation process.

Keywords: Android apps · Web authentication · Password cracking

1 Introduction

As the prevailing of mobile smart devices, entry of web authentication is migrating from browsers to mobile apps. Many apps, however, do not implement secure authentication process and thus are vulnerable to typical web attacks such as password cracking. Therefore, the security of mobile web is significantly weakened compared to that of traditional web. With millions of apps released nowadays, it is essential to shed light on the status quo of how mobile web authentication processes are handled by those apps.

Previous studies has revealed many aspects of vulnerable implementation in Android apps that affect remote web authentication, especially on cryptographic misuses [2,4] and insecurity of user and session authentication [3]. However, they either rely on manual reverse engineering or only concern about simple vulnerabilities such as using hard-coded key. In this paper, we mainly consider two general types of web authentication vulnerabilities exposed by mobile apps concerning mobile web authentication implementations.

One is the transport channel downgrade vulnerability. Services serving for both mobile and web apps are most likely to use the same identity database, since maintaining separate databases is profitless and resource-consuming. But they

© Springer International Publishing Switzerland 2016
F. Li et al. (Eds.): APWeb 2016, Part II, LNCS 9932, pp. 483–487, 2016.
DOI: 10.1007/978-3-319-45817-5_51

would offer different interfaces when dealing with users from different platforms, out of the consideration of optimizing user experience and usability. That is, the **authenticator**[1] sent to server is the same for both mobile and web apps, while the transport channel of the authenticator can be diverse. The less secure channel that mobile app uses can result in the authenticator leakage, which opens a door for attackers to attack the web authentication. Web authentication security is completely compromised if the transformation process is weak enough such that an attacker can reveal the corresponding password simply knowing the authenticator. If the password can not be deduced from the authenticator at once, with the knowledge of the transformation process obtained by app analysis, an attacker can still launch an off-line password brute-forcing attack, trying all possible passwords until the output of the transformation process matches the objective authenticator, which manifests the correct password.

The other vulnerability is the CAPTCHA bypass defect. When an attacker has no access to a user's authenticator, she may actively launch a password guessing attack, pretending to be the user and trying possible passwords until getting a correct response from server. Since this is a well-known on-line attack, a countermeasure involving CAPTCHAs is proposed. Nevertheless, for mobile apps, the enforcement of this measure can be omitted, which makes this kind of attack possible again. With the knowledge of the transformation process, the corresponding authenticators that may be accepted by the server are successfully generated.

2   Inferring Authentication Process via App Analysis

As described above, the key factor to successfully launch both the off-line and on-line brute-forcing/password guessing attack is the ability of transforming the guessed passwords to authenticators. If the generated authenticator matches the objective, the password is cracked. In this section, we introduce the way how to get a sketch of the authentication process via app analysis.

We need to analyze apps to get an overview of what apps have done to the password. And we only concern about the authentication process part of an app. Therefore, our general idea of analyzing an app is (1) running the sign-in process, (2) logging method call traces and specified parameters, and (3) getting the sketch of the transformation process by mapping the input-output relationship and profiling critical method calls. Figure 1 depicts the analysis process. We use appium [1] to automatically run the app's login process, filling in the pre-chosen username and password. Then an instrumented Dalvik VM on which the app is running is used to log down the method call trace and specified method parameters. At last, we extract these logs and abstract the semantic information of the transformation process.

---

[1] During user authentication, an app typically receives a password, encodes or encrypts it, and sends the result together with other data to a remote web server. We let authenticator denote the result for the rest of this paper.

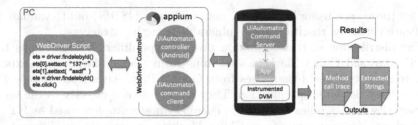

Fig. 1. The process of app analysis.

3   Testing Results

Using the method introduced in Sect. 2, we analyzed 100 top free apps from Google play and 200 top free apps from China Android Market. We successfully analyze 209 apps[2] and classify them into four categories and nine types according to the type of transformation process, as listed in Table 1. The category basically matches discoveries found in [3].

Table 1. Summary of the analysis results.

Type of Transformation Process		China Android Market			Google Play Market		
		HTTP	HTTPS[a]	HTTPS[b]	HTTP	HTTPS[a]	HTTPS[b]
Trivial Transformation	Plaintext	44	19	8	1	8	42
	Encoding	2	0	0	0	1	0
Hash	One-time MD5	21	8	5	0	0	0
	Fixed-salt MD5	6	0	0	0	0	0
	Multi-time hash	3	0	0	0	0	0
Symmetric Encryption	AES/DES with hard-coded key	24	7	2	0	0	0
	AES/DES with randomly generated key	1	0	0	0	0	0
Asymmetric Encryption	RSA/ECB/ NoPadding	3	2	0	0	0	0
	RSA/ECB/ PKCS1Padding	2	0	0	0	0	0
	Sum	106	36	15	1	9	42

[a] apps using HTTPS connection and inadequately validating the certificates.
[b] apps using HTTPS connection and correctly validating the certificates

Overall, 149 apps (71.3 %) fail to provide secure user authentication. Nearly all apps from Google Play use the same type of authentication process, while apps from China Android Markets use different types of transformation process.

---

[2] Other apps are either packed or involved with native APIs, in which case manual intervention is needed.

And for those apps using HTTPS, 45 out of 92 apps (48.9 %) fail to validate the certificate correctly, resulting apps vulnerable to active attackers.

The distribution of transformation process type differs between apps from Google play and China Markets, and apps from China Markets are exposed to more security threats than ones from Google Play. Apps from Google Play seem to obey the same specification. They send password directly to the server. The security totally relies on a secure channel, mostly implemented as HTTPS connections. However, apps from China Markets are completely different. On one hand, most apps send the authentication material through HTTP. On the other hand, apps try to protect password by making transformations (security by obscurity), which, as we will show in Sect. 4, basically enforce no security.

Case Study. *DangDang* is the Chinese most popular online bookstore. While its web entry sends user's password in plain text via a secure TLS channel, its Android client sends the authentication information without any protection.

In terms of CAPTCHA Bypass, on the website of *Meilishuo*, which is the largest fast fashion e-commerce platform in China, a group of pictures are presented for users to click and rotate each of them until all the pictures are in the right position when a user logs in. This is a pretty strong approach to defend against password guessing attacks. Nevertheless, the entry that its mobile app provide totally breaks this defense. It neither presents any CAPTCHAs nor limits the times a user can submit the authentication information, making the online password guessing attack unimpeded. This weakness also appears in *Sogou*, *Jiayuan*, *Mia* etc.

4   Password Cracking Cost Measurement

To intuitively demonstrate the consequences of password cracking enabled by the knowledge of password transformation process, we recruit 20 volunteers to randomly use those 209 apps and launch the authentication process in a controlled network environment. We collect 180 authenticators and conduct the password cracking attack.

The attack cost of some transformation type is trivial, such as encoding and symmetric encryption, since password can be directly calculated from authenticator by inversing the transformation process. As for hash and asymmetric encryption type, cracking is needed. We let two GPU cards (Telsa K20c, 4800 MB global memory, 2496 CUDA cores, 706 MHz clock rate; Quadro K620, 2047 MB global memory, 384 CUDA cores, 1124 MHz clock rate) work together to crack hash. And we use an Intel Xeon Processor E5-2643v3 with 20M smart cache and 3.4 GHz base frequency CPU to run the RSA calculation. Finally, we successfully recover 84.4 % of the passwords in 2 days.

Acknowledgments. This work is supported by the Major program of Shanghai Science and Technology Commission (15511103002).

References

1. Appium automation for apps. http://appium.io/. Accessed 20 Apr 2016
2. Bugiel, S., Davi, L., Dmitrienko, A., Fischer, T., Sadeghi, A., Shastry, B.: Towards taming privilege-escalation attacks on android. In: NDSS (2012)
3. Cai, F., Hao, C., Yuanyi, W., Yuan, Z.: Appcracker: widespread vulnerabilities in user and session authentication in mobile apps. In: IEEE Mobile Security Technologies. IEEE (2015)
4. Fahl, S., Harbach, M., Muders, T., Baumgärtner, L., Freisleben, B., Smith, M.: Why eve, mallory love android: an analysis of android ssl (in) security. In: Proceedings of the ACM Conference on Computer and Communications Security, pp. 50–61. ACM (2012)

K-th Order Skyline Queries
in Bicriteria Networks

Shunqing Jiang[1], Jiping Zheng[1,2][✉], Jialiang Chen[1], and Wei Yu[1]

[1] College of Computer Science and Technology,
Nanjing University of Aeronautics and Astronautics, Nanjing, China
{jiangshunqing,jzh,chenjialiang,echoyu}@nuaa.edu.cn
[2] School of Computer Science and Engineering, University of New South Wales,
Sydney, Australia

Abstract. We consider the problem of k-th order skyline queries in bicriteria networks. Our proposed k-th order skyline queries consider distance, time preferences thus having two kinds of skyline queries, named Distance/Time Optimal k-th Order Skyline Queries (DO-kOSQ/TO-kOSQ). We design algorithms for the two kinds of skyline queries in bicriteria networks based on incremental network expansion method and further develop maximum distance/time restriction strategies to improve the efficiency of the algorithms. Experimental results show efficiency and effectiveness of our proposed methods to answer k-th order skyline queries in real road networks.

Keywords: Skyline query · Bicriteria networks · Euclidean distance

1 Introduction

With the increasing complexity of traffic conditions in road networks, the nearest destination cannot be necessarily reached in the fastest time. The traditional Nearest Neighbor (NN) or k Nearest Neighbor (kNN) queries in a single cost criterion (Euclidean distance or travel-time) spatial network are often strongly restrictive [4]. In this paper, we propose the concept of k-th order skyline queries [1,3] in bicriteria networks where every edge labeled with two kinds of cost criterion: Euclidean distance and travel-time, and the value of travel-time for each edge is a function of the departure time. For a query point q in bicriteria networks, the attributes of each facility in the networks are the sum of lengths and travel-time of the edges along a same path and the selected path could be a Distance Optimal Path (DOP, the shortest path between q and this facility) or a Time Optimal Path (TOP, the fastest path between q and this facility). A k-th order skyline result consists of the subset of the conventional facilities which is been dominated by at less k facilities [3]. And the k-th order skyline queries will be divided into Distance Optimal k-th Order Skyline Query (DO-kOSQ) and Time Optimal k-th Order Skyline Queries (TO-kOSQ) by the selected path. In the rest paper, we will present our techniques for these two kinds of k-th order skyline queries and conduct experiments to show the efficiency and effectiveness of our proposed methods.

© Springer International Publishing Switzerland 2016
F. Li et al. (Eds.): APWeb 2016, Part II, LNCS 9932, pp. 488–491, 2016.
DOI: 10.1007/978-3-319-45817-5_52

2   *K*-th Order Skyline Computation in Bicriterion Networks

2.1   Basic Solutions for DO-*k*OSQ and TO-*k*OSQ

Given a query point q and departure-time t, the objective of the basic solutions of DO-*k*OSQ (TO-*k*OSQ) (termed *DO-kOSQ algorithm* (*TO-kOSQ algorithm*)) is to perform the distance cost (travel-time cost) expansions and calculate the travel-time (road distance) of the expanded facilities concurrently from q.

For *DO-kOSQ algorithm*, We maintain a priority queue Q to store the expanded vertices and facilities. Each point p of Q in the form of $(p, TD(P_s(q \to p)), TT(P_s(q \to p)), AT(P_s(q \to p)))$, and the elements in Q are given by the increasing order of $TD(P_s(q \to p))$. $P_s(q \to p)$ is the shortest path (P_s) between q and p. For the convenience of narration, we use (TD_p, TT_p, AT_p) for $(TD(P_s(q \to p)), TT(P_s(q \to p)), AT(P_s(q \to p)))$. $TD(P)$ and $TT(P, t)$ are the travel-distance and travel-time of path P respectively which are computed by Eqs. 1 and 2. AT_p represents the arrive time of p from q with departure-time t along the shortest path, $AT_p = t + TT_p mod(T)$ (T is time domain). The AT_p is used to determine the travel-time of adjacent edges of the end node p.

$$TD(P) = \sum_{(u,v) \in P} l_{u,v} \tag{1}$$

$$TT(P, t) = \sum_{(u,v) \in P} c_{u,v}(t_u), \tag{2}$$

where t_u is the arriving time of node u, $t_v = t_u + c_{u,v}(t_u)$.

The algorithm of DO-*k*OSQ takes three parameters as the input, i.e., query location q, departure-time t and an order k. *DO-sky(q, t)* is used to store the skyline query results. We first insert the tuple $(q, TD_q = 0, TT_q = 0, AT_q = t)$ generated by q into the priority queue Q. Next, we take a loop starting at de-queue the first element from Q and expand it. The de-queued element is termed as (u, TD_u, TT_u, AT_u) and u can be either a road networks vertex or a facility. We compare u to all the points that stored in *DO-sky(q, t)* to determine its skyline order. If u's skyline order exceeds to k (i.e., there are at least k points in *DO-sky(q, t)* dominating u), we do nothing for this loop and process the next loop. If u's skyline order is less than k: if u is a facility point, then, u is a skyline point, we can insert (u, TD_u, TT_u) into *DO-sky(q, t)* directly. If u is a vertex, we expand vertex u. For each u's non-visited adjacent vertex v and the each facility p (if exists) on edge (u, v), we calculate $TD_{v(or\ p)}, TT_{v(or\ p)}, AT_{v(or\ p)}$, and insert $(v(or\ p), TD_{v(or\ p)}, TT_{v(or\ p)}, AT_{v(or\ p)})$ into the Q. The above process is repeated until Q is empty.

For *TO-kOSQ algorithm*, we perform the incremental network expansion by the travel-time cost criterion of each edge, and the time one arrives at the edge entry determines the travel-time on that edge. The form of each element in priority queue Q is denoted as $(p, TT(P_f(q \to p)), TD(P_f(q \to p)), AT(P_f(q \to p)))$ and sorted according to the value of $TT(P_f(q \to p))$. $P_f(q \to p)$ is the fastest path between q and facility p. The other operations for TO-kOSQ are similar to DO-kOSQ as shown above.

2.2   Improved Solutions for DO-kOSQ and TO-kOSQ

To speed up the processing of basic solutions of DO-kOSQ and TO-kOSQ, we exploit a distance/time restriction value to avoid expanding operations of unnecessary vertices. The improved version of the above two algorithms is named *DO-kOSQ$^+$ algorithm* and *TO-kOSQ$^+$ algorithm*, respectively.

For *DO-kOSQ$^+$ algorithm*, we first get the maximum speed V_{max} of a bicriteria network. We find that, let $\{p_1, p_2, ..., p_k\}$ be the k nearest points of q according to the travel-distance, and t_{max} be $\max\{TT_{p_1}, TT_{p_2}, ..., TT_{p_k}\}$. Then, for each facility p that belongs to *DO-sky(q, t)*, the shortest distance between p and q is less than $V_{max} \times t_{max}$. So, after acquiring the value $V_{max} \times t_{max}$, we needn't visit those points the shortest distance between which and q is larger than $V_{max} \times t_{max}$ for DO-kOSQ. In order to get $V_{max} \times t_{max}$, we don't need to answer kNN query early. We can get the k nearest neighbor points and acquire t_{max} in the processing of DO-kOSQ.

For *TO-kOSQ$^+$ algorithm*. First we should acquire the V_{min} of the network. Second, $d_{max} = \max\{TD(P_f(q \to p_1)), TD(P_f(q \to p_2)), ..., TD(P_f(q \to p_k))\}$ is acquired when we expand the road network by travel-time. Then, we have all of the points belonging to TO-kOSQ result of q where the fastest travel-time between each point p and q is less than d_{max}/V_{min}. Similar to *DO-kOSQ$^+$* *algorithm*, we needn't visit those points the fastest travel-time between which and q is large than d_{max}/V_{min} for TO-kOSQ.

3   Experiments

Experimental Setup. We use the road map of San Francisco Road which consists of 174,956 vertices and 223,001 edges. The generating method of the time function cost of each edge is similar as [2] and the data storage mode can be seen in [5]. For each experiment, we generate 20 distinct time-dependent road networks and execute 20 queries randomly selected for each network.

Effect of the Number of Facilities. As shown in Fig. 1, when the number the facilities becomes denser (varying from $1K$ to $20K$ and the default number of facilities is $10K$), the process time of these algorithms decreases since the expansions of four algorithms need to consider little edges before next facilities are found. *DO-KOSQ$^+$ (TO-kOSQ$^+$) algorithm* is more efficient than *DO-kOSQ (TO-kOSQ) algorithm*, but the efficiency is also been clearly influenced by the number of facilities. This is because *DO-KOSQ$^+$ (TO-kOSQ$^+$) algorithm* has non-preprocessing step, a lot of vertices and facilities need to be traversed in the process of the two algorithms.

Effect of k. Figure 2 shows the performance of the two kinds of k-th order skyline queries for various values of k (the value of k varying from 1 to 20 and the

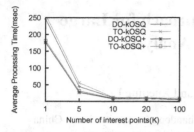

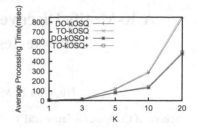

Fig. 1. CPU cost when the number of facilities increases

Fig. 2. CPU cost when k increases

default value is 10). It is obvious that as the value of k increases, algorithms will expand more vertices and facilities. Therefore, the processing time will increase. Similar observations hold for *DO-kOSQ$^+$ algorithm* and *TO-kOSQ$^+$ algorithm*.

4 Conclusion

In this paper, we address two kinds of k-th order skyline queries. We formalize these queries and design algorithms for their processing. Experiments show that our methods to answer k-th order skyline queries are efficient and effective in real road networks. Future work aims at investigating more appropriate paths to calculate the attributes of each facility.

Acknowledgment. This work is partially supported by Natural Science Foundation of Jiangsu Province of China under grant No. BK20140826, the Fundamental Research Funds for the Central Universities under grant No. NS2015095, the Funding of Graduate Innovation Center in NUAA under grant No. KFJJ20151606.

References

1. Börzsöny, S., Kossmann, D., Stocker, K.: The skyline operator. In: ICDE, pp. 421–430 (2001)
2. Costa, C.F., Nascimento, M.A., Macêdo, J.A.F.D., Machado, J.: A*-based solutions for knn queries with operating time constraints in time-dependent road networks. In: MDM, pp. 23–32 (2014)
3. Huang, X., Jensen, C.S.: In-route skyline querying for location-based services. In: Kwon, Y.-J., Bouju, A., Claramunt, C. (eds.) W2GIS 2004. LNCS, vol. 3428, pp. 120–135. Springer, Heidelberg (2005)
4. Mouratidis, K., Lin, Y., Yiu, M.L.: Preference queries in large multi-cost transportation networks. In: ICDE, pp. 533–544 (2010)
5. Papadias, D., Zhang, J., Mamoulis, N., Tao, Y.: Query processing in spatial network databases. In: VLDB, pp. 802–813. VLDB Endowment (2003)

A K-Motifs Discovery Approach for Large Time-Series Data Analysis

Yupeng Hu, Cun Ji, Ming Jing, and Xueqing Li[(✉)]

School of Computer Science and Technology, Shandong University, Jinan, China
{huyupeng,jicun,jingming,xqli}@sdu.edu.cn

Abstract. Motif discovery is a method for finding some previously unknown but frequently appearing patterns in time series. However, the high dimensionality and dynamic uncertainty of time series data lead to the main challenge for searching accuracy and effectiveness. In our paper, we propose a novel k-motifs discovery approach based on the Piecewise Linear Representation and the Skyline index, which is superior to traditional R-tree index. As the experimental results suggest, our approach is more accurate and effective than some other traditional methods.

Keywords: Time series · Data mining · K-motifs · Indexing and retrieval

1 Introduction

The potential value in time series data has long been watched closely by data analysis and mining. Motif has been defined as some previously unknown but frequently appearing patterns in time series [1].

While Looking for the top K-motifs of large time series data, we face several vital issues: the high dimensionality of the data and the efficiency of similarity calculation between different subsequences [2]. Likewise, we need to consider the optimization of the spatial indexes of time series.

Existing algorithms for finding motifs are mainly based on some dimensionality reduction methods and indexing techniques for similarity searching [3, 4]. Zhou proposed a time series segmentation based on series importance point (PLR_SIP) [5], which considers to employ the local extreme points defined by Pratt and Fink [6]. SonNT et al. propose a novel method for finding motifs by using the Middle Points and Clipping (MP_C) and skyline index [7]. The MP_C method can represent the sequence and reduce the high dimension more efficiently but less accurately. Especially in the treatment of some volatile data, the middle points are difficult to represent the temporal features compared with the extreme points of the sequence.

The main contributions of this paper are as follows: (1) We propose a novel K-Motifs finding approach by combining the PLR_IDP and Skyline index. (2) We define the lower-bounding region for PLR_IDP and utilize Skyline for indexing time series data. (3) The experimental results verify the effectiveness of our approach.

© Springer International Publishing Switzerland 2016
F. Li et al. (Eds.): APWeb 2016, Part II, LNCS 9932, pp. 492–496, 2016.
DOI: 10.1007/978-3-319-45817-5_53

2   Methodology

Our proposed method mainly consists of the following two parts: piecewise linear representation with important data points (PLR_IDP), skyline index for PLR_IDP.

PLR_IDP is a novel time series data representation method in our previous work [8], which considers not only the global characteristics and the global fitting error of the time series, but also the fitting error of the single point in a certain segment.

An example of the IDP identification is described in the Fig. 1. Figure 1(a) and (b) describe how to identify the point with the greatest vertical distance to the fitting line. Figure 1(b) and (c) describe how to find the point whose single point fitting error is more than half of the threshold of the fitting error segment.

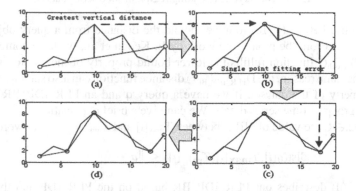

Fig. 1.   The steps of the TP identification process

Figure 2(a) shows a group of time series as an example, Fig. 2(b) shows the skylines bounding regions defined by the raw data. Evidently, the bounding regions describe in Fig. 2(b) are minimum, however the straightforward method is impractical because of its higher computing cost. Figure 2(c) shows the traditional data bounding regions by R-tree index, whose overlaps between the minimum bounding rectangles (MBRs) are too large to affect the search efficiency. Therefore, only the IDPs selected by PLR_IDP should be stored for each MBR and the bounding region called PLR_IDP Bounding Region (PLR_IDP_BR), is defined as follows: There is a group T of time series n of length m. $T = (t_1, t_2, \ldots, t_n)$, the PLR_IDP_BR of T specifies a two-dimensional rectangle enveloped by the top and bottom skylines [9].

$$R = \left(T_{\max}, T_{\min}\right) \tag{1}$$

where

$$T_{\max} = \left\{t_{1\max}, t_{2\max}, \cdots, t_{m\max}\right\}$$
$$T_{\min} = \left\{t_{1\min}, t_{2\min}, \cdots, t_{m\min}\right\} \tag{2}$$

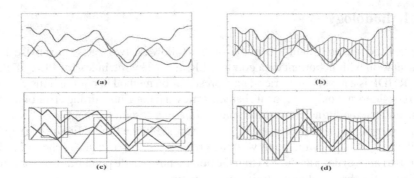

Fig. 2. The Skyline bounding region of time series data

We can use $LB_{Keogh}()$ [10] to lower bound the distance from a query object to an PLR_IDP_BR. From the pioneering work by E. Keogh et al., we have come to know that the $LB_{Keogh}()$ function fulfills the lower bound property of the whole time series data. For the same reason, the $D_{PLR\_IDP\_BR}()$ distance function can also satisfies the lower bound property of the dataset. If we have a query q and an PLR_IDP_BR R which contains a group of time series data x. We can give a brief explanation:

for any $x \in R$, the value of x[i] is between $T_{max}[i]$ and $T_{min}[i]$. In other words,

$$\text{dist}(q[i], T_{max}[i], T_{min}[i]) \leq \|q_i - x_i\| \tag{3}$$

Figure 2(d) describes our PLR_IDP_BR based on the PLR_IDP and the skyline index.

3   Experimental Evaluation

In this section, we evaluate the performance of our k-motifs discover methods. In order to accomplish the experiment, we select some kinds of typical time series datasets, which include Electrocardiograph dataset (ECG) provided by the UCR homepage [11], the monitoring data of Jinan municipal steam heating system(JMSHSD), the monitoring data of the Dong Fang Hong satellite (DFHSD). All of the experiments compared on ECG, JMSHSD and DFHSD, which contain no less than 100,000 data records.

We compare PLR_IDP with Skyline (PLR_IDP&Skyline), PLR_SIP with R-tree index (PLR_SIP&R) and MC_P with Skyline (MC_P&Skyline) by the running times on the above datasets in Fig. 3(a) describes the running times of the three algorithms normalized by the baseline algorithm (PLR_SIP&R). Obviously, in the different data-sets, the running time of our proposed method is shorter than the two other algorithms, especially in the JMSHSD and DFHSD datasets, which contain plentiful volatile data.

In Fig. 3(b), we focus on the DFHSD dataset and specify the motif length from 64 to 1024. We can see that the running times for all of the tree algorithms increase when the length of motif becomes large, however the cost of our method increases at a slower rate, which means that the PLR_IDP&Skyline provides a more accurate representation

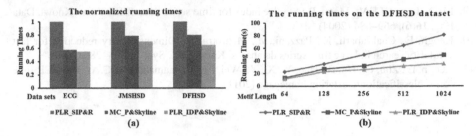

Fig. 3. The Skyline bounding region of time series data

for the temporal features and a more powerful pruning ability for K-Motifs finding than the rest of methods, when the dataset is more volatile.

4 Conclusion

In this paper, our k-motifs finding algorithm based on PLR_IDP and Skyline index ameliorates the existing methods by considering the dimensionality reduction and the optimized index on the whole time series dataset. Experimental results demonstrate the performance of the proposed method.

There are also some limitations of our method. Due to the high data volume of the industry time series data, our approach is inevitable to encounter the performance bottleneck, the maximum capacity of dataset which can be processed is no more than 80 GB, which may hurt the overall experiment performance to some extents. We will attempt to address these problems in the future.

References

1. Lin, J., Keogh, E., Patel, P., Lonardi, S.: Finding motifs in time series. In: The 2nd Workshop on Temporal Data Mining. The 8th ACM International Conference on Knowledge Discovery and Data Mining, Edmonton, Alberta, Canada, pp. 53–68 (2002)
2. Hailin, L., Chonghui, G.: Survey of feature representations and similarity measurements in time series data mining. Appl. Res. Comput. **30**, 1285–1291 (2013)
3. Mueen, A., Keogh, E., Zhu, Q., Cash, S., Westover, B.: Exact discovery of time series motifs. In: Proceedings of SIAM International Conference on Data Mining, pp. 473–484 (2009)
4. Chiu, B., Keogh, E., Lonardi, S.: Probabilistic discovery of time series motifs. In: The 9th International Conference On Knowledge Discovery And Data Mining, pp. 493–498 (2003)
5. Dazhuo, Z., Minqiang, L.: Time series segmentation based on series importance point. Comput. Eng. **34**, 14–16 (2008)
6. Pratt, K.B., Fink, E.: Search for patterns in compressed time series. J. Image Graph. **2**, 86–106 (2002)
7. Son, N.T., Anh, D.T.: Discovery of time series k-motifs based on multidimensional index. J. Knowl. Inf. Syst. **46**(1), 59–86 (2016)
8. Cun, J., Shijun, L., Chenglei, Y.: The 20th IEEE International Conference on Computer Supported Cooperative Work in Design, Nanchang, Jiangxi, China (2016, in press)

9. Li, Q., Lopez, I.F.V., Moon, B.: Skyline index for time series data. IEEE Trans. Knowl. Data Eng. **16**(6), 669–684 (2004)
10. Keogh, E., Chakrabarti, K., Pazzani, M., Mehrotra, S.: Dimensionality reduction for fast similarity search in large time series databases. Knowl. Inf. Syst. **3**, 263–286 (2001)
11. Keogh, E., Zhu, Q., Hu, B., Hao, Y., Xi, X., Wei, L., Ratanamahatana, C.A.: The UCR time series classification/clustering homepage (2011)

User Occupation Prediction on Microblogs

Xia Lv[1], Peiquan Jin[1,2(✉)], and Lihua Yue[1,2]

[1] School of Computer Science and Technology, University of Science
and Technology of China, Hefei 230027, China
jpq@ustc.edu.cn
[2] Key Laboratory of Electromagnetic Space Information,
Chinese Academy of Sciences, Hefei 230027, China

Abstract. User occupation plays an important role in many applications such as personalized recommendation and targeted advertising. However, user occupations on microblogging platforms are usually unavailable to public due to personal privacy. This opens an interesting problem, i.e., how to predict user occupations on microblogging platforms. In this paper, we propose a framework for extracting user occupations on microblogs. In particular, we implement a number of classification models and devise various sets of features for predicting user occupations, and devise an occupation-oriented lexicon to generate the training data. The experimental results show that the proposed lexicon-based method can achieve higher accuracy compared with traditional models.

Keywords: Occupation prediction · Feature extraction · Word embedding

1 Introduction

Microblog platforms like Sina Weibo enable users to receive and send messages, share information, and post short notices [1, 2]. Microblog has become an important source to obtain user information including age, occupation, summary of interests, and other relevant attributes like political preferences [1]. Among them, user occupation is conducive to optimize commercial applications such as precise advertising push and personalized recommendation. Accordingly, user occupation detection on microblog platforms has many potential business values.

Extracting information from microblogs has been studied widely in recent years. For example, previous work in [2] used the spatial information [3] and the temporal information [4] to extract events from microblogs. There are already some efforts concentrating on mining users' profile information [5–7]. A multi-source integration framework concentrates more on building feature set of content model and network information [5], and latent feature representation such as word clusters and embedding is used to classify occupations, but they only generate full-text semantic features without comparison to classical models [6]. So far, most of them are related to "occupation field". An occupation field indicates an area, which is less specific than an occupation. Thus, it is not sufficient for real applications. For example, "entertainment" is an "occupation field", but we may wonder whether a user is an actor/actress or a singer.

© Springer International Publishing Switzerland 2016
F. Li et al. (Eds.): APWeb 2016, Part II, LNCS 9932, pp. 497–501, 2016.
DOI: 10.1007/978-3-319-45817-5_54

Therefore, in this paper we focus on predicting user specific occupation on microblogging platforms like Sina Weibo. We address it by leveraging available information such as observable digit content, user behavior habit, custom tag, and linguistic messages of user. In the process, several classical models and lexicon-based feature selection method are respectively used to generate feature sets. Then, we compare several different classification methods on different feature sets. The main contributions of the paper are as follows:

(1) We build a framework for *Microblog User Occupation Prediction*. Our framework screens out 8 occupations and can achieve high precision up to 87.12 %. These occupations contain writer, reporter, lawyer, photographer, actor, singer, doctor and dietitian. We integrate message contents with three linguistic models, and compare it with lexicon-based selection method.

(2) An occupation-oriented lexicon adapting semantic similarity and word importance is built using an iterative approach. We use the occupation-oriented lexicon to simplify the feature selection step and reduce the dimension of feature.

2 Framework of User Occupation Prediction

We randomly collect 8000 users' ID according to "verified reason" from Sina Platform, and then fetch every user's latest 500 microblogs and custom tags by web crawler. Our 8 occupations are as follows: *writer, reporter, lawyer, photographer, actor, singer, doctor*, and *dietitian*. The framework is shown in Fig. 1.

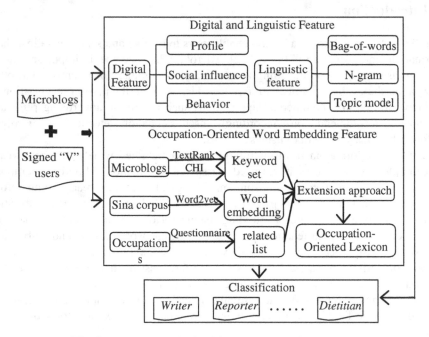

Fig. 1. Framework of microblog user occupation prediction

In the first part, we extract the digital and linguistic features. Digital features contain three types: basic profile feature, social influence and behavior feature. Gender, province, messages count and favorites count ("like" count) are the only four basic profile features. The social influence is evaluated by two-fold: one is the number of followers, friends and mutual fans, and the other one is reflected by comments or retweets. The behavior features like average number of topics and average number of messages per day are considered. Moreover, linguistic content information contains user name, description, custom tag and messages posted by user; it encapsulates the main topic of user type as well as the user's lexical usage. We explore various linguistic content features based on three linguistic models: bag-of-words, n-gram and topic model. *Chi-square (CHI)* and *term frequency–inverse document frequency (TFIDF)* are applied to select and represent words respectively. In this paper, we derive the unigram and bigrams of the text. *Latent Dirichlet Allocation (LDA)* and *Biterm Topic Model (BTM)* [8] are both used to give a generative probabilistic model for the text corpus.,

Next, we first use *word2vec* to train the corpus and get a model of word embedding to express similarity between words. And then we screen keyword set with *TextRank* and *CHI*. In this part, *CHI* aims to select feature words, which measures the degree of the independence between the feature and categories. *TextRank* is applied to solve keyword extraction and it is tasked with the automatic identification of terms that best describe the subject of a document. Moreover, we invite several persons to perform our questionnaires that ask for related words they first thought about different occupation, these related words are used as an input of our extension approach. Finally, we use the combination to get occupation-oriented word embedding features.

3   Experimental Results

We exploit 8 occupations to label 8000 Sina Weibo "*V*" identification users, over which 70 % are used as the training data. The corpus of *word2vec* contains 69.3 billion effective words. We adapt *Skip-gram* model and window size of 8 to train corpus with different word representation size. The *TextRank* and *CHI* are applied on all messages of labeled 8000 users, and we get 22.75 thousand keywords and 9030 words selected by the *CHI* value.

As shown in Table 1, with the increasing of the topic size, the precision of prediction keeps growing until the topic size is 30, and then has a slight decreasing trend. Thus, we use 30 topics for the next task. We apply 300 dimensions of word embedding to represent remaining words, and integrate all vector representations by computing sum value of each dimension. Table 2 shows the accuracy of classification. We can see that only 2 rounds of extension steps can extract appropriately lexicon words. In addition, we compare three machine learning methods: a linear SVC classifier with *L2* regularized logistic regression, a *Support Vector Machine* classifier with linear kernel, and an ensemble classification of *Random Forest*. The performance of word embedding with different dimensions using 3530 words lexicon and traditional feature set are shown in Table 3. It shows that the accuracy of traditional feature set is 77.92 % when we use all the features and the random-forest model. The *LR* and linear *SVM* models get

Table 1. Performance w.r.t. topic size (%)

	Size				
Meric	20	25	30	35	40
Precision	71.26	72.54	75.09	74.97	74.29
Recall	70.54	71.92	75.54	74.62	73.67
F-measure	70.90	72.23	74.81	74.80	73.98

Table 2. Accuracy w.r.t. lexicon size (%)

Lexicon size	Accuracy
1537	77.54
3530	82.50
6480	80.62

Table 3. Performance on different feature set

	Classifier		
Feature	LR	SVM	Random Forest
Digital and Linguistic-24116	74.67	75.21	77.92
Lexicon-200	80.21	79.96	66.21
Lexicon-400	83.62	83.58	68.21
Lexicon-600	84.58	85.46	67.29
Lexicon-800	85.42	86.21	68.62
Lexicon-1000	86.08	87.12	67.46

similar accuracy, while the *Random Forest* has worse performance when using word embedding. In general, we can see that combining the occupation-oriented lexicon with word embedding helps to achieve higher accuracy compared with traditional feature sets.

4 Conclusion

Detecting user occupation from microblog platform is an important issue for information extraction on short texts. This paper presents a framework for classifying 8 occupations based on Sine Weibo. We implement a number of classification models and devise an occupation-oriented lexicon to improve the accuracy. The experimental results show that the lexicon-based features with lower dimension achieve higher accuracy compared to traditional models.

Acknowledgements. This work is supported by the National Science Foundation of China (61379037 and 71273010).

References

1. Pennacchiotti, M., Popescu, A.: A machine learning approach to Twitter user classification. In: ICWSM, pp. 281–288 (2011)
2. Zheng, L., Jin, P., Zhao, J., Yue, L.: A fine-grained approach for extracting events on microblogs. In: Decker, H., Lhotská, L., Link, S., Spies, M., Wagner, R.R. (eds.) DEXA 2014, Part I. LNCS, vol. 8644, pp. 275–283. Springer, Heidelberg (2014)

3. Zhang, Q., Jin, P., Lin, S., Yue, L.: Extracting focused locations for web pages. In: Wang, L., Jiang, J., Lu, J., Hong, L., Liu, B. (eds.) WAIM 2011. LNCS, vol. 7142, pp. 76–89. Springer, Heidelberg (2012)
4. Jin, P., Lian, J., Zhao, X., Wan, S.: TISE: a temporal search engine for web content. In: IITA, vol. 3, pp. 220–224 (2008)
5. Huang, Y., Yu, L., Wang, X., Cui, B.: A multi-source integration framework for user occupation inference in social media systems. World Wide Web 18(5), 1247–1267 (2015)
6. Preoţiuc-Pietro, D., Lampos, V., Aletras, N.: An analysis of the user occupational class through Twitter content. In: ACL, pp. 1754–1764 (2015)
7. Tu, C., Liu, Z., Sun, M.: PRISM: profession identification in social media with personal information and community structure. In: Zhang, X., Sun, M., Wang, Z., Huang, X. (eds.) SMP 2015. CCIS, vol. 568, pp. 15–27. Springer, Heidelberg (2015)
8. Yan, X., Guo, J., Lan, Y., et al.: A biterm topic model for short texts. In: WWW, pp. 1445–1456 (2013)

Industry Full Paper

Combo-Recommendation Based on Potential Relevance of Items

Yanhong Pan, Yanfei Zhang, and Rong Zhang[✉]

Institute for Data Science and Engineering, Software Engineering Institute,
East China Normal University, Shanghai, China
{51151500037,51151500076}@ecnu.cn, rzhang@sei.ecnu.edu.cn

Abstract. Combo recommendation expects to recommend a collection
of products to users in a Groupon way. The representative application is
combo recommendation in the travel industry, which is also called package recommendation and may include different landscapes according to
the inherent features. Compared with traditional recommendation scenario, combo recommendation has the following characteristics: (1) sparsity: information for combos is much less than that for individual items;
(2) collectivity: every combo is composed of multiple individual products
with different features; (3) diversity: products composed of combos may
have different features; (4) relevance: products inside combos have some
kind of potential relevant. Traditional recommendation algorithms may
perform poor for they consider nothing about these four characteristics
in the models. Aiming at improving performance of combo recommendation, our work proposes a novel combo recommendation algorithm
called RBM-CR based on the Restricted Boltzmann Machine. RBM-CR
algorithm takes advantage of users' consumption histories to derive the
correlations among products by mapping from visible features to hidden features, and to profile users and combos by those hidden features.
Finally, experiments on real dataset verify effectiveness and accuracy of
our algorithm.

1 Introduction

With the development of E-commerce websites, the number of products and users
is growing faster and faster. Users are willing to comment on products and to
share the shopping experience after using the products. The reviews and ratings
express the preferences of users and opinions on products which can help potential users to select the products. In the meantime, these reviews and ratings have
been used in the recommendation systems to profile users and products, which
includes collaborative filtering algorithms [4,7,14], content-based recommendation algorithms [11], and model-based algorithms [6,15]. According to users'
consumption histories, the traditional recommendation algorithms learn users'
preferences to recommend products to users. However, these traditional recommendation algorithms are mainly designed for recommending a single product
called an item so it can't adapt to the various types of consumption patterns.

© Springer International Publishing Switzerland 2016
F. Li et al. (Eds.): APWeb 2016, Part II, LNCS 9932, pp. 505–517, 2016.
DOI: 10.1007/978-3-319-45817-5_55

For instance, in order to attract more users, E-commerce launches Groupon consumption, which sales collections including multiple individual products. The new consumption style has been applied in many industries, such as tourist industry and catering industry. Tourists usually expect to be served with some combos, which are also called travel packages in the tourist industry and consist of some landscapes with the same geographic location and open season. More and more restaurants offer some combos that include dishes with lower price. Usually, the new products in combos can be paid more attention. Different from item recommendation, the data model for combo recommendation has the following issues that should be considered:

- Sparsity: The consumption information of combos is much less than that of individual products. The reason is that Groupon has a short history as a new consumption style.
- Collection: Combos are composed of multiple different individual products. The recommendation models should consider combination characteristics of products.
- Diversity: Products combined in the same combo are complementary. For instance, a combo maybe have some sweet dishes and some spicy dishes in catering industry. Diversity has a contradiction with item recommendation algorithms which take advantage of the similarity to recommend products to users.
- Relevance: Products in the same combo are relevant. The relevance between products in combos is reflected in different industries: combo contains a series of representative landscapes along an itinerary in travel industry. And combo recommendation for dishes needs to mix vegetables with meat and balance taste in catering industry.

Ignoring data characteristics introduced above, the traditional recommendation algorithms do not work well for combo recommendation. This paper aims at proposing an approach for combo recommendation which considers the above characteristics. Under the background of application in the catering industry, every combo is made up of multiple dishes. We design and implement a novel Combo Recommendation algorithm based on Restricted Boltzmann Machine (RBM-CR). In our approach, we make the following contributions:

1. RBM-CR uses mixed reviews as dataset, which consists of Groupon reviews and non-Groupon reviews, to solve the data sparsity problem. One single dish in a combo is also in non-Groupon menu. Therefore, the information of a dish can be obtained from comments of non-Groupon services. As shown in Fig. 1, the analysis of topic model indicates that the probabilistic distribution of potential theme in Groupon data is extremely similar to that in non-Groupon data. This result clearly shows that review contents are similar and mixed data can be used as data source for combo recommendation.
2. Based on RBM [10], model, RBM-CR algorithm analyzes the user preferences and mines the correlations among combo dishes. Then we calculate the predicted rating that target users give to each dish in combo, multiplied by the weight coefficient to obtain the overall score.

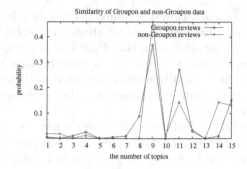

Fig. 1. Data analysis diagram of Groupon and non-Groupon reviews

3. The experimental results show that the experimental result of RBM-CR is better than that of baselines in terms of accuracy.

2   Related Work

Based on the way to generate the combos, combo recommendation can be divided into two types. The first one expects to recommend a set of products for users according to their personal customization. As mentioned in [13], personalized travel package recommendation adopts the greedy strategy to generate a sequence of points of interest (POIs) for users. Because of a wide range of products, formulating such travel packages is an NP hard problem. The second type is to recommend predefined combos to users according to the latent relationship among products. For example, travel industry takes area, season, price and characteristics of POIs into consideration to travel package. Work [2,5,8,9] adopt the tourist-area-season topic (TAST) model which is used to analyze tourists travel log to profile uses and then recommends travel package to users. Since travel package service information is still of small size, in work [12], it combines the available additional context information to make the analysis data set rich. By treating additional context information as the collection of the feature Cvalue pairs, it proposes an Object-oriented travel package recommendation algorithm, using the Objected-Oriented Topic Model (OTM) and the Objected-Oriented Bayesian Network (OBN).

Existing combo recommendation algorithms are mostly used in the tourism industries, and it is not adaptive to other application domains, such as restaurant services. For travel package generation, it considers much more about prices, time intervals, and locations which are the inherent characteristics of POIS. These characteristics are obviously domain sensitive. In addition, work [5,9,12,13] use LDA model to analyze the probability distributions of potential topics in travel logs, and combine with collaborative filtering algorithms to perform travel package recommendation. In restaurant service, every combo is made up of multiple dishes. These combo services are generally consume-able by customers with cheaper price. However combo generation relies on the inner correlations among dishes, which may consider about the combination of hot, cold, spice or sweet foods. Comparing with travel package recommendation in travel industry, combo recommendation shall focus more on user profiles instead of customization requirement. Additionally, although topic-based analysis can get the interest of users, to evaluate the extension of the preference is also important for recommendations.

Considering about these problem, we propose a new recommendation model for restaurant Combo recommendation combing the review comments together with ratings. In other words, we try to profile user favorites by mining latent topics in comments and evaluate the influence of topics to user choices by ratings.

3   RBM-CR Recommendation Model

In this section, we propose a method based on Restricted Boltzmann Machine named RBM-CR applying to combo recommendation. The major challenges are to calculate the influence of dishes on combos ratings to users and to mine the potential relevance among dishes inside combos facing the scarce of dataset (Table 1).

Our method is divided into four parts: *Data Preprocessing, Features Converting and Mapping, Model Training* and *Combo Recommendation*. (1) Data Preprocessing (Sect. 3.1), we take advantage of natural language processing technology to get the rating information to dishes from users by analyzing user reviews. (2) Features Converting and Mapping (Sect. 3.2), our method applies RBM model to build mapping from visible units with the user-dishes matrix to hidden units with potential topics. (3) Model Training (Sect. 3.3), we obtain parameters by training data of visible units. (4) Combo Recommendation (Sect. 3.4), we design the algorithm to recommend top-k combos to target users by training the parameters.

Table 1. Notations definition

Symbol	Description
$U = \{u_i\}, I = |U|$	U is the set of users, the size is I
$D = \{d_j\}, J = |D|, D^l = \{d_j = 0|1\}$	D is the set of dishes, the size is J, the restaurant l has the dish d_j when d_j equals to 1
$C = \{c_m\}, M = |C|$	C is the set of combos, the size is M
$R = \{r_{ij}\}$	R is user-dish matrix, r_{ij} is the rating to the dish d_j from the user u_i
$R^P = \{r_{ij}^P\}$	R is user-dish predictive rating matrix, r_{ij}^P is the predictive rating to the dish d_j from the user u_i
$R^P = \{r_{im}^P\}$	R is user-combo predictive rating matrix, r_{im}^P is the predictive rating to the combo c_m from the user u_i
$S = \{s\}, s\epsilon[1,5]$	S is set of discrete values describing users rate dishes

3.1   Data Preprocessing

We first use an open source tool named Ansj[1] to split reviews into sentences. If the sentence contains of the dish' name, we then map sentences to dishes in combos. After the mapping between sentences and dishes, we score the dishes according to the emotional polarity of words inside sentences redreference $NTUSD$[2] and DAD[3]. If the sentence contains strong emotional adverbs such as greatly, excellently and so on, the rating is more far away from the average. We then can obtain the rating r_{ij} to dish d_j from user u_i as calculated in Eq. 1. In such a way, we obtain user-dishes rating matrix which is used to train model.

$$
r_{ij} = \begin{cases}
1 & \text{if the review contains strong emotional negative words} \\
2 & \text{if the review only contains negative words} \\
3 & \text{if the review only contains no emotional words} \\
4 & \text{if the review only contains positive words} \\
5 & \text{if the review contains strong emotional positive words}
\end{cases} \tag{1}
$$

Considering the above two challenges, We adopt the SoftMax structure to represent visible units of RBM model. As shown in Fig. 2, the SoftMax structure has five units of values to represent the ratings [1–5], from top to bottom, the visible units valued from 1 to 5. Every time there is only one unit in active state valued "1", and the other four units are inactive valued "0". In Fig. 2, the small black squares represent activated state, and the other white block units represent inactivated states. For example, if rating r_{ij} is scored 5, one of the rating states in RBM model can be expressed as $\{0, 0, 0, 0, 1\}$, in which the fifth unit is activated, and the other four "0"s are inactive. In addition, the SoftMax structure may have missing units, so that it represent the rating missings.

3.2   Features Converting

RBM model uses visible features to train feature distribution on hidden units. It assumes visible units have the multinomial distribution given the hidden units, represented as Eq. 2, it assumes hidden units have the Bernoulli distribution given the visible units, represented as Eq. 3.

$$
P(r_{ij} = s|H) = \frac{exp(b_{js} + \sum_{k=1}^{K} h_{ik} w_{jks})}{\sum_{s' \in S} exp(b_{js'} + \sum_{k=1}^{K} h_{ik} w_{jks'})} \tag{2}
$$

$$
P(h_{ik} = 1|R) = sigmod(b_k + \sum_{d \in \rho_i} w_{jk r_{ij}}), \tag{3}
$$

where the two dimensional hidden units H corresponds to matrics $U \times K$, which is user-hidden feature mapping relationship. h_{ik} is the kth hidden unit for user u_i. Visible units V is three dimensional matrix, which is user-dish-rating matrix

---

[1] http://www.ansj.org/.

[2] NTUSD: http://www.datatang.com/data/44317/.

[3] NTUSD: http://pan.baidu.com/s/1sjoqp1z/.

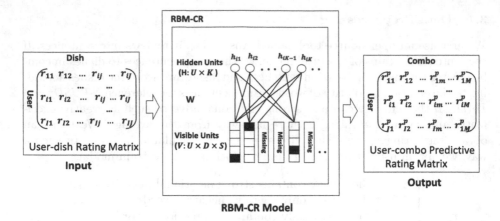

Fig. 2. Framework of our method

$U \times D \times S$ to represent the user profile. If $r_{ij} = s$, it means that user u_i rates dish d_j by s^{th} level and s^{th} cell in the corresponding visible unit is blacked. W is also a three dimensional matrix $D \times K \times S$, which is a global parameter to adjust the degree of user preferences for different hidden features. w_{jks} weights rating the pair of dish d_j and the score s for the kth hidden unit. The visible bias b_{js} is a two dimensional matrix $D \times S$, which represents the visible bias of the score s to the dish d_j and is used to adjust estimated deviations from visible units. The hidden bias b_k, size of K, is a one-dimensional vector, which is used to adjust estimated deviations of hidden units to users. ρ_i is the collection dishes rated by the user u_i.

Based on Eqs. 2 and 3, using user-dish rating history, it is easy to get user-hidden factor relationship on iterative calculation which is described in details in next section.

3.3   Model Training

RBM-CR model takes an iterative method to train user profiles. It trains the model to get user preference using user rating collections $\rho_i = \{r_{ij}\}$, which is the set of rated dishes by user u_i. Users are also expected to be represented by K hidden features such as $u_i = \{h_{ik} = 1|0\}(1 \leqslant k \leqslant K)$. The hidden factor is equivalent to a latent topics, and if the kth hidden factor has an influence on user ratings, $h_{ik} = 1$, or else, $h_{ik} = 0$. We can use the weights W between V and H in RBM-CR model to adjust the degree of users preference represented by hidden units. Since each user corresponds to a RBM-CR model, though the same dish has the same rating from different users, the bias and the connecting weights are identical to users for the status of all hidden units may not be equivalent. Model training details can be found in Algorithm 1.

In Algorithm 1, Line 2 to 8 is a full iterative process for all users. In the iterative process, RBM-CR algorithm uses rating data and preference of every

---

Algorithm 1. User preference model training

Input:

    User-dish rating matrix $R = \{r_{ij}\}$, the set of combos $C = \{c_m\}$ and adjusting parameters including the penalty parameter $\lambda = 0.001$, the momentum $\mu = 0$ and learning rate $\eta = 0.001$.

Output:

    Training parameters in $RBM - CR$ model includes weights W, visible bias b_{js} and hidden bias b_k

1: Initialize parameters;
2: Obtain ρ_i;
3: According to Eq. 3, we calculate $P(h_{ik}^0 = 1 | R^0)$ to obtain H^0;
4: According to Eq. 2, we calculate $P(r_{ij}^1 = s | H^0)$ to obtain R^1;
5: Repeat Step 2, we calculate $P(h_{ik}^1 = 1 | R^1)$ to obtain H^1;
6: Update weights:

$$\delta w_{jks} = P(h_{ik}^0 | r_{ij}^0) \times r_{ij}^0 - P(h_{ik}^1 | r_{ij}^1) \times r_{ij}^1;$$
$$\Delta w_{jks}^{n+1} = \mu \times \Delta w_{jks}^n + \eta \times (\delta w_{jks} - \lambda \times w_{jks}^n);$$
$$w_{jks}^{n+1} = \Delta w_{jks}^{n+1} + w_{jks}^n;$$

7: Update visible bias:

$$\delta b_{js} = r_{ij}^0 - r_{ij}^1;$$
$$\Delta b_{js}^{n+1} = \mu \times \Delta b_{js}^n + \eta \times (\delta b_{js} - \lambda \times b_{js}^n);$$
$$b_{js}^{n+1} = \Delta b_{js}^{n+1} + b_{js}^n;$$

8: Update hidden bias:

$$\delta b_k = P(h_{ik}^0 | r_{ij}^0) - P(h_{ik}^1 | r_{ij}^1);$$
$$\Delta b_k^{n+1} = \mu \times \Delta b_k^n + \eta \times (\delta b_k - \lambda \times b_k^n);$$
$$b_k^{n+1} = \Delta b_k^{n+1} + b_k^n;$$

9: Based on Eqs. 4, 5 and 6, calculate predictive rating from users to dishes.
10: Traverse the users set U, calculate RMSE value in test dataset;
11: If $RMSE^{n+1} < RMSE^n$, repeat Step 2 to 9. If not, iterate over.

---

user to update global parameters including weights W, visible bias b_{js} and hidden bias b_k. The initial values for W, b_{js} and b_k are value "0" respectively, which are updated iteratively. Line 3 to 5 takes Constrastive Divergence (CD) [1,3,10] methods for sampling visible and hidden units to speed up training [?], which may make the model more feasible. The sampling results of CD methods are R^0, H^0, R^1 and H^1, which are used for updating W, b_{js} and b_k during each round of calculation based on individual users from Line 6 to 8. During the updating processing, we have the penalty parameter λ, the momentum μ and learning rate η to adjust the importance of the difference (δw_{jks}, δb_{js} and δb_k) calculated from the two round of sampling process to current parameters computation (W, b_{js} and b_k). After updating weights W, visible bias b_{js} and hidden bias b_k, we can calculate user preference distribution on hidden units as Eq. 4, the probability of predictive rating s to dish d_j for any target user u_i as Eq. 5 and finally the predicative ratings r_{ij}^p for u_i to d_j as Eq. 6. After having the predictive rating matrix for users U, we compare the $RMSE^{n+1}$ value with the previous round of $RMSE^n$ values based on predictive rating matrix. Our algorithm will stop only when $RMSE^{n+1} > RMSE^n$.

Algorithm 2. Combo recommendation algorithm

Input:

The target user u_i, training parameters in $RBM - CR$ model includes weights W, visible bias b_{js} and hidden bias b_k,

Output:

Recommendation list $C_{(top-K)} = \{top - k(C)\}(C_{(top-K)} \in C)$

Based on Eqs. 4, 5 and 6 Implement an mean field to calculate the expectation to every dish from the target user;

2: Repeat step 1 for every dish inside the combo;

Obtain $w_j^m = \frac{\sum_{u_i \in \rho(j)} r_{ij}}{\sum_{d_j \in c_m} \sum_{u_i \in \rho(j)} r_{ij}}$;

4: Calculate $r_{im}^p = \sum_{d_j=1 \cup d_j \in c_m} r_{ij}^p \times w_j^m$;

Repeat step 1 to 5 for every combo;

6: Sort r_{im}^p to obtain top-k combos;

$$P_{ik} = b_k + \sum_{d_j \in \rho_i} w_{jk} r_{ij} \tag{4}$$

$$P(r_{ij} = s|P_{ik}) = \frac{exp(b_{js} + \sum_{k=1}^{K} P_{ik} w_{jks})}{\sum_{s' \in S} exp(b_{js'} + \sum_{k=1}^{K} P_{ik} w_{jks'})} \tag{5}$$

$$r_{ij}^p = \sum_{s \in S} s \times P(r_{ij} = s|P_{ik}), \tag{6}$$

3.4   Combo Recommendation

RBM-CR model profiles users by the preference distribution on hidden units, so that it builds potential relevance between users and dishes. Then user-dish rating can be easily calculated relying on model parameters including weights, visible bias and hidden bias. Listed in Algorithm 2 detailedly.

In Algorithm 2. In Line 1 to 5, it recommends the top-k combos for the target user u_i. In Step 1 to 2, we take advantage of mean filed method to gain user-dish matrix based on model parameters including weights, visible bias and hidden bias. In Step 3 to 4, we will convert user-dish predictive rating matrix into user-combo predictive rating matrix through the weighting coefficient. The reason of considering weighting coefficient is that different dishes in the same combo have different degree of influence on the combo. For example, the main course is more important than the dessert in the same combo. In Step 3, we obtain the weight coefficient through accounting the proportion of the dish rating summation $\sum_{u_i \in \rho(j)} r_{ij}$ in dishes consisting of the combo rating summation $\sum_{d_j \in c_m} \sum_{u_i \in \rho(j)} r_{ij}$, where w_j^m represents weight of the dish d_j in the combo c_m and $\rho(j)$ represents collections of users who rated the dish d_j. In Step 4, we use predictive rating of every dish to multiply by weight coefficient of the dish to gain the predictive rating of combo r_{im}^p. In Step 5, we traverse the set of combos $C = \{c_m\}$ to get predictive ratings of all combos from the target user u_i. In Step 6, we sort the predictive ratings of all combos to obtain the top-k combos for the target user u_i.

4 Experiments

4.1 Data Set

In this paper, we gathered review data from *Dianping*[4], which is the biggest restaurant review site in China. As shown in Table 2, we crawled the messages of 95,260 restaurants in Shanghai area. We remove the restaurants which don't provide combos and obtain the dataset including 775,878 reviews from 2838 restaurants providing combos. Each review contains information, such as review text, user ID, restaurant ID, an overall rating, and three aspect ratings containing taste, surroundings and service. All the ratings take values from 1 to 5. According to the statistical information, these selected restaurants involve 3306 combos and 25,000 dishes. And we gain the number of Groupon reviews 121,349. Each combo contains information, such as multiple dishes, the original price, the discount, the current price and the number of people.

Table 2. Experimental data

Experimental data	Quantity	Experimental data	Quantity
The number of restaurants	95260	The number of combos	3306
The number of restaurants providing combos	2838	The number of dishes	25000
The number of reviews	775878	The number of combo reviews	121349

4.2 Contrast Tests

We compare our system with three baseline systems based on the previous studies. The baseline systems are described as follows:

BaseLine1: We adopt collaborative filtering algorithm to recommend combos as items rather than dishes collections to target users. We obtain the user-combo matrix as the input of BaseLine1. Then, we calculate the similarity between combos to gain predictive combos ratings for target users. Finally, we recommend the top-k recommendation list to target users.

BaseLine2: The same as the BaseLine1, the BaseLine2 also applies collaborative filtering algorithm to combo recommendation. Different from the BaseLine1, it treats combos as collections so that it considers dishes' influence on the whole combo. Therefore, we obtain user-dish rating matrix and combo-dish weight matrix as the input of the BaseLine2 to calculate the similarity between dishes. The output is the same as the BaseLine1.

BaseLine3: We use tourist-area-season topic ($TAST$) model mentioned in work [5] in the BaseLine3. Considering that the application background of this article is combo recommendation in the catering industry, we adopt mixed review data rather than travel data as training data. Therefore, we slightly change the $TAST$ model to apply to combo recommendation.

---

[4] Dianping: http://www.dianping.com.

4.3 Evaluation Index

- **Root Mean Square Error (RMSE) and Mean Absolute Error (MAE):** They are two of accuracy evaluation indexes of recommendation algorithms. And they have two advantages. One is that calculation method of RMSE and MAE is simple and understood easily. Because the RMSE value and MAE value are respectively unique in each algorithm, the other is that we are intuitive to distinguish the difference of values between two algorithms. The smaller $RMSE$ and MAE value are, the higher the accuracy of recommendation algorithm will be. Represented as Eqs. 7 and 8.

$$RMSE = \sqrt{\frac{1}{c} \sum_{m=1}^{c} \left(r_{im} - r_{im}^p \right)^2} \tag{7}$$

$$MAE = \frac{1}{c} \sum_{m=1}^{c} |(r_{im} - r_{im}^p))|, \tag{8}$$

where c is the number of combos rated by users in test data, r_{im} is real rating to the combo c_m from the user u_i mined from Groupon data. r_{im}^p is predictive rating to the combo c_m from the user u_i.
- **Position:** Position is used to measure the accuracy of recommendation algorithms. We obtain the exact position of purchased combos in the top-k recommendation list. In a general way, if combos in the recommendation list have been adopted, recommended effect will be better.
- **Coverage:** Coverage can be defined as the proportion of the user's consumed combos accounted for the recommendation list, which is particularly important in the recommendation system. The higher coverage value is, the better personal service of recommendation system will be.
- **Significance Test:** We use the significance test to test whether the differences between the samples and the totality are significant. Shown as Eqs. 9 and 10.

$$t = \frac{|\overline{r^p} - u_0|}{s^* \div \sqrt{n}} \tag{9}$$

$$s^* = \sqrt{\frac{1}{n-1} \times \sum_{u_i \in U} \sum_{m=1}^{M} \left(r_{im}^p - \overline{r^p} \right)^2}, \tag{10}$$

where test samples are respectively test sample and predictive sample data. $\overline{r^p}$ is the average of the predictive data. u_0 is the average of test data. s^* is the standard variance of predictive data. n is the sample quantity.

4.4 Experimental Results

In this paper, Table 3 is evaluated results of $RMSE$ and MAE in three different types of dataset. RBM-CR refers to our proposed method, and BaseLine1, BaseLine2 and BaseLine3 are baselines. The TestSet1 is a collection of active combos rated by plenty of users. The TestSet2 is a collection of inactive combos rated by few users less than 6. And the Testset3 is complete dataset including 12,477 rating records. Seen in experimental results in Table 3, $RMSE$ and MAE of the $RBM - CR$ algorithm are superior to baselines. The TestSet1 is so ascendant that it is normal that the experimental results under Testset1 are more accurate than under Testset2. The reason why the $RBM - CR$ algorithm's recommended results are more accurate is that it takes the potential relationship among dishes in combos into consideration based on RBM model.

Table 3. The experimental results of RMSE and MAE in different dataset

Dataset	Experiment	RMSE	MAE	Dataset	Experiment	RMSE	MAE	Dataset	Experiment	RMSE	MAE
Testset1	BaseLine1	0.942	0.720	Testset2	BaseLine1	1.001	0.795	Testset3	RBM-CR	0.972	0.750
	BaseLine2	0.921	0.710		BaseLine2	0.970	0.761		RBM-CR	0.930	0.744
	BaseLine3	0.900	0.700		BaseLine3	0.930	0.760		RBM-CR	0.917	0.735
	RBM-CR	0.760	0.570		RBM-CR	0.860	0.724		RBM-CR	0.840	0.704

Table 4 is experimental evaluation results of position and coverage on condition that top-k is equal to 50 in all test data, as well as the improvement of position and coverage comparing to BaseLine1 experiment. To some degree, the experiment of $RBM - CR$ algorithm on position improves substantially comparing with BaseLine1. In this experiment, the specific definition of coverage is the proportion of the user's actual consumption combos accounted for the recommendation list. The top 50 of monthly sales are supposed to the baseline of coverage experiment, which is selected from combos sales in November 2015. As is known to all, the higher coverage shows that the actual recommended results are better. The experiment of coverage makes a significant improvement comparing with baselines.

We select sample size $n = 500$ and assume $H_0 : \bar{x} = u_0$. In the case of unknown variance, we use t-test whose acceptable domain is $|t| < t_{1-\alpha/2}(n - 1), (\alpha = 0.002)$. And we obtain $t_{0.975}(500) = 3.107$ by looking up $table^5$. We calculate the values such as $\bar{x} = 3.141, u_0 = 3.246, s = 0.774$ by predictive samples to obtain $t = 3.015$. Therefore, experimental results accept the original hypothesis, which is regarded as meeting the significance test.

Table 4. Experimental results of position and coverage

Experiment name	Position	Improved position	Coverage	Improved coverage
Top 50 of monthly sales	-	-	0.1539	-
BaseaLine1	23.12	-	0.1551	0.779 %
BaseaLine2	22.4166	3.042 %	0.1670	8.512 %
BaseaLine3	22.0357	4.689 %	0.1588	3.183 %
$RBM - CR$	19.3516	16.299 %	0.1729	12.345 %

---

[5] Table: http://wenku.baidu.com/view/83bca9d1195f312b3169a5ad.html.

5   Conclusion

In this paper, we propose that package recommendation apply to new business field defined as combo recommendation in the catering industry. We first introduce inherent features of combo data in detail, which includes sparsity, collection, diversity and relevance. Next, we explain our method based RBM model to deal these features. RBM-CR algorithm is divided into two main parts: training model and combo recommendation. Training model part is able to mine potential relevance among dishes and relationship from dishes to users. Considering the different dishes with different influences on the same combo, we use the mean filed method and weight coefficient to calculate the predictive rating for target users in the combo recommendation. Finally, our method has good effects in the real dataset. For future work, there are several ways in which this research could be extended. Making our algorithm more availability, we plan to speed up training the model. Considering the traffic problem, we further plan to add geographic location of restaurants as a new factor in the combo recommendation. In addition, we plan to apply our combo recommendation algorithm to Groupon recommendation in the other business field.

Acknowledgment. This work is partially supported by National Science Foundation of China under grant (No. 61232002 and No. 61402180), National Science Foundation of Shanghai (No. 14ZR1412600), and Shanghai Agriculture Science Program (2015) Number 3-2. The corresponding author is Rong Zhang.

References

1. Carreira-Perpinan, M.A., Hinton, G.: On contrastive divergence learning. In: AISTATS, vol. 10, pp. 33–40. Citeseer (2005)
2. Dhanumjaya, V.S., Gollapalli, A.R.: A novel model and efficient topic extraction for travel package recommendation. IJRCCT 4(12), 2011–2014 (2015)
3. Hinton, G.E.: Training products of experts by minimizing contrastive divergence. Neural Comput. 14(8), 1771–1800 (2002)
4. Li, D., Chen, C., Lv, Q., Shang, L., Zhao, Y., Lu, T., Gu, N.: An algorithm for efficient privacy-preserving item-based collaborative filtering. Future Gener. Comp. Syst. 55, 311–320 (2016)
5. Liu, Q., Chen, E., Xiong, H., Ge, Y., Li, Z., Wu, X.: A cocktail approach for travel package recommendation. IEEE Trans. Knowl. Data Eng. 26(2), 278–293 (2014)
6. Pichardo-Lagunas, O., Herrera-Alcántara, O., Arroyo-Figueroa, G. (eds.): Advances in Artificial Intelligence and Its Applications. LNCS, vol. 9414. Springer, Heidelberg (2015)
7. Polatidis, N., Georgiadis, C.K.: A multi-level collaborative filtering method that improves recommendations. Expert Syst. Appl. 48, 100–110 (2016)
8. Liu, Q., Ge, Y., Li, Z., Chen, E.: Personalized travel package recommendation. In: International Conference on Data Mining (2011)
9. Ranjane, R.S., Vidhya, R.: Travel package recommendation system
10. Salakhutdinov, R., Mnih, A., Hinton, G.E.: Restricted boltzmann machines for collaborative filtering. In: Proceedings of the Twenty-Fourth International Conference (ICML) Machine Learning, Corvallis, Oregon, USA, pp. 791–798, 20–24 June 2007

11. Subercaze, J., Gravier, C., Laforest, F.: Real-time, scalable, content-based twitter users recommendation. Web Intell. **14**(1), 17–29 (2016)
12. Tan, C., Liu, Q., Chen, E., Xiong, H., Wu, X.: Object-oriented travel package recommendation. ACM Trans. Intell. Syst. Technol. (TIST) **5**(3), 43 (2014)
13. Zhang, C., Liang, H., Wang, K., Sun, J.: Personalized trip recommendation with poi availability and uncertain traveling time. In: Proceedings of the 24th ACM International on Conference on Information and Knowledge Management, pp. 911–920. ACM (2015)
14. Zhang, M., Guo, X., Chen, G.: Prediction uncertainty in collaborative filtering: enhancing personalized online product ranking. Decis. Support Syst. **83**, 10–21 (2016)
15. Zhou, X., He, J., Huang, G., Zhang, Y.: SVD-based incremental approaches for recommender systems. J. Comput. Syst. Sci. **81**(4), 717–733 (2015)

Demo Paper

OPGs-Rec: Organized-POI-Groups Based Recommendation

JiaPeng Li$^{(\boxtimes)}$, Yanxia Xu$^{(\boxtimes)}$, and Lei Zhao$^{(\boxtimes)}$

School of Computer Science and Technology, Soochow University, Suzhou, China
trajepl@gmail.com, xyx.edu@gmail.com, zhaol@suda.edu.cn

Abstract. With development of urban modernization, a large number of Organized POI Groups (OPGs), such as multipurpose buildings, business streets and shopping malls, scatter over the city which have a great impact on people's lives and urban planning etc. Nowadays recommender system is based on Point-of-Interest (POI) which plays an important role in ours lives. However, it is well-known that there exists no work about groups of POIs which serves in recommendation. In this paper, we propose an OPGs-Rec, a novel OPGs-Based recommender system that recommends OPGs extracted from sets of POIs to target users by their locations and needs. We will demonstrate step by step how use OPGs to recommend valuable information to target users.

Keywords: OPGs recommendation · Organized POI groups · Density-based clustering

1 Introduction

With the proliferation of spatial data, lots of research work on spatial recommendation has been presented. The development of urban modernization fosters a large number of hot spots, such as multipurpose buildings, business streets and shopping malls. One of these hot spots is larger than a single POI and smaller than a functional region [1]. Intuitively, a large hot spot is defined as Oganized POIs Group (OPG) [2] has a set of POIs which locate close to each other. The OPGs not only reflect the distribution of the center of city, but also have a great effect on recommendation. If an user plans a series of activities (e.g., shopping, watching a movie or having dinner), the existing POIs-Based recommendation usually recommend the best POI according to evaluation functions. Users will receive a set of nearest POIs which cover all of user's requirements [3] or a set of related POIs with higher ranking [4]. However, people have their individual requirements and the existing POIs-Based recommendation cannot meet all of people's requirements simultaneously. Besides, an enormous number of people are unclear about their requirements until they have a visit to the POI recommended. For example, a woman plans to buy a pair of shoes. But she is unclear about what kind of shoes she wants until she has a shopping. There are also so many other possibilities after buying a pair of shoes, Maybe she suddenly wants

© Springer International Publishing Switzerland 2016
F. Li et al. (Eds.): APWeb 2016, Part II, LNCS 9932, pp. 521–524, 2016.
DOI: 10.1007/978-3-319-45817-5_56

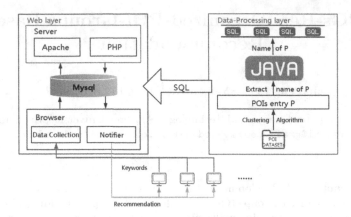

Fig. 1. System overview

to buy a skirt to match her new shoes. Obviously, the OPGs-Based recommendation can offer user more choices than traditional POIs-Based recommendation. Hence, recommending a shopping mall flexibly is better than recommending several shoes shops according to evaluation functions because the users can choose the best one just taking a little time.

To recommend a set of Organized POIs Groups, this paper proposes a novel system OPGs-Based Recommender System (OPGs-Rec). It will be shown that when an user plans a series of activities, recommendation based OPGs is more efficient and the accuracy does not obviously decrease. The solution to discovering OPGs from lots of POIs is a variant of density-based clustering algorithm, i.e. DBSCAN [2]. After clustering POIs, we get a complete file R which contains a set of OPGs $O = < O_1, O_2, O_3...O_n >$. Each OPG contains a set of POIs which locate close to each other, i.e. $O_i = < P_1, P_2, P_3...P_n >$. There is $P_i = < LOG, LAT, M, L, I >$ which indicates geographical longitude, geographical latitude, address of POI, tags of POI and ID of OPGs in order. For these sets of P_i, we extract the name information of OPGs from sets of $P_i.M$ which group by value of $P_i.I$ for further recommendation.

2   System Overview

As shown in Fig. 1, our recommender system OPGs-Rec has two layers: Web architecture layer and Data-Processing layer. The front-end of our recommender system is designed with a good interaction and presentation. The backstage of OPGs-Rec provides an interface for interacting with Mysql storing a series of dataset that processed by Data-Processing layer. Each party has several components and their functionalities are described as follows.

Web Layer: The front end of website is designed with a good interaction and presentation. As you can see, in Fig. 2, there are three regions forming our page: keyword-select, result table and the map display. When you visit our page, you

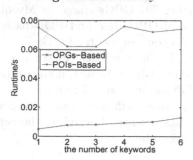

Fig. 2. OPGs classify

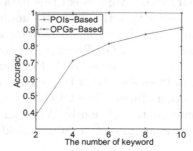

Fig. 3. All OPGs in Beijing 2014

Fig. 4. Efficiency

Fig. 5. Accuracy

can delimit the scope with a circle when you click the button is at upper right corner. Then you can select the keywords got from google POI classification[1] in keyword-select region. Finally, the backstage returns OPGs related all keywords you input. The feedback results are shown ordered by the linear addition of distance from circle center and the evaluation ranking. Note that, if you do not delimit the scope, this page will raise a warning.

Generally, the data entry $IN = (L, R, K)$ created in front-end is considered as input data of our recommendation. $IN.L$ and $IN.R$ is the center and radius of circle. $IN.K$ is the keywords that users can select. The output data depends on these three elements is represented as the $OUT = (ID, Info)$ indicates the ID and information of OPGs. Every OPG in OUT is displayed as a covering in the map. The audience could act as users and visit our website in http://ada. suda.edu.cn/opg/.

Data Processing Layer: In this layer, we process the original POIs dataset using a variant of density-based clustering algorithms i.e. DBSCAN using hybrid similarity which different from traditional DBSCAN. (The details can be seen in [2]). In this way, we can get a set of OPGs $O = (O_1, O_2, O_3...O_n)$ mentioned above. Then we extract the name information of OPGs. As you can see, in Fig. 3, there are 701 OPGs clustered from original POIs dataset.

System Implementation: The Web layer is implemented as a B/S pattern based on Apache2.4, php7 and Baidu Map API[2]. For Data-Processing layer,

---

[1] http://lbs.juhe.cn/data.php?PHPSESSID=acunetixsessionfixation.

[2] http://developer.baidu.com/map/reference/index.php.

we use OpenJDK8 Runtime Environment and OpenJDK 64-Bit Server VM to processing POIs dataset and output sql file to insert keywords and some supplementary information to Mysql. The version of Mysql is Ver 15.1 Distrib 10.1.12-MariaDB, for Linux (x86-64).

3   Performance

This section, we conduct experiment using the POIs data in Beijing (2014) which has 490317 POIs. We process this dataset getting 701 OPGs stored in Mysql. Figure 4 shows that the different running time of OPGs-Rec and POIs-based when there are different numbers of keywords. As you can see, the OPGs-Rec is more efficient than POIs-Based recommendation. In Fig. 5, with the increasing of keywords, the accuracy of OPGs-Rec increases obviously. In Table 1, it shows the number of POIs and OPGs that recommended depending on different distances from location of users. The OPGs-Rec greatly increases efficiency and accuracy of recommendation obviously. When an user plans a series of activities, he can find all he wants in several OPGs quickly and accurately.

Table 1. POI and OPGs search results

	500 m	1000 m	1500 m	2000 m	2500 m	3000 m
POI	2318	5399	9080	14281	19226	21575
OPG	2	3	3	7	7	7

References

1. Yuan, J., Zheng, Y., Xie, X.: Discovering regions of different functions in a city using human mobility and POIs. In: KDD 2012, pp. 186–194 (2012)
2. Xu, Y., Liu, G., Yin, H., Xu, J., Zheng, K., Zhao, L.: Discovering organized POI groups in a city. In: Liu, A., Ishikawa, Y., Qian, T., Nutanong, S., Cheema, M.A. (eds.) DASFAA 2015 Workshops. LNCS, vol. 9052, pp. 223–226. Springer, Heidelberg (2015)
3. Cao, X., Cong, G., Jensen, C.S., Ooi, B.C.: Collective spatial keyword querying. In: SIGMOD Conference, pp. 373–384 (2011)
4. Chen, W., Zhao, L., Jiajie, X., Liu, G., Zheng, K., Zhou, X.: Trip oriented search on activity trajectory. J. Comput. Sci. Technol. **30**(4), 745–761 (2015)

A Demonstration of Encrypted Logistics Information System

Huakang Li, Xinwen Zhang, Yitao Yang, and Guozi Sun$^{(\boxtimes)}$

Institute of Computer Technology,
Nanjing University of Posts and Telecommunications,
Nanjing 210023, China
{huakanglee,sun}@njupt.edu.cn

Abstract. Recently, some illegal delivery persons sale customers information caused the fraud and criminal cases because of the visibility of logistics papers. Traditional information encryption technology packaging all logistic information is difficult for couriers to sort the parcels. In this paper, a multiplayer encryption technology based on QR code was proposed to protect the customers privacy information without disabling the normal logistic process.

Keywords: Logistic information · Multiplayer encryption · QR code

1 Introduction

With the development of Internet and e-commerce, Logistics Information System (LIS) has become an indispensable tool for logistics express industry. In LIS, the express list contains some personal information for receiving, transiting and delivering parcels conveniently.

Some illegal delivery persons sale customers information caused the fraud and criminal cases in recent years. The British Postal Service Management Committee [1] writes all the terms of the postal security to the license, and it is enforced by law. The logistics staff jobs in the USA must provide a social security number [2]. To some extent, these methods can protect the user's privacy information, but there is still a risk of privacy leakage [3].

The logistics staff could store home address as the file name using ciphertext instead of the express list, in the Android mobile phone [6]. However, there are some limitations in the practical applications: (1) Do not use express: it is difficult to be compatible with the existing LIS; (2) This method has not yet fully realized the privacy protection of personal information. In [7], the authors introduced a personal logistics information protection mechanism to store ciphertext in a database with strict management. However, Customer representatives will contact the receiver to realize logistics transit, and it has not good compatibility with existing LIS. Therefore, we propose a multiplayer encryption technology, named Logistic Information Privacy Protection System (LIPPS), using QR code to protect the customers privacy information without disabling the normal logistic process.

© Springer International Publishing Switzerland 2016
F. Li et al. (Eds.): APWeb 2016, Part II, LNCS 9932, pp. 525–528, 2016.
DOI: 10.1007/978-3-319-45817-5_57

2   System Architecture Design

LIPPS is an architecture of mobile terminal-verification cloud platform (Fig. 1). The system structure of LIPPS is introduced as following:

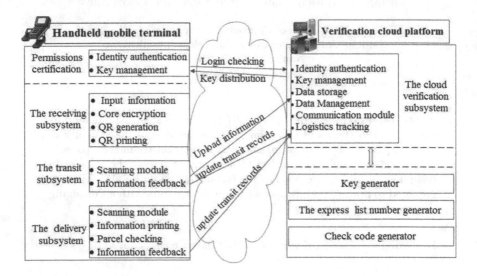

Fig. 1. LIPPS system architecture design.

(1) The cloud verification subsystem is the foundation of the whole LIPPS, which includes identity authentication, key management, data management, logistics records tracking, and information communications. This subsystem completes logistics business processes through a series of data transmission and control operations.

(2) The receiving subsystem is system terminal, including the personal logistics information input, core information encryption, QR code generation, and QR code printing. It also provides temporary personal information storage to store to the verification cloud platform within the network connected environment.

(3) The transit subsystem is the kernel of logistics business process which includes QR code scanning, authorization decryption, and logistics records update. The logistics staff obtains the appropriate authorization using login certification, and scans QR code to get readable logistics information with requirements. The system automatically updates the logistics records of the current node to the next logistics node.

(4) The delivery subsystem is LIPPS output including the QR code scanning, parcel information printing, parcel check, and information feedback. Delivery staff scans the QR code to obtain the final node address and prints prompt information. Then, the verification cloud platform sends prompt information

and check code information to the receiver via SMS module. Finally, the parcel checking function provides the accuracy verification of users parcel information.

3   Information Encryption and Decryption

Encryption operation is the process of encrypting personal logistics information in handheld terminal, as shown in Fig. 2(a). According to the logistics platform actual transit mechanism, the personal logistics information (the top information in Fig. 2(c)) is divided into n piece $\{M_1, M_2, ..., M_n\}$. The handheld terminal uses the public key $Auth\_i\_pubkey$ to encrypt M_i and outputs the segment encryption C_i. The ciphertext $C = \{C_1, C_2, ..., C_n\}$ is encoded to QR code (the above QR code in Fig. 2(c)) using a special separator distinguish method.

The decryption operation is the process of obtaining the next node distribution information to complete logistics transit with requirements, as shown in Fig. 2(b). The logistics staff scans QR code to obtain the ciphertext information, and uses the private key $Auth\_i\_prikey$ to decrypt the ciphertext to obtain the current job information such as '$Add: BCity$' in the middle of Fig. 2(c). The system will provide little information as possible to dispatch for the final courier (the bottom QR code in Fig. 2(c)).

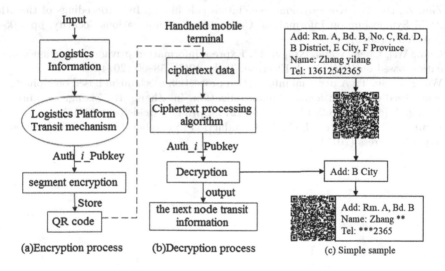

Fig. 2. Encryption and decryption process.

4   Conclusion

In this paper, a Logistics Information Privacy Protection System based on encrypted QR code (LIPPS) was demonstrated. According to the logistics business process, we designed the hierarchical authorization mechanism to solve the contradiction between information encryption and logistics transit. The logistics information is encrypted and stored in the QR code. The logistics staff can scan QR code to decrypt the destination address of the next node using the handheld mobile terminal. Therefore, this system can solve the problem of privacy protection in the logistics transit process, and has a good compatibility with existing LIS.

Acknowledgement. This work was supported by the Foundation of Nanjing University of Posts and Telecommunications (Grant No. NY213085 and No. NY214069), the NSFC (No. 61502247, 11501302, 61502243), Natural Science Foundation of Jiangsu Province (BK20140895,BK20130417).

References

1. Zongjie, G.: Research on the legal issues of universal postal service. Jinan J. **34**(10), 95–100 (2012)
2. Berghel, H.: Identity theft, social security numbers, and the web. Commun. ACM **43**(2), 17–21 (2000)
3. http://www.informationweek.com/news/internet/social_network/219401268
4. Qian, W., Xingyi, L.I.: Express information protection application based on K-anonymity. Appl. Res. Comput. **31**(2), 555–557 (2014)
5. Zhu, Z., Du, W.: K-anonymous association rule hiding. In: Proceedings of the 5th ACM Symposium on Information, Computer, Communications Security, pp. 305–309. ACM (2010)
6. Qian, Wei, Chen, Wang, Xingyi, Li: Express information privacy protection application based on RSA. Appl. Electron. Tech. **40**(7), 58–60 (2014)
7. Wang, L., Su, T.: A personal information protection mechanism based on ciphertext centralized control in logistics informatization. In: Zhang, R., Zhang, Z., Liu, K., Zhang, J. (eds.) LISS 2013, pp. 1207-1212. Springer, Heidelberg (2015)
8. Kraft, J.S., Washington, L.C.: An Introduction to number Theory with Cryptography. CRC Press (2016)

PCMiner: An Extensible System for Analysing and Detecting Protein Complexes

Danyang Xiao, Jia Zhu$^{(\boxtimes)}$, Yong Tang, Lingxiao Chen, and Jingmin Wei

School of Computer Science, South China Normal University, Guangzhou, China
{danyangxiao,jzhu,ytang,chenlingxiao,weijingmin}@m.scnu.edu.cn

Abstract. Protein complexes detection is a hot topic in bioinformatics. More and more researchers tend to detect the protein complexes from the Protein-Protein interaction networks. Since there are many approaches for detecting protein complexes, it will be great to integrate all approaches into a system, which can help researchers focus on analyzing. In this paper, we introduce PCMiner, an extensible system that integrates some state-of-arts protein complexes detection algorithms with distributed design and visualization technology. The system also provides application interfaces for researchers to implement their own approaches.

Keywords: Protein complexes detection · Graph clustering

1 Introduction

Protein-Protein Interaction (PPI) network [1] is a large scale graph, which formed by the PPI data. As Shown in Fig. 1, each point in the graph represents a protein, and each edge represents the interaction between two proteins. Protein complex detection aims to detect the complexes (clusters) from the graph, which means this is a clustering problem. Though there are many approaches being proposed over the past few years, few studies focus on proposing some system, which can provide a platform that integrates existing protein complexes detection algorithms. In addition, if the PPI data is massive, the running time of current protein complexes detection algorithms on a single machine (standalone mode) will be unacceptable. To overcome above described challenges, we introduce PCMiner, a scalable system aims to help the research of protein complexes detection. Our system' contributions are described as follows:

- The system provides a convenient experimental environment by integrating the whole protein complexes detection procedure.
- The PCMiner integrates some approaches such as MCODE, MCL, DPClus, COACH [1] and SNMF [2], and provides application interfaces for researchers to implement their own approaches.
- The PCMiner is being designed in a distributed way, which all algorithm can be run in the Spark^1. Experiments shows that distributed algorithms' performance run on the PCMiner are better when the PPI data is massive.

---

^{1} http://spark.apache.org/.

© Springer International Publishing Switzerland 2016
F. Li et al. (Eds.): APWeb 2016, Part II, LNCS 9932, pp. 529–532, 2016.
DOI: 10.1007/978-3-319-45817-5_58

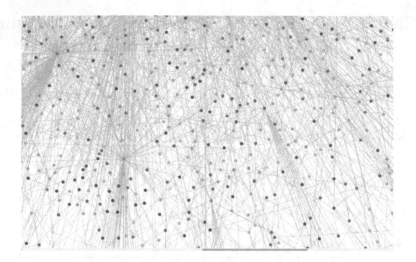

Fig. 1. Protein-protein interaction network

2   System Overview

PCMiner is designed using browser/server mode, as shown in Fig. 2(a). The server invokes the core module to detect complexes and visualizes the results to browser after the algorithm stops, the core module stores the clustering information into database. The description for main modules of PCMiner are shown below:

Interaction Module: This module provides a management interface to configure the system parameter and allows users to interact with PCMiner. Before performing an experiment, users can select a protein complex detection algorithm and set the model parameters. Interaction module also responses the analyses request from the user and display the detection/clustering results.

Core Module: Core module integrates the whole protein complex detection procedure. As shown in Fig. 2(b), when a detection task runs, this module will pre-process the data, then execute the users' selected algorithm, and store the result in database and analyze it. After analyzing, our system will visualize the result. This module also provides a few protein complexes detection algorithm libraries to invoke.

Distributed Module: This module supports the system to run detection algorithms on distributed architecture. As show in Fig. 2(b), Our algorithm libraries consists of two part: one is stand alone algorithms, the other is distributed algorithms. Since Spark [3] performs better in iterative computation, and it provides some libs such as mllib, graphX, we adapt a few algorithms to run on Spark including an algorithm we proposed previously, which showed better performance than other approaches. The pseudocode is outlined in Algorithm 1, and more technical details about SNMF can be found in [2].

Algorithm 1. DistributedSNMF

Input: G=(V,E);
Output:S(a set of clusters, each cluster contains one or more proteins).
Begin:
  transform G to r   //r is Spark'data structure:RDD
  transform R to m   //m is Spark'data structure:BlockedMatrix
  initialize the SNMF's parameters
  convergence = false //check the status
  while convergence = false **do:**
    x = distributedMatrixMultiply(m) //x is cluster member matrix
    error = computeMatrixLoss()
    if isConvergence(error):
      convergence = true
    end if
  end while
  S = getClustersFromMatrix(m,x)   //get clusters from the matrix
End

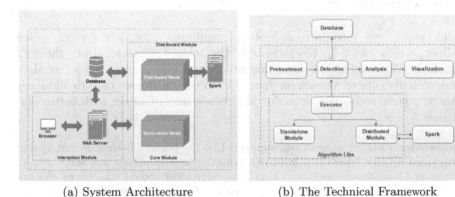

(a) System Architecture   (b) The Technical Framework

Fig. 2. System architecture and technical framework

3   System Demonstration

Optional Algorithms: As shown in Fig. 3(a), PCMiner' interaction interface shows a list of algorithms we have implemented. When users upload the data file(different PPI) and input some parameters of the algorithms, the PCMiner prepare for running. In addition, some experiments' information maybe show during the system running.

PPI Visualization: As shown in Fig. 3(b), PCMiner shows the whole PPI in the browsers. The protein complexes will distribute color randomly based on clusters. PCMiner also demonstrates the evaluate result of the experiments. Since PPI network is a scale-free network, how to show all details about the PPI network in a screen is a big problem. PCMiner provides a function that user

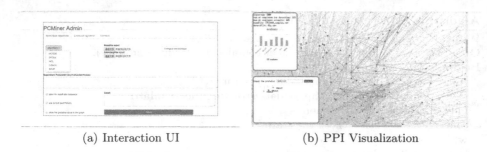

(a) Interaction UI   (b) PPI Visualization

Fig. 3. Interaction UI and visualization results

could search a detected protein complex by protein'name. If the complex exists, it will be displayed in a child window.

4   Conclusions

In this paper, we introduce a protein complexes detection system called PCMiner, which can help researchers in the related areas to study the PPI network. PCMiner integrates detecting procedures and implements a few algorithms with distributed version on Spark [4]. In the future, we plan to add more advanced protein complexes detection algorithms into PCMiner with better performance.

Acknowledgments. This work was supported by the Natural Science Foundation of Guangdong Province, China (No. 2015A030310509), the Public Research and Capacity Building in Guangdong Province, China (No. 2016A030303055), the major Science and Technology Projects of Guangdong Province, China (No. 2016B030305004, No. 2016B010109008, No. 2013B090800024), the National High Technology Research and Development Program of China (863, No. 2013AA01A212).

References

1. Li, X., Wu, M., Kwoh, C.K., et al.: Computational approaches for detecting protein complexes from protein interaction networks: a survey. BMC Genomics **11**(Suppl. 1), 1–19 (2010)
2. Zhu, J., Wu, X., Huang, C., et al.: A graph based clustering approach for protein complexes detection using symmetric non-negative matrix factorization (2016). (Online First)
3. Antoine, H., David, A.: Distributed graph layout with Spark, pp. 271–276 (2015)
4. Li, J., Li, D., Zhang, Y.: Efficient distributed data clustering on Spark. In: IEEE International Conference on CLUSTER Computing, pp. 504–505 (2015)

MASM: A Novel Movie Analysis System Based on Microblog

Xingcheng Wu, Jia Zhu$^{(\boxtimes)}$, Yong Tang, Rui Ding, Xueqin Lin,
and Chuanhua Xu

School of Computer Science, South China Normal University, Guangzhou, China
{xingchengwu,jzhu,ytang,ding,xqLin,chxu}@m.scnu.edu.cn

Abstract. With the big data technology booms, it plays an increasingly important role in many areas including movie analysis. This paper designs and implements a novel movie analysis system based on Microblog called MASM, which not only can help audiences to decide which movies they would like to watch but also make producers understand the public's reaction to a certain movie quickly. In this demonstration, we show four functions of MASM for movies, namely, keywords cloud, sentiment analysis, movie correlation network and movie box office prediction.

Keywords: Sentiment analysis · Keyword extraction · Movie network · Box office prediction

1 Introduction

Microblog[1], as an online social networking service that enables users to send and read short messages, has become an essential part of daily life. More and more people tend to comment movies through Microblog, which can express people's real feeling for different movies. We extract the keywords of hundreds of movies from Microblog to give more useful information to audience. We perform the sentiment analysis for the extracted information. Through this way, not only can help audience to choose the movie they are interested, but also help the movie production company gains the audience reaction of their movie quickly.

In addition, box office, as an important indicator of the movie's quality, has attracted more and more attentions from ordinary people. Since they have their own taste of the protagonists, our prediction model can give box office prediction for a movie based on users' preference.

Generally speaking, our system firstly fetches the movie data from some related websites. We then build a movie correlation network based on the data we fetched, and audiences can click a movie in the network to check the attributes (such as protagonists and directors), which may be related to other movies, or click one of the movie attributes to see other related movies. Based on the network, we can extract multiple features of those information and model those data

---

[1] http://www.weibo.com.

© Springer International Publishing Switzerland 2016
F. Li et al. (Eds.): APWeb 2016, Part II, LNCS 9932, pp. 533–536, 2016.
DOI: 10.1007/978-3-319-45817-5_59

to predict the box office with a classification model by replacing the protagonist. In other words, this function can help a director to pick an actor at some levels. In the following sections, we will introduce our system in more details.

2   System Overview

In this section, we introduce the architecture of our system shown in Fig. 1.

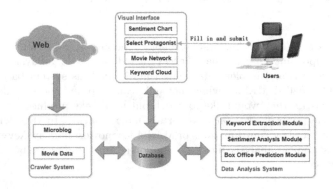

Fig. 1. System framework

Crawler System: We obtain data sources such as movie title, genre, directors and issuing company through a self-developed crawler system, which adopts with the latest MapReduce[2] distributed framework. Firstly, we put the movie titles into the waiting queue lists, then crawl the data we need from Microblog, and other related movie websites. After regular filtering and cleaning processes, all the related useful movie data are stored in a database.

Keyword Extraction Module: This module extracts keywords from every Microblog by using a semantic-based keyword extraction(SKE) algorithm Wang and Huai [2012]. Based on the frequency statistics of all keywords, we select the 50 highest frequent keywords of all movies. The results is used for the keyword cloud and the sentiment analysis later.

Sentiment Analysis Module: In this module, after the word segment processing for every message in the Mircoblog, we divide each message into positive emotion and negative emotion using the theory proposed by Liu [2013], and calculate the percentage of each type of emotion.

Box Office Prediction Module: With the above three modules, we can use these information combine for box office prediction. Firstly, based on the relationship between some of the properties (including actors, directors, distribution

---

[2] https://en.wikipedia.org/wiki/MapReduce.

companies, type, production companies) of movies, we construct a movie correlation network as shown in Fig. 2(c). We then construct features for each movie based on the network. We apply the two-step filter feature selection method proposed by Ding et al. [2016], which can remove the redundancy features and filter out the relevance features. Last but not least, we then apply our previously proposed multiple classifiers model (MCM) to perform prediction, more technical details and experimental results of MCM can be found at Zhu et al. [2015].

3   Demonstration

In this section, we introduce the demonstration plan of our system. When users access our system, the search results are as shown in Fig. 2 after entering a movie name. Next, we will introduce each function one by one.

Keyword Cloud Results: As shown in Fig. 2(a), through this module, users can not only see the audience's reaction to the movie, such as good and praise, but also can understand some of the most relevant information with the movie.

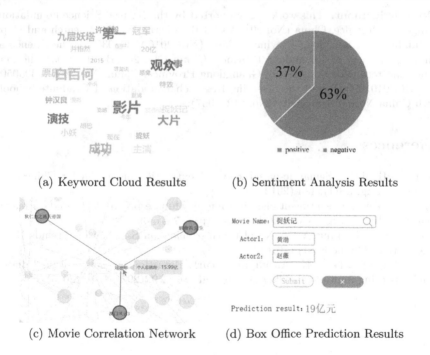

(a) Keyword Cloud Results   (b) Sentiment Analysis Results

(c) Movie Correlation Network   (d) Box Office Prediction Results

Fig. 2. Demonstration Details

Sentiment Analysis Results: When you search for a new movie, the evaluation of audiences will have an impact on whether or not you should watch it. In this module, we get the results of the sentiment analysis based on the information we extracted from Microblog. As shown in Fig. 2(b), users can understand the appraisal of audience's preferences for the movie.

Movie Correlation Network: For a movie, people usually pay more attentions to the attributes such as protagonist, director, type, box office and so on. As shown in Fig. 2(c), users can click on any nodes in this network and see the information associated with it, e.g., how many movies did the actor being a protagonist.

Box Office Prediction Results: The choice of the protagonist usually has a great influence on the box office. In this module, users can replace the protagonist with other favorite actors and make a box office prediction as shown in Fig. 2(d).

4   Conclusions

In this paper, we introduce a novel movie analysis system based on Microblog(MASM). By using our MASM, audiences can get suggestion to choose the movie they are interested. At the same time, the system can provide some support for the movie production company in the choice of actors.

Acknowledgement. This work was supported by the Natural Science Foundation of Guangdong Province, China (No. 2015A030310509), the Public Research and Capacity Building in Guangdong Province, China (No. 2016A030303055), the Science and technology project of Guangdong Province, China (No. 2015B010109003), the major science and technology projects of Guangdong Province, China (No. 2016B030305004, No. 2016B010109008) and the Scientific Research Foundation of Graduate School of South China Normal University (No. 61272067).

References

Wang, L., Huai, X.: Semantic-based keyword extraction algorithm for Chinese text. Comput. Eng. **38**(1), 1–4 (2012)

Liu, D.: Research on sentiment classification of Chinese micro blog based on machine learning. Int. J. Digit. Content Technol. Appl. **7**(3), 395 (2013)

Ding, R., Zhu, J., Tang, Y.: A Novel Feature Selection Strategy for Friends Recommendation (Online First) (2016)

Zhu, J., Xie, Q., Zheng, K.: An improved early detection method of type-2 diabetes mellitus using multiple classifier system. Inf. Sci. **292**(292), 1–14 (2015)

A Text Retrieval System Based on Distributed Representations

Zhe Zhao, Tao Liu$^{(\boxtimes)}$, Jun Chen, Bofang Li, and Xiaoyong Du

School of Information, Renmin University of China, Beijing, China
{helloworld,tliu,chenjun2013,libofang,duyong}@ruc.edu.cn

Abstract. Most text retrieval systems are essentially based on bag-of-words (BOW) text representations. Despite popularity of BOW, it ignores the internal semantic meanings of words since each word is treated as an atomic unit. Recently, distributed word and text representations become increasingly popular in NLP literatures. They embed syntactic and semantic information of words and texts into low-dimensional vectors, thus overcome the weaknesses of traditional BOW representations to some extent. In this paper, we implement a text retrieval system that are totally supported by distributed representations. Our new system no longer relies on the matchings of words in queries and texts, but uses semantic similarity to judge if a text is relevant to a query and to what extent, which provides better user experience compared with traditional text retrieval systems.

Keywords: Text retrieval · Distributed text representation · Hierarchical paragraph vector

1 Introduction

Bag-of-words (BOW) is a simple but surprisingly powerful approach for text representation. Nowadays mainstream text retrieval systems such as Lucene[1] still rely on BOW representations. In traditional text retrieval systems, the relevance of query and text is essentially based on words matchings. And the heuristic weighting schemes and smoothing techniques are usually required to obtain better performance [3]. In our new text retrieval system, all objects (include word, text, etc.) are represented by low-dimensional vectors, each dimension of which represents high-level semantics rather than the occurrence of concrete word. The degree of relevance between query and text is measured by distance between their distributed representations.

Paragraph Vector (PV) is a popular approach for generating distributed text representations (embeddings), where text embeddings are trained to be useful to predict words in the texts [1]. PV models are trained in unsupervised framework. However, in many cases, label (supervised) information is available such

---

[1] http://lucene.apache.org/.

© Springer International Publishing Switzerland 2016
F. Li et al. (Eds.): APWeb 2016, Part II, LNCS 9932, pp. 537–541, 2016.
DOI: 10.1007/978-3-319-45817-5_60

as the news categories and hashtags in Twitter. To exploit this kind of knowledge, we extend traditional PV by adding label layer upon it, which is called Hierarchical Paragraph Vector (HPV). HPV regards each label as a pseudo text and its 'words' is the texts that process the corresponding label. As a result, the framework of HPV is illustrated in Fig. 1. HPV takes supervised information into consideration and can generate better text representations. When labels are not available, traditional PV is used for training in our system.

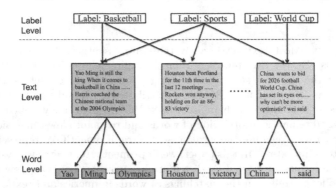

Fig. 1. Framework of hierarchical paragraph vector. Arrows denote 'predict'. Labels are treated as pseudo texts and their embeddings are trained to predict texts.

2   Overview Framework

Our system consists of two components: (1) Off-line component, where words, texts and labels are embedded into low-dimensional vectors. (2) On-line component, where queries are embedded into low-dimensional vectors and compared with trained text embeddings to return the ranked texts list. The framework of the system is illustrated in Fig. 2.

A. Off-line component: Firstly, word embeddings and parameters in neural network are pre-trained by external large-scale datasets such as Wikipedia. This component is included mainly for two reasons: (1) word embeddings trained in external large-scale datasets are 'universal' features, which already capture syntactic and semantic information of words well and can be used for many NLP tasks directly. (2) when the query is composed of words that never occur in our dataset (a very common case), directly utilizing word embedding trained in external large-scale dataset can provide satisfying results.

Texts and their labels are then fed into Hierarchical Paragraph Vector model. Embeddings and parameters in neural model are fine-tuned to make labels and texts to be useful to predict their corresponding texts and words.

B. On-line component: When a query is provided, we embed it into the same semantic space with text embeddings. Specifically, an embedding is initialized randomly and trained to be useful to predict the words in the query.

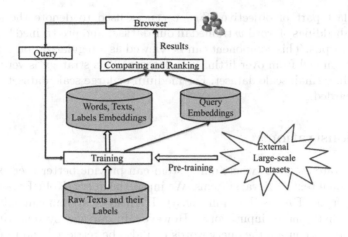

Fig. 2. System overview

During the training, parameters in the model are fixed. When training process is finished, the query embedding is compared with texts embeddings and top k texts are retrieved for users.

3   Key Technologies

In this section, we give a detailed discussion about HPV. Firstly some notations are established. Texts collection is denoted by $C = \{t_1, t_2, ..., t_{|C|}\}$ and i_{th} text in C is denoted by $t_i = \{w_{i1}, w_{i2}, ..., w_{i|t_i|}\}$. V is vocabulary set and $|WN|$ is the number of words in the whole collection. Local context of word w is denoted by $w^{context}$. Label set is denoted by $L = \{l_1, l_2, ..., l_{|L|}\}$. T_i is the collection of texts that possess label l_i. The training objective of our system consists of four components:

$$\sum_{i=1}^{|C|} \sum_{j=1}^{|t_i|} logP(w_{ij}|t_i) + \sum_{i=1}^{|WN|} logP(w_i|w_i^{context})$$

$$+ \sum_{i=1}^{|L|} \sum_{t_k \in T_i} logP(t_k|l_i) \qquad (1)$$

$$+ \sum_{w_k \in V} Sim\_reg(w_k)$$

where conditional probability $P(|)$ is defined by negative sampling softmax [2]. The first two components is just the objective of PV, where target words are predicted by texts and local contexts respectively [1]. The third component introduces supervised information into the models by maximizing the conditional probabilities of texts given their labels. By taking the label information into consideration, we can not only improve the quality of text embeddings, but also obtain the trained label embeddings.

In the last part of objective, *Sim_reg(w)* is used to denote the similarity between embeddings of word w trained in our dataset and pre-trained by external large-scale corpus. This component can be viewed as a regularization term which prevents the model from over-fitting our dataset. This strategy is very effective when handling small-scale dataset. For medium or large scale dataset, this part can be discarded.

4   Demonstration

We demonstrate scenarios where our system can provide better user experience than traditional text retrieval systems. We input the query 'Kobe Bryant', a basketball player in NBA (as shown in Fig. 3). Traditional system can only retrieve those texts that contain input words. However, in our new system, the related texts that do not contain the query words can also be retrieved and ranked basically according to the semantic similarities with the query: News about NBA is ranked at relatively top positions, the following is sports news and in turn followed by news in other fields.

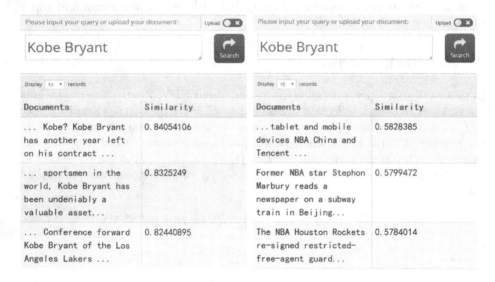

Fig. 3. The left snapshot shows the top 3 texts in response to the query. The texts that contain the key words are given very high similarities with query and are ranked at high positions. As we scroll down, our system returns texts that are related to the query but do not contain query words, which is shown in the right snapshot.

Acknowledgements. This work is supported by the Fundamental Research Funds for the Central Universities, the Research Funds of Renmin University of China No. 14XNLQ06.

References

1. Le, Q.V., Mikolov, T.: Distributed representations of sentences and documents. In: ICML 2014, Beijing, China, 21–26 June 2014, pp. 1188–1196 (2014)
2. Mikolov, T., Sutskever, I., Chen, K., Corrado, G.S., Dean, J.: Distributed representations of words and phrases and their compositionality, pp. 3111–3119 (2013)
3. Mogotsi, I.C., Christopher, D.M., Prabhakar, R., Hinrich, S.: Introduction to Information Retrieval, 482 p. Cambridge University Press, Cambridge, (2008). Inf. Retr. **13**(2), 192–195 (2010). ISBN: 978-0-521-86571-5

A System for Searching Renting Houses Based on Relaxed Query Answering

Jianfeng Du$^{(\boxtimes)}$, Kunxun Qi, and Can Lin

Guangdong University of Foreign Studies, Guangzhou 510006, China
jfdu@gdufs.edu.cn

Abstract. In this paper, a system for searching renting houses is presented. The system integrates Semantic Web technologies to provide house information to people who want to rent houses. It gathers data from multiple house renting websites and traffic websites to construct a graph database. It accepts a natural language query for searching renting houses, then translates it into a set of conjunctive queries and then to a set of Cypher queries upon the graph database, and finally evaluates the Cypher queries to return answers to the given query. The system can also relax the translated conjunctive queries to recommend more answers. All the returned answers are shown with explanations.

1 Motivation and Novelty

In recent years, the price of houses is rising quickly in China. More and more people prefer to rent houses rather than buy houses. A number of house renting websites have emerged in China to help people rent houses. Existing house renting websites such as GanJi (http://gz.ganji.com/fang/) and FangTianXia (http://zu.gz.fang.com/) provide search facilities based on information retrieval. These websites usually offer house information to users by keyword matching. They do not work correctly with natural language queries raised by users and do not provide explanations for returned answers to a user query.

In this paper, we present a system that accepts natural language queries and shows answers with explanations. It has the following novelties compared to existing house renting websites.

1. The system integrates data from multiple house renting websites and traffic websites. Moreover, it exploits the Baidu map information to discover nearby bus stops, metro stations and other points of interest for renting houses. All the above data are stored in a graph database, which provides efficient accessing to the stored data by employing graph-based join operations.
2. The system is able to process a natural language query for searching renting houses. More precisely, it translates the given natural language query into a set of conjunctive queries and then to a set of Cypher queries that can directly be evaluated in the graph database.
3. To recommend more relevant renting houses to users, the system relaxes the translated conjunctive queries to more related conjunctive queries.

© Springer International Publishing Switzerland 2016
F. Li et al. (Eds.): APWeb 2016, Part II, LNCS 9932, pp. 542–546, 2016.
DOI: 10.1007/978-3-319-45817-5_61

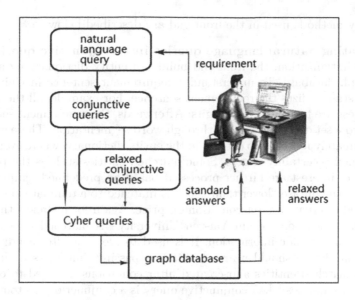

Fig. 1. The front-end of the proposed system

4. All answers including both standard and relaxed ones are shown with explanations. These explanations can tell why a standard answer matches the given
   query and how a relaxed answer differs from the given query.

2   Descriptions of the Proposed System

The proposed system is composed of a back-end and a front-end. The back-
end gathers data from multiple house renting websites and traffic websites. It
also gathers the Baidu map information to discover nearby points of interest
for a renting house. All the above integrated data are stored in a Neo4j [4]
graph database so as to support efficient accessing. Actually, this back-end is
almost the same as the back-end proposed in [1] except that it stores data in
Neo4j rather than in MySQL. The front-end of the proposed system, shown in
Fig. 1, is completely different from the front-end proposed in [1] on the following aspects. Firstly, it translates a given natural language query into a set of
conjunctive queries rather than accepts a conjunctive query directly. Secondly,
to recommend more relevant renting houses to users, it relaxes the translated
conjunctive queries to more related queries rather than applies abductive conjunctive query answering (ACQA) [1]. This way is more efficient since ACQA
is computationally hard. Finally, it translates all obtained conjunctive queries
into Cypher [2] queries other than SQL queries. The Cypher queries can directly
be evaluated in the Neo4j graph database, yielding both standard answers and
relaxed answers, where standard answers come from originally translated conjunctive queries and relaxed answers come from relaxed conjunctive queries.

The key methods used in the front-end are described below.

Translating natural language queries into conjunctive queries. During the initialization, the front-end builds an entity dictionary by accessing the graph database. To understand a requirement sentence in Chinese, the system standardizes the requirement sentence. For example, all the numbers in Chinese are translated into digits. Afterwards, the requirement sentence is split into a set of Chinese words through word segmentation. These words are approximately matched to entities in the entity dictionary, where every match is assigned a certainty degree. Some matched entities such as the price and a point of interest are further processed based on predefined templates. For example, the phrase "lower than 2000" is matched to a template about price. It is used to create a filter condition on price. Again for instance, the phrase "has transportation to Sun Yat-sen University" is matched to a template about nearby traffic information. It is used to create a filter condition that a resulting house should near a traffic line passing Sun Yat-sen University. All the matched entities and created filter conditions are used to construct conjunctive queries [1]. A conjunctive query is a conjunction of atoms of four forms including $\tau(x)$, $\alpha(x,y)$, $\delta(x,y,z)$ and $\phi(x,y)$. $\tau(x)$ means that the type of x is τ. $\alpha(x,y)$ means that the value of the attribute α of x is y. $\delta(x,y,z)$ means that the value of the distance attribute δ between x and y is z. $\phi(x,y)$ means that the comparison relation between x and y is ϕ, where ϕ can be EQ (=), NEQ ($\neq$), GE ($\geq$), GT ($>$), LE ($\leq$) and LT ($<$). For example, the requirement sentence "has a price lower than two thousand yuan and has transportation to Sun Yat-sen University" can be translated to the following conjunctive query, where an entity is defined to be near another entity if the distance between them is not greater than 500 m.

$$Q(x) = \text{House}(x) \wedge \text{Price}(x,y) \wedge \text{LT}(y, 2000) \wedge \text{Distance}(x, z, d_1) \wedge \text{Stop}(z)$$
$$\wedge \text{LE}(d_1, 500) \wedge \text{HasStop}(w, z) \wedge \text{TrafficLine}(w) \wedge \text{HasStop}(w, u)$$
$$\wedge \text{Stop}(u) \wedge \text{Distance}(u, \text{SunYatSenUniverity}, d_2) \wedge \text{LE}(d_2, 500).$$

A natural language query can be translated to multiple conjunctive queries since the matching is done approximately. Only conjunctive queries that have a high total certainty degree are kept for further processing.

Relaxing conjunctive queries. The front-end employs the "goal replacement" approach proposed in [3] to relax every translated conjunctive query. It replaces some predicates in a translated conjunctive query with more general predicates, yielding relaxed conjunctive queries. For example, to relax a conjunctive query that has an atom whose predicate is MetroLine, the predicate MetroLine is replaced with a more general predicate TrafficLine.

Translating conjunctive queries into Cypher queries. To evaluate conjunctive queries over the constructed Neo4j graph database, every conjunctive query is translated into a Cypher [2] query according to the following mappings.

$$\tau(x) \overset{\text{map}}{\Longrightarrow} (x) - [: \text{TYPE}] \rightarrow (n \ \{\text{ID} : \tau\})$$

$$\alpha(x,y) \overset{\text{map}}{\Longrightarrow} (x) - [: \alpha] \rightarrow (y)$$

$$\alpha(x,y) \wedge \phi(y,w) \overset{\text{map}}{\Longrightarrow} (x) - [: \alpha] \rightarrow (y) \ \text{WHERE} \ \phi(y.\text{ID}, w)$$

$$\delta(x,y,z) \wedge \phi(z,w) \overset{rmmap}{\Longrightarrow} (x) - [d : \delta] \rightarrow (y) \ \text{WHERE} \ \phi(d.\text{value}, w)$$

For example, the aforementioned conjunctive query $Q(x)$ is translated into the following Cypher query.

> MATCH $(x) - [: \text{TYPE}] \rightarrow (n_1 \ \{\text{ID} : \text{House}\}), (x) - [: \text{Price}] \rightarrow (y)$,
> $(x) - [d_1 : \text{Distance}] \rightarrow (z), (z) - [: \text{TYPE}] \rightarrow (n_2 \ \{\text{ID} : \text{Stop}\})$,
> $(w) - [: \text{HasStop}] \rightarrow (z), (w) - [: \text{TYPE}] \rightarrow (n_3 \ \{\text{ID} : \text{TrafficLine}\})$,
> $(w) - [: \text{HasStop}] \rightarrow (u), (u) - [: \text{TYPE}] \rightarrow (n_4 \ \{\text{ID} : \text{Stop}\})$,
> $(u) - [d_2 : \text{Distance}] \rightarrow (n_5 \ \{\text{ID} : \text{SunYatSenUniversity}\})$
> WHERE $y.\text{ID} < 2000$ and $d_1.\text{value} \leq 500$ and $d_2.\text{value} \leq 500$
> RETURN x, y, z, w, u.

Generating explanations for answers. Every (standard or relaxed) answer is presented to users with an explanation. The explanation corresponds to a natural language description for the conjunctive query that the answer comes from. Roughly speaking, the explanation is generated from the conjunctive query by separating different aspects for the answer variable and by instantiating variables with matched entities in the graph database. For example, the following explanation is generated for an answer c to the aforementioned conjunctive query $Q(x)$, where c_y, c_z, c_w and c_u are entities instantiated from the returned variables in the translated Cypher query.

> c has a price c_y, and is near stop c_z which is connected to stop c_u
> through traffic line c_w, where c_u is near SunYatSenUniversity.

3 What Will Be Demonstrated?

We have implemented a prototype system available at http://115.28.29.106:8080/Search/. The following key features will be shown in the demonstration.

1. The system is effective in processing natural language queries for renting houses. A given query is allowed to contain requirement information on house facilities, nearby traffic routes and nearby points of interest.
2. The efficiency in answering naturally language queries is rather high.
3. The answers are derived from different house renting websites. Every data item about an answer is shown with its information source.
4. Relaxed answers have explicit associations with the given query and can be used to recommend more relevant renting houses.

Acknowledgements. This work is partly supported by NSFC 61375056.

References

1. Du, J., Wang, S., Qi, G., Pan, J.Z., Qiu, C.: An abductive CQA based matchmaking system for finding renting houses. In: Pan, J.Z., Chen, H., Kim, H.-G., Li, J., Horrocks, I., Mizoguchi, R., Wu, Z., Wu, Z. (eds.) JIST 2011. LNCS, vol. 7185, pp. 394–401. Springer, Heidelberg (2012)
2. Holzschuher, F., Peinl, R.: Querying a graph database language selection and performance considerations. J. Comput. Syst. Sci. **82**(1), 45–68 (2016)
3. Inoue, C.S.K.: Negotiation by abduction and relaxation. In: AAMAS, p. 242 (2007)
4. Webber, J.: A programmatic introduction to Neo4j. In: SPLASH, pp. 217–218 (2012)

TagTour: A Personalized Tourist Resource Recommendation System

Tian Han, Yongjian Liu, and Qing Xie$^{(\boxtimes)}$

School of Computer Science and Technology,
Wuhan University of Technology, Wuhan 430070, China
{hant,felixxq}@whut.edu.cn, liuyj626@163.com

Abstract. The resource recommendation is an important service for tourists when they are traveling in an unfamiliar city. A tourist can request the information of food, entertainment and accommodation for her reference, so a recommendation system providing good user experience is highly desirable, which can effectively stimulate the tourism industry. However, the previous recommendation systems usually face the cold-start problem, and the keyword-based querying approach is too inefficient to guarantee the user experience. Considering the limitations above, this demo will introduce an interactive recommendation system named TagTour, which utilizes the tag transfer technique and the users' interaction to provide more accurate and personalized recommendation results, so as to improve the user experience. In this demo, we will introduce the important components and the relevant techniques of TagTour system, and then demonstrate the system performance by some actual resource recommendation cases.

Keywords: Recommendation system · Interactive recommendation · Tag transfer · Collaborative filtering

1 Introduction

Tourism is now a fast developing industry, with the rapidly increasing number of tourists nowadays. When a tourist is traveling in an unfamiliar city, she is always keen to know the information of food, entertainment and accommodation in a convenient way. Therefore, a well-designed recommendation service with good user experience is necessary and desirable, which can effectively guarantee a pleasant tour. Naturally, the tourists are expecting the recommendation system to provide precise and personalized service. However, there is a big gap between the tourists and the local resources. Since the tourists are usually new to the destination city, they have no idea about the resources with distinctive local features, while the systems cannot fully recognize the tourists' personal preference, especially for those new users, because they usually express little information of their interests, i.e., the *cold-start* problem. Therefore, it is difficult to effectively provide specific recommendation for the tourists.

© Springer International Publishing Switzerland 2016
F. Li et al. (Eds.): APWeb 2016, Part II, LNCS 9932, pp. 547–551, 2016.
DOI: 10.1007/978-3-319-45817-5_62

Considering the situation above, how to build a bridge to fill the gap between the tourists and the distinctive local resources is an essential problem, which requires the recommendation system to effectively manage the local resources, and explore the characteristics of different tourists for personalized recommendation. In this work, we propose a resource recommendation system named **Tag-Tour**, which provides a tag-based solution to explore the characteristics of different resources and tourists, and adopts an interactive method to dynamically adjust the recommendation results for the tourist's specific requirements.

To achieve the desirable recommendation performance, there are some challenges to address. First, tagging technique is widely accepted to describe the characteristics of resource items, but it is difficult to automatically assign different tags to individual resources without priori knowledge. Second, since the tourists are unfamiliar to the local resources, how to precisely match them to the resources related with their personal preferences is a critical problem. Tag-Tour system addresses the above challenges by proposing *tag transfer* concept and leveraging the knowledge collected from local people to achieve the tag assignment, and the resource matching is performed by an interactive process for precise recommendation.

2   System Description

Figure 1 shows the overall framework of TagTour system, which is developed based on the API of Baidu Map[1]. There are two main components: the location-based recommendation module will provide the ranked candidates according to the user's current location for quick recommendation; if the user primely selects an item or chooses to further specify her preference by some tags, the tag-based recommendation module will function to recommend refined candidates based on collaborative filtering mechanism. Then we will introduce the system components in details with relevant techniques.

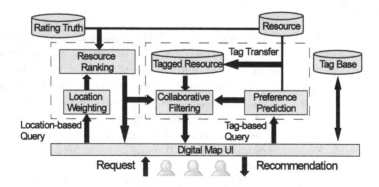

Fig. 1. System overview.

[1] http://lbsyun.baidu.com/.

Location-Based Recommendation Module. Primely, TagTour maintains two-level POI categorization of all resource items, and allows the user to select a category as request submission, e.g., "Tourist Attractions - Museum". For all resource items under a category, we evaluate their ranking based on the associated vote counts and rates, which are provided and verified by the digital map service. In order to remove the influence from different vote counts, we propose to apply Bayesian average [2] to evaluate the resource rating. For each category, all resource items will be ranked based on the rating, and the results will be stored for utilization in recommendation process.

When a request is submitted, the location-based recommendation module will perform a range query to retrieve a list of ranked resource candidates within a specified range of the user's current location. To consider the reachability, each candidate's rating score will be weighted according to its distance to the user's current location, so as to adjust the ranking and recommend the most convenient and valuable items to the user. For the results in this recommendation round, the user can select an item immediately, or choose to submit more information to the system for more precise recommendation service. At this stage, the tag-based recommendation module will take effect.

Tag-Based Recommendation Module. We adopt the tagging technique for users to provide more information due to its flexibility. However, the tag-based service usually suffers from the sparsity problem [1], so for the efficiency and users' convenience, we maintain a tag base consisting of well-defined tags for each resource category. The resource items will be assigned with different tags for more precise recommendation.

If the user chooses to submit more information for alternative choices, the tag base will provide her the tag set to check. The user can select those tags most suitable to describe her preference, and/or provide additional free tags to specify her demand. With the user's tag information, the tag-based recommendation module will carry out collaborative filtering [3] to find the best matching items from the tagged resource set. We adopt the normalized TF-IDF to generate the resource feature for similarity comparison. It should be noticed that the additional free tags specified by the user will be assigned more weight for the selection of refined candidates. After retrieving the refined candidates, they will be returned to the user according to their location-based ranking.

Now the problem is, how to assign the tags to different resource items initially? We propose a tag transfer solution, which leverages the power of local residents, since they well understand the local tourist resource. Initially, each item's tag information is empty. For a local resident, if she chooses some tags and then selects a resource item, it means these tags can probably describe this item. Therefore, if thousands of people are involved into this procedure, the tag information for each resource item can be learned during the interaction with local users in a statistical way. Based on this intuition, TagTour is primely tested among local residents, and the resource items will be assigned with different tags automatically. During the recommendation process, we also keep updating the tag base and the tag assignment for each resource item with sufficient tag-use statistics.

3   Demonstration

In this demonstration, we will exemplify the system functions by recommending the tourist resources in Wuhan City, and the resource items are retrieved from the Baidu Map API. Before the demonstration, we have initialized the tag information of all resource items by tag transfer, with the system tested among the volunteers.

The user can simulate her position in the city by locating on the digital map, and the system will provide the general recommendation near her, as shown in Fig. 2(a). The user can select an item, or choose to add some labels to express her specific preference. Figure 2(b) shows the interface through which the user can submit her preferred labels, and the system will return refined recommendation results for the user. If the user determines her choice, the system will display the details of the chosen item, with a suggested route guiding the user to the destination. Also some recommendations similar to the chosen item will be provided for alternative choice. At all stages, the user can add or change labels freely to establish real time interaction.

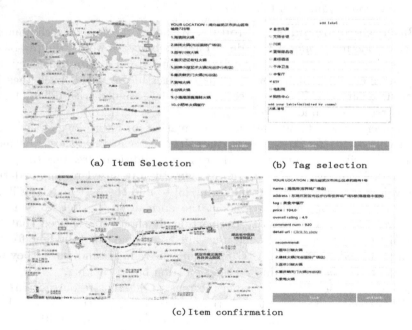

(a) Item Selection   (b) Tag selection

(c) Item confirmation

Fig. 2. Demonstration of the system workflow.

References

1. Djuana, E., Xu, Y., Li, Y., Sang, A.: A combined method for mitigating sparsity problem in tag recommendation. In: Proceedings of International Conference on System Sciences, pp. 906–915 (2014)

2. Kruschke, J.K., Aguinis, H., Joo, H.: The time has come Bayesian methods for data analysis in the organizational sciences. Organ. Res. Meth. **15**, 722–752 (2012)
3. Linden, G., Smith, B., York, J.: Amazon.com recommendations: item-to-item collaborative filtering. IEEE Internet Comput. **7**(1), 76–80 (2010)

A Chronic Disease Analysis System Based on Dirty Data Mining

Ming Sun, Hongzhi Wang[(✉)], Jianzhong Li, Hong Gao, and Shenbin Huang

Departmemt of Computer Science and Technology,
Harbin Institute of Technology, Harbin, China
{mingsun,wangzh,lijzh,honggao,huangshenbin}@hit.edu.cn

Abstract. With the rapid progress in data mining techniques, more and more systems are facing to the analysis of chronic disease because of the convenience for doctors and patients. However, low-quality data seriously leads to low-quality analysis results which may cause one's life lost. Even though many efforts have been made to enhance data quality, there always exists the data which we cannot get the exact value. Motivated by this, we develop a chronic disease analysis system adopting the mechanism that combines data cleaning and fault-tolerant data mining. In our system, we conduct a complete data mining of raw dirty data set and integrate the analysis of some kinds of chronic disease which is different to just analysis for a single disease. Moreover, our system also provides a platform for training and testing a new medical data set which is more convenient for users who do not know data mining well.

1 Introduction

With the widespread of medical needs of the big data, its unique challenge arrests more attentions. Redundancy in large-scale data declines the effectiveness and low quality brings about severe problem that the fatally wrong analysis result affects people's life. Statistics shows that $13.6\% - 81\%$ of the critical data in American Medical Information System is incomplete or obsolete [1]. Every year in America, the number of deaths due to medical errors may be high as $98,000$ [2]. However, existing medical data mining systems fail to tolerate dirty data [3–5].

To overcome these drawbacks, we develop a chronic disease analysis system that could tolerate low-quality medical data. In our system, dirty data is known inaccurate data and missing value. We add data cleaning and dirty data mining mechanisms. Firstly we repair dirty data whose true values could be found with high certainty. For the data unable to be repaired determinedly, we develop dirty data mining approaches that could discover high-quality knowledge from low-quality data. It is possible because our approach attempts to minimize the negative effect of dirty data.

The paper is organized as follows. In Sect. 2, we overview the system to preprocess data first and then train a chronic disease model with them in brief. Then, we introduce key technologies in Sect. 3, after which we discuss the plan of demonstrations in Sect. 4.

© Springer International Publishing Switzerland 2016
F. Li et al. (Eds.): APWeb 2016, Part II, LNCS 9932, pp. 552–555, 2016.
DOI: 10.1007/978-3-319-45817-5_63

2   System Overview

In this section, we introduce the architecture of our system. The organization of the system is shown in Fig. 1. The system has 2 modules.

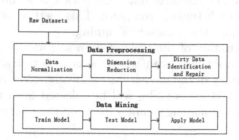

Fig. 1. System overview

In Data Preprocessing Module, raw data sets are fit into a uniform form suitable to be mined, and then dirty data will be identified and we repair data whose truth values could be found with high certainty. In Data Mining Module, in case of the dirty data which cannot be repaired in the module above affects the result much, we adopt a fault-tolerant mechanism to train a model to determine whether a user suffered from any chronic disease. The key technologies of these two modules will be introduced in next section.

3   Key Technologies

In this section, we introduce key technologies in our system including data preprocessing techniques and dirty data mining strategies.

3.1   Data Preprocessing

Before mining, we preprocess data with the order that first normalizing, then reducing the dimensions, and at last repairing dirty data as far as possible. We normalized all the data to be between 0 and 1 which is suitable for the train of neural network in data mining module. In the part of dimension reduction, we develop two methods combined which are missing value filter and low variance filter. Missing value filter reduces the dimension with a large number of missing values since such column may be useless in high possibility. Low variance filter reduces columns with small variances due to their little impact on data mining results. The column which is selected by both methods will be reduced. Then we adopt Baysian network to identify the uncertainty of missing value and inaccurate data. For the data with low uncertainty, the method in [6] will be performed to repair them. For the data with high uncertainty, we will adopt a fault-tolerant mechanism to mine the dirty data set which will be introduced in Sect. 3.2.

3.2 Data Mining

In the data mining module, we choose neural network as the mining technique since any continuous function can be approximated with a neural network with three layers at least. Via training the model with different views of the dirty data set, We can tolerant the dirty data. Each view is a complete share of the dirty data set where each tuple is complete. This method not only guarantees the effectiveness but also the accuracy of mining.

In case of the limitation of the scale of data set, we adopt a mechanism to divide structured data set into two parts. One is used to train and the other is used to test the accuracy of the model. We randomly select n samples from the data set k times, that is, nk samples will be selected from the data set to be the testing set and the remaining samples make up the training set. Accuracy is defined as follows:

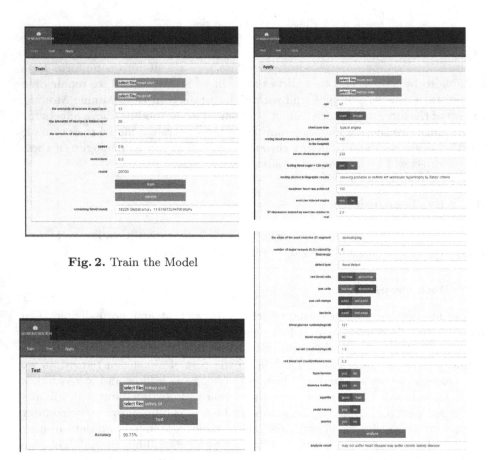

Fig. 2. Train the Model

Fig. 3. Test the Model

Fig. 4. Apply the model

$$accuracy = \frac{1}{k} \sum_{i=1}^{k} (1 - (\frac{n}{100} \sum_{j=1}^{n} |y(j) - d_j|)) \tag{1}$$

which $y(j)$ is the analysis result of model we trained for the jst sample while d_j is the reality. In our system, we set n and k as 10.

4 Demonstration

In this section, we introduce demonstration plan of our system. The features of our system will be shown in three aspects.

In the training page as shown in Fig. 2, the user can set different values of some parameters to adjust the global error of model. This page can also show that the remaining rounds to be trained and current global error which enhances the experience for users. In the testing page (Fig. 3), users can obtain the accuracy of the model after choosing the testing set and the model. In the apply page shown in Fig. 4, users enter all the information needed to be recorded, then acquire the key result which is just a direction of diagnosis and treatment, never the ultimate judgement, in spite of which, it reduces the workload of doctors.

Acknowledgement. This paper was partially supported by National Sci-Tech Support Plan 2015BAH10F01 and NSFC grant U1509216,61472099,61133002 and the Scientific Research Foundation for the Returned Overseas Chinese Scholars of Heilongjiang Provience LC2016026.

References

1. Yeast, J.: Missing prenatal records at a birth center: a communication problem quantified (2005)
2. Kohn, L.T., Corrigan, J.M., Donaldson, M.S., et al.: To err is human: building a Safer Health System (2000)
3. Knaus, W.A., Zimmerman, J.E., Wagner, D.P., Draper, E.A., Lawrence, D.E.: Apache-acute physiology, chronic health evaluation: a physiologically based classification system. Crit. Care Med. **9**, 591–597 (1981)
4. McCullough Jr., J.P.: Treatment for chronic depression: cognitive behavioral analysis system of psychotherapy (CBASP) (2003)
5. Holden, R.J., Valdez, R.S., Schubert, C.C., Thompson, M.J., Hundt, A.S.: Macroergonomic factors in the patient work system: examining the context of patients with chronic illness. Ergonomics (2016)
6. Jin, L.: Research on Missing Value Imputation of Incomplete Data. Harbin Institute of Technology, Harbin (2013)

ADDS: An Automated Disease Diagnosis-Aided System

Zhentuan Xu, Xiaoli Wang$^{(\boxtimes)}$, Yating Chen, Yangbin Pan,
Mengsang Wu, and Mengyuan Xiong

Software School of Xiamen University, Xiamen 361005, China
{1023243863,120903400,1004567472,423848331,1163864969}@qq.com,
xlwang@xmu.edu.cn

Abstract. This demo paper describes an automated disease diagnosis-aided system in healthcare (ADDS). The system contains four main components: medical knowledge base, data processing, data analytics, and interactive crowd-based feedback. First, we build a semantic-rich knowledge base using practical clinical data that are collected from several collaborated hospitals. Different from existing knowledge bases, we focus on building the personalized knowledge graph for each patient. Second, we develop an automated integration platform for patients to upload and manage their health data. The data are integrated into their clinical data, and then used as input for the data analytics tool. The automated diagnosis results are retuned and visualized to the corresponding patients and doctors. If the doctors have questions about the results, they can use our provided interactive feedback system to contact the patients. The feedbacks from expert doctors are used to improve the accuracy of the data analytics tool.

Keywords: Automated disease diagnosis-aided system · Medical knowledge base · Data analytic tool · Clinical data · Health data

1 Introduction

With the development of information age in healthcare industry, more and more healthcare data is being collected and stored into databases [1]. How to manage such large amount of data for supporting various medical applications has taken much attention from the research community. Many researchers have focused on using advanced information technology for healthcare, and the disease diagnosis-aided system is one of the most important problems (e.g., [2–7]).

As far as we know, there is no mature automated disease diagnosis-aided system, which can have practical use in healthcare. Existing systems can be classified into two categories: the computer-aided system (e.g., [2–4]) and the crowdsourcing based system [7]. The earlier computer-aided systems often suffer from very low accuracy for predicting the diagnosis results, as they use machine learning algorithms that are sensitive to the dynamic environment. To overcome this limitation, a most current system is proposed based on crowdsourcing [7]. This system introduces a new direction for improving machine learning results with the human interactive feedback. However, it also has limitation for practical use as they use out of date clinical data for prediction. To collect the real-time healthcare data from patients, several systems have focused on

© Springer International Publishing Switzerland 2016
F. Li et al. (Eds.): APWeb 2016, Part II, LNCS 9932, pp. 556–560, 2016.
DOI: 10.1007/978-3-319-45817-5_64

using sensor techniques (e.g., [5, 6]). However, such systems are currently used only for the old man remote monitoring systems.

In this paper, different from existing systems, we develop a novel system, denoted by ADDS, based on sensor techniques and crowdsourcing. First, we build personalized patient knowledge graph using their real clinical data. Second, we use sensors to collect a patient's real-time health data, and online integrate them into the patient's clinical knowledge graph. After that, we analyze the patient's most current health data, and provide diagnosis results to the doctors in the crowdsourcing system. The doctors can make final decision and send results to the patient, or interact with the patient if they have doubts about the predication results. Our main contribution can be summarized as follows.

- We build a personalized knowledge graph for each patient by using their real clinical data. Compared with existing systems, our system could help to provide more customized and accurate service for each patient.
- We integrate the patient's real-time health data into the historical clinical data for providing more accurate diagnosis predication results. This can overcome the out of date limitation for existing prediction systems by using only the historical data.
- We develop a doctor-patient interaction system based on crowdsourcing. This helps to further refine the machine predication results with human feedback. It is also very convenient for doctors to interact with the patients without using the real limited healthcare resources. This also potentially helps to reduce the rising healthcare costs of patients and improve the quality of healthcare.

2   System Architecture

Figure 1 shows our system framework with five main modules. It is implemented as a J2EE project developed by Eclipse and deployed on the Apache Tomcat.

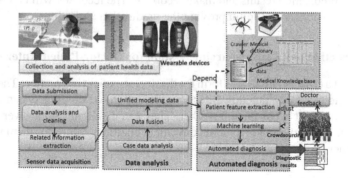

Fig. 1. System architecture

2.1   Collection and Analysis of Human Health Data

Personalized sensor is used to collect real-time health data for each patient. The collected data will be directly sent to ADDS for preliminary analysis, and the patient's profile data is extracted and integrated into the corresponding clinical data. It is noteworthy that the clinical data are collected from several collaborated hospitals.

2.2   Construction of Patients' Personalized Knowledge Graphs

The accuracy of ADDS relies heavily on the comprehensiveness of the medical knowledge base. We construct the personalized knowledge graph for each patient by using the historical clinical data. We extract the related entities and the relationships between entities according to the medical dictionaries and medical knowledge base crawled from several professional medical web sites.

2.3   Data Fusion and Unified Storage

The historical clinical data and the real-time health data are fused and modeled as one unified profile graph for each patient. Similarly, we process the real-time health data by extracting the related entities and the relationships between entities using the medical knowledge base and the constructed patient knowledge graph.

2.4   Data Analysis and Automated Disease Diagnosis

Both the clinical data and the health data of the patient are used as the input for the data analytics tool. The automated diagnosis predication results are directly sent to the doctor-patient interaction system. According to the results, warning messages are also sent to the patient if new disease condition is detected. The doctors in the crowdsourcing system can judge the predication results and make feedback. The feedback will be further used as input of the analytics tool for improving the predication algorithms

2.5   Construction of a Crowdsourcing System for Doctor-Patient Interaction

We implement a crowdsourcing platform for the doctor-patient interaction. The platform contains all the necessary functionalities in a typical crowdsourcing system. Both the doctor and the patient act as the job worker. They can distribute and answer questions in the system. We also provide a more convenient chatting feature for them to interact with each other by chatting.

3   Demonstration

We introduce demonstration plan of our system. Figure 2 shows the patient knowledge graph, which includes the related diseases, medicines, laboratory tests and symptoms. When you put the mouse on attachment, it will show the relationship between two

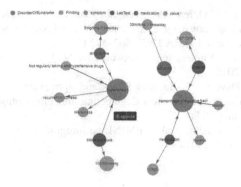

Fig. 2. Patient knowledge graph

Fig. 3. Crowdsourcing questions   **Fig. 4.** Interactive platform

entities. When an expert logins to the crowdsourcing system, he or she can see all the questions that are relevant to their expertise areas. Each question contains three radio options of "yes", "no", and "uncertain", as shown in Fig. 3. Moreover, we provide a convenient platform for doctors and patients to interact with each other via chatting, as shown in Fig. 4.

Acknowledgment. This work was supported by the Fundamental Research Funds for Central Universities (201601000679).

References

1. IBM big data for healthcare. http://www.ibm.com
2. Phuong, N.H.: Fuzzy set theory and medical expert systems: survey and model. In: Bartosek, M., Staudek, J., Wiedermann, J. (eds.) SOFSEM 1995. LNCS, vol. 1012, pp. 431–436. Springer, Heidelberg (1995)
3. Friedman, C., Alderson, P.O., Austin, J.H., Cimino, J.J., Johnson, S.B.: A general natural-language text processor for clinical radiology. JAMIA **1**(2), 161–174 (1994)

4. Savova, G.K., Masanz, J.J., Ogren, P.V., Zheng, J., Sohn, S., Schuler, K.K., Chute, C.G.: Mayo clinical text analysis and knowledge extraction system (cTAKES): architecture, component evaluation and applications. JAMIA 17(5), 507–513 (2010)
5. Alemdar, H., Ersoy, C.: Wireless sensor networks for healthcare: a survey. Comput. Netw. 54(15), 2688–2710 (2010)
6. Dimitrievski, V.P., Davcex, D.: Security issues and approaches in WSN. Department of Computer Science, Faculty of Electrical Engineering and Information Technology, Skopje, Republic of Macedonia (2011)
7. Ling, Z.L., Tran, Q.T., Fan, J., et al.: GEMINI: an integrative healthcare analytics system. PVLDB 7(23), 1766–1771 (2014)

Indoor Map Service System Based on Wechat Two-Dimensional Code

Chen Guo[✉] and Qingwu Hu

School of Remote Sensing Information Engineering, Wuhan University, Wuhan, China
{496474070,1261704852}@qq.com

Abstract. Along with the rapid development of urbanization and commercial economy, the size of buildings become bigger and bigger, and the internal structure tend to be more complex. Addressing, navigating and providing location service indoors have a good prospect in several areas such as business development, three-dimensional traffic navigation and public safety. Aiming at the requirement of more efficient and convenient indoor map service, WeChat two-dimensional code technology is introduced in this paper. The paper puts forward methods to compressing map data and storing it in the two-dimensional code effectively, and techniques to decoding two-dimensional code and displaying the map. A prototype system was developed and tested with 3 indoor maps and the method is proved to be feasible and promising.

Keywords: Indoor location service · Indoor map · Two-dimensional code · Map compression and encoding · Decoding

1 Introduction

With the widespread of GPS and outdoor positioning technology, it's convenient to get accurate position of outdoor objects and access to location service by GPS. While in the case of GPS signal barrier, it's difficult to address and navigate indoors. Thanks to development of mobile positioning technology and widespread of smartphones, indoor location service using indoor positioning technology and indoor maps are spreading. Many indoor location service methods have been developed. For instance, the Assisted Global Positioning System utilizes the mobile network and positioning server to make up the influence caused by GPS signal barrier [1]. The infrared positioning technology utilizes the miniature infrared emitter and infrared receiving station spreading all over the building to calculate user's position [2]. Ultrasonic positioning technology is similar with the infrared positioning. Ultrasonic label and receiver are used to get the position information [3]. However, above indoor location service methods are all limited by the positioning accuracy and high cost, and can't apply widely.

Wechat is a popular social platform currently developed by Tencent and Wechat two-dimensional code is a new kind of information carrier designed for realizing extra functions for Wechat, such as adding friends, Wechat Payment etc. Wechat two-dimensional code has the advantages of small size, large capacity and wide popularity. Lin proposed a method of indoor map service based on Wechat two-dimensional code, an

© Springer International Publishing Switzerland 2016
F. Li et al. (Eds.): APWeb 2016, Part II, LNCS 9932, pp. 561–564, 2016.
DOI: 10.1007/978-3-319-45817-5_65

order number of indoor map was stored in the code and when users scan the code, the server respond and send the corresponding map to the user through mobile network [4]. It's a kind of online map service and the speed of accessing location information is largely limited to server respond ability and network transmission speed, and the server is required to have a certain level of storage capacity.

Therefore, the paper propose a new indoor location service method, using Wechat two-dimensional code to store compressed indoor map directly and realizing the decoding and map reconstruction process on mobile device.

2   Framework

Instead of storing the link or an order number of an indoor map into the Wechat two-dimensional code, in the method presented, the compressed indoor map itself is stored in the code. Figure 1 shows the frame of the method presented.

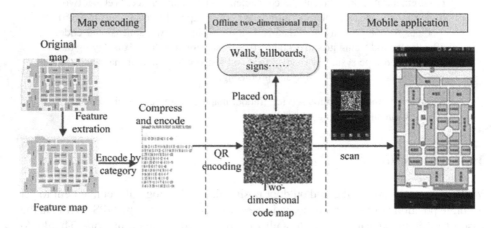

Fig. 1. The frame of indoor map service based on Wechat two-dimensional code

As is shown in Fig. 1, the map service based on Wechat two-dimensional code is divided into 3 parts: map encoding, two-dimensional code map generation and mobile application. In the map encoding part, typical elements of the map are extracted and then compressed into a character string along with the metadata of the map such as size, name, latitude and longitude, etc. Then the string is compressed to meet the capacity of two-dimensional code. In the second part, the string is transferred into a two-dimensional code map through QR encoding process and then placed on carriers like walls, billboards, signs or any other physical carrier. In the third part, a mobile application with the function of scanning two-dimensional code and reconstructing the original map is developed. When user scans the two-dimensional code map with the application, the original map will be reconstructed and displayed on the device.

3   System Test

A prototype system is developed to test the feasibility of the method presented. The prototype system contains two parts running on windows platform and android platform respectively. The first part is for map compression and two-dimensional code generation, and the second part is an application on android platform for scanning code and reconstruct and display the map. Three typical indoor maps are chosen for the experiment. Experiments on windows platform are run on a 2.60 GHz processor with 4 GB memory, and experiments on android platform are on a SONY Lt29i smartphone.

3.1   Indoor Map Encoding Experiment

Utilizing the method proposed to compress and encode the experimental map, the size of the character string and its compression ratio against the original map image is shown in Table 1, along with a contrast with RAR compression.

Table 1. Comparing the size of string before and after compression and the efficiency of method presented with RAR

Data	Size (B)	Method presented		RAR compression	
		Compressed size (B)	Ratio	Compressed size (B)	Ratio
D1[a]	66791	2066	32.3	66673	1.0002
D2[b]	14529	691	21.0	13528	1.0740
D3[c]	64319	1240	51.9	64158	1.0025

[a]D1 is the first floor of Qunguang Square in Wuhan.
[b]D2 is the T1 public zone of Capital Airport in Beijing.
[c]D3 is the Hangzhou railway station.

The result suggests the ascendancy of utilizing the method proposed for compression is obvious against direct RAR compression. And the size of string after compression is limited to 2 KB, meets the condition for storing in a WeChat two-dimensional code. The map encoding algorithm proposed is feasible for generating a character string fit the WeChat two-dimensional code capacity.

3.2   Two-Dimensional Code Map Decoding and Map Reconstruction Efficiency Experiment

Utilizing the Android application to decode the two-dimensional code map and display the original map, the result is shown in Table 2.

Table 2 shows the time consumed for scanning the two-dimensional code map and decoding it and display the original map. The result suggests with the increase of data size in two-dimensional code map, time consumed of the application also increased significantly. However, the response time is always kept within a few seconds, the efficiency of decoding and display meets the practical application requirements.

Table 2. Comparing the time consumed on scanning and decoding of different data

Data	Size (B)	Scan time (s)	Decode time (s)	Title time (s)
D1[a]	2066	2	2	4
D2[b]	691	1	1	2
D3[c]	1240	2	1	3

[a]D1 is the first floor of Qunguang Square in Wuhan.
[b]D2 is the T1 public zone of Capital Airport in Beijing.
[c]D3 is the Hangzhou railway station.

4   Conclusions

The paper introduces an indoor map service method based on WeChat two-dimensional code. In contrast with the conventional indoor location service approaches, it's more efficient and convenient, and independent of network. The prototype system developed in this paper realizes the whole process. The experimental results demonstrate the method proposed is feasible and efficient, and has a prospect in the field of business development, three-dimensional traffic navigation and public safety, etc.

References

1. Anwar, A.K., Loannis, G., Pavlidou, F.N.: Indoor location tracking using AGPS and Kalman filter. In: Workshop on Positioning, Navigation and Communication, pp. 177–181 (2009)
2. Emilsson, E., Rydell, J.: Chameleon on fire – thermal infrared indoor positioning. In: Position, Location and Navigation Symposium-Plans, pp. 637–644 (2014)
3. Gonzalez Hernandez, J.R., Bleakley, C.J.: Accuracy of spread spectrum techniques for ultrasonic indoor location. In: International Conference on Digital Signal Processing, pp. 284–287 (2007)
4. Zefei, L.: Construction of a two-dimensional barcode API for library location service based on WeChat public platform. Libr. Inf. Serv. **58**(16), 138–142 (2014)

Factorization Machine Based Business Credit Scoring by Leveraging Internet Data

Ge Zhu and Lin Li[✉]

Wuhan University of Technology, Wuhan, China
ziwuyoulin@foxmail.com, cathylilin@whut.edu.cn

Abstract. Small business accounts for an important part in national economy. However, there are many difficulties for small business to get loan from big banks. Although big banks have begun to accept small loan over the last decades, the threshold is relatively too high. Therefore, many banks have to search help from outside vendors that provide small business credit scoring service, which in return increases their costs. Existing small business credit scoring systems ignore the Internet information, especially Web. This paper introduces a factorization machine model to predict credit scores for small firms. In this model, we combine firms' basic information with Internet data. Experimental results show that our result is better than traditional linear regression. A demo system is given in the end.

1 Introduction

Traditional banks tend to loan to big companies with good financial statements and credit because of low risk, but the condition of small firms is just on the contrast. Small business usually cannot afford the cost of making certified and audited financial statements, in addition, banks need to assess the credit of small businesses themselves or buy credit scoring service from other companies, which increase the costs of banks [3]. But facing challenges from Internet finance, traditional merchant banks begin to expand their business to small business loan [5].

There are already some available business models on SBCS, but what they have in common is that they ignore the Internet data. Small businesses create a lot of information in the Internet age [4]. In this paper, we will introduce a factorization machine based scoring by combining basic company data with Internet data to improve traditional SBCS models. To make it more visualized, we have developed a web system to demonstrate the results of our model.

This research was undertaken as part of Project 15BGL048, 2015AA015403, 2015BAA072 and 61303029 and supported by Hubei Key Laboratory of Transportation Internet of Things.

F. Li et al. (Eds.): APWeb 2016, Part II, LNCS 9932, pp. 565–569, 2016.
DOI: 10.1007/978-3-319-45817-5_66

2   Our Approach

In general, our scoring approach can be divided into two parts: online part and offline part. Online part provides a query entry as shown in Sect. 3. Offline part completes jobs like data collection, data cleaning, data storage and model training.

2.1   Data Collection and Feature Extraction

(1) Data Collection: All data comes from the Internet, including company's basic data, online transaction data, product comments, and etc. Because the whole data could be very huge, we choose to collect firm's data in each province, one after another. We have collected approximately all firms' data in Jiangsu province of China so far, based on which we carry out our following research.

(2) Data Cleaning: Our collected data set is usually noisy. For example, the values of some fields are null or empty. We need to filter out these kinds of invalid data, or set null or empty field a default value.

(3) Data Storage: After the steps above, we get cleaned data set. Then we need to store them into database. We divide the data set into several fields corresponding to database fields.

(4) Feature Extraction: At present, we have 12 kinds of features extracted from Internet data. They are registered capital, place of business, date of establishment, business scope, shareholder information, branches, change information, Transaction information, administrative sanction, operation anomalies, serious offences and spot check information. We choose these features based on the assumption that they will affect on the final credit score of a company.

2.2   FM Model Training

To train our model, we use factorization machine (FM) because FMs can work with any real valued feature vector [2]. Its equation is as follows:

$$\hat{y} := \omega_0 + \sum_{i=1}^{n} \omega_i x_i + \sum_{i=1}^{n} \sum_{j=i+1}^{n} \langle V_i, V_j \rangle x_i x_j \qquad (1)$$

where

$$\langle V_i, V_j \rangle := \sum_{f=1}^{k} v_{i,f} \cdot v_{j,f} \qquad (2)$$

In our model, ω_0 can be seen as an base score that all firms share, $\sum_{i=1}^{n} \omega_i x_i$ shows each feature's effect on the final score, where x_i represents the ith feature and ω_i is its weight, and the last part demonstrates interactions between features.

We can treat this as a regression problem, so linear regression is our baseline. Here, we try the above FM model that considers the interactions between different features and use the software tool libFM [1].

Table 1. RMSE's trend with the growth of pairwise interactions

Number of pairwise interactions	REMS
0	1.3795
1	0.9265
2	0.8690
3	0.8183
4	0.7498
5	0.7473
6	0.7318
7	0.6996
8	0.6904
9	0.6812
10	0.6845
11	0.6703
12	0.6703

There is no credit score information at hand, so we have discussed with financial experts and got some rules for manual scoring. The data size of training set is not large, there are about 50000 manual scores. To avoid over fitting problems, we randomly divide these artificial data into 5 parts on average, choose four of them as training set and the remaining one as test set for 5 cross validation. Finally, root-mean-square error (RMSE) is calculated. At this evaluation proccdure, we let the number of factors used for pairwise interactions (namely the last part of the equation) changes from 0 to 12, the final result is showed in Table 1.

In Table 1, we see that as number of pairwise interactions grows, RMSE gradually falls. When number grows larger than 8, the change of RMSE becomes slowly, so we choose 8 pairwise interactions.

2.3   Experimental Results

If we do not consider the final part in the equation of FMs, it is the traditional linear regression and its RMSE is exactly shown at the first point in Table 1. As shown above, the FM model is better than linear regression in terms of RMSE.

2.4   Discussions

To ensure persuasion, we will evaluate our experiment from several aspects.

(1) Credibility of Data: As mentioned above, all firms' information comes from the Internet. More specifically, firms' essential information comes from National Enterprise Credit Information Publicity System which is an authoritative official website. The comment information of a company comes from 1688.com, a sub

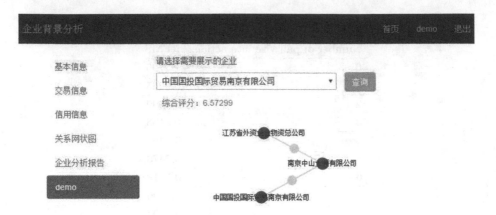

Fig. 1. Credit score and connection graph

site of Alibaba. Maybe the comments are to some extent subjective, but they still reveal something. In contrast, the overall evaluation is more persuasive.

(2) Approach Evaluation: We choose cross validation to avoid over fitting problems, and to find an appropriate parameter, we let the number of pairwise interactions changes from 0 to 12 and get 8 as final number by checking their RMSE. Compared with standard linear regression, our result shows improvement. So we believe our model has a better performance than traditional approaches.

3   Our Demo

Our demo provides query function, and users can search for firms and see their credit scores online. Figure 1 shows the html page of our credit scoring system, and a score is given. In our model, we set the score range from 0 to 10. A larger score indicates a better credit. We can also see a graph of several enterprises which are connected to one another. This is our connection graph. Two connected firms have at least two common shareholders. We create this graph to show the potential relationship between enterprises.

4   Conclusion

In this paper, we present a credit scoring system. We take information that firms create on the Internet into consideration. To train the model, we select the most important 12 features. The number of pairwise interactions between features is considered in model training and finally credit scores are predicted. Because we generate firms' credit score with our approach, it is unavoidable that the credit scores may be influenced by subjective factors. In the future, we consider to corporate with some authority departments like banks.

References

1. Rendle, S.: Factorization machines with libFM. Acm Trans. Intell. Syst. Technol. **3**(3), 219–224 (2012)
2. Rendle, S.: Factorization machines. In: The 10th IEEE International Conference on Data Mining, pp. 995-1000. IEEE Computer Society (2010)
3. Berger, A.N., Scott Frame, W: Small business credit scoring and credit availability. Federal Reserve Bank of Atlanta, pp. 5–22 (2005)
4. Eisenbeis, R.: Recent developments in the application of credit scoring techniques to the evaluation of commercial loans. IMA J. Math. Appl. Bus. Ind. **7**(4), 271–290 (1996)
5. Berger, A.N., Udell, G.F.: The Importance of Bank Organisational Structure, Finance and Economics Discussion Series with number 2001-36

OICRM: An Ontology-Based Interesting Co-location Rule Miner

Xuguang Bao, Lizhen Wang$^{(\boxtimes)}$, and Meijiao Wang

School of Information Science and Engineering, Yunnan University,
Kunming, Yunnan, China
bbaaooxx@163.com, lzhwang@ynu.edu.cn

Abstract. Spatial co-location rule mining plays an important role in spatial data mining. The usefulness of co-location rules is strongly limited by the huge amount of discovered rules. To overcome this drawback, several methods were proposed in the literatures. However, being generally based on statistical information, most of these methods do not guarantee that the extracted co-location rules are interesting for the user. Thus, it is crucial to help the decision-maker with an efficient processing step to reduce the number of co-location rules. This demonstration presents OICRM, an interactive system to discover interesting co-location rules based on the ontology. First, the ontology is used to improve the integration of user knowledge; next, a powerful formula sub-system is designed to easily represent domain's background and constraint knowledge; finally, OICRM has an interactive post-processing step (the secondary mining) to reduce the number of rules furthermore.

1 Introduction

Co-location rule mining, first introduced in [1], is considered as one of the most important tasks in Knowledge Discovery in Spatial Databases. Co-location rules are generated from co-location patterns. A spatial co-location pattern represents a subset of spatial features whose instances are frequently located in spatial neighborhoods. Among sets of features in a spatial database, co-location rule mining aims at discovering implicative tendencies that can be valuable information for the decision-maker.

A co-location rule cr is defined as the implication $cr: C_1 \rightarrow C_2(p, cp)$ where C_1 and C_2 are co-location patterns and $C_1 \cap C_2 \neq \Phi$, p is a number representing the prevalence measure of the co-location pattern $C_1 \cup C_2$ and cp is a number measuring conditional probability. Given a prevalence threshold min_prev and a conditional probability threshold min_cond, cr will be discovered if $p \geq min\_prev$ and $cp \geq min\_cond$.

The size of a co-location c is the number of spatial features in c. For a k-size co-location c, it will generate as much as $2^k - 3$ co-location rules. Thousands of co-location rules can be extracted from a spatial database of several features and several co-location patterns.

To overcome this drawback, several methods were proposed in the literatures such as co-location pattern concise representations [2], redundancy reduction [3], and post processing [4]. However, most of these methods do not guarantee that the extracted co-location rules are interesting for the user, since rule interestingness strongly

© Springer International Publishing Switzerland 2016
F. Li et al. (Eds.): APWeb 2016, Part II, LNCS 9932, pp. 570–573, 2016.
DOI: 10.1007/978-3-319-45817-5_67

depends on user knowledge and goals. A co-location rule could be interesting to one user but not to another. Thus, the co-location rule-finding methods should be imperatively based on a strong interactivity with the user.

OICRM (Ontoloy-based Interesting Co-location Rule Miner) is a system to discover interesting co-location rules interactively. In the Semantic Web field, ontologies are considered as the most appropriate representation to express the complexity of the user knowledge [5]. OICRM extends the ontology specification language as a rule schema, and provides three kinds of operators applied on the rule schema to express the user's interestingness precisely, and then OICRM can discover the interesting co-location rules for the user.

2 System Overview

As Fig. 1 shows, our system contains three major sub-systems: formula sub-system, mining sub-system and re-filter sub-system. In pre-processing step, formula sub-system uses the ontology to integrate the user's knowledge and requires the user to provide formulas based on his/her knowledge using rule schema and operators; Mining sub-system employs an improved co-location rule-finding method to mine co-location rules from the spatial database constrained by the formulas provided by the formula sub-system. The mined co-location rules are called coarse co-location rules, although they satisfy the constraint of formulas, they are required to be filtered because of the possible huge amount of them. In post-processing step, re-filter sub-system supplies two co-location rule filters to remove the redundant co-location rules in order to reduce the number of co-location rules without loss.

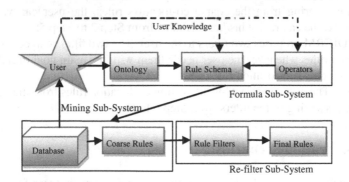

Fig. 1. OICRM description

3 Interactive Processes

The OICRM framework proposes to the user two interactive processes of co-location rule discovery. Taking into account his/her feedback, the user is able to revise his/her expectations in function of immediate results.

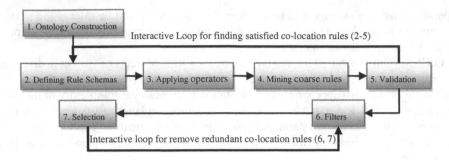

Fig. 2. Interactive processes description

As Fig. 2 shows, several steps are suggested to the user in the framework as follows:

1. *ontology construction*—starting from the spatial database or existing ontologies, the user develops an ontology based on the features of the database;
2. *defining Rule Schemas*—the user expresses his/her goals and expectations concerning the co-location rules that he/she wants to find;
3. *choosing the right operators*— three kinds of operators are provided for the user to apply on the rule schema: $P(RS)$ eliminates all co-location rules matching RS; $C(RS)$ confirms all co-location rules matching RS; $U(RS)$ proposes to filter co-location rules with a surprise effect for the user: co-location rules unexpected regarding the antecedent U_p, co-location rules unexpected regarding the consequent U_c, and co-location rules unexpected regarding both sides U_b;
4. *mining coarse co-location rules*—based on the formulas generated by Step 3, OICRM mines the coarse rules from the spatial database;
5. *validation*—staring from the coarse co-location rules, the user can validate the results or he/she can revise his/her formulas from Step 2 to Step 5;
6. *filters*—OICRM proposes two *filters* to the user. The two filters can be applied over co-location rules whenever the user needs them with the main goal of reducing the number of coarse co-location rules and
7. *selection*—The user can select his interesting co-location rules from the final results or revise his setting of the filters from Step 6 to Step 7.

4   Demonstration Scenarios

OICRM is well encapsulated with a friendly interface, what the user faces is only a simple UI. We use part of the data from points of interests in Beijing to show the Demonstration of OICRM.

Figure 3 shows the interface of OICRM. Figure 3(a) shows the main function of OICRM; Fig. 3(b) shows the tree visualization of the ontology used in OICRM; Fig. 3(c) shows the formulas provided by the user according to the user's interestingness; Fig. 3(d) shows the runtime information during the mining of coarse co-location rules; Fig. 3(e) takes the coarse co-location rules as input and proposes two filters interactively with the user to reduce the number of coarse co-location rules with no loss.

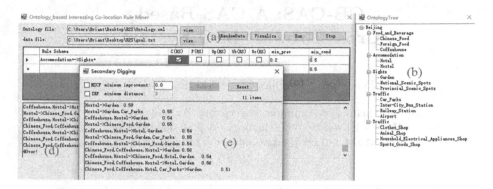

Fig. 3. Demonstration of OICRM

In Fig. 3(c), it can be seen that the Rule Schema provided by the user is RS: $Accommodation+\rightarrow Sights+$, and the selected operator is $C(RS)$, which means that the user wants to know when there exist accommodations, the probability of the existence of sights. Figure 3(e) shows that only eleven rules are discovered for the user, which can reduce the burden of the user effectively.

Acknowledgements. This work was supported in part by grants (No. 61472346, No. 61262069) from the National Natural Science Foundation of China and in part by a grant (No. 2016FA026, No. 2015FB149, and No. 2015FB114) from the Science Foundation of Yunnan Province.

References

1. Huang, Y., Shekhar, S., Xiong, H.: Discovering co-location patterns from spatial data sets: a general approach. IEEE Trans. Knowl. Data Eng. (TKDE) **16**(12), 1472–1485 (2004)
2. Wang, L., Zhou, L., Lu, J., et al.: An order-clique-based approach for mining maximal co-locations. Inf. Sci. **179**(2009), 3370–3382 (2009)
3. Xin, D., Shen, X., Mei, Q., et al.: Discovering interesting patterns through user's interactive feedback. In: 12th ACM SIGKDD International Conference on Knowledge Discovery and Data Mining, pp. 773–778 (2006)
4. Bao, X., Wang, L., Fang, Y.: OSCRM: a framework of ontology-based spatial co-location rule mining. J. Comput. Res. Dev. **52**(Suppl), 74–80 (2015)
5. Marinica, C., Guillet, F.: Knowledge-based interactive postmining of association rules using ontologies. IEEE Trans. Knowl. Data Eng. **22**(6), 784–797 (2010)

CB-CAS: A CAS-Based Cross-Browser SSO System

Peng Gao, Yongjian Liu, and Qing Xie$^{(\boxtimes)}$

School of Computer Science and Technology,
Wuhan University of Technology, Wuhan 430070, China
{gaop1024,felixxq}@whut.edu.cn, liuyj626@163.com

Abstract. With the internal business structures of enterprise becoming more and more complex and the increasing applications, SSO (Single Sign On) has been increasingly applied to integrate enterprise business and information resources. We only need to login once by using SSO to access a variety of applications within the enterprise. This will greatly reduce the burden of the user's memory, and improve the management efficiency of enterprises because administrators only need to manage one set of employee information. However, currently available SSO systems have a common flaw, that they do not support cross-browser. If the browser compatibility of enterprise applications is not sufficient to enable multiple systems operating within the same browser, the function of SSO will not be able to show, so building a cross-browser SSO system is particularly necessary. In this demo we will introduce a Cross-Browser SSO system based on CAS. We first introduce the core technique of our system, and then show the system operating procedures.

Keywords: SSO system · Cross browser · CAS

1 Introduction

Nowadays, modern enterprises established a variety of systems such as financial system, personnel management system and OA system. The design methods and development techniques of these systems are usually different, which leads the enterprise to form multiple sets of user information database and a plurality of user authentication systems. To improve the work efficiency, enterprises raised the demand of establishing a unified database of user information and user authentication system, and the SSO technique has been designed to meet the demand. If SSO system is deployed in a plurality of trusted applications, users only need to login one of these systems, and there is no need to login again when accessing any other systems. It centralizes the distributed multiple user information databases to manage, and each system needs to access a unified user authentication system for authentication. As the user information is stored centrally, administrators only need to operate once when managing user information, without separately accessing multiple systems, which greatly improves the work efficiency.

F. Li et al. (Eds.): APWeb 2016, Part II, LNCS 9932, pp. 574–578, 2016.
DOI: 10.1007/978-3-319-45817-5_68

However, the application systems within an enterprise are usually built in different periods and different environments, and browser compatibility of each application system is different, so when building SSO system, this situation should be taken into account so as to make the SSO system supporting cross-browser operations. However, conventional SSO technique stores user's authentication information in the HTTP Cookie [2], which can not be shared between browsers. In this demo, we will introduce CB-CAS: a cross-browser SSO system based on CAS, which is currently the best open source SSO protocol. CB-CAS utilizes Flash Cookie which is more powerful than browser HTTP Cookie, and extracts the cookie from the browser to create a cross-browser SSO.

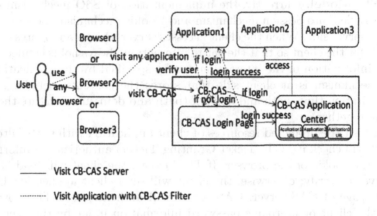

Fig. 1. System workflow.

2   System Overview

Figure 1 shows the system workflow, and users can operate the CB-CAS SSO system in two ways: Visit CB-CAS Server and Visit Application with CB-CAS Filter:

- Visit CB-CAS Server: users type CB-CAS Server URL directly in the browser (assuming the first-time visit), and access the polymerization center of enterprise application system (CB-CAS Application Center), which contains all client applications with CB-CAS Filter, by entering the username and password in the CB-CAS Login Page for identity verification. Users can access enterprise applications by clicking these applications within a URL link. At this time, no matter for which CB-CAS-related system or in which browser users visit, they will be already logged in as the role of verified users, until the cache of identification information in this PC is invalid (the cache will be expired at default cache expiration time or when exit buttons in these systems being clicked).

- Visit Application with CB-CAS Filter: users type the URL of an application with CB-CAS Filter in the browser (assuming the first-time visit), and it will be redirected to CB-CAS Login Page for identity verification since the users are not logged in. After verifying successfully, it will be redirected back to the application which users wanted to visit at first as the role of verified users. Then no matter for which application with CB-CAS Filter or in which browser the user visits, the user will be already logged in until the cache of identification information in this PC is invalid.

CB-CAS is a cross-browser SSO system that is based on CAS SSO model and introduces Flash Cookie caching mechanism.

CAS SSO Model. Currently, the implementation of SSO mechanism can be roughly divided into Session mechanism and Cookie mechanism. Session mechanism is at server side. When the client accesses server, the server creates a unique SessionID for the client so that the entire interaction always holds connected status, and information of the interaction can be specified by the applications [1]. Cookie mechanism is at client side that stores the contents including: name, value, expiration time, path and domain (path and domain constitute the scope of Cookie together).

Users visit the protected resources of client application with CAS Filter, and the client will check if TGT (Ticket Granting Ticket) authorization information exists in the cookie of the browser. If TGT exists, the client will send it to the CAS server to verify; otherwise, the client will send client application URL to the login page of CAS server. CAS server will authenticate TGT information sent by the client or username-password information input by the user. If the authentication is successful, CAS server will redirect to the client and return successful verification information; or else the user is prompted to re-enter username and password. At this point, if users are the first time to authenticate, the client will store the TGT information into the browser cookie, and users can access the protected resources of any application with CAS Filter.

Flash Cookie Caching Mechanism. CB-CAS SSO system introduces Flash Cookie which is more flexible than conventional HTTP Cookie caching mechanism to realize cross-browser SSO. Flash Cookie is a client-shared storage technology controlled by Flash Player, which is realized through SharedObject that provides a method to store the local data. These local data are like HTTP Cookie stored in a Web browser on the client machine, and usually be Flash Cookie. Compared with HTTP Cookie, the advantages of Flash Cookie are:

- cross-browser: No matter how many different versions of browser installed on a user's computer, Flash Cookie can make all browsers sharing the same Cookie;
- larger capacity: Flash Cookie can accommodate up to 100 KB of data, while a standard HTTP Cookie can only accommodate 4 KB;
- privacy: Flash Cookie is stored in different locations in users computer, making it difficult to find.

Flash Cookie can be managed just like HTTP Cookie, and users can disable or remove Flash Cookie in the Flash Player of their computer, so Flash Cookie is secure. Meanwhile, CB-CAS system provides an exit button to log out of system and makes user authentication information stored in Flash Cookie out of date, so there is no need for users to worry about leakage of their information.

3 Demonstration

In this demonstration, we will show the function of CB-CAS. Two simple applications are created to simulate the enterprises application system: *Application A* developed by .NET and *Application B* developed by Java Web. We embed CB-CAS Filter into the two applications, and the filter will be valid when visiting these two applications.

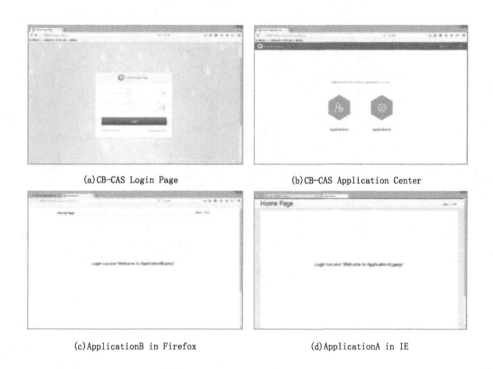

(a) CB-CAS Login Page   (b) CB-CAS Application Center

(c) ApplicationB in Firefox   (d) ApplicationA in IE

Fig. 2. Visit CB-CAS Server.

Primely, we type CB-CAS Servers URL in Firefox to enter CB-CAS Login Page as shown in Fig. 2(a), and then input username and password to verify. After verifying successfully, we enter CB-CAS Application Center as shown in Fig. 2(b). In this page, there are two applications we can use. When we click *Application B*, it is found that *Application B* has been logged in as shown in Fig. 2(c). To verify the function of cross-browser, we type the URL of *Application A* in the IE browser, and it will be seen that *Application A* is also in status of login as shown in Fig. 2(d), which means the cross-browser function works.

References

1. Ayadi, I., Serhrouchni, A., Pujolle, G., Simoni, N.: Http session management: architecture and cookies security. In: Proceedings of Conference on Network and Information Systems Security, pp. 1–7 (2011)
2. Kormann, D.P., Rubin, A.D.: Risks of the passport single signon protocol. Comput. Netw. Int. J. Comput. Telecommun. Netw. **33**, 51–58 (2000)

A Demonstration of QA System Based on Knowledge Base

Zhenjiang Dong[1], Hong Chen[1], Jingqiang Chen[2], Huakang Li[2], and Tao Li[2(✉)]

[1] Cloud Computing and IT Product Operation ZTE Corporation,
Nanjing 210012, China
[2] Institute of Computer Technology, Nanjing University of Posts
and Telecommunications, Nanjing 210023, China
`towerlee@njupt.edu.cn`

Abstract. QA service is one important service for Telecommunication, while it's very difficult because of the variety of question formations. This paper designs and implements an automatic QA system based on knowledge base. We first construct a knowledge base using the Entity-Attribute-Value model. We collect the entities, the corresponding attributes and values from a corpus in Telecommunication domain. We analyze customers' questions, extract entities and attributes from the questions, get the question type to retrieve answers from the knowledge base. The results show our system work well especially for multi-questions.

Keywords: QA system · Knowledge base · Entity · Multi-questions

1 Introduction

Automatic Question Answer (QA) system is useful and necessary for Telecommunication domain. Customers' questions are analogous and repetitive and answered automatically. Most of the questions are about the services and the telephone fare, so a knowledge base can be built to model the knowledge in telecommunication domain.

In telecommunication domain, most of the customers' questions are like "how to attend to the Bestpay service" or "what is the fare rate of the Easyown service". The two questions are about "Bestpay service" and "Easyown service" of China Telecommunications company. The two services both has the properties of fare, rate of flow, short message, etc. And each property has its corresponding values. Accordingly, we can build a knowledge base about the services to answer the questions by retrieving properties and values automatically. Therefore, to develop such a QA system, we need to deal with the following two key problems.

The first problem is to build a knowledge base in Telecommunication domain to store complex structured and unstructured information used by a computer system [1]. It has been studied for long history, and there are many

© Springer International Publishing Switzerland 2016
F. Li et al. (Eds.): APWeb 2016, Part II, LNCS 9932, pp. 579–582, 2016.
DOI: 10.1007/978-3-319-45817-5_69

techniques to describe a knowledge base [2–4]. In our system, we use the Entity-Attribute-Value model (EAV for short) to represent services in Telecommunication domain. EAV is a data model to encode, in a space-efficient manner, entities where the number of attributes that can be used to describe them is potentially vast [5]. EAV model is suitable for our QA system. Represent the services in telecommunication domain, and the properties are the attributes of the services. We build EAV model in Telecommunication domain from FAQ and China Mobile 10000 manual question answer.

The second problem is to analyze customers' questions. We focuses on customers' questions about services in the Telecommunication domain. We first get the question type of the questions, and normalize the questions. We then extract the entities and attributes from the questions. Finally, we get the values from the knowledge base and construct the answers which are then returned to askers

In the following, we give the architecture of the QA system, describe each part and give the experiment results.

2   System Architecture Design

Our QA system is called TELEQASYS. Figure 1 shows the architecture of the system. It comprises three module: Knowledge Base Creating Module, Question Processing Module, Answer Creating Module.

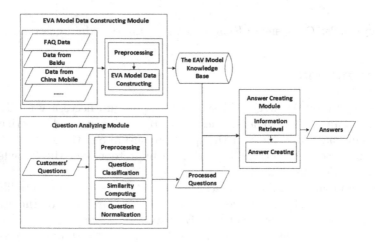

Fig. 1. The architecture of TELEQASYS

Knowledge Base Creating Module aims to build the EAV model of services in Telecommunication Domain from the corpus collected from FAQ data of China Mobile, the data crawled from the Baidu search engine, and etc. We first pre-process the collected text data through noise cleaning, sentence splitting, word segmentation, POS tagging and named entity recognition. The tools we used are StanfordCoreNLP and HaNLP. HaNLP is to process Chinese language. The excel-format text data provided by China Mobile is semi-structured. We collect entities, attributes and values from this data, and save the results in xml format. New entities and corresponding attributes are collected from crawled text data, and are updated periodically. The relationships among entities and attributes are also discovered. The relationships between entities include the co-occurrence relation and the sequential relation. The relationships between entities and attributes are belonging relations. We use Semantic Link Network model [6] to describe the relationships. Reasoning can be carried out over a semantic link network based on the rules on relations. The EAV model is stored in xml format.

Question Processing Module aims to process customers' questions and to automatically give answers by retrieve information from the EAV model knowledge base. Firstly, we preprocess the questions which are input by customers through the input interface. We split complex question into simple questions. A complex question usually contain two or more simple questions. Secondly, we extract the entities and the attributes from the questions based on the EAV model knowledge base. Customers often use different words to refer to same entities, so a thesaurus is used to transform the entities and the attributes into the ones in the knowledge base. The thesaurus is built in advance using the corpus provided by China Mobile. Thirdly, we normalize the questions. One question can be asked in different forms. For example, "What is the fare rate of Bestpay service" and "Can you tell me the fare rate of Bestpay service" share the same meaning. However, "Who is Bush's son" and "Whose son Bush is" have different meanings. To solve the problem, we first compute the similarity between two questions using Morphological similarity and Structure similarity, and then compute the semantic similarity based on the dependency tree. The questions are normalized through the two steps. Based on the processed questions in Question Processing Module, we get what does the customer ask.

Answer Creating Module aims to give answers to questions. Firstly, we compare the questions with the ones in FAQ through similarity computing. If a similar question exists, we return the corresponding answers in FAQ. Secondly, if similar questions does not exist, we need to create the answers. Based on the results in the Question Processing Module, we retrieve attributes and corresponding values from the EAV model knowledge base. For example, if customers ask "What is Bestpay service", we can return its fare rate, flow rate and etc. If they ask "What is the fare rate of Bestpay service", we can return its fare rate. We can return just answers, and the answers need not to be complete sentences.

3   Experiment

Figure 2 demonstrates one experimental result that user submitted tow questions at once. Our system separated the two question and the second question is complemented with the Knowledge base with context information. More demonstration could be referred with 'http://bigdata.njupt.edu.cn/project1/'.

Fig. 2. One demonstration result for online TELEQASYS.

4   Conclusion

This paper designs and implements an automatic Question-Answer system in Telecommunication Domain. The system comprise three parts: Knowledge Base Creating Module, Question Processing Module and Answer Creating Module. Firstly, we use Entity-Attribute-Value model to build the knowledge base and store the data in xml format. Secondly, we process the customers' questions into computer understandable format by entity extracting and question normalization. Finally, we create answers through answer searching in FAQ or attributes and values retrieval in EAV model knowledge base.

References

1. Green, C., Luckham, D., Balzer, R., Cheatham, T., Rich, C.: Report on a knowledge-based software assistant. In: Readings in Artificial Intelligence, Software Engineering, pp. 377–428. Morgan Kaufmann (1986). Accessed 1 Dec 2013
2. Bao, J., Duan, N., Zhou, M., et al.: Knowledge-based question answering as machine translation. Cell **2**(6) (2014)
3. Shenghuo, Z., Tao, L., Zhiyuan, C., Dingding, W., Yihong, G.: Dynamic active probing of helpdesk databases. In: Proceedings of the 34th International Conference on Very Large Data Bases (VLDB), pp. 748–760 (2008)
4. Wang, D., Li, T., Zhu, S., Gong, Y.: iHelp: an intelligent online helpdesk system. IEEE Trans. Syst. Man Cybern. Part B **41**(1), 173–182 (2011)
5. Marenco, L., Tosches, N., Crasto, C., Shepherd, G., Miller, P.L., Nadkarni, P.M.: Achieving evolvable web-database bioscience applications using the EAV, CR framework: recent advances. J. Am. Med. Inform. Assoc. **10**(5), 444–453 (2003). Epub 4. PubMed PMID 12807806; PubMed Central PMCID: PMC212781., June 2003
6. Zhuge, H.: The Knowledge Grid: Toward Cyber-Physical Society, 2nd edn. World Scientific, Singapore (2012)

An Alarming and Prediction System for Infections Disease Based on Combined Models

Jiahong Li, Hongzhi Wang[✉], Shengqiang Zhang, Xiangyu Gao,
Ziqi Qu, and Shenbin Huang

Department of Computer Science and Technology,
Harbin Institute of Technology, Harbin, China
lijh0226@163.com, qzq92900@126.com,
{zhangsq5829,1137744315}@qq.com,
{huangshenbin,wangzh}@hit.edu.cn

Abstract. With the continuous improvement of the economy and the medical technology, people's living conditions and sanitation facilities have been greatly improved. However, the human infectious disease with a higher prevalence can cause sudden public health incidents. They are more difficult to control and can easily cause public panic disorders. Early identification of infectious disease out breaks and takes prompt and effective measures in a timely, and can greatly reduce morbidity and mortality of infectious disease and loss of property. Therefore, the monitoring of infectious disease, early detection of infectious disease outbreaks of dangerous diseases to make early alarming and prediction is the focus of attention and research. Motivated by this, we develop the alarming and prediction system for infection diseases. Such system integrates the data collected by hospital, pharmacies, and other infectious disease control monitoring system as well as Internet public opinion. And then through data processing, mining analysis, monitoring public opinion, as we as the integration of BP artificial neural network modes, SIR model and the complex network model, the system provides early Alarming and Prediction and forecasting functions of infectious diseases.

Keywords: Public opinion mining · BP artificial neural model · SIR model · Complex network model

1 Introduction

In the field of public health, Alarming and Prediction is defined as that in the absence of adequate dose and the lake of determination of causality-reflect the relationship under the evidence, to take measures to promote a method that is before adjusting behavior or prevent the occurrence of environmental threats. Through the collection, collation and analysis of information about the infectious disease, assess time trends and the degree of harm before the event occurs, alarm before or early in the event, so that the responsible departments and incident target population can respond in a timely manner to prevent infection or to reduce the risk of the diseases.

© Springer International Publishing Switzerland 2016
F. Li et al. (Eds.): APWeb 2016, Part II, LNCS 9932, pp. 583–587, 2016.
DOI: 10.1007/978-3-319-45817-5_70

Existing alarming and prediction systems mainly use the information from social media [1]. Even though for some case, they are effective. Due to the bias of the information from social media, they fail to predict the infectious disease accurately and completely. Motivated by this, we develop a novel infectious diseases alarming and prediction system, which use sufficient data sources including not only social media but also medical records and inspection reports integrated from hospital, pharmacies, and other infectious disease control monitoring system.

For the accurate prediction, our system extracts features from multiple data sources, and obtains preliminary results of the number of population in the area, to determine the proportion of various groups. By the statistics of the areas between the crowd flow information, the system calculates the number of people in a certain area in the next period of time. Thus, it is determined whether a region will be the trend of the epidemic situation of infectious disease outbreak. The area which exceeds the alarming threshold is issued an alarming and prediction signal. With such signal, the government and medical mechanisms are advised to take the necessary action to control the epidemic spread.

This system has the following characteristics:

Accuracy: The system uses large amounts of data from public network, data mining in order to achieve accurate analysis and forecasting.

Based on combined models: Our system involves a novel model as a combination of multiple models for accurate prediction. Such a model has a good portability, and can be applied in other areas, such as predicting advertising point and the trend of network public opinion.

User-friendly: The system provides users with a simple interface, which is convenient and easy to understand.

2 System Architecture

The architecture of our system is shown in Fig. 1.

Our system has four modules:

- Data obtaining and sorting: Our system collects data from multiple data sources including micro-blog and existing information systems. It extracts keywords data from website automatically by web collector. It can find the most important part of data by using probability model and pattern information.
- Alarming: When the keywords are extracted, the type of infectious disease is determined, and then the prediction is performed based on them. A threshold is set for a certain type of infectious disease for each city. The initial numbers of infected, susceptible infected people are estimated with artificial neural network. If the number of infected people reaches the threshold, the alarming will be given.
- Prediction: Because of population movements, starting from the alert point, infectious disease spreads out through the city transportation network gradually. Our system will use SIR model algorithm [2] of the people who are infected, susceptible or recovered from the disease in each city adjacent to transport network. These data

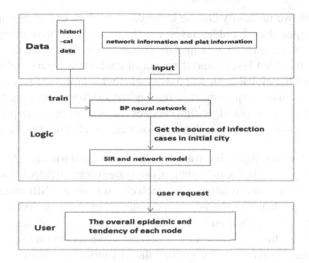

Fig. 1. System architecture

are updated every day. The system can also predict the propagation tendency of various infectious diseases throughout the topological network, as well as the propagation tendency in a single city, then analyze and give the corresponding disease control advise.

- User interface: User can know the visualized output from algorithm in the interface and appoint the cities.

3 Key Technologies

Our system aims on life-long immunity diseases, likes mumps for example. It integrates data from hospitals, drugstores and CDC. These data are stored in the local database. Then a precise BP neural network is trained using he data in the local database. After training, this system will monitor and mine Internet public opinion information and integrate data from disease control and detection system. These data are converted to the input data of the BP neural network in order to predict the epidemic situation of infectious diseases in the surrounding cities. In addition, our system provides a friendly interactive interface for users to acquaint the epidemic situation.

Infection Source City Discovery Based on BP Neural Network. There is no unified method of determining the layer number of the BP neural network. However, most people think that comparing with one hidden layers, BP neural network with two hidden layers are easier to fall into local optimal solution which may cause errors in prediction [3]. Therefore, we choose three-layer BP neural network including one input layer, one hidden layer and one output layer. Initially, we set each connection weight a random number between -1 and 1 and set up an error function e, quarter value ε and maximum training times M. we randomly choose k samples and their expected output, compute the input and output of each hidden layer, and the partial derivative of error function e to each output layers. And then we adjust the connection weight using the output of each output layers and hidden layers, and the input of each input layers and hidden layers. Next, it computes the global error and judges whether it meets the requirements. If the error meets the accuracy requirement or reaches the maximum training time, the algorithm finishes. Otherwise, our algorithm chooses the next training sample and excepted output, trains it again. Finally, we obtain the prevalence of illness of the city.

Prediction of Epidemic Situation Based on the Combined Model. According to city personnel exchanges and traffic information, our system draws a directed weighted graph to describe the roads between cities. Then each city is used as a SIR-model calculating space. The total number N of the area is a constant during the period of disease transmission and the time is measured in day. We set the proportion of healthy persons and patients at the initial time is $s_0(s_0 > 0)$ and $i_0(i_0 > 0)$. We set the number of out-migrant r_0 at the initial time as zero. Then we obtain the SIR model as follows.

$$\begin{cases} \dfrac{ds}{dt} = -\lambda si, & s(0) = s_0 \\ \dfrac{di}{dt} = \lambda si - \mu i, & i(0) = i_0 \\ \dfrac{dr}{dt} = \mu i, & r(0) = 0 \end{cases}$$

Then the numerical solution is obtained by numerical method. The $s(t) \sim i(t)$ and $s \sim i$ curve of the SIR model are obtained. Next we obtain the overall situation of the epidemic spread of the topological map and the development of the epidemic situation in the city by using this curve. Therefore, we can take effective control measures in time. Then, we model the net migration rate of population movement as a matrix and use the population of illness as an input vector

4　Demonstration

The map shown on the interface includes all of the cities in algorithm network. If a city suffer reach the threshold, Alarming and Prediction signal will be generated automatically. Users can input to query one of the cities, and be shown with four days of the epidemic situation of infectious diseases in future.

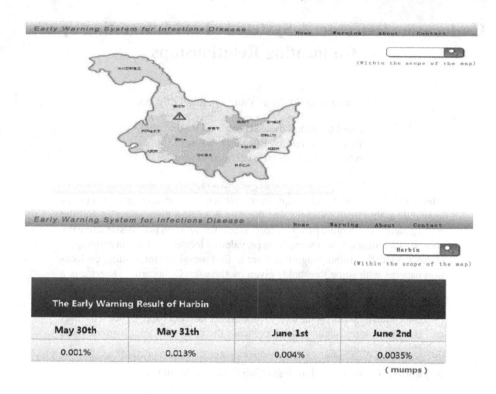

In the process of system to complete, we have done a lot of demand survey, to design the basic functions of the system. This system has a friendly user interface and clear function, but also need further development, to strengthen the user interaction and prediction accuracy.

Acknowledgement. This paper was partially supported by National Sci-Tech Support Plan 2015BAH10F01 and NSFC grant U1509216, 61472099, 61133002 and the Scientific Research Foundation for the Returned Overseas Chinese Scholars of Heilongjiang Province LC2016026.

References

1. 陈碧云, 高立冬, 陈长等. 我国传染病预警研究及工作现况. Pract. Prev. Med. **21**(12) (2014)
2. Yu, L.: To predict the number Research of Epidemic Spread Model. 计算机仿真, April 2007
3. Hou, R.: Application of back propagation artificial neural network in forecasting of infectious diseases. Science of Travel Medicine, June 2008

Co-location Detector: A System to Find Interesting Spatial Co-locating Relationships

Xuguang Bao, Lizhen Wang$^{(\boxtimes)}$, and Qing Xiao

School of Information Science and Engineering,
Yunnan University, Kunming, Yunnan, China
bbaaooxx@163.com, lzhwang@ynu.edu.cn

Abstract. Data Mining develops from original transactional data to current spatial data, this trend indicates that the data is getting more complex and the mining algorithms require better performances. Co-location patterns describe the subsets of features whose instances are prevalently located together in geographic space. Co-location mining algorithms are to find prevalent (interesting) co-location patterns with some thresholds given by the user. Co-location Detector is a system which improves the join-less algorithm and optimizes some details, it owns friendly interactive interface and good operational experiences, visualizes the co-location patterns for the user to process the next decision, besides, the user can change his input parameters to compare the results in order to mine more valuable information.

Keywords: Co-location · Join-less · User decision · Visualize

1 Introduction

The rapid growth of spatial data and widespread use of spatial databases indicate the need for the discovery of spatial knowledge. Spatial data mining is the process to discover interesting and previously unknown, but potentially useful patterns from spatial databases. Extracting interesting co-location patterns is more difficult than extracting the corresponding patterns from traditional numeric and categorical data due to the complexity of spatial data types, spatial relationships and so on.

Co-location patterns describe the subsets of features (objects) whose instances are prevalently located together in geographic space. Extracting interesting and useful co-location patterns from massive spatial datasets is important in many application domains, such as ecology, public health and homeland defense, etc.

Suppose F is a set of spatial features, S is a set of their instances and R is a spatial neighbor relationship over S. If the Euclidean metric is used as the neighbor relationship R, two spatial instances are neighbors if the ordinary distance between them is no more than a given distance threshold d. A co-location pattern c ($c \subseteq F$) is a subset of spatial features whose instances form cliques under the neighbor relationship frequently. In co-location pattern mining, participation index (PI) is often used as a measure of the prevalence of a co-location pattern. Given a minimum prevalence threshold min_prev, a co-location c is a **prevalent co-location** if $PI(c) \geq min\_prev$ holds.

© Springer International Publishing Switzerland 2016
F. Li et al. (Eds.): APWeb 2016, Part II, LNCS 9932, pp. 588–591, 2016.
DOI: 10.1007/978-3-319-45817-5_71

Co-location mining plays an important role in spatial data mining. [1] proposed a join-based co-location mining algorithm similar to apriori-gen [2]. [3] gave a join-less co-location mining algorithm to reduce expensive join operations used for finding co-location instances in join-based by materializing the neighbor relationships of a spatial dataset.

We present a Co-location Detector System based on join-less to find prevalent co-location patterns and try our best to reduce the cost of time and space. In this system, the user can load data from file and watch the data at any time. Also, it provides a visualized graph for the user to see the neighbor relationships of any instance. Besides, the user can stop the process and change his parameters setting to restart the progress in order to get more useful information.

2 System Overview

Our system is performed as follows: First, collecting data and setting parameters. Second, generating candidates and mining prevalent co-location patterns size by size until there is no candidate generated, and then outputting the prevalent co-location patterns per size. Finally, the co-location rules are generated.

As shown in Fig. 1, there are five modules in Co-location Detector. In storage module, the original data are read and shown for the user. Besides, the data and configuration modified by the user can be saved to out-side files or databases through storage module. Data module receives data and configuration module receives configuration from storage module, if there is no data or configuration at all, default configuration is used. After data generated, graph module will run automatically to draw graph based on data and configuration. Method module runs improved join-less algorithm using the data and configuration to mine prevalent co-location patterns, when the mining process finished, it shows the user the prevalent co-location patterns and rules and the certain instances. The user can modify his configuration at any time, interrupt current running method, change data or configuration and restart the algorithm.

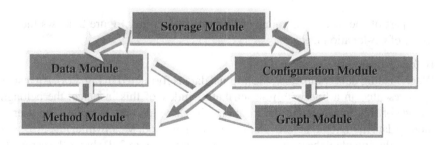

Fig. 1. Modules of Co-location-Detector

3   Challenges and Approaches

In this section, we discuss our implementation details coping with some major challenges in Co-location Detector.

– **Multi-thread Programming**
  Co-location Detector uses multi-thread programming to improve operates experiences for the user. When the thread for generating data finished, a new thread will start to draw graph. When the user wants to start the main mining algorithm, a main thread will start to run the main algorithm, during this main thread, some information about progress, status, current prevalent patterns, instances and rules need to be printed. If printing information once generating a prevalent pattern, it will take quite a long time on UI operations. Thus, Co-location Detector delays the printing until the prevalent patterns with the same sizes generated.
– **Generating neighborhood relationships**
  A new method is proposed to generate neighborhood relationships. First, the whole space is divided into several $d * d$ cells, where d is a distance threshold. Then every instance has its unique cell and every cell has some instances. In order to find a certain instance's neighbors within d, the method only locates its cell A and find the instances around A, that is, the instances in the squares which are A's 8 directions (north, northeast, east, southeast, south, southwest, west, northwest) and A self.
– **Drawing graph efficiently**
  If the number of instances is too big to show all of the instances in the graph clearly, it is completely useless to waste so much time on drawing them. Thus, an approximate method is proposed to show instances. The system just collects all instances of each feature and calculates their average location as per feature's location. Besides, the user can check any instance's neighbors by selecting an instance and the graph will dynamically refresh to show its relations.

4   Demonstration Scenarios

We use part of the data from points of interests in Beijing. Figure 2 shows the main interface of Co-location Detector.

– **Before running the main method**
  When loading data completely, the original data is formatted in Fig. 2(a). An instance is represented in a line with four attributes: index of this instance, the belonging feature of this instance and the location with X value and Y value. The user can modify data in Fig. 2(a) before running the main method. Meanwhile, Fig. 2(b) shows the distribution of all the instances of the input data. Figure 2(d) shows the parameters set by the use, the user can modify these parameters at any time, and the next running on the data will be based on the current configuration.
– **While running the main method**
  Figure 2(c) shows the run-time information while running the main method including progress, discovered frequent co-location patterns and rules. Status bar shows more

detail information and percentage of current operation. If the user wants to restart a new method, he/she just interrupts the current method and runs the method again at any time.

– **After running the main method**

The main method takes the data in Fig. 2(a) and mines the prevalent co-location patterns based on the parameters in Fig. 2(d). The final results are shown in Fig. 2(c). In Fig. 2(c), it can be seen that the maximal size of the mined co-locations is five, {Car Parks, Hotel, Hostel, Chinese Food, Garden} is one of the 5-size prevalent co-location patterns which means that the five features are frequently located together in this area.

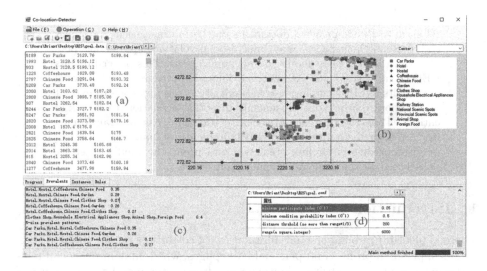

Fig. 2. Demonstration of Co-location Detector

Acknowledgements. This work was supported in part by grants (No. 61472346, No. 61262069) from the National Natural Science Foundation of China and in part by a grant (No. 2016FA026, No. 2015FB149, and No. 2015FB114) from the Science Foundation of Yunnan Province.

References

1. Huang, Y., Shekhar, S., Xiong, H.: Discovering co-location patterns from spatial data sets: a general approach. IEEE Trans. Knowl. Data Eng. (TKDE) **16**(12), 1472–1485 (2004)
2. Agrawal, R., Skrikant, R.: Fast algorithms for mining association rules. In: The 20th International Conference on Very Large Data Bases, pp. 487–499 (1994)
3. Yoo, J.S., Shekhar, S.: A joinless approach for mining spatial collocation patterns. IEEE Trans. Knowl. Data Eng. (TKDE) **18**(10), 1323–1337 (2006)

KEIPD: Knowledge Extraction and Inference System for Personal Documents

Zhaoyang Lv[1], Yuanyuan Liu[1], and Xiaohui Yu[1,2(✉)]

[1] School of Computer Science and Technology, Shandong University, Jinan, China
{zhylv,yuanyliu}@mail.sdu.edu.cn
[2] School of Information Technology, York University, Toronto, ON, Canada
xyu@sdu.edu.cn

Abstract. Public personal documents on the Internet, such as resumes and personal homepages, may imply social relationships among people, which is of great value in various applications. This paper presents KEIPD, a system to extract and infer knowledge from personal documents. KEIPD employs a tree-similarity based approach to extract information from personal documents to obtain a relational network of entities. Then the inference of social relationships can be transformed into a link prediction problem. KEIPD implements some popular unsupervised predictors for link prediction and prune the candidate entity pairs based on the domain-dependent constraint.

1 Introduction

There is plentiful public personal information of celebrities on the Internet, e.g. resumes, personal profiles and personal homepages, which may imply social relationships among the celebrities. For example, two people may be schoolmates if they have been studied in the same university during an overlapped time period. This information can be organized as a social network to support community discovery, most influential nodes discovery and other researches. Compared with traditional social networks, it has some distinguished characteristics: First, links in this network may represent various types of relationships (e.g. schoolmates and colleagues) rather than homogeneous relationships. Second, the network is more realistic where the links are deduced from the factual experiences of people instead of the interaction data of users via a social application. Third, the formation of a link is sensitive to time as the example described above.

The construction of such a social network can be viewed as a two-step process. We can build a relational network by extracting events from personal documents where nodes represent main entities in the documents, including the person, the organizations he belongs to, the locations of these organizations, etc. Then the social network can be regarded as a view of the relational network after predicting the link between arbitrary person-person pair. Challenges to implement such a system can be concluded as follows: the information unit in a personal document is an event rather than a binary relation, which is more complicated to extract; how to infer knowledge properly on a network embedded with heterogeneous

© Springer International Publishing Switzerland 2016
F. Li et al. (Eds.): APWeb 2016, Part II, LNCS 9932, pp. 592–595, 2016.
DOI: 10.1007/978-3-319-45817-5_72

nodes and links. KEIPD employs a tree-similarity based approach to extract events. For link prediction, unsupervised predictors. The system is based on a considerably mature graph database.

2 System Overview

Figure 1 shows the overview of KEIPD. We will introduce the details of the information extraction module and knowledge inference module in this section.

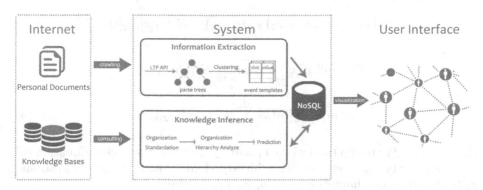

Fig. 1. System overview

2.1 Information Extraction

According to the classical entity types proposed in the Message Understanding Conference (MUC) and the characteristics of personal documents, we consider three entity types here: *Person*, *Organization* and *Location*.

Event Template. Take resume documents for example. A resume displays some fixed classes of events by time order, among which the most typical event is membership, as shown in Table 1. Events in the same class correspond to a common predefined template.

Table 1. The template of membership events

Slot	Note
Type	Type of membership: employment/education/member
Person	Participant
Organization	Organization name
Role	The role of the participant in the organization. For employment, role is the position; for education, role is "student"
Time	Start time and end time of this membership

Tree-similarity based method. We adapt the method from [3] to perform event extraction. It is assumed that sentences describing the same type of events own similar parse tree structures. First, we refer to a integrated natural language processing tool, LTP [1], to preprocess the text through Named Entity Recognition (NER) and Dependency Parsing tasks. The results of these two tasks are merged to constitute a NE-tagged parse tree where key attributes of the nodes include the word, part-of-speech tagging, the results of NER and Dependency Parsing.

Then, the parse trees are clustered with the tree-similarity function:

$$K(T_1, T_2) = m(r_1, r_2) * s(r_1, r_2) + K_c(r_1[\mathbf{c}], r_2[\mathbf{c}]) \tag{1}$$

where

$$K_c(p_1[\mathbf{c}], p_2[\mathbf{c}]) = \arg\max_{\mathbf{a}, \mathbf{b}} K(p_1[\mathbf{a}], p_2[\mathbf{b}]) \tag{2}$$

$$K(p_1[\mathbf{a}], p_2[\mathbf{b}]) = \sum_{i=1}^{l} K(p_1[\mathbf{a}_i], p_2[\mathbf{b}_i]) \tag{3}$$

Here, T_1 and T_2 are two trees where r_1 and r_2 are their root nodes. Equation (3) is the similarity function over two arbitrary children node sequences $p_1[\mathbf{a}]$ and $p_2[\mathbf{b}]$. Due to space limitations, see [3] for more details.

We adjust the match function $m(r_1, r_2)$ and the node-similarity function $s(r_1, r_2)$ to suit our data:

$$m(p_i, p_j) = \begin{cases} 0 & p_i.relate = p_j.relate \\ 1 & otherwise \end{cases} \tag{4}$$

$$s(p_i, p_j) = \begin{cases} 0.2 & p_i.ne \neq p_j.ne \\ 0.5 & p_i = O, p_j = O, p_i.pos \neq p_j.pos \\ 0.8 & p_i = O, p_i.pos = p_j.pos \\ 1.0 & p_i \neq O, p_i.ne = p_j.ne \end{cases} \tag{5}$$

The weight in Eq. (5) is assigned empirically according to the discriminative ability of the feature types.

The calculation of similarity starts from leaf nodes and goes up to the root employing a dynamic programming algorithm. We summarize syntactic rules manually for different clusters to fill the corresponding event template.

2.2 Knowledge Inference

Online Knowledge Bases. Considering the complexity of natural language, we process the relational network with the assistance of some online knowledge bases. The hierarchical characteristics of entities belonging to *Location* and *Organization* are key factors for link prediction. Therefore, we crawl an external knowledge base about fine-grained regionalism in China which contains more

Table 2. Unsupervised predictors for link prediction

Predictor	Explanation
Preferential attachment	$\|\Gamma(x)\| \cdot \|\Gamma(y)\|$
Common neighbors	$\|\Gamma(x) \cap \Gamma(y)\|$
Jaccard's coefficient	$\dfrac{\|\Gamma(x) \cap \Gamma(y)\|}{\|\Gamma(x) \cup \Gamma(y)\|}$
Adamic/Adar	$\sum_{z \in \Gamma(x) \cap \Gamma(y)} \dfrac{1}{log\|\Gamma(z)\|}$
Shortest path	The length of the shortest path between x and y
Katz	$\sum_{l=1}^{L} \beta^l \cdot \|paths_{x,y}^{(l)}\|$

than 700K locations. For organizations, we refer to an online encyclopedia[1] to normalize their names and design a simple algorithm to infer the hierarchy by analyzing prefix relations.

Link Prediction. Besides the predictors shown in Table 2, we also experiment with *rooted PageRank* and *PropFlow*, see [2] for more details. As demonstrated in Sect. 1, the formation of a link is strongly dependent on the time attributes, so we prune the candidate entity pairs before prediction using the time constraint.

3   Demonstration Scenarios

There are about 15K personal documents of politicians crawled from *People*[2] as source data. The system will be demonstrated via two types of query operations:

(1) Point query. Given a specific person as a query condition, the system will return related people with corresponding relationships.
(2) Path query. Given two specific people as query conditions, the system will return all the paths/the shortest path between them.

References

1. Che, W., Li, Z., Liu, T.: LTP: a Chinese language technology platform. In: Proceedings of the 23rd International Conference on Computational Linguistics: Demonstrations. pp. 13–16. Association for Computational Linguistics (2010)
2. Davis, D., Lichtenwalter, R., Chawla, N.V.: Multi-relational link prediction in heterogeneous information networks. In: 2011 International Conference on Advances in Social Networks Analysis and Mining (ASONAM), pp. 281–288. IEEE (2011)
3. Zhang, M., Su, J., Wang, D., Zhou, G., Tan, C.-L.: Discovering relations between named entities from a large raw corpus using tree similarity-based clustering. In: Dale, R., Wong, K.-F., Su, J., Kwong, O.Y. (eds.) IJCNLP 2005. LNCS (LNAI), vol. 3651, pp. 378–389. Springer, Heidelberg (2005)

---

[1] http://baike.baidu.com.
[2] http://www.people.com.cn.

Author Index

Bao, Xuguang II-406, II-570, II-588

Cai, Xiangrui II-434
Chai, Yale II-449
Chao, Han-Chieh I-581
Chen, Haiming II-368
Chen, Hong I-169, II-454, II-579
Chen, Hongmei II-406
Chen, Jialiang II-488
Chen, Jingqiang II-579
Chen, Jun II-317, II-537
Chen, Ling I-66
Chen, Lingxiao II-529
Chen, Meng I-480
Chen, Min II-343
Chen, Qun II-289
Chen, Wei I-28, I-329
Chen, Xiaoying II-469
Chen, Yating II-556
Chen, Zhiyong I-220
Cheng, Jiajun II-464
Cheng, Shaoyin II-410
Cheng, Xueqi I-268
Chow, Chi-Yin I-104
Cui, Xingcan II-214
Cui, Zhiming II-162

Dey, Kuntal I-342
Ding, Bin I-359
Ding, Chen I-519
Ding, Rui II-533
Ding, Yue II-98
Ding, Zhaoyun II-464
Dong, Guozhong I-245
Dong, Zhenjiang II-579
Dou, Peng I-304
Du, Jianfeng II-542
Du, Sizhen I-304
Du, Xiaoyong I-41, I-420, I-555, II-537

Fang, Junhua II-3
Feng, Ling I-431
Feng, Shi I-567, I-594

Feng, Yong II-174
Feng, Zhiyong II-444
Fournier-Viger, Philippe I-581
Fu, Peng I-531
Fujita, Hideyuki II-459

Gan, Wensheng I-581
Gao, Hong II-239, II-251, II-552
Gao, Hua II-402
Gao, Hui II-30
Gao, Peng II-574
Gao, Xiangyu II-583
Gao, Yunjun I-383, II-17, II-227
Ge, Bin II-469
Ge, Wei II-343
Geng, Shaofeng II-30
Gu, Caidong II-162
Gu, Dawu II-483
Guo, Chen II-561
Guo, De II-214
Guo, Hao II-410
Guo, Mengyu II-68

Hameurlain, Abdelkader II-355
Han, Jianhua I-468
Han, Shanshan II-251
Han, Shuai I-292
Han, Shumin II-201
Han, Tian II-547
Han, Yanbo I-116
He, Jun I-555
He, Qinming II-227
He, Su II-464
He, Xiaofeng I-15
Hitzler, Pascal II-429
Hong, Liang I-383
Hong, Shenda II-56, II-389
Hong, Xiaoguang I-220
Hongyan, Tang II-186
Hou, Xiaoyun I-41, I-420
Hu, Qinghua I-506
Hu, Qingwu II-561
Hu, Qingyang II-227
Hu, Yanli II-469

Hu, Yue   I-256
Hu, Yupeng   II-492
Huang, Changqin   I-371
Huang, Hao   I-383, II-227
Huang, Jianbin   II-398, II-402
Huang, Liusheng   I-53, II-276
Huang, Lu   I-555
Huang, Shanshan   II-150
Huang, Shenbin   II-251, II-552, II-583
Huang, Weijing   I-329
Huang, Xiu   I-256
Huang, Yihua   II-343
Hung, Patrick C.K.   II-263

Ji, Cun   II-492
Ji, Wendi   I-395, II-263
Jia, Xiaolin   II-398, II-402
Jiang, Fan   I-316, II-410
Jiang, Jing   I-196
Jiang, Shunqing   II-488
Jiang, Tao   II-289
Jin, Hai   I-543
Jin, Peiquan   I-129, II-497
Jin, Yuanyuan   II-263
Jing, Ming   II-492

Kang, Hong   II-68
Kaushik, Saroj   I-342
Ke, Bingqing   I-79
Kou, Yue   II-201, II-414

Lamba, Hemank   I-342
Lan, Yunshi   I-196
Leung, Carson K.   I-316
Li, Bofang   I-41, II-537
Li, Cuiping   I-169, II-454
Li, Guohui   I-153
Li, Hong   I-53
Li, Hongyan   II-56, II-389
Li, Huakang   II-525, II-579
Li, Hui   I-220
Li, Jiahong   II-583
Li, Jianjun   I-153
Li, Jianzhong   II-239, II-251, II-552
Li, JiaPeng   I-28, II-521
Li, Jingyuan   I-268, I-280
Li, Juanru   II-483
Li, Lin   II-565
Li, Qi   I-431

Li, Qing   I-104
Li, Shanshan   I-232
Li, Tao   II-579
Li, Weibang   II-289
Li, Weiwei   II-385
Li, Wenzhong   II-124
Li, Xueqing   II-492
Li, Yaping   I-456
Li, Yeting   II-368
Li, Yijia   I-292
Li, Yijin   II-439
Li, Yueyang   II-464
Li, Yuqiang   II-385
Li, Zhanhuai   II-289
Li, Zhengbo   I-468
Li, Zhixu   I-28
Lian, Defu   I-359
Liao, Xiangke   I-232
Lim, Ee-Peng   I-196
Lin, Bin   I-232
Lin, Can   II-542
Lin, Hailun   I-531
Lin, Huaizhong   II-17
Lin, Jerry Chun-Wei   I-581
Lin, Lanfen   I-444
Lin, Xueqin   II-533
Lin, Yiyong   II-389
Lin, Zheng   I-531
Lini, Chen   II-98
Liu, An   I-28, I-104
Liu, Bo   I-543
Liu, Bozhong   I-66
Liu, Chang   II-429
Liu, Chen   I-116
Liu, Chunyang   I-66
Liu, Guanfeng   II-162
Liu, Hao   I-141
Liu, Hongyan   I-555
Liu, Hui   II-483
Liu, Li   I-280
Liu, Peng   II-150
Liu, Qing   I-456
Liu, Sisi   II-111
Liu, Tao   I-41, I-420, II-537
Liu, Xiaodong   I-232
Liu, Xiaoqing   I-456
Liu, Yang   I-92, I-480, II-85
Liu, Yongjian   II-111, II-543, II-574
Liu, Yuanyuan   II-592
Liu, Zhihui   II-444

Liu, Ziyan   I-220
Lu, Dongming   II-17
Lu, Hongtao   I-208
Lu, Sanglu   II-124
Lu, Xiaorong   II-276
Lu, Yuan   I-268, I-280
Luo, Changyin   I-153
Luo, Wenyi   I-129
Luo, Yonghong   II-419
Lv, Lei   II-85
Lv, Xia   II-497
Lv, Ze   II-398
Lv, Zhaoyang   II-592

Ma, Jun   I-408, II-150
Ma, Lintao   II-150
Ma, Yu-Jia   II-474
Masada, Tomonari   II-420
Meng, Dan   I-531
Meng, Fanshan   II-239
Morvan, Franck   II-355
Moumen, Chiraz   II-355
Mutharaju, Raghava   II-429

Nagar, Seema   I-342
Nie, Min   I-359
Nie, Tiezheng   II-201, II-414

Ohmori, Tadashi   II-459

Pan, Haiwei   I-292
Pan, Jing   I-506
Pan, Yangbin   II-556
Pan, Yanhong   II-505
Peng, Feifei   II-368
Peng, Zhaohui   I-220
Peng, Zhiyong   I-383

Qi, Guilin   II-429
Qi, Kunxun   II-542
Qian, Tieyun   I-383, II-227
Qian, Weining   I-183
Qiao, Fengcai   II-464
Qin, Hongchao   II-478
Qiu, Weidong   I-66
Qiu, Yuan   II-459
Qiu, Zhen   II-56
Qu, Ziqi   II-583

Rao, Guozheng   II-444
Ren, Yan   I-268

Shang, Shuo   I-3
Shao, Jie   I-79, I-256
Shen, Derong   II-201, II-414
Shen, Fumin   I-256
Shen, Yao   II-276
Shi, Hong   I-506
Shi, Huike   II-419
Shi, Xiaohua   I-208
Shi, Xuanhua   I-543
Shintani, Takahiko   II-459
Shu, Jiwu   II-305
Song, Guojie   I-304
Sun, Guozi   II-525
Sun, Heli   II-398, II-402
Sun, Hui   II-454
Sun, Ming   II-239, II-552

Takasu, Atsuhiro   II-420
Tang, Yong   I-371, II-529, II-533
Tong, Jia   II-186
Tu, Jiaqi   I-444

U, Leong Hou   II-17

Wang, Binbin   II-305
Wang, Changping   II-317
Wang, Chao   II-393, II-434, II-449
Wang, Chaokun   II-317
Wang, Chengyu   I-15
Wang, Daling   I-567, I-594
Wang, Dong   II-98
Wang, Guoren   II-478
Wang, Hao   I-141, I-431, II-317
Wang, Hongzhi   II-239, II-251, II-552, II-583
Wang, Hui   II-464, II-483
Wang, Jianxin   I-53
Wang, Jing   I-444
Wang, Jun   I-493
Wang, Lizhen   II-406, II-570, II-588
Wang, Meijiao   II-570
Wang, Qisen   II-124
Wang, Shaoqing   I-169
Wang, Song   I-383
Wang, Tengjiao   I-329
Wang, Wei   I-245
Wang, Weiping   I-531

Wang, Wendy Hui   II-439
Wang, Xiaoli   II-556
Wang, Xiaoling   I-395, II-137, II-263
Wang, Xiaotong   II-3
Wang, Xin   II-444
Wang, Xiongbin   I-116
Wang, Yan   II-385
Wang, Yang   I-594, II-276
Wang, Yaqi   I-567
Wang, Yongheng   II-30
Wang, Yongjin   II-150
Wang, Yuanzhuo   I-268, I-280
Wang, Zheng   I-169
Wei, Feng   II-410
Wei, Jingmin   II-529
Wei, Jinmao   I-493
Wei, Zhewei   I-3
Wen, Ji-Rong   I-3
Wen, Yanlong   II-330, II-434, II-449
Wenlong, Shao   II-186
Wu, Bu-Xiao   I-519
Wu, Jian   II-162
Wu, Meng   II-56
Wu, Mengsang   II-556
Wu, Shanshan   II-17
Wu, Tianzhen   II-454
Wu, Xingcheng   II-533

Xia, Fan   I-183
Xia, Hu   I-359
Xiao, Danyang   II-529
Xiao, Jing   I-371, I-519
Xiao, Lin   I-456
Xiao, Qing   II-588
Xiao, Weidong   II-469
Xie, Qing   II-111, II-547, II-574
Xie, Xiaoqin   I-292
Xin, Xin   II-98
Xiong, Mengyuan   II-556
Xu, Chuanhua   II-533
Xu, Hengpeng   I-493
Xu, Huachun   I-359
Xu, Jiajie   II-162
Xu, Jingyang   I-408
Xu, Xiefeng   II-162
Xu, Yang   I-53
Xu, Yanxia   II-521
Xu, Zhentuan   II-556
Xue, Yuanyuan   I-431

Yan, Qian   II-227
Yang, Fan   II-474
Yang, Jing   I-371
Yang, Lei   I-359
Yang, Nan   I-456
Yang, Wei   I-53, II-276
Yang, Wu   I-245
Yang, Yajun   I-506
Yang, Yang   I-79, I-256
Yang, Yitao   II-525
Yang, Zhifan   II-419
Yao, Kai   I-153
Ye, Xiaojun   II-317
Yin, Li'ang   I-468
Yin, Zhilei   II-289
Ying, Chen   II-227
Ying, Li   II-186
Ying, Yuanxiang   II-389
Yong, Yang   II-186
Yu, Chengcheng   I-183
Yu, Donghai   I-92
Yu, Ge   I-567, I-594, II-201, II-414
Yu, Jianye   I-280
Yu, Wei   II-488
Yu, Xiaohui   I-92, I-480, II-85, II-214, II-592
Yu, Yong   I-468
Yuan, Chunfeng   II-343
Yuan, Fengcheng   I-531
Yuan, Xiaojie   II-68, II-330, II-393, II-449
Yuan, Ye   II-478
Yue, Lihua   I-129, II-497

Zhang, Bei   II-174
Zhang, Bo   II-414
Zhang, Chengqi   I-66
Zhang, Chong   II-469
Zhang, Cong   I-141
Zhang, Detian   I-104
Zhang, Dongxiang   I-79
Zhang, Haiwei   II-330
Zhang, Haiyang   II-43
Zhang, Hao   I-316
Zhang, Heng   I-444
Zhang, Jinbo   II-389
Zhang, Kai   II-137, II-263
Zhang, Lamei   I-329
Zhang, Linlin   I-480
Zhang, Mengqi   I-196
Zhang, Rong   I-15, II-3, II-505
Zhang, Shengqiang   II-583

Zhang, Xiao   II-124
Zhang, Xiaolan   II-368
Zhang, Xiaowang   II-444
Zhang, Xinwen   II-525
Zhang, Yanfei   II-505
Zhang, Yi   I-79
Zhang, Yifei   I-567, I-594
Zhang, Ying   I-395, II-393, II-419
Zhang, Yuanyuan   II-483
Zhang, Zhenguo   II-330, II-434
Zhang, Zhiqiang   I-292
Zhao, Dapeng   II-137, II-263
Zhao, Gansen   I-371
Zhao, Kankan   I-169
Zhao, Lei   I-28, II-521
Zhao, Liang   I-431
Zhao, Pengpeng   II-162
Zhao, Qingjie   I-141
Zhao, Xue   II-393

Zhao, Zhe   I-41, I-420, II-537
Zhao, Zhiqiang   II-402
Zheng, Hai-Tao   II-43
Zheng, Jiping   II-488
Zhou, Aoying   I-15, I-183, II-3
Zhou, Guomin   I-15
Zhou, Kailai   II-454
Zhou, Yiping   I-395
Zhou, Yu   II-398
Zhou, Zhangquan   II-429
Zhou, Ziwei   II-414
Zhu, Feida   I-196, I-245, II-478
Zhu, Ge   II-565
Zhu, Jia   I-371, I-519, II-529, II-533
Zhu, Meiling   I-116
Zhu, Shunzhi   I-3
Zhu, Yue-Sheng   II-474
Zou, Xueqing   II-464

Printed in the United States
By Bookmasters